Handbook of Research on the Impacts, Challenges, and Policy Responses to Overtourism

Cláudia Ribeiro de Almeida
University of Algarve, Portugal

Alfred Quintano
Independent Researcher, Malta

Moisés Simancas
University of La Laguna, Spain

Raquel Huete
University of Alicante, Spain

Zélia Breda
University of Aveiro, Portugal

IGI Global
DISSEMINATOR OF KNOWLEDGE

A volume in the Advances in Hospitality, Tourism, and the Services Industry (AHTSI) Book Series

Published in the United States of America by
IGI Global
Business Science Reference (an imprint of IGI Global)
701 E. Chocolate Avenue
Hershey PA, USA 17033
Tel: 717-533-8845
Fax: 717-533-8661
E-mail: cust@igi-global.com
Web site: http://www.igi-global.com

Library of Congress Cataloging-in-Publication Data

Names: Almeida, Cláudia Ribeiro de, 1971- editor.
Title: Handbook of Research on the Impacts, Challenges, and Policy Responses to Overtourism / Cláudia Ribeiro de Almeida, Alfred Quintano, Moisés Simancas, Raquel Huete, and Zélia Breda, editors.
Description: Hershey, PA : Business Science Reference, 2020. | Includes bibliographical references and index. | Summary: "This book is a collection of research on best practices and legislation solutions for the management of tourism destinations suffering from overtourism, tourismophobia, or antitourism movement issues"-- Provided by publisher.

Identifiers: LCCN 2019037702 (print) | LCCN 2019037703 (ebook) | ISBN 9781799822240 (hardcover) | ISBN 9781799822264 (ebook)
Subjects: LCSH: Overtourism--Case studies.
Classification: LCC G156.5.O94 I67 2020 (print) | LCC G156.5.O94 (ebook) | DDC 910.68--dc23
LC record available at https://lccn.loc.gov/2019037702
LC ebook record available at https://lccn.loc.gov/2019037703

This book is published in the IGI Global book series Advances in Hospitality, Tourism, and the Services Industry (AHTSI) (ISSN: 2475-6547; eISSN: 2475-6555)

British Cataloguing in Publication Data
A Cataloguing in Publication record for this book is available from the British Library.

For electronic access to this publication, please contact: eresources@igi-global.com.

Advances in Hospitality, Tourism, and the Services Industry (AHTSI) Book Series

Maximiliano Korstanje
University of Palermo, Argentina

ISSN:2475-6547
EISSN:2475-6555

Mission

Globally, the hospitality, travel, tourism, and services industries generate a significant percentage of revenue and represent a large portion of the business world. Even in tough economic times, these industries thrive as individuals continue to spend on leisure and recreation activities as well as services.

The Advances in Hospitality, Tourism, and the Services Industry (AHTSI) book series offers diverse publications relating to the management, promotion, and profitability of the leisure, recreation, and services industries. Highlighting current research pertaining to various topics within the realm of hospitality, travel, tourism, and services management, the titles found within the AHTSI book series are pertinent to the research and professional needs of managers, business practitioners, researchers, and upper-level students studying in the field.

Coverage

- Destination Marketing and Management
- Customer Service Issues
- Leisure & Business Travel
- Tourism and the Environment
- Travel Agency Management
- Cruise Marketing and Sales
- Casino Management
- Sustainable Tourism
- Service Training
- Service Management

IGI Global is currently accepting manuscripts for publication within this series. To submit a proposal for a volume in this series, please contact our Acquisition Editors at Acquisitions@igi-global.com or visit: http://www.igi-global.com/publish/.

The Advances in Hospitality, Tourism, and the Services Industry (AHTSI) Book Series (ISSN 2475-6547) is published by IGI Global, 701 E. Chocolate Avenue, Hershey, PA 17033-1240, USA, www.igi-global.com. This series is composed of titles available for purchase individually; each title is edited to be contextually exclusive from any other title within the series. For pricing and ordering information please visit http://www.igi-global.com/book-series/advances-hospitality-tourism-services-industry/121014. Postmaster: Send all address changes to above address.

Titles in this Series

For a list of additional titles in this series, please visit:
https://www.igi-global.com/book-series/advances-hospitality-tourism-services-industry/121014

Handbook of Research on Smart Technology Applications in the Tourism Industry
Evrim Çeltek (Gaziosmanpasa University, Turkey)
Business Science Reference • © 2020 • 430pp • H/C (ISBN: 9781799819899) • US $295.00

Global Opportunities and Challenges for Rural and Mountain Tourism
Devkant Kala (University of Petroleum and Energy Studies, India) and Satish Chandra Bagri (Hemvati Nandan Bahuguna Garhwal University, India)
Business Science Reference • © 2020 • 354pp • H/C (ISBN: 9781799813026) • US $225.00

Organizational Behavior Challenges in the Tourism Industry
Şule Aydin (Nevsehir Haci Bektas Veli University, Turkey) Bekir Bora Dedeoglu (Nevsehir Haci Bektas Veli University, Turkey) and Ömer Çoban (Nevsehir Haci Bektas Veli University, Turkey)
Business Science Reference • © 2020 • 524pp • H/C (ISBN: 9781799814740) • US $225.00

Handbook of Research on Social Media Applications for the Tourism and Hospitality Sector
Célia M.Q. Ramos (University of Algarve, Portugal) Cláudia Ribeiro de Almeida (University of Algarve, Portugal) and Paula Odete Fernandes (Polytechnic Institute of Bragança, Portugal)
Business Science Reference • © 2020 • 477pp • H/C (ISBN: 9781799819479) • US $295.00

New Trends and Opportunities for Central and Eastern European Tourism
Puiu Nistoreanu (The Bucharest University of Economic Studies, Romania)
Business Science Reference • © 2020 • 338pp • H/C (ISBN: 9781799814238) • US $225.00

Managing, Marketing, and Maintaining Maritime and Coastal Tourism
Mónica Morais de Brito (CEGOT, Portugal & Instituto Politécnico de Beja, Portugal) Alvaro Dias (Universidade Lusófona de Humanidades e Tecnologias, Portugal & ISCTE, Instituto Universitário de Lisboa, Portugal) and Mafalda Patuleia (Universidade Lusófona de Humanidades e Tecnologias, Portugal)
Business Science Reference • © 2020 • 325pp • H/C (ISBN: 9781799815228) • US $225.00

Global Developments in Healthcare and Medical Tourism
Sudip Paul (North-Eastern Hill University, India) and Sharad Kumar Kulshreshtha (North-Eastern Hill University, India)
Medical Information Science Reference • © 2020 • 341pp • H/C (ISBN: 9781522597872) • US $245.00

IGI Global
DISSEMINATOR OF KNOWLEDGE

701 East Chocolate Avenue, Hershey, PA 17033, USA
Tel: 717-533-8845 x100 • Fax: 717-533-8661
E-Mail: cust@igi-global.com • www.igi-global.com

Editorial Advisory Board

List of Contributors

Table of Contents

Section 5
Case Studies: Natural and Rural Areas

Detailed Table of Contents

This chapter deals with and underscores sustainable policy responses towards overtourism, which is becoming an increasingly important issue for the survival of tourist destinations. Based on the current literature, this chapter proposes a wide range of policy responses that can be adapted by a variety of elements in tourist destinations, ranging from stakeholders to local residents. As a consequence, the measures and strategies in response to the overtourism phenomenon can expand our understanding of the concept.

This chapter conceptualises the notion of overtourism by providing a systematic literature review mapping the existing research knowledge. The mapping exercise benefits from textual data from academic articles on overtourism. The study focuses on measures to counteract overtourism from a marketing perspective. Since tourism is about customers' experiences, understanding overtourism and its aspects is crucial at a time the market growth calls for measures to control this phenomenon. Using online databases and NVIVO 12 software, 66 articles were selected for bibliographic content analysis. Findings highlight the need to work the concept of overtourism in a holistic approach encompassing five dimensions: social, political & governance, marketing & customer experience, economic, and environmental. Results permitted to identify other dimensions; overtourism, symbolic, ethical-moral, cultural. Managing overtourism is challenging because whereas tourism is about customers' experiences, for policy makers overtourism encompasses strategies with collective effects.

Luciana Aparecida Barbieri da Rosa, Federal University of Santa Maria, Brazil
Maria Carolina Martins-Rodrigues, CIEO, University of Algarve, Portugal
Tais Pentiado Godoy, Federal University of Santa Maria, Brazil
Luana Inês Damke, Federal University of Santa Maria, Brazil
Clandia Maffini Gomes, Federal University of Santa Maria, Brazil

Studies on overtourism have been growing in the last decade. This article aims to understand the characteristics of the international scientific literature that links the theme Smart Cities in the Web of Science and Scopus databases, from 1998 to 2018. The established methodology was a bibliometric survey. The sample was composed of 24 articles selected in WOS and Scopus. The survey results show that the year with the most publications was 2018. The most relevant article is: "Understanding and overcoming the negative impacts of tourism in city destinations: a conceptual model and strategic framework" by A. Postma, & D. Schmuecker, published in the Journal of Tourism Futures in 2017 with 10 citations. Thus, future studies should expand the search for articles on the topics in other bases as well as national and international scientific events.

Alejandro Mantecón, University of Alicante, Spain
María Velasco, Complutense University of Madrid, Spain

This chapter discusses the factors that have led to the emergence of expressions of criticism toward tourism. This review serves to frame the original contribution of this text: a theoretical model that clarifies the defining features of the main attitudes towards tourism. Merton's model is here adjusted for the analysis of a new relationship between social ends and economic means. In this case, the end is economic progress. The way is the tourism, conceived as a massive social phenomenon. The relation between goals and means generates tensions. Its management derives in strategies of adaptation that include different ways of identification or discussion. The five types of adaptation of the new model are useful for addressing subject positions, political discourses, or attitudinal dispositions towards tourism. To illustrate this typology a purposive sampling of news on the tourismphobia has been selected, with no statistical generalization reflecting the constituent elements of each of the types: legitimization, innovative criticism, resignation, radical criticism, and subversive utopia.

Hio Kuan Lai, Faculty of Economics, University of Algarve, Portugal
Patrícia Pinto, Faculty of Economics, Research Centre for Tourism, Sustainability, and Well-Being (CinTurs), University of Algarve, Portugal
Pedro Pintassilgo, Faculty of Economics, CEFAGE, University of Algarve, Portugal

The concept of overtourism has been popularly discussed in recent years, upon local residents starting to recognize the tourism disturbances in some high-profile destinations. For a sustainable tourism development, it is crucial to investigate potential impacts of overtourism towards the local residents living in the destination. Thus, this chapter employs the existing academic work on quality of life (QoL)

In the last few years, the number of visitors in historic cities has grown, resulting in a situation labeled 'overtourism'. In these tourist-historic cities, tourism is one of the main local economic foundations. This chapter asks whether the social carrying capacity of the host community has been exceeded, through a case study in Toledo, Spain, a UNESCO World Heritage Site since 1986. The methodology employed is a survey to residents in the city. The main findings show that the residents do not perceive the overtourism situation in the city as something serious. However, some impacts such as the historic centre becoming a museum for tourists, or the traffic congestion derived from overcrowding in both the traffic and pedestrian flow are clearly perceived by the host community. It is the residents in the historic centre those who manifest a more negative opinion of touristic development. This chapter may be of interest to academics, decision makers, and those responsible for tourism in historic cities.

This chapter reflects upon the reality associated with running sports events in protected areas, in particular at the Peneda Gerês National Park (PNPG), with a particular focus on the Trans Peneda-Gerês (TPG), and understands its impacts within a sustainable tourism development perspective. With the diversity of landscapes and natural resources, and its characteristic villages, PNPG, as many other protected areas in Portugal, has a strong potential for Outdoor Sports/Tourism, offering some of Portugal's best running trails. As such, a considerable number of running events are currently being hosted. Based on a research process which includes both questionnaires and interviews, the authors propose to identify the profile and behavior of 2019 TPG edition's participants, and to assess local stakeholders' perception of economic, environmental, and social impacts, and to contribute to the discussion from a sustainable development perspective of the creation and management of tourism-related sports events, namely running, in protected areas.

Foreword

The socio-cultural and environmental impacts caused by the excessive tourist demand have been investigated since the 90's of the last century, in the tourist destinations of the coast, namely in southern Europe and in the historical cities, such as Venice, places where the growth in tourist demand has led to exceeding the carrying capacity limits. The concept of sustainable tourism, which emerged during this period, aimed to make the preservation of the sites compatible with the interests of visitors.

The pressure generated by the steady increase in tourist flows over the last twenty years (with the exception of years of severe economic crisis) coupled with the absence of effective planning and development policies to counter steep growth has led to anti-tourism movements, particularly in the most visited cities as Venice, Barcelona or Amsterdam. The numerous scientific literature on overtourism, over the last decade, has emerged as a reflection on how to find the best solutions for turning mass tourism destinations into sustainable places.

This book provides an important effort to capture some of these new developments. The four sections correspond to the core themes of the theme under analysis. They include a first section devoted to concepts and methodologies for overtourism analysis followed by three more filled with case studies - in urban areas, coastal areas, and historical centres and rural areas.

Section one, present concepts and methodologies for overtourism analysis, focuses the researcher on literature review strategies and appropriate research methods that include assessing impacts on residents' quality of life, seeking de-marketing strategies for overtourism and the conceptualization of new theoretical frameworks to analyse attitudes towards tourism.

The remaining three sections contain a wide range of case studies, approaches, contexts, research methods and techniques, policy responses and strategies implemented or recommended by the authors, appropriate to each of them.

The contents of its four sections make this book a useful resource for students and researchers.

Ana Maria Ferreira
University of Évora, Portugal

Ana Maria Ferreira *is a Senior Fellow at the CIDEHUS - Interdisciplinary Centre for History, Culture and Societies.*

Preface

The concept of overtourism gained popularity at the start of the 21st century, as some reports started emerging of negative reactions to tourism flows in several major destinations. Local residents started to be conscious of the negative impact of tourism in their own communities, bringing to light some of the main issues. The term and concept of overtourism entered in daily discourse, in political debate, in media reports and in business lobbies agendas.

In terms of academic literature, this is an important concept, most notably in studies related to overcrowded cities. This is due, among other factors, to what the Spanish media refers to as turismofobia, a negative backlash to the touristification of spaces and elements that until now, not used for tourism, as well as to the lack of planning or poor management of the phenomenon.

"Overcrowding", "saturation" or "congestion" of a tourist destination, area, space or even just a localized place, are pejorative descriptions that refer to the negative impact of tourism on the perceived quality of life of local residents and/or the experience of visitors.

Labelling a tourism area as "saturated" underscores the existence of a negative impact, understood as a change seen as undesirable by tourists; thus, it increases the difference between the perceived value and the expected value of the model developed. This is because overtourism has a decisive influence on how the tourist perceives the quality or the perceptual quality of a particular destination. Using the definition of the psychological carrying capacity of tourists, we can consider a coastal tourism area saturated, from the supply-side point of view in terms of accommodation capacity or when it reaches a minimum level of tourist satisfaction and the tourist's interest (tolerance) is low.

Very high levels of occupancy make tourists feel uncomfortable resulting in a 'psychological' carrying capacity. This is because the installations, facilities, equipment and infrastructure become overcrowded, in terms of 'physical' carrying capacity and the degradation of environmental quality of the area resulting in 'ecological' carrying capacity. A feeling arises among visitors that there are too many tourists, accommodations spaces and/or constructed space in a 'perceptual' carrying capacity. Finally, the local residents or 'host community' begin to manifest their rejection of the tourists through a 'social' carrying capacity. The causation of visitor-resident irritants was already researched and intimated in the seminal work by Doxey (1975).

The perception that a tourism area suffers from accommodation saturation can lead tourists to feel less comfortable and lead to loss of enjoyment, causing them to lose interest in the destination and not return or even to advise other potential tourists not to visit. In this way, it influences the image of the tourism area and the destination's brand and attractiveness. This could generate a negative perception of the place, with the consequent impact on the key indicators of tourism (number of overnights, average stay, etc.) and business profitability (occupancy rates, average daily revenue per available and occupied room, etc.). In other words, it can cause the destination to become less competitive (economic carrying capacity).

Excessive saturation puts a tourism area in a decisive, critical and highly uncertain situation, accompanied by symptoms of degradation/exhaustion. If no action is taken, this can lead to the final stages of degeneration, with the consequent decline or decay of the destination, with tourists and tour operators choosing alternative destinations that are not saturated or damaged. This typically occurs towards the tail end of the tourism destination life cycle (Butler, 1980).

Furthermore, estimating the optimal limit of overtourism is not an end in and of itself, but rather an instrument that can provide continuous, relevant information to those in charge of making decisions related to planning and managing mature tourist destinations. Therefore, it is a management concept that reflects the objectives of managers or those responsible for satisfying tourists and local residents.

The book editors as well as the various contributing authors aim to reach out to all those seeking a reference work on the topic of overtourism. In particular it is aimed at tourism stakeholders including but not limited to policy makers, government officials, academic researchers and practitioners. It is particularly aimed at whoever is trying to understand the primary impacts caused in the host communities, as well as hopefully offering some solutions to the problematic by evaluating, analysing and monitoring of overtourism in different realities and geographic locations.

Handbook of Research on the Impacts, Challenges, and Policy Responses to Overtourism is organized in five sections with 19 chapters incorporating case studies from different geographical regions and realities, thus helping to understand the phenomenon of overtourism, presenting some solutions and measures on how to control it.

Section 1 focuses on concepts and methodologies for overtourism analysis presented in six chapters.

Chapter 1, "Developing Sustainable Policies in Response to Overtourism," proposes a wide range of policy responses that can be adapted by a variety of elements in tourist destinations.

Chapter 2, "Overtourism: A Systematic Review of Literature," conceptualises the notion of overtourism by providing a systematic literature review mapping the existing research knowledge.

Chapter 3, "Methods of Published Articles on Overtourism: Analysis and Reflections of Scopus and Web of Science (1998-2018)," brings to light the discussion of the international scientific literature that links the theme.

Chapter 4, "Beyond Tourismphobia: Conceptualizing a New Framework to Analyse Attitudes Towards Tourism," presents factors that led to the emergence of expressions of irritation towards tourism; the authors conceptualize a new framework to analyse attitudes towards tourism.

Chapter 5, "Overtourism: Impacts on Residents' Quality of Life – Evidence From Macau," details the existing academic work on quality of life concepts to examine the correlations between residents' perception of tourism's impact to them in material, community, emotional, and health and safety life domains.

Chapter 6, "Demarketing: A Marketing Framework for Overtourism," focuses on the deconstruction of the situation of overtourism in different parts of the world, and the use of the concept of demarketing as a strategy for sustainable tourism.

Section 2 presents four case studies of overtourism in urban areas.

Chapter 7, "Overtourism in Megalopolises: The Case of St. Petersburg, Russia," discusses overtourism in St. Petersburg, with some review of the major methods of preventing and overcoming overtourism employed in 18 different megapolis, in order to identify the cures that are more suitable for St. Petersburg.

Chapter 8, "Bye Bye Lisbon: Tourism Gentrification Impacts on Lisbon´s Inner-City Housing Market," is an analysis of the impacts of tourism gentrification in Lisbon's inner-city housing market, with focus on the driving forces of austerity urbanism and the neoliberal turn in urban policy, which fuels the process of tourism gentrification, in the last decade, in historic neighbourhoods.

Chapter 9, "Touristificacion in the Central Market of Valencia: Fact or Fiction?" presents a diagnostic situation of Valencia Central Market and its consumers' perception in order to understand the main transformations that took place. This chapter establishes the debate between society and tourism considering the tourism process and its impact on cultural heritage, initially aimed at locals.

Chapter 10, "The Impact of Vacation Rentals on Housing: State of the Art and Proposals," discusses one of the impacts of excessive tourism: the increase in the price of residential homes and rental properties in the city of Malaga, which create a social and academic debate about public policy proposals to limit the growth of the phenomenon.

Section 3 contains four case studies of overtourism situations in coastal tourism areas.

Chapter 11, "The Collaborative Lodging in Puerto Vallarta: Who Wins and Who Loses?" is an analysis on who actually receives the benefits of collaborative accommodation, as well as the negative impacts that it generates in Puerto Vallarta (Mexico), through an analysis of the behaviour of the offer, as well as in-depth interviews with the registered hosts on Airbnb.

Chapter 12, "The Development of a Scale to Measure Tourism-Phobia: An Exploratory Case of Residents in Majorca," is an attempt to understand the negative aspects which overtourism is causing in the perception of both native and foreign residents on Majorca (Spain), as well as to proposing a scale developed to measure tourism-phobia.

Chapter 13, "Vacation Rentals, Tourism, and International Migration: Gentrification in Las Palmas de Gran Canaria (Spain) From a Diachronic Perspective," discusses the growing presence of vacation rentals and international residential migrations in some Spanish Mediterranean and island cities, with the analysis of some conflicts inherent to tourism gentrification in Las Palmas de Gran Canaria.

Chapter 14, "Alternative Tourism Strategies to Enhance Local Sustainable Tourism Development: The ALTER-ECO Project in Gandia (Spain)," exposes the ALTER-ECO project in Gandia (Spain); it proposes monitoring, public Wi-Fi and gamification as three strategies for dispersing tourist load.

Section 4 focuses on historical centres, with three chapters analysing the emergence of overtourism and theirs effects on the quality of life of the residents, conservation of the tourist resources and on the degree of tourist satisfaction with the tourist experience.

Chapter 15, "Lisbon Historic Quarters: Identity vs. Overtourism," addresses the debate regarding the policy responses of the Lisbon Council in the framework of an integrated urban rehabilitation in Lisbon's historic quarters and the Lisbon Council Strategy.

Chapter 16, "In Search of Overtourism Indicators in Urban Centres," reflects on the need to define overtourism indicators that allow dimensioning the phenomenon and its impacts, as well as its application in the historic centre of Madrid, given the absence of reference values and operational difficulties to obtain data.

Chapter 17, "Perceptions and Opinions of the Host Community Regarding Overtourism in the Tourist-Historic City: A Case Study in Toledo (Spain)," evaluates the perceptions and opinions of the host community about overtourism in Toledo (Spain). The chapter is an analysis of the impact of an historic centre becoming a museum for tourists or traffic congestion resulting from overcrowding in both traffic and pedestrian flow clearly perceived by the host community.

Section 5 contemplates two case studies of overtourism in rural areas.

Chapter 18, "Perceived Impacts of Running Events in Protected Areas: The Case of Trans Peneda-Gerês at Peneda Gerês National Park, Portugal," reflects the impact of running sports events in protected areas, in particular at the Peneda Gerês National Park.

Chapter 19, "Outdoor Solutions for the Seasonal Concentration of Tourism Demand in Northern Portugal: An Integrated Approach Based on the Gini Index," focuses on the development of a Gini Index to analyse the distribution of tourism seasonality in the Northern Portugal and current public control measures in practice and suggests the implementation of different and more effective policy measures.

The book editors believe that readers will be made aware of the overtourism problematic not only in a conceptual manner but also due to the case studies presented throughout the book chapters. To know, to understand and to discuss the topic is nowadays more important than ever, mainly for those that need to make decisions and establish goals in order to create and manage sustainable tourism destinations that involve different stakeholders in an integrated management approach and above all, measures to avoid overtourism and the main negative impacts that it causes.

Cláudia Ribeiro de Almeida
University of Algarve, Portugal

Alfred Quintano
Independent Researcher, Malta

Moisés Simancas
University of La Laguna, Spain

Raquel Huete
University of Alicante, Spain

Zélia Breda
University of Aveiro, Portugal

Acknowledgment

The editors would like to acknowledge the help of all the people involved in this project and, more specifically, to the authors and reviewers that took part in the review process. Without their support, this book would not have become a reality.

To all the authors, our sincere gratitude to their contributions, time and expertise.

To all the reviewers, our appreciation to their valuable contributions that allowed the improvement of quality, coherence and content presentation of the chapters.

To Professor Ana Maria Ferreira, for her kind words in the foreword.

And to all the others that directly or indirectly helped us during the project, with some recommendations and technical work in order to bring to light this new publication.

To all… THANKS!

Cláudia Ribeiro de Almeida
University of Algarve, Portugal

Alfred Quintano
Independent Researcher, Malta

Moisés Simancas
University of La Laguna, Spain

Raquel Huete
University of Alicante, Spain

Zélia Breda
University of Aveiro, Portugal

Section 1

Concepts and Methodologies for Overtourism Analysis

Chapter 1
Developing Sustainable Policies in Response to Overtourism

Kadir Çakar
Mardin Artuklu University, Turkey

ABSTRACT

This chapter deals with and underscores sustainable policy responses towards overtourism, which is becoming an increasingly important issue for the survival of tourist destinations. Based on the current literature, this chapter proposes a wide range of policy responses that can be adapted by a variety of elements in tourist destinations, ranging from stakeholders to local residents. As a consequence, the measures and strategies in response to the overtourism phenomenon can expand our understanding of the concept.

INTRODUCTION

The tourism industry is considered to be an economic engine as it increases employment, income and tax revenue while boosting regional development. On the other hand, however, it has some adverse effects on the environment and society, which pave the ways in which host and guest communities are destroyed (McKercher, 1993). Today, tourism has many problems stemming from a negative growth pattern that is inconsistent with sustainable tourism policies and/or goals (Higgins-Desbiolles, 2018b).

It is, therefore, these negative impacts of tourism that have tarnished the sustainability of tourist destinations and engendered the presence of many new concepts, such as anti-tourism, tourismphobia (Higgins-Desbiolles, 2018a), responsible and sustainable tourism, or more concretely responsustable tourism (Mihalic, 2016), and overtourism (Koens et al., 2018), to explain the current problems. These include the emergence of activist groups at a local level defending anti-tourism movements against the increasing number of travellers visiting tourist destinations (Seraphin et al., 2018).

In response to the question of how can tourism be maintained more sustainably, the aim of this chapter is to offer ways in which destinations can overcome the issue of overtourism. In light of this aim, the chapter also focuses on proposing some alternative ways for tourist destinations to cope with the problem of overtourism.

DOI: 10.4018/978-1-7998-2224-0.ch001

DEFINING OVERTOURISM

The contemporary debate on overtourism and the related term, the anti-tourist movement, has led to a redefinition of tourism development and planning issues (Panayiotopoulos & Pisano, 2019). It has also drawn attention due to its adverse impact on society and the environment, stemming from overcrowding, which is closely linked to and derived from mass tourism (Jacobsen et al., 2019; Koens et al., 2018; Martín Martín et al., 2018). The concept of overtourism is described as "the phenomenon of a popular destination or sight becoming overrun with tourists in an unsustainable way" (Collins Online Dictionary, 2019). This concept is closely related to a carrying capacity that is exceeded by destinations hosting an excessive number of visitors (Cheer et al., 2019), and is associated with tourism growth (Gonzalez et al., 2018).

By bringing a distinct perspective, Cheer, Milano and Novelli (2019) put forwards the approach of temporal overtourism. According to the authors, the temporal overtourism concept refers to a seasonal overcrowding problem that mostly emerges during peak holiday seasons or periods.

In their study, Namberger, Jackisch, Schmude and Karl (2019) noted two main types of resident typology towards the overtourism phenomenon, namely, the "mass tourism avoider" and the "tourism sympathizer". The authors reported that travellers representing the "mass tourism avoider" type often prefer to stay in less crowded places as compared to tourists who fall within the typology of "tourism sympathizer".

As such, the concern of overtourism is one of the main manifestations of growth (Higgins-Desbiolles et al., 2019). Recently, in several European destinations (e.g., Venice, Rome, Amsterdam, Florence, Berlin, Lisbon, Palma de Mallorca, etc.), due to overtourism, anti-tourism movements or tourismphobia, local residents initiated measures in response to the overflow of travellers visiting their areas (Burgen, 2019). Particularly in Venice, local residents have made many protests against tourism as they believe that tourists devastate their city (Higgins-Desbiolles, 2017). Similarly, in Barcelona, as a result of the excessive amount of visitors in European tourism destinations such as Venice, anti-tourism movements have emerged protesting against tourists (Hughes, 2018). The issue of overtourism, for both local people and travellers, has also been seen in New York City, Reykjavik, the Isle of Skye, Koh Phi Phi, Thailand and Palawan, the Philippines (Oklevik et al., 2019). All the issues deriving from destinations hosting an excessive number of tourists reveal that the overtourism phenomenon is a problem potentially faced by mature destinations.

SUSTAINABLE TOURISM VS. OVERTOURISM

The increasing interest both in sustainable tourism development and sustainable tourism has been associated with an equally growing concern about the social and biophysical impacts of tourism (McCool & Lime, 2001). On the other hand, local residents, whose attitudes towards improving tourism through the promotion of sustainable tourism development is a key factor (Eusébio et al., 2018), as they are regarded as the most important part of the system through which tourism is provided (Muler-Gonzalez et al., 2018), have been negatively affected due to the adverse impact of tourism. For this reason, the discourse on carrying capacity has been given more attention alongside overtourism (Leung et al., 2018). However, even though they evoke an identical meaning, the concepts of overtourism and carrying capacity are different from each other. The overtourism problem has diverse negaitve impacts on carrying capacity, which can be divided into several distinct forms (Namberger et al., 2019, p. 5):

The physical carrying capacity refers to the maximum number of tourists that any particular tourist destination can host.

The environmental or ecological carrying capacity addresses the number of tourists who can visit a destination without destroying the environment or the ecosystem.

The economic carrying capacity identifies the degree to which the economic damage emanating from tourism becomes conspicuous, negatively influencing the cost of living for residents, and thereby resulting in tourism being no longer acceptable to local residents.

The infrastructural carrying capacity depicts the number of tourists that can be held by a destination's predominating infrastructure.

The perceptual carrying capacity highlights the number of tourists above which the quality of the tourist's experience deteriorates, although sometimes they do not notice that they are creating the problem themselves.

The social carrying capacity outlines the number of tourists above which the social and cultural changes driven by tourism are no longer acceptable to local residents.

The pros and cons of the tourism industry are still contested and its impacts are multilayered, consisting of social, cultural, political and environmental issues, along with economic ones (Pizam & Milman, 1986). According to McKercher (1993), tourism has some economic benefits as well as detrimental effects on host communities, cultures and environments. He presents some fundamental truths about tourism (p. 7):

As an industrial activity, tourism consumes resources, creates waste and has specific infrastructural needs.

As a consumer of resources, tourism has the ability to overconsume.

Tourism, as a resource-dependent industry, must compete for scarce resources to ensure its survival.

Tourism is a private sector-dominated industry, with investment decisions being based mainly on profit maximisation.

Tourism is a multi-faceted industry and, as such, it is almost impossible to control.

Tourists are consumers, not anthropologists.

Tourism is entertainment.

Unlike other industrial activities, tourism generates income by importing clients rather than by exporting its products.

Given the statements mentioned above, overall one can claim that tourism as an industry has adverse effects as well as economic benefits. Further, most importantly, when compared to other industries, in considering the second point it is more consistent with, and closely related to, overtourism. On the other hand, the seventeen Sustainable Development Goals (SDGs), developed by the World Tourism Organization (UNWTO), can potentially be used as a tool to ensure the sustainability of tourism in a more effective and efficient way. Among these goals, SDG 11 and SDG 12 deal with the problem of overtourism in destinations (Boluk et al., 2017). Overall, it can be said that the SDGs, with 17 goals and 169 targets, are in a better position to serve not only sustainable tourism but also to help cope with overtourism as well.

DIMENSIONS AND CHALLENGES OF OVERTOURISM

Phi (2019) identifies the causes of overtourism that have substantial and long-lasting effects on the phenomenon:

Causes of overtourism, centring on tourists: This refers to the increasing number of tourists, including people who travel mostly to popular destinations, which subsequently precipitates overtourism. The main problem is engendered by tourists who exert a negative impact on the environment, communities and cultures (Panayiotopoulos & Pisano, 2019).

Impacts of overtourism, centring on locals: This theme addresses the negative impact of tourism from the point of view of local residents, who are disturbed by overcrowding problems in destinations. Therefore, this theme is also related to the concepts of tourismphobia and anti-tourism protest or movements, which are growing against tourism and tourists due to issues such as the overuse of public places, congested streets and the rising cost of living.

Centring on cities or destinations: Today, the issue of overtourism is mainly encountered in major urban centres or cities, where overcrowding is seen as highly problematic, while in the past attention was predominantly given to protected areas, in terms of tourism and visitor management of overcrowding (Leung et al., 2018). The major cities facing overtourism are Barcelona, Venice, Dubrovnik and Paris (Cheer et al., 2019; Oklevik et al., 2019). The common element among these cities is that their tourism industry is heavily driven by the cruise industry, which thus leads to overtourism.

Referring to the growth agenda, centring on the travel industry: In this arena, the discourse on growth takes up a significant amount of space in the overtourism debate, by underscoring the economic contributions made by tourism as being part of neoliberalism. This stance supports the view that overtourism can be addressed only by one symptom of the current problem (Higgins-Desbiolles, 2019a) and is thus supported by the SDGs and dealt with in the agenda of the UNWTO, which covers continued growth that is vital for providing sustainable tourism while tackling the negative impacts of overtourism (Boluk et al., 2019).

As described above, the negative impacts of overtourism are multiple and wide-ranging, affecting a series of elements operating or placed within the tourism-value system. As such, five types of challenges regarding overtourism are addressed by the World Travel & Tourism Council and McKinsey & Company (2017):

Alienated local residents: Here, the impact on stakeholders is largely addressed. Local residents are opposed to the negative impact of tourism, such as increasing rents, the displacement of locals and local retail businesses, noise, and the shifting neighbourhood attributes and outflow of economic tourism benefits.

A degraded tourist experience: This refers to the negative effects on tourists from the overcrowding of cities. In many tourist destinations, the tourist experience itself can suffer owing to queues, crowding and disturbances caused by overcrowding, which can lead to increasing visitor displeasure.

An overloaded infrastructure clearly signals negative effects in the economic arena. Considering the infrastructure benefiting tourists is shared with a series of essential non-tourism activities, such as commerce, health and transport, the pressure placed on such infrastructure can adversely affect not only visitors, but also local people and businesses. Furthermore, the consumption patterns of some resources, such as water, and the production of waste by visitors add to local overconsumption and pollution.

Negative effects on nature underline the fact that visitors contribute to the overuse of natural resources, such as water and forests, waste pollution, and damage to flora and fauna.

Threats to culture and heritage: Overcrowding can undermine a destination's spiritual and physical integrity, and crowds can lead to inadequate security measures, resulting in areas suffering from vandalism (Kuščer, & Mihalič, 2019, p. 6).

Taking into consideration the negative impact and challenges of overtourism stemming from growth, destinations' reactions in terms of remedies in response to overtourism should be effective and include the proactive management of visitors, rather than ending tourism altogether, as stated by the UNWTO (2017): "growth is not the enemy, it is how we mange it".

SUSTAINABLE POLICY RESPONSES IN COPING WITH OVERTOURISM

In facing the issue of overtourism, there are two possible methods for tourist destinations to adopt: either stop promoting and maintaining tourism activity altogether or develop effective tourism policies and strategies in response to overtourism and tourismphobia. Some measures and recommendations adopted by destinations faced with overtourism, to be applied by public and private sector representatives, can be summarised as follows (Higgins-Desbiolles, 2018a; Seraphin et al., 2018, p. 2):

- Repression plus education of tourists,
- Teach visitors how to appreciate the city,
- Provide authentic tourist experiences (eat and drink with locals),
- TUI Group urged clients to choose destinations other than Spain,
- Increase taxes,
- Respect campaigns,
- Restriction or tightening of Airbnb activities,
- Promote a policy based on "local-hood" tourism, and
- Use demarketing strategies or deflect places of interest in order to address the increasing tensions of overtourism.

On the other hand, based on a social cooperative and collaborative model as a community-centred initiative, Fairbnb has been developed by Amsterdam and Venice destinations in response to the challenge of the devastating and disruptive model of Airbnb (Higgins-Desbiolles, 2019). The new platform has been designed to ensure visitor accommodation services put the priorities of locals first over profit, and by preserving local values in terms of authentic, sustainable and enriched travel experiences (https://fairbnb.coop/, 2019).

Additionally, the World Travel & Tourism Council and McKinsey & Company (2017) introduced some possible strategies that tourist destinations may initiate in order to cope with the issue of overtourism:

Spread out visitors over time: Many destinations face degradation due to an imbalance of visitors during the high season, called 'temporal overtourism', which is often seen in Mediterranean destinations during the summer (Cheer et al., 2019). One response of destinations to the negative impact of mass crowding that destroys infrastructure is to develop strategies to spread the tourist load, thus supporting the sustainability of the tourism industry.

Spread visitors across sites: Conveying visitors from hot spot locations to low-density areas can help spread tourists more evenly across residential and under-visited sites, preventing congestion in overcrowded locations.

Adjust pricing to balance supply and demand: Pricing might also be a useful way to provide a balance between supply and demand while tackling the overtourism problem. However, this strategy may lead to the likelihood of decreasing the number of travellers visiting any particular tourist destination due to the price increase. Moreover, it also brings up the issue of elitism, while ensuring domestic tourists are still able to reach their own heritage sites.

Regulate accommodation supply: Some destinations may try to directly control the supply of tourism accommodation. In doing so, the overtourism problem can be effectively managed and controlled by mature destinations facing overcrowding. For instance, in Barcelona, where the use of Airbnb accommodation is widely used, even taking precedence over hotels on peer-to-peer accommodation platforms (Gutiérrez et al., 2017), there is an effort to constrain Airbnb rentals by officials (Higgins-Desbiolles, 2018a).

Limit access and activities: In the case of overtourism reaching a critical stage, another response of destinations could be the limiting or banning of certain activities. For example, Dubrovnik is determined to limit the number of cruise ships entering the city at any time, while the Thai government closes the popular Maya Beach on Phi Phi Island for four months of every year to ensure the sustainability of sea life (Higgins-Desbiolles, 2018a).

In addition to the measures described above, more recently a list of destinations affected by overtourism that travellers should avoid has been released by the media (Groundwater, 2019). Furthermore, some key strategies and measures addressing the increased growth in visitors have also been developed by the UNWTO (2018):

- Promote the dispersal of visitors within the city and beyond,
- Promote time-based dispersal of visitors,
- Stimulate new visitor itineraries and attractions,
- Review and adapt regulation,
- Enhance visitors' segmentation,
- Ensure local communities benefit from tourism,
- Create city experiences that benefit both residents and visitors,
- Improve city infrastructure and facilities,
- Communicate and engage with local stakeholders,
- Communicate and engage with visitors, and
- Set monitoring and response measures.

Taking the policies developed in response to overtourism into account, it can be said that the strategies developed by the United Nations contain more comprehensive, broad and inclusive measures. It can also be emphasised that strategies and measures prioritise a longer-term management approach.

CONCLUSION

More recently, mature destinations are faced with an overtourism problem due to hosting an excessive number of visitors that exceeds the carrying capacity of those cities. This problem is seen as crucial since the tourism industry relies heavily on the interactions between host residents and tourists, which have become distorted due to the number of visitors exceding the carrying capacity of the destination (Cheung & Li, 2019), thus leading to anti-tourism movements. This stance confirms that local residents

Table 1. Strategies and measures to address visitor growth in cities

Strategies	Measures
Strategy 1 Promote the dispersal of visitors within the city and beyond	• Host more events in less visited parts of the city and its surroundings. • Develop and promote visitor attractions and facilities in less visited parts of the city and its surroundings. • Improve the capacity of and time spent at attractions. • Create a joint identity of the city and its surroundings. • Implement a travel card for unlimited local travel. • Mark entire city as inner-city to stimulate visitation of less visited parts.
Strategy 2 Promote time-based dispersal of visitors	• Promote experiences during off-peak months. • Promote dynamic pricing. • Stimulate events in off-peak months. • Set timeslots for popular attractions and/or events aided by real-time monitoring. • Use new technologies (apps and others) to stimulate dynamic time-based dispersal.
Strategy 3 Stimulate new visitor itineraries and attractions	• Promote new itineraries at the city entry points and throughout the visitor's journey, including at tourist information centres. • Offer combined discounts for new itineraries and attractions. • Produce city guides and books highlighting hidden treasures. • Create dynamic experiences and routes for niche visitors. • Stimulate the development of guided tours in less visited parts of the city. • Develop virtual reality applications to famous sites and attractions to complement onsite visits.
Strategy 4 Review and adapt regulation	• Review opening times of visitor attractions. • Review regulation on access for large groups to popular attractions. • Review regulation on traffic in busy parts of the city. • Ensure visitors use parking facilities at the edge of the city. • Create specific drop-off zones for coaches in suitable places. • Create pedestrian-only zones. • Review regulation and taxation on new platform tourism services. • Review regulation and taxation on hotels and other accommodation. • Define the carrying capacity of the city and of critical areas and attractions, etc. • Consider an operator's licence system to monitor all operators, etc. • Review regulation on access to certain areas of the city for tourist-related activities.
Strategy 5 Enhance visitors' segmentation	• Identify and target visitor segments with lower impact according to the specific city context and objectives. • Target repeat visitors. • Discourage visitation of the city of certain visitor segments.
Strategy 6 Ensure local communities benefit from tourism	• Increase the level of employment in tourism and strive to create decent jobs. • Promote the positive impacts of tourism, create awareness and knowledge of the sector amongst local communities. • Engage local communities in the development of new tourism products. • Conduct an analysis of supply-demand potential of the local communities and promote their integration in the tourism value chain. • Improve quality of infrastructure and services considering residents and visitors. • Stimulate development of impoverished neighbourhoods through tourism.
Strategy 7 Create city experiences that benefit both residents and visitors	• Develop the city to fit with the residents' needs and desires and consider tourists as temporary residents. • Develop tourism experiences and products that promote the engagement of residents and visitors. • Integrate visitor facilities within local festivities and activities. • Create and promote local city ambassadors. • Promote art and culture initiatives, such as street art, to provide fresh perspectives on the city and expand visitation to new areas. • Extend opening times of visitor attractions.

continues on following page

Table 1. Continued

Strategies	Measures
Strategy 8 Improve city infrastructure and facilities	• Create a city-wide plan for a well-balanced, sustainable traffic management. • Ensure that major routes are suitable for extensive tourism activity and that secondary routes are available at peak times. • Improve urban cultural infrastructure • Improve directional signage, interpretation materials and notices. • Make public transport better suited for visitors. • Set up specific transport facilities for visitors during peak periods. • Provide adequate public facilities. • Create safe cycling routes and stimulate bicycle rentals. • Set up specific safe and attractive walking routes. • Ensure that routes are suitable for the physically impaired or elderly visitors in line with accessible tourism principles. • Safeguard quality of cultural heritage and attractions. • Ensure cleaning regimes fit with tourism facilities and with peak times.
Strategy 9 Communicate and engage with local stakeholders	• Ensure that a tourism management group (including all stakeholders) is set up and is regularly convened. • Organise professional development programmes for partners, etc. • Organise local discussion platforms for residents. • Conduct regular research amongst residents and other local stakeholders. • Encourage locals to share interesting content about their city on social media. • Communicate with residents about their own behaviour. • Unite disjointed communities.
Strategy 10 Communicate and engage with visitors	• Create awareness of tourism impact amongst visitors. • Educate visitors on local values, traditions and regulations. • Provide adequate information about traffic restrictions, parking facilities, fees, shuttle bus services, etc.
Strategy 11 Set monitoring and response measures	• Monitor key indicators such as seasonal fluctuations in demand, arrivals and expenditures, patterns of visitation to attractions, visitor segments, etc. • Advance the use of big data and new technologies to monitor and evaluate tourism performance and impact. • Create contingency plans for peak periods and emergency situations.

Source: UNWTO, 2018, pp. 8–9

are now more consciously aware of their environment and interested in protecting and increasing their quality of life rather than the income they receive from tourism (Seraphin et al., 2018). In response to this problem, destinations have begun to take some necessary measures, on their own initiative, by introducing useful regulation. However, the strategies and measures of the UNWTO towards overtourism appear to be very detailed and inclusive when compared to those policies and strategies adopted by destinations themselves. They also cover proactive strategies in response to city growth, which leads to the overcrowding that subsequently engenders the overtourism problem. The present chapter, based on a literature review, highlights the need for a current understanding of the overtourism phenomenon. Thus, the reasons behind the social tensions amongst host communities and tourists emerging from overtourism need to be further explored by future research studies providing empirical results.

REFERENCES

Boluk, K., Cavaliere, C. T., & Higgins-Desbiolles, F. (2017). Critical thinking to realize sustainability in tourism systems: Reflecting on the 2030 sustainable development goals. *Journal of Sustainable Tourism*, *25*(9), 1201–1204. doi:10.1080/09669582.2017.1333263

Boluk, K., Cavaliere, C. T., & Higgins-Desbiolles, F. (2019). A critical framework for interrogating the United Nations Sustainable Development Goals 2030 Agenda in tourism. *Journal of Sustainable Tourism*, *27*(7), 847–864. doi:10.1080/09669582.2019.1619748

Burgen, S. (2018). 'Tourists go home, refugees welcome': Why Barcelona chose migrants over visitors. Retrieved 16 May, 2019, from https://www.theguardian.com/cities/2018/jun/25/tourists-go-home-refugees-welcome-why-barcelona-chose-migrants-over-visitors

Cheer, J. M., Milano, C., & Novelli, M. (2019). Tourism and community resilience in the Anthropocene: Accentuating temporal overtourism. *Journal of Sustainable Tourism*, *27*(4), 554–572. doi:10.1080/09669582.2019.1578363

Cheung, K. S., & Li, L. H. (2019). Understanding visitor–resident relations in overtourism: Developing resilience for sustainable tourism. *Journal of Sustainable Tourism*, *27*(8), 1197–1216. doi:10.1080/09669582.2019.1606815

Collins Online Dictionary. (2019). Overtourism. Retrieved May 16, 2019, from https://www.collinsdictionary.com/submission/19794/Overtourism

Eusébio, C., Vieira, A. L., & Lima, S. (2018). Place attachment, host–tourist interactions, and residents' attitudes towards tourism development: The case of Boa Vista Island in Cape Verde. *Journal of Sustainable Tourism*, *26*(6), 890–909. doi:10.1080/09669582.2018.1425695

Fairbnb.coop. (2019). Community Powered Tourism. Retrieved October 5, 2019, from https://fairbnb.coop/

Groundwater, B. (2019). Ten destinations to avoid in 2019. Retrieved May 16, 2019, from https://www.stuff.co.nz/travel/themes/109938533/ten-destinations-to-avoid-in-2019

Gutiérrez, J., García-Palomares, J. C., Romanillos, G., & Salas-Olmedo, M. H. (2017). The eruption of Airbnb in tourist cities: Comparing spatial patterns of hotels and peer-to-peer accommodation in Barcelona. *Tourism Management*, *62*, 278–291. doi:10.1016/j.tourman.2017.05.003

Higgins-Desbiolles, F. (2017). 'Sustainable Tourism' is not working – here's how we can change that. Retrieved May 16, 2019, from https://theconversation.com/sustainable-tourism-is-not-working-heres-how-we-can-change-that-76018

Higgins-Desbiolles, F. (2018a). Why Australia might be at risk of 'overtourism'. Retrieved May 16, 2019, from https://theconversation.com/why-australia-might-be-at-risk-of-overtourism-99213

Higgins-Desbiolles, F. (2018b). Sustainable tourism: Sustaining tourism or something more? *Tourism Management Perspectives*, *25*, 157–160. doi:10.1016/j.tmp.2017.11.017

Higgins-Desbiolles, F. (2019). Rethinking tourism so the locals actually benefit from hosting visitors. Retrived October 5, 2019, from https://theconversation.com/rethinking-tourism-so-the-locals-actually-benefit-from-hosting-visitors-116066

Higgins-Desbiolles, F., Carnicelli, S., Krolikowski, C., Wijesinghe, G., & Boluk, K. (2019). Degrowing tourism: Rethinking tourism. *Journal of Sustainable Tourism*, *27*(12), 1926–1944. doi:10.1080/09669582.2019.1601732

Hughes, N. (2018). 'Tourists go home': Anti-tourism industry protest in Barcelona. *Social Movement Studies*, *17*(4), 471–477. doi:10.1080/14742837.2018.1468244

Jacobsen, J. K. S., Iversen, N. M., & Hem, L. E. (2019). Hotspot crowding and over-tourism: Antecedents of destination attractiveness. *Annals of Tourism Research*, *76*, 53–66. doi:10.1016/j.annals.2019.02.011

Koens, K., Postma, A., & Papp, B. (2018). Is overtourism overused? Understanding the impact of tourism in a city context. *Sustainability*, *10*(12), 4384. doi:10.3390u10124384

Kuščer, K., & Mihalič, T. (2019). Residents' Attitudes towards Overtourism from the Perspective of Tourism Impacts and Cooperation – The Case of Ljubljana. *Sustainability*, *11*(6), 1823. doi:10.3390u11061823

Leung, Y. F., Spenceley, A., Hvenegaard, G., & Buckley, R. (Eds.). (2018). *Tourism and Visitor Management in Protected Areas: Guidelines for Sustainability. Best Practice Protected Area Guidelines Series No. 27*. Gland, Switzerland: IUCN. doi:10.2305/IUCN.CH.2018.PAG.27.en

Martín Martín, J., Guaita Martínez, J., & Salinas Fernández, J. (2018). An analysis of the factors behind the citizen's attitude of rejection towards tourism in a context of overtourism and economic dependence on this activity. *Sustainability*, *10*(8), 2851. doi:10.3390u10082851

McCool, S. F., & Lime, D. W. (2001). Tourism carrying capacity: Tempting fantasy or useful reality? *Journal of Sustainable Tourism*, *9*(5), 372–388. doi:10.1080/09669580108667409

McKercher, B. (1993). Some fundamental truths about tourism: Understanding tourism's social and environmental impacts. *Journal of Sustainable Tourism*, *1*(1), 6–16. doi:10.1080/09669589309450697

Mihalic, T. (2016). Sustainable-responsible tourism discourse – Towards 'responsustable' tourism. *Journal of Cleaner Production*, *111*, 461–470. doi:10.1016/j.jclepro.2014.12.062

Muler Gonzalez, V., Coromina, L., & Gali, N. (2018). Overtourism: Residents' perceptions of tourism impact as an indicator of resident social carrying capacity – Case study of a Spanish heritage town. *Tourism Review*, *73*(3), 277–296. doi:10.1108/TR-08-2017-0138

Namberger, P., Jackisch, S., Schmude, J., & Karl, M. (2019). Overcrowding, Overtourism and Local Level Disturbance: How Much Can Munich Handle? *Tourism Planning & Development*, *16*(4), 452–472. doi:10.1080/21568316.2019.1595706

Oklevik, O., Gössling, S., Hall, C. M., Steen Jacobsen, J. K., Grøtte, I. P., & McCabe, S. (2019). Overtourism, optimisation, and destination performance indicators: A case study of activities in Fjord Norway. *Journal of Sustainable Tourism*, 1–21. doi:10.1080/09669582.2018.1533020

Panayiotopoulos, A., & Pisano, C. (2019). Overtourism dystopias and socialist utopias: Towards an urban armature for Dubrovnik. *Tourism Planning & Development*, *16*(4), 393–410. doi:10.1080/21568316.2019.1569123

Phi, G. T. (2019). Framing overtourism: A critical news media analysis. *Current Issues in Tourism*, 1–5. doi:10.1080/13683500.2019.1618249

Pizam, A., & Milman, A. (1986). The social impacts of tourism. *Tourism Recreation Research*, *11*(1), 29–33. doi:10.1080/02508281.1986.11014414

Seraphin, H., Sheeran, P., & Pilato, M. (2018). Over-tourism and the fall of Venice as a destination. *Journal of Destination Marketing & Management*, *9*, 374–376. doi:10.1016/j.jdmm.2018.01.011

UNWTO. (2017). UNWTO & WTM Minister' Summit Overtourism: "growth is not the enemy, it is how we manage it". Retrieved October 5, 2019, from https://www2.unwto.org/unwto-wtm-ministers-summit-2017

World Tourism Organization (UNWTO); Centre of Expertise Leisure, Tourism & Hospitality; NHTV Breda University of Applied Sciences; and NHL Stenden University of Applied Sciences. (2018). 'Overtourism'? – Understanding and Managing Urban Tourism Growth beyond Perceptions, Executive Summary, UNWTO, Madrid, Doi:10.18111/9789284420070

World Travel & Tourism Council (WTTC) and McKinsey & Company. (2017). Coping with Success: Managing Overcrowding in Tourism Destinations. Retrieved August 14, 2019, from https://www.wttc.org/priorities/sustainable-growth/destination-stewardship/

Chapter 2
Overtourism: A Systematic Review of Literature

Fatima Lampreia Carvalho
https://orcid.org/0000-0002-1189-8843
University of Algarve, Portugal

Manuela Guerreiro
https://orcid.org/0000-0002-6398-9712
University of Algarve, Portugal

Nelson Matos
University of Algarve, Portugal

ABSTRACT

This chapter conceptualises the notion of overtourism by providing a systematic literature review mapping the existing research knowledge. The mapping exercise benefits from textual data from academic articles on overtourism. The study focuses on measures to counteract overtourism from a marketing perspective. Since tourism is about customers' experiences, understanding overtourism and its aspects is crucial at a time the market growth calls for measures to control this phenomenon. Using online databases and NVIVO 12 software, 66 articles were selected for bibliographic content analysis. Findings highlight the need to work the concept of overtourism in a holistic approach encompassing five dimensions: social, political & governance, marketing & customer experience, economic, and environmental. Results permitted to identify other dimensions; overtourism, symbolic, ethical-moral, cultural. Managing overtourism is challenging because whereas tourism is about customers' experiences, for policy makers overtourism encompasses strategies with collective effects.

DOI: 10.4018/978-1-7998-2224-0.ch002

INTRODUCTION

The main goal of this chapter is to find a common ground underlying the several dimensions of overtourism and explain how the many facets of overtourism interact in theoretical and practical terms. This literature review aims to provide the basis for the conceptualisation of *resilient destinations* that are supplied with information on how to prevent the negative impacts of tourism. A secondary goal of this chapter is to identify potential gaps in the literature concerning overtourism. This chapter offers therefore three main contributions for new scholarship on tourism studies: (i) It scrutinises key concepts for a discussion of overtourism based on a systematic literature review; (ii) It indicates main absences in research concerning overtourism and (iii) it explores possible areas for collaboration among policy makers, business managers and travel operations in tackling overtourism.

Resilient destinations that are robust enough to cope with the ongoing dynamisms of society depend on, robust assessment tools able to detect signs of overtourism, and support strong *stakeholder negotiation*. (Figure 1).

With basis on the literature, there is evidence that destination robustness depends on stakeholder cooperation as a major enabler on overtourism policy making. This chapter seeks therefore to elaborate an argument on how the several dimensions of overtourism, found in the academic literature (1997-2020) can contribute to the conceptualisation of *resilient destinations* that are robust enough to cope with the ongoing dynamisms of society (Hartman, 2018, p.7). Studies on building resilient destinations are associated with debates on *optimal tourism* (Morteza, Seddiq, Sharareh, and Jamal, 2016; Wang, Li, & Li, 2013; Shafiee et al, 2019), optimal tourism clustering (Gao, Zhang, Lu, Wu and Du, 2018). and with optimal crowding levels (Jacobsen, Iversen, and Hem, 2019) as well as with new models for *Smart tourism* destinations (Shafiee, Ghatari, Hasanzadeh, Jahanyan, 2019). Moreover, new methodologies to support the management of tourism using a competitive, smart, and sustainable approach have demonstrated that the competitive advantage of a destination originates also from its managerial efforts and the ability for optimal resource allocation (Wang, Li, & Li, 2013; Shafiee et al, 2019, p. 289). Yet progress towards optimal tourism and resilient destinations cannot progress without deep examination of empirical and theoretical studies of overtourism.

Literature in tourism studies provides several case studies of how overtourism impacts on the customers' experience (Fotiadis, 2016; Zhang et al, 2017), natural parks (Prakash, Perera, Newsome, Kusiminda and Walker, 2019), and at the sea and beaches (Saveriades, 2000; Andriotis, 2005; Marrero Rodríguez, 2006; Capo et al., 2007; Cole, 2012; Ordiz, 2017). Moreover, city overtourism is already notorious for residents' complaints about places like Berlin, Paris, London, Amsterdam, Venice, Rome, or Florence (Wall, 1997; Rees, 1997; Russo 2002; Garrigós, Narangajavana, and Palacios, 2004; Fistola and La Rocca, 2017; Lerario and Di Turi, 2018; Martí Martín et al., 2018). Not only the tourism experience in the city context but also in other areas, such as rural areas (Royo et al., 2009; Shin et al., 2017), natural sanctuaries and mountains (Steven et al., 2011; Luque-Gil et al., 2018) kindle debate on the matter.

Overtourism now poses a great challenge to tourism policy makers, private sector and academics aiming to propose responses to the social discomfort. The problem to stakeholder cooperation and involvement arises because apart from destination residents, there are also two sides to every narrative on overtourism: From the perspective of governments, travel operators and tourists, on one side, and from private businesses and entrepreneurs, on the other side. Tourism boosts economies, employs thousands of people, enriches businesses and pays for important public services. But overtourism may increase housing prizes in popular tourist areas.

Figure 1. Overtourism literature tree map. Source: 84 Academic Journals 1997- 2019
Source: The authors content analysis of literature with NVIVO 12 plus.

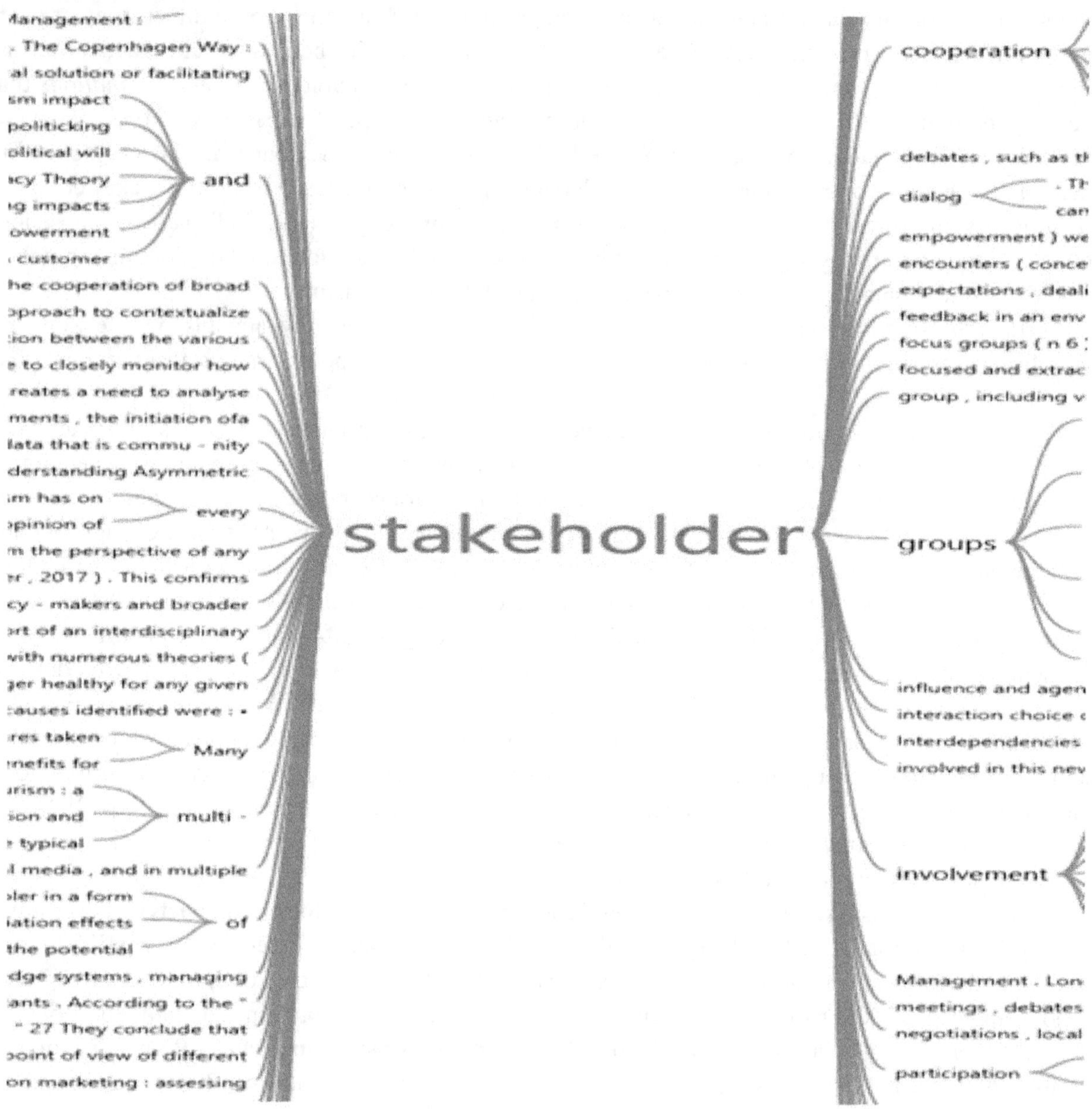

Tourism puts a strain on already scarce resources and the negative side of the overflow of tourists led destinations, regions and tourism areas to consider de-marketing. Using the coin analogy, previously referred to as the Janus-faced character of the tourism industry in Cuba (Sanchez and Adams, 2008), is helpful to think whether de-marketing would benefit tourism businesses, mitigate unpleasant experiences for customers in overcrowded venues. But it is disputable whether *de-marketing* is an efficient deterrent against overtourism (Séraphim Sheeran and Pilato, 2018 b). But there is not a one-size-fits-all solution for overtourism and therefore solutions need to be made to fit each local context (Koens, Postma, Papp,2018, p. 8-10).

Cheer, Milano and Novelli (2019: 561-2) questioned how practicable the pursuit of sustainable tourism is in an era in which human beings have become the primary agents of change on a planetary scale. This new era may undermine the resilience of tourism dependent communities. Several authors (Bec, McLennan & Moyle, 2016; Calgaro, Lloyd, Hoves, 2014; Lew & Cheer, 2018) anticipated a new approach to overtourism whilst adopting *resilience* as a framework to assess responses to perturbations emerging from the tourism system. Such responses should recognize that people and their institutions are integral components of ecological systems, and it makes sense to tackle the root causes of social and environmental issues rather than perpetually react to disaster and crisis events (Cretney, 2014, p. 636).

In the same vein, Kuscer and Mihalic (2019, p.3) combined the existing sustainability–responsibility tourism framework (from academic research with industry research on overtourism) to propose stakeholder cooperation based on the overtourism risk monitoring model. Their model of Sustainable-Responsible Tourism (SRT) relies on an awareness of full sustainability and on sustainability ethics, supported by environmental education, knowledge and values, with full awareness about sustainability issues on the part of all stakeholders on both the demand and supply sides. It is therefore crucial investigate the direct and indirect relationships among overtourism impact factors in the several thematic dimensions found in tourism literature.

theoretical Background

Certain authors described the overtourism as a multidimensional and complex phenomenon, which has expanded out in the context of wider societal and city developments (Koens, Postma and Papp, 2018). The original research debate on overtourism is coherent with concerns over the natural and social environments from the early 1970s regarding its costs and benefits of tourism (Young, 1973) which initiated debates on environmental ecology in tourism studies (Kaspar, 1973, p. 139), termed environmental ecology. Later, Swiss ecologist Jost Krippendorf (1984) challenged the sense of mass tourism in his book entitled "Vacation People" ("Ferienmenshen" in German), translated as "Holiday Makers" in English (Krippendorf, 1987, 1999) and began the search for alternatives. Tourism critics in the 1980s called for "...more responsibility for the effects of travel and behaviour on host environments, both physical and human" (Butler, 1995, p. 5). This interest in more responsibility led to so-called alternative tourism forms and concepts which have been given many names, such as alternative, soft, quality, eco, responsible, minimum impact tourism, green and ethical tourism, with all of them representing an alternative to the mainstream mass tourism. The latter has become environmentally, socially, ethically and politically intolerable (Mihali, 2006; Swarbrooke, 1999). In general, global environmental concern culminated in "Our Common Future", as defined in the Brundtland report (WCED, 1987), which brought the global environmental debate and the notion of sustainability to the forefront of global and local social and political thinking and agendas.

The current debates pervading the overtourism debate proved helpful to clarify the need for a desired tension between consumerist behaviours, post-consumerist attitudes as the basis for new policymaking able to control overtourism. The contention here is that, on one hand, a consumerist vision of tourism in late Capitalist societies breed insatiable hedonistic impulses and economic pressure to increase tourism profits. On the other hand, post consumerist vision of tourism (Boyle, Wilson, and Dimmock, 2015; Hales and Jennings, 2017; Phi, Whitford, Dredge and Reid, 2017; Higgins-Desbiolles, 2018; Higgins-

Desbiolles, Carnicelli, Krolikowski, Wijesinghe and Boluk, 2019) encouraged the re-establishment of a moral order of consumption via: (a) the symbolic management of destination image; (b) social management of tourists' image; (c) political activism and social protests to stop overtourism and (d) increased global environment anxiety to protect fragile ecosystems.

Previous studies seem to propose that the concept of crowding is more closely connected to user's expectations than to actual level of density (Andereck and Becker, 1993; Ditton, Fedler and Graefe, 1983; Shelby, Heberlein, Vaske and Alfano, 1983). Remarkably, hotspot crowding does not evoke purely negative emotions; it also stimulates positively directed sentiments and positive reactions that can add value to places reflecting a "collective gaze" (Urry, 1990). Overall, these places can have a social density that can enable positive (e.g., fun) and negative (e.g., discomfort) atmosphere among vacationers due to space violations and loss of interest and uniqueness (Jakobsen, Iversen and Hem, 2019; Podilchak, 1991).

The expectancy theory explains that tourists tend to hold a temporary belief that an act will be followed by a determined result. There is therefore a weak relationship between crowding perception and actual encounter level. People who participate in recreational activities wait for a reward and when an expectancy is frustrated visitors and residents alike can "...experience an increased level of loss of control, perceived lack of space, invasion of personal space, or unwanted interference, even though the actual number of encounters is not high" (Luque-Gil et al. 2018, p. 94). The stimulus overload theory applied by social psychologists to urban areas explain that an overload of stimuli associated with overtourism can be negative because the individual is not able to cope with the high number and frequency of stimuli. (Andereck and Becker, 1993).

The overtourism problematic has a strong sociological component as it implies an overuse of the resources, infrastructure, or facilities of a destination or parts of it (Koens et al. 2018). *Tourism Towards 2030* (2011) (a broad research project in continuation of United Nations World Tourism Organization's work in the area of long-term forecasting) presented evidence that the number of international tourists arrivals worldwide should reach 1.8 billion by 2030, which represents an average estimation of 3.3% per year within the next decade (UNWTO, 2016).

Study design

This chapter relied on secondary data collection for the literature review. Concept-driven literature review is a process of research and evaluation of available literature on an area of scientific knowledge (Bem, 1995; Webster and Watson, 2002; Boote and Beile,2005; Volgger and Huang, 2019), covering different research traditions and study ðelds. The present work adopted mapping literature review to show the associations between different literature and help to identify issues such as proximity and connections in terms of ideas and findings. The purpose was to conceptualize the notion of "overtourism" and describe the "state of the art" pinpointing the cause and consequences of "overtourism" (e.g., Volgger, 2018; Goodwin, 2017; Oskam and Boswijk, 2016; Veiga, Santos, Águas and Santos, 2018; Cruz and Zaragoza, 2019).

Data collection of academic articles took place between May and October 2019 with the assistance of EndNote X9. The latter is a software for automatic generation of citations and bibliographic references. Its integration with a word processor allows citations to be inserted as articles are produced. Researchers with an interest in overtourism used this software to collect data and perform data analysis of academic sources and online reviews (Cheer, Milano and Novelli, 2019; Papathanassis, 2019a).

The starting point for this literature review was scoping, planning, researching, selecting and evaluating academic articles, book chapters and PhD theses. In the scoping phase, the researchers formulated a clear and objective research question: considering the last 20 years of approaches to overtourism and overcrowding in Tourism Studies publications, how the many facets of overtourism interact in theoretical and practical approaches to the matter? Can a literature review with focus on the period 1997-2019, provide the basis for the conceptualisation of *resilient destinations* able to prevent the negative impacts of tourism and support strong *stakeholder negotiation*? What potential gaps exist in the literature selected for a systematic analysis on overtourism?

The study design also contemplated the impact of overtourism on customer experience management and on destination management. On this concern, the researchers located previous reviews on those topics (Gonzalez, Coromina and Galí, 2018; Jakobsen, Iversen and Hem, 2019; Luque-Gil et al., 2018; Perkumienė, Pranskûnienė, 2019; Volgger and Huang, 2019). The researchers undertook work to map the literature and highlight thematic emphases and associations in the overtourism debate. A descriptive or mapping literature review (Fink, 1999; Chiou, 2009; Lingard, 2017) became the method of choice to assemble selected textual data on overtourism as well as in other contexts that are relevant to studies in tourism.

Sampling and Data Collection Procedures

This systematic literature review process involved six steps mentioned in previous literature on research methodologies (Templier and Paré, 2015):

(1) framing the research question(s) and objective(s)

The first research objective of this chapter is to perform a content analysis of 84 tourism academic sources, from 1997 to 2019, to explore how the several dimensions of overtourism interrelate, thus supporting new conceptualisation of *resilient destinations* that are robust enough to cope with the ongoing dynamisms of society, as identified in empirical case studies about overtourism

The second research objective is to identify potential gaps in the literature establishing therefore promising areas for future research on stakeholder behaviour in specific areas of overtourism studies.

(2) searching the existing literature in data bases

Tourism researchers recognized the critical role that academic journals play in academic scholarship (Xiao and Smith, 2006). Some suggested that "research journals are ..the major showcase of research in the field" (van Doren, Koh and McCahill, 1994:308) and for this reason over the past decades tourism researchers have been devoting more attention to publish in academic journals (e.g., Cheng, Li, Petrick and O'Leary, 2011; Hall et al., 2004; Jogaratnam, Chon, et al., 2005).

To locate pertinent research, an academic article search was performed in February 2018, May 2019 and October 2019. The authors used the Web of Science, a website that allows subscription-based access to multiple databases with comprehensive citation data for different academic disciplines. Access to academic articles relied on the Online Knowledge Library (b-on) which provides unlimited and permanent access to research institutions and higher education. This library gives access to Social Sciences data bases such as Business Source Complete (EBSCO) and ProQuest databases (ProQUest, 2019). The search on the Online Knowledge Library (b-on) was supplemented with the exploration of full text via the Scopus and Web of Science databases.

(3) screening literature for inclusion

In the planning phase of literature search using EndNote X9 the keywords to be used in the literature search terms were decided considering alternative words to "overtourism", synonyms, singular, plural and derivative words, such as "over-crowded", and generalisations. Between February 2019 and October 2019, the searches returned with 497 hits (Table 1).

Most of the 48 academic articles on "Overtourism" found via the Web of Science bibliographic management were classified published under the categories 'Hospitality leisure sport tourism' (31 or 64.5%) followed by 'Green sustainable science technology' (17 or 35.4%), 'Environment sciences' (12 or 25%) 'Environmental studies' (12 or 25%), 'Economics' (2 or 4% and 'Management' (2 or 4%).

Selection criteria of 10 bibliographic sources was based on what the Web of Science core collection results identified as 'highly cited in the field' such as Perkuniene, D. and Perkuniene, R. (2019) Overtourism: Between the Right to Travel and Residents' Rights and Gürsoy (2019) - Beauty and the Beast: A Fairy Tale of Tourismphobia. Further selection criteria were the inclusion of a collection of works from Web of Science with high rate of citations. Works selected fluctuated between 104 cited references (Martín Martín, Martínez and Fernández, 2018) to 83 cited references (Milano, 2018) with a minimum of 80 cited references (Novy, 2019).

Other bibliography such as Moreno-Izquierdo, Ramón-Rodríguez & Such-Devesa, 2018 (69 references), Koens, K., Postma, A. and Papp, B., 2018 (66 cited references), Postma and Schmuecker, 2017 (60 cited references), Sarantakou & Terkenli, 2019 (51 cited references) were included in this analysis for being relevant case studies in overtourism in destinations of great interest.

In the planning phase, the most common areas of identification of "overtourism" debates as well as criteria for inclusion and exclusion of academic articles was the authors' ability to conceptualise overtourism across various dimensions to inform stakeholders on overtourism management. Other key planning aspects were the units of observation, as well as time factor and data aimed. At the screening

Table 1. Overtourism bibliography search

Source name	Search words	Date	Hits
B-on	• Overtourism	14-05-2019	157.
	• Overtourism control	14-05-2019	38.
	• Overtourism governance	14-05-2019	30.
	• Overtourism control	14-05-2019	65.
ProQuest E-Book Central Springer	•Overtourism	14-05-2019	12.
Scopus	Tourism Leisure and hospitality Management journals	16-05-2019	113.
Web of Science	Tourism Management Journal: "Overtourism"	16-05-2019	2.
Emerald Insight	Overtourism	09- 07-2019	32.
Web of Science	Overtourism - all sources	22-10-2019	48

Source: (Authors, 2019)

phase the authors decided to keep academic journal articles, retaining also book chapters, two conference proceedings, and two magazines. The date aimed was for the period 1997- 2019.

(4) assessing the quality and relevance of primary studies.

In their planning steps, directives and protocols which helped in the elaboration and presentation of the systematic reviews were consulted (e.g., AMSTAR -Assessment of Multiple Systematic Reviews by Shea, Grimshaw, Wells, Boers, Andersson, Hamel, Porter, Tugwell,, Moher, and Bouter, 2007). The authors also benefited from the Preferred Reporting Items for Systematic Reviews and Meta-Analyses (PRISMA) and the PRISMA Flow Diagram 2009- PRISMA Statement by Moher, Liberati, Tetzlaff, Altman and PRISMA Group (2009), cited in 46 of the 320 related articles (Google search, 2019). Tao, Li, Zhou, Moher, Ling and Yu (2011) noted that in a sample of 146 journals publishing systematic reviews, the PRISMA Statement was mentioned frequently by authors, more specifically in 27% (40/146) of the journals; more often in general and internal medicine journals (7/14; 50%) than in specialty medicine journals (33/132; 25%) (Figure 2).

Figure 2. PRISMA Flow Diagram - Overtourism study
Source: Adapted from Moher, Liberati, Tetzlaff, Altman, and PRISMA Group,2009)

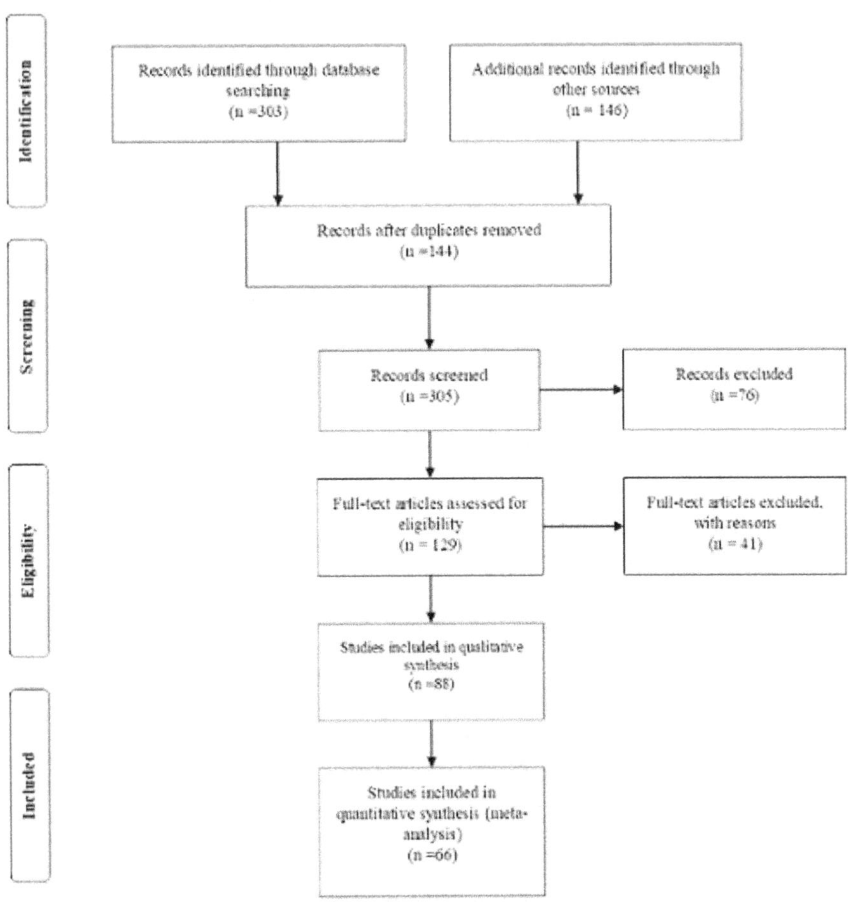

(5) extracting data

At a later stage, all selected journal contents with direct reference to "overtourism" were codified and quantified using software NVIVO 12 Plus, able to import bibliography gathered via EndNote X9 software. Such as other literature reviews, the present was organized chronologically, thematically, and methodologically (Litchfield, 2008). With basis on the above search, 497 articles were screened to select 84 journal articles, book chapters and editorials for in depth analysis. A descriptive or mapping literature review was the chosen method (King and He, 2005; Paré et al., 2015). Published literature on overtourism provided a database from which any interpretable trends were identified.

(6) analysing data

Nvivo 12 Plus software NVivo (QSR International) has been the most used software package (Sotiriadou, Le, Brouwers, 2014). NVivo requires the researcher to code the data and to develop themes or categories, to derive the list of codes and rules for attaching these to the data. Thus, one can argue that the data analysis is subjective. Yet, in the present study, the categories (overtourism dimensions) were derived from previous literature to demonstrate, via content analysis, what dimensions of overtourism prevail in the literature and how those dimensions are related. This work evolved across four phases, namely, conceptualization, planning, execution, and writing this chapter for dissemination. In the planning section, the researchers have assessed each source with basis on: (1) author's credentials, arguments supported by evidence; (2) Objectivity and the use of pertinent information to prove a point; (3) Persuasiveness; (4) Value concerning whether the work ultimately contribute in a significant way to explain the subject.

Content Analysis

A content analysis executed on 84 relevant journal articles and book chapters with focus on overtourism, published between 1997 and 2019 facilitated an analysis of the main dimensions of overtourism in the tourism literatures. The software NVIVO 12 Plus supported a thematic organisation of sources enabling an overview of the subject, issue or theory under consideration, along with the objective of clarifying how overtourism affects customer experience. Based on first insights given by the literature, the researchers created a grid of categories or nodes in NVIVO (Figure 2). The latter allowed the categorization of 90 PDF files stored in Nvivo 12 across the major categories and subcategories, with the purpose to analyse the frequency of topics associated with overtourism.

Results and Discussion

Content analysis results of academic articles on overtourism revealed that the topic has 9 dimensions. As seen in Figure 2, overtourism studies have a strong political and social dimension to it. (Table 2).

The dimensions most salient in journal articles are the social dimension (316 references, 34%) followed by a political and governance dimension (193 references, 20,8%) and marketing and customer experience dimension (153 references, 16,5%), with 662 (71,2%) references out of a total of 930 (100%). Broad approaches to overtourism in the selected literature (65 references, 7%) include 15 allusions to overtourism in a lengthy document requested by the European Parliament's on Transport and Tourism

Table 2. NVIVO Nodes and sub-nodes

Node and Sub-node	Description
1. **Overtourism**	Generic mentions to overtourism, overtourism origin, factors & implications, overtourism as a multidimensional concept, new conceptual models to tackle overtourism. The tourism challenge for destinations is to become robust enough to endure perturbations and flexible enough to recover or to re-develop/re-invent itself. (Hartman, 2018, p.2)
Sub-nodes	1.1. Tourism Evolutionary phenomenon little change; 1.2. Overtourism Origin, Factors & Implications; 1.3. Overtourism multidimensional concept; 1.4. New conceptual models to tackle overtourism; 1.5. Myths on overtourism.
2. **Ethical Dimension**	The responsible tourism argument as an aspect of overtourism theory in which it produces advice on how to continuously implement the sustainability concept and move from market-value-led and environmentally laissez-faire tourism towards more environmental- and social-value-driven responsible tourism" (Mihalic, 2016, p. 461).
Sub-nodes	2.1. Responsibility for overtourism; 2.2. Responsibility for Business Model; 2.3. Overtourism limit; 2.4. Overtourism and corruption; 2.5. Ethics of tourism growth; 2.6. Corporate social responsibility; 2.7. Best practices; 2.8 Accountability for overtourism externalities.
3. **Social Dimension**	Mentions to the concept of mutual gaze (Maoz, 2006) which can be useful to study social problems affecting tourists. Problematic behaviours reported by the global media (Pile, 2017; Zhang, Pearce, Chen, 2019, p. 71). Reference to optimal levels of accommodation density and the need to consider the level of satisfaction of the tourists (Simancas Cruz and Zaragoza, 2019)
Sub-nodes	3.1. Accommodation saturation; 3.2. Carrying Capacity; 3.3. Community & tourism enterprise relations; 3.4. Conflict Management; 3.5. Gentrification related to Overtourism; 3.6. Overtourism as a Social problem; 3.7. Overtourism impact on work force; 3.8. Overtourism unsatisfaction; 3.9. Resident community Sentiment and tourist visitor relations; 3.10. Seasonality and overtourism; 3.11. Social and Psychological impacts of tourism; 3.12. Social media and overtourism; 3.13. Sustainability of Tourism Model; 3.14. Sustainable Responsible tourism; 3.15. Tourism and Migration the clash UNWTO & IOM; 3.16. Tourism De-growth.
4. **Political & Governance Dimension**	Governance of overtourism. Anti-tourism activists define overtourism as the excessive growth of visitors leading to overcrowding in areas where residents suffer the consequences of temporary and seasonal tourism peaks. Claims that overtourism is harming the landscape, damaging beaches, putting infrastructure under enormous strain (Milano, Novelli & Cheer, 2019).
Sub-nodes	4.1. Anti-tourism movements and debates; 4.2. Careless policy making; 4.3. Custom made policies; 4.4. Destination Management Organisations; 4.5. Destination planning and management; 4.6. Indicators of overtourism; 4.7. Mobility Perspective human rights; 4.8. Models of urban democracy; 4.9. Multi-entry and relaxation permit policy; 4.10 Policy to reduce issues from overtourism & crowding; 4.11. Political activism; 4.12 Progressive Tourism Development; 4.13. Pro-growth policy; 4.14. Resistance to colonisation; 4.15. Right to life and residency; 4.16. Rights of travellers & local populations; 4.17. Risk Monitoring Model; 4.18. Stakeholder theory; 4.19. Statutes and Regulation for Tourism; 4.20. The Neoliberal model & perspective; 4.21 Tourism policy research.
5. Economic Dimension	The connections between modern tourism to development, which encompasses growing number of new destinations. In this view (by WTO) these dynamics have turned tourism into a key driver for socio-economic progress
Sub-nodes	5.1 Benefits of tourism growth for cities; 5.2. Capitalist economy system; 5.3. Consumerist dynamic; 5.4. Low cost flights and overtourism; 5.5. Markets; 5.6. Particularity and Uniqueness in overtourism analysis; 5.7. the Sharing economy; 5.8. Tourism economic impact on local community; 5.9. Tourism Growth Incongruencies.
6. **Symbolic Dimension**	The social nature of tourism consumption, and "the destination's ability to 'say something' about the visitor…can play a significant role in tourism" (Corneo and Jeanne, 1999). By managing a destination's meaning, it may be constructed to be fashionable to niche segments. Being fashionable can help manage overtourism by making the destination popular
Sub-nodes	6.1. Fashionable destinations; 6.2. Interactionism perspective; 6.3. Myths and overtourism; 6.4. Threats to culture and Heritage.

continues on following page

Table 2. Continued

Node and Sub-node	Description
7. **Environmental Dimension**	The notion of sustainable tourism in response to overtourism converges with discourses on threats to global environment and its moderation in "Our Common Future", as outlined in the Brundtland report (WCED, 1987). Other key document brought the notion of sustainability to the forefront of global and local social and political agendas. (Goodwin, 2017).
Sub-nodes	7.1. Geographic density of touristic accommodation; 7.2. Climate change and overtourism; 7.3. Destination Landscapes; 7.4. Eco-friendly behaviour. 7.5. Good practices by destination managers; 7.6. Overcrowding and Pollution; 7.7. Sustainability in Tourism; 7.8. Threat to biodiversity; 7.9. Threat to natural environment; 7.10. Tourism and unsustainability; 7.11. Walkability and overtourism.
8.**Marketing and CE Dimension**	Lessons from Customer Experience theory which can refine the study of overtourism due to the diversity of disciplinary approaches to CEM in the hospitality context. These approaches demonstrate the importance of bringing together marketing, human resources and technology to manage customer experience providing a detailed understanding of the mechanisms for managing CE in the hospitality industry (Kandampully, Zhang and Jaakola, 2018).
Sub-nodes	8.1. Visitor & Market segmentation; 8.2. Tourism management; 8.3. Online reputation and Overtourism; 8.4. Marketing Strategies against sustainable development; 8.5. Management of Service interactions; 8.6. Learning from customers and innovating; 8.7 Innovation and Technology; 8.8. Impact of Externalities on Customer Experience; 8.9. Hotspot crowding & agglomeration; 8.10. Film reputation and overcrowding; 8.11. Destination Management; 8.12. Destination Marketing and Branding & SIT; 8.13. De-Marketing; 8.14. Customer satisfaction constructs; 8.15. Customer satisfaction and loyalty; 8.16. Customer Experience; 8.17. Ambidextrous Management strategy for overtourism.
9. Cultural Dimension	Mentions to cultural tourism and overtourism (Richards, 2018). Boniface (2013) signalled problems with the overcrowding of World Heritage Sites, a phenomenon linked with the idea of 'overtourism'. The problems being encountered with the conservation of tangible heritage and the growing desire of tourists for new experiences (Du Cros, 2012)
Sub-nodes	9.1 Sustainability of cultural Tourism; 9.2. Cultural Consumption; 9.3. Cultural production; 9.4. Cultural tourism and creative economy; 9.5. Cultural Tourism and overcrowding; 9.6. Culture and overtourism; 9.7. Definition of cultural tourism; 9.8. Destination culture; 9.9. Developments in cultural tourism; 9.10. Economic effects of cultural tourism; 9.11. Loss of authenticity & heritage mismanagement; 9.12. Loss of cultural identity; 9.13. Mindscapes and overtourism; 9.14. Overcrowding of World Heritage Sites; 9.15. Paradoxes of cultural tourism; 9.16. Relationship Culture-Consumption; 9.17. Sustainability of cultural Tourism; 9.18. Synergies between tourism and culture; 9.19. Travel related cultural distance tolerance; 9.20. Turismophobia from perspective of culture.

Source: The authors NVIVO analysis, 2019.

(TRAN) (Peeters, Gössling, Klijs, Milano, Novelli, Dijkmans, Eijgelaar, Hartman, Heslinga, Isaac, Mitas, Moretti, Nawijn, Papp and Postma, 2018), with the following results (Figure 3).

Amongst the nine thematic dimensions explored in this study the five most salient are: (1) Social dimension on overtourism (340 references, 30%), (2) Political and Governance dimension (262 references, 23%), (3) Marketing and CE Dimension (155 references, 13,5%), (4) Economic dimension (85 references, 7,4%), and (5) Environmental dimension (105 references, 9,2%) . Together these dimensions represent a total of 947 references, 83,18% of all references. As such a more exhaustive content analysis delved into those five dimensions.

Overtourism Five Main Dimensions

The social dimension of overtourism and the role of stakeholders towards effective sustainable tourism management

The *social dimension of overtourism* (317 references), the most discussed amongst all, incorporates notions such as "Resident community sentiment and tourist visitor relations" (87), "Tourism De-growth"

Figure 3. Content Analysis: Overtourism dimensions and number of references. Source: 84 Academic Journals 1997- 2019
Source: (Authors, 2019)

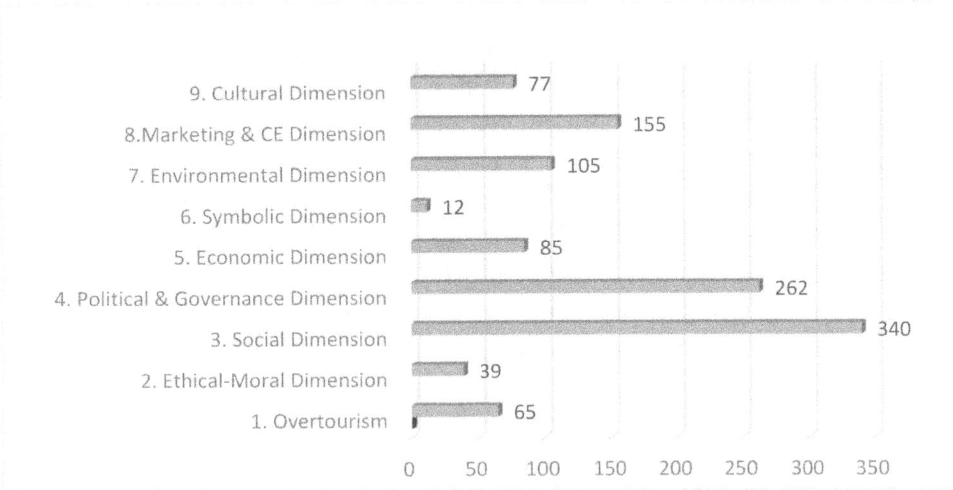

(50 references), "Social and Psychological impacts of tourism" (36), "Carrying Capacity" (30), as well as "Sustainable responsible tourism" (20). Other relevant aspects in narratives concerning the social dimension of overtourism are: "Community and tourism enterprise relations", "Conflict Management", and "Gentrification related to overtourism. (Figure 4).

Plichta (2018, p.58) proposed that "the dynamic development of tourism….is becoming a problem that requires a broad look and ways to solve it". As said, "the balance of benefits from the point of view of different stakeholder groups, including residents, can be positive if they cooperate actively in managing

Figure 4. Content Analysis: Overtourism Social dimension and number of references. Source: 84 Academic Journals, 1997- 2019
Source: (Authors, 2019)

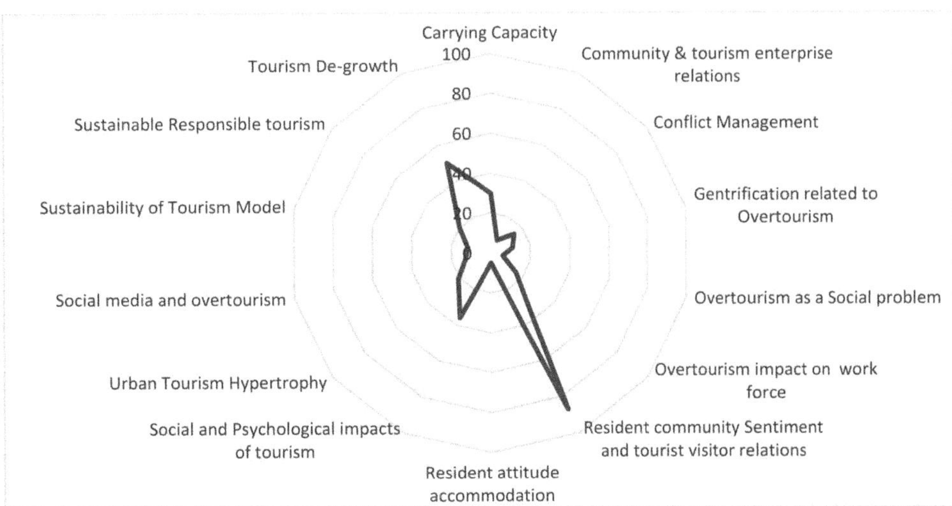

this problem in a longer period" (p.58). Muler, Coromina and Galí (2018, p.277) analysed the value of residents' perceptions of tourism as an indicator of social carrying capacity levels in a heritage town in Spain concluding that carrying capacity is multidimensional as environmental, economic, psychological and perceptual factors need to be considered. Yet those aspects depend on "the particular concerns of the stakeholders involved" (Garrigós Simón et al., 2004). Stakeholder perspectives vary within and between tourism places and there is "no one-size-fits-all solution" (Goodwin, 2017, p.4). In this view, the public authorities play a key part in managing tourism and studies of tourism resorts since that at issue is the ability of local government to manage tourism. Local government need to determine "where the balance is struck between the interests of different stakeholders in tourism, with different target markets, and the other stakeholders in the community" (Goodwin, 2017, p.9).

Koens, Postma and Papp et al (2018) noted that whilst *carrying capacity* continues to be a popular concept to appreciate the negative consequences of tourism, but the usefulness of this perspective is questioned. The main issue is the notion of carrying capacity's focus on tourism numbers, which brings with it that negative effects are equated with mass tourism or increasing visitor numbers. This hinders a common understanding between different stakeholders of the speciðc nature of the problem(s) within the local context. Alternative perspectives, such as the levels of acceptable change framework (LAC), provide greater nuance. The LAC framework and similar impact-based approaches has shifted emphasis from numbers to perceived beneðts and disadvantages (Koens et al, 2018, p.3)

Jordan, Spencer and Prayag (2019, p.223) drew attention to practical aspects of tourism impacts, emotions and stress proposing that a concerted effort by multiple stakeholders (local government, community leaders, and NGOs) must address the multitude of challenges facing communities. Therefore, a holistic rather than ad-hoc approach to community tourism planning and development is necessary for residents' continued well-being. Richards (2018, p. 18) added the development of new technologies also creates a need to analyse stakeholder relationships, particularly in terms of who beneðts from and who pays for the considerable investment required (Tscheu & Buhalis, 2016). Those examples show the wealth of connection between stakeholders' engagement and the social aspects of tourism management.

Walmsley (2017) studied overtourism and underemployment looking at the implications of overtourism on three groups of stakeholders: employees, employers and tourists. Aiming to assist policy makers and destination management/marketing organisations to ensure that tourism development addresses the needs of stakeholders, this paper highlighted that "the dilemma is that the costs are not borne by the businesses that retain profits, but largely by the local community primarily who endure the worst of overtourism's excesses" (Timothy, 2007).

Political and Governance Issues

The second most relevant aspect of overtourism found in the literature was the *political and governance dimension* (262 references) with attention to "Stakeholder theory" (44), "Policy to reduce issues from overtourism and crowding" (38), "Destination Management Organisations" (13), "Statutes and Regulation for Tourism" (23), "Anti-tourism movements and debates" (12), "Multi-entry and relaxation permit policy" (12), and "Indicators of overtourism" (11). (Figure 5).

Challenging overtourism from the perspective of policy making and governance is a difficult project because while governance is a forthright concept, tourism policy is very uncertain (Joppe, 2018). Aall and Koens (2019) focused on stakeholder theory and overtourism noting that the UNWTO (United nations

Figure 5. Content Analysis: Overtourism Political and Governance dimension and number of references. Source: 84 Academic Journals, 1997- 2019
Source: (Authors, 2019)

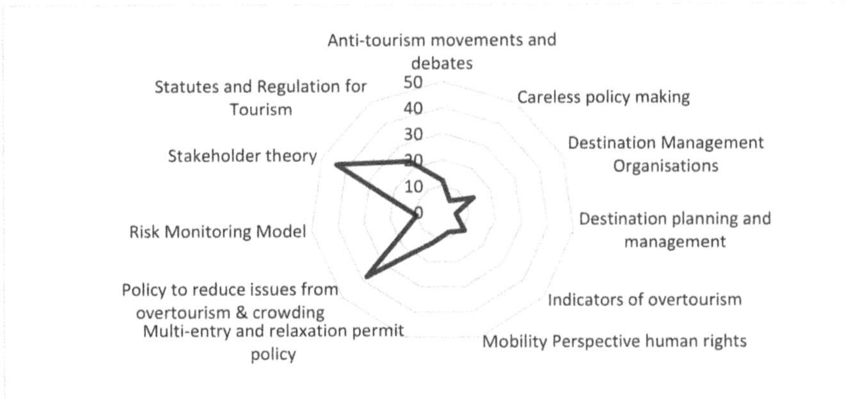

World Tourism Organization) already established that sustainable tourism should aim at a harmonious and productive relationship between host communities, visitors and the environment.

Koens, Melissen, Aall, and Mayer (2019) declared in "The Smart City Hospitality Framework" that collaboration represents a social dilemma, because tourism is discussed on different hierarchical levels of policy with limited communication between stakeholders. Aall and Koens (2019, p.5) reminded that the overarching question of whether one should develop sustainable urban tourism raises a normative question, it depends on the values of the stakeholders addressing it. That is to say, the development of sustainable tourism depends on a critical mass of stakeholders sharing the same viewpoint. On the other hand, Kucer and Mihahic (2019, p.3) delivered a sustainable-responsible tourism (SRT) model demonstrating that sustainable responsible tourism must be based on an awareness of full sustainability awareness on the part of all stakeholders on both the demand and supply sides.

On the topic "Policy to reduce issues from overtourism and crowding" Koens, Postma and Papp (2018, p.1) presented results from a qualitative investigation among 80 stakeholders in 13 European cities to clarify what overtourism entails in policy terms. The authors concluded that the number of stakeholders in the cities who advocate a need to curb tourism growth and increase regulation is on the rise.

Marketing and Customer Experience Issues

The third most relevant dimension of overtourism in the referred literature was the *marketing and customer experience dimension* with 153 references in 31 documents giving salience to "Hotspot crowding and agglomeration" (33) and "de-marketing" (23 references), followed by "visitor and market segmentation" (14 references) and "destination management" (14 references), amongst others (Figure 6).

Only few articles with an interest in marketing and customer experience discussed the impact of stakeholder negotiation in destinations affected by overtourism. Looking at seasonal overcrowding in Australia, Cheer, Milano and Novelli, (2019, p.563) explored the contrasting views between those stakeholders (Local tour operators) who rely on tourism and the so-called out-of-towners (Local tourism association members) who own holiday houses and play a large role in decision-making by rejecting

Figure 6. Content Analysis: Overtourism Marketing and Customer Experience dimension and number of references. Source: 84 Academic Journals, 1997- 2019
Source: (Authors, 2019)

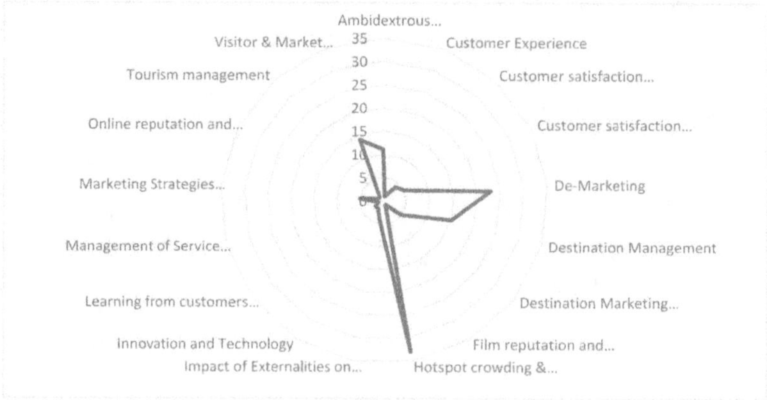

expansion. Séraphin, Gowreesunkar, Zaman, Bourliataux-Lajoin (2019, p.222) approach to community-based festivals as a tool to tackle tourismphobia acknowledged that engaging tourists in local festivals may fuel further anti-tourism movements. Medway, Warnaby and Dharni (2010, p.125) carried research involving interviews with place-marketing stakeholders who were identified as operating in contexts where place-demarketing activities might be appropriate.

It is therefore relevant to know the arguments selected in the above content analysis to facilitate an evidence base crafting of new strategies to manage overtourism in diverse contexts (e.g., destinations), specially with reference to its relationship with the CE approach. In this perspective, authors have proposed to "**redirecting tourism ñows to areas usually not visited** by tourists, such as street art tours, which could "contribute to reinforcing social cohesion in destinations" (Veiga, Santos, Águas and Santos, 2018, p. 670). Barrera and Cesario (2017) for example, demonstrated that the characteristics of a tourism destination (e.g., the Algarve region, Portugal) are rated and assessed differently in peak (summer) and off-peak (autumn/winter) seasons, thus contributing for tourists' satisfaction and experience at the destination in various ways. Jacobsen, Iversen and Hem (2019) recognised that whereas overtourism in public spaces generate political responses from residents, the situation in private paid touristic venues is different

Only few articles with an interest in marketing and customer experience discussed the impact of stakeholder negotiation in destinations affected by overtourism. Medway, Warnaby and Dharni (2010, p.125) carried research involving interviews with place-marketing stakeholders who were identified as operating in contexts where place-demarketing activities might be appropriate. Interviewees were representatives (or employees) at local authorities, tourist sites, national parks and voluntary/pressure groups with an interest in how space is managed and marketed. The authors proposed that demarketing effort and intervention is not always welcome by all stakeholders, especially if the message itself is perceived as being over-alarmist (p.132)

Economic Issues

A content analysis of the Scholarly articles selected with reference to 'overtourism' and economy revealed the presence of a significant debate on the matter relating to markets (25), tourism growth incongruencies (11) and tourism economic impact on local community (4), the sharing economy (2), and the consumerist dynamic (1), amongst others, as part of a total of 79 references in 21 files. Scholarly articles by Butler (2018), Papathanassis, (2017b) and Bresson Lagossah (2011) emerged as the most relevant for a discussion on economy and overtourism. Some of those articles focused on the broken promise of tourism, questioning - "To what extent do the benefits of increasing tourism-generated income reach the local economies and communities?" (Papathanassis, 2017, p. 290-91).

Other authors linked overtourism to a technical revolution that changed and influenced the tourism market, since the tourists have more information, and bargain power to create their own tourism product (Boivin and Taguay, 2019). The sharing economy is one of such examples of technology influencing many tourism services – from transportation (e.g., Uber) to accommodation (e.g., Airbnb) (Peeters et al., 2018).

Environmental Issues

A content analysis of the Scholarly articles selected with reference to 'overtourism' and 'environment' identified 73 references in 19 files, with sub-node 'accommodation density and saturation (17) as the most emphasised topic, followed by 'overcrowding and pollution'(9) and sustainability in tourism (8). In the context of the overtourism debate, "cruises prominently feature as key drivers of overcrowding and pollution (Séraphin et al, 2018; Stefanidaki and Lekakou, 2014, Papathanassis, 2019, p. 3). The contribution of cruise tourism in port destinations is being increasingly questioned. "Overcrowding of heritage attractions and tourism infrastructure, limited cruiser-spending on land (Papathanassis, 2016b; Larsen et al., 2013; Klein, 2009) and environmental pollution (i.e. emissions and waste) all affect the life of locals as well as the satisfaction of overnight tourists" (Papathanassis, 2017, p. 108). Studies aligned with the sustainability paradigm also extended the perspective of overtourism to coastal tourism. Focusing on the Canary Islands, Spain, Simancas Cruz and Saragosa (2019, p.2) hypothesised that "tourism density is one of the main causes of overtourism".

With focus on city breaks, in addition to the undoubted benefits of growing tourism, Plitcha, (2018, p. 58) noted that "Krakow is beginning to feel the impact of this factor on many areas, such as environmental pollution". Another study limited to Barcelona revealed that the impacts of overtourism, as perceived to a larger extent by local citizens, "are the overcrowding of infrastructures and public spaces…more waste generation, environmental deterioration, or overuse of resources, among others" (Martin, Martinez and Fernandez, 2018, p. 4) As the tourist activity creates numerous interactions with groups that pursue different interests, the Stakeholder Theory proposed by Freeman (1984) would serve as "a good model to improve the organization of the activity derived from the collaborative economy within the tourism industry, given that in order to reduce the negative impacts derived from the interactions…" from the point of view of the public policy-makers and the private sector, it would be important to identify the stakeholders, get to know the way they interact with each other, and the potential conñicts in order to work jointly in the planning process" (Martín, Martinez and Fernandez, 2018, p. 13). More importantly, stakeholders should envisage their marketing strategy for minimize and reduce overtourism environmental problems and potential social-economic conflicts.

CONCLUSION

The systematic literature review performed allowed to identify nine dimensions of overtourism, which can reflect common grounds regarding this concept. The most relevant dimensions of overtourism studies are *social, political and governance, marketing and customer experience, economic, and environmental.*

The major findings in this literature review are of two kinds: (1) On a theoretical level, overtourism avoidance involves trade-offs between competing priorities for tourists, residents, tourism operators, destination residents and local governments. Managing overtourism not only depends on a desired dialogue between consumerist and post-consumerist values, but it depends on the values of the stakeholders addressing it. Those stakeholders are the key drivers of the network of actors who debate, decide and implement tourism related strategies and policies from a holistic perspective. The development of overtourism management policies depend on a critical mass of stakeholders sharing the same viewpoint. (2) On an empirical level, it is now clear that a wide range of tourism studies provide policy makers with the tools to carry their assessments.

After this literature review of overtourism literature, the role of stakeholders was found to be central in overtourism literature because sustainable tourism addresses the needs of visitors, the industry and the environment. There is a prevalent view in the literature researched, voiced by Aall and Koens (2019) that governments are hesitant to insist on changes in the tourism sector, and with the privatization of services that dominate governance, regulations are not welcome. The fragmentation of the tourism sector increased, now comprising networks where some players have greater power than others, tourism transpires economic interests, ethical beliefs of stakeholders and the effectiveness of lobbies. In this context pervaded by interest groups, stakeholders need to be aware that cooperation is a social dilemma because tourism is discussed on different hierarchical levels of policy with limited communication between stakeholders. Stakeholders need to be aware that conñicting values, priorities and interests are unavoidable (Aall & Koens, 2019). But as Jordan, Spencer and Prayag (2019, p.223) suggested, there is at least some convergence on the value of holistic methods for stakeholder involvement in overtourism management. In this context, overtourism can be at least discussed via a concerted effort by multiple stakeholders (local government, community leaders, and NGOs) to address the multitude of challenges facing communities and the tourism activity. In this attempt, stakeholders should avoid ad-hoc approaches to the community tourism planning and development and concentrate on residents' continued well-being.

The real value this literature review adds to knowledge is its ability to articulate mentions to stakeholder theory and empirical studies of stakeholder behaviour with the overtourism debate and with tools that researchers in the field of tourism developed to explain and to manage overtourism in the last 20 years.

FUTURE RESEARCH DIRECTIONS

Future research for the academy should consider the identified nine dimensions of overtourism, to assess and explore with future in-depth stakeholder management of overtourism. Instead of a disperse and ill-defined study of overtourism, the present study enables to identify core dimensions which can help to frame the concept within those areas making easier to find common grounds and future research gaps. Also, for the industry, this study highlights once more previous views on the need to find solutions to prevent the increase of conflicts and tensions between tourism and residents of tourism destinations. Policies and strategies have been implemented to reduce this dichotomy in tourism, as found in the literature review, but very few studies have a wider view of the phenomena, helping destination managers to consider new dimensions besides those related with socio-economic, environmental and political. Findings suggest that the dimension marketing and customer experience, can be helpful for destination managers to reduce, change and re-direct tourism demand and tourism flow.

ACKNOWLEDGMENT

This research was funded by national funds through the FCT- Foundation for Science and Technology under the project UID/SOC/04020/2019.

REFERENCES

Aall, C., & Koens, K. (2019). The Discourse on Sustainable Urban Tourism: The need for Discussing More Than Overtourism. *Sustainability*, *11*(15), 4228. doi:10.3390u11154228Andereck, K. L., & Becker, R. H. (1993). Perceptions of carry-over crowding in recreation environments. *Leisure Sciences*, *15*(1), 25–35. doi:10.1080/01490409309513184

Andriotis, K. (2005). Community groups' perceptions and preferences to tourism development. Evidence from Crete. *Journal of Hospitality & Tourism Research (Washington, D.C.)*, *29*(1), 67–90. doi:10.1177/1096348004268196

Barrera, A. P., & Cesario, M. (2017). Factors influencing the choice of the Algarve region as a tourist destination: Does season matter? *International Journal of Tourism Research*, 1–10. doi:10.1002/jtr.2207

Bec, A., McLennan, C. L., & Moyle, B. D. (2016). Community resilience to long-term tourism decline and rejuvenation: A literature review and conceptual model. *Current Issues in Tourism*, *19*(5), 431–457. doi:10.1080/13683500.2015.1083538

Boivin, M., & Tanguay, G. A. (2019). Analysis of the determinants of urban tourism attractiveness: The case of Québec City and Bordeaux. *Journal of Destination Marketing & Management*, *11*, 67–79. doi:10.1016/j.jdmm.2018.11.002

Boniface, P. (2013). *Managing quality cultural tourism*. London, UK: Routledge. doi:10.4324/9780203424230

Brakus, J. J., Schmitt, B., & Zarantonello, L. (2009, May). Brand Experience: What Is It? How Is It Measured? Does It Affect Loyalty? *Journal of Marketing*, *73*(3), 52–68. doi:10.1509/jmkg.73.3.052

Bresson, G., & Lagossah, K. (2011, February). Crowding-out effects of cruise tourism on stay-over tourism in the Caribbean: Non-parametric panel data evidence. *Tourism Economics*, *17*(1), 127–158. doi:10.5367/te.2011.0028

Butler, R. (1995). Introduction. In R. Butler, & D. Pearce (Eds.), *Change in Tourism: People, Places, and Processes* (pp. 1–11). New York: Routledge.

Butler, R. (2018). Sustainable Tourism in Sensitive Environments: A Wolf in Sheep's Clothing? *Sustainability*, *10*(6), 1789. doi:10.3390u10061789

Calgaro, E., Lloyd, K., & Dominey-Howes, D. (2014). From vulnerability to transformation: A framework for assessing the vulnerability and resilience of tourism destinations. *Journal of Sustainable Tourism*, *22*(3), 341–360. doi:10.1080/09669582.2013.826229

Capo, J., Font, A., & Nadal, J. (2007). Dutch disease in tourism economies: Evidence from the Balearics and the Canary Islands. *Journal of Sustainable Tourism*, *15*(6), 615–627. doi:10.2167/jost698.0

Castells, M. (1986). *La ciudad y las masas: sociologia de los movimientos urbanos*. Madrid, Espana: Alianza Editorial.

Cheer, J. M., Milano, C., & Novelli, M. (2019). Tourism and community resilience in the Anthropocene: Accentuating temporal overtourism. *Journal of Sustainable Tourism*, *27*(4), 554–572. doi:10.1080/09669582.2019.1578363

Cheng, C., Li, X. R., Petrick, J. F., & O'Leary, J. T. (2011). An examination of tourism journal development. *Tourism Management*, *32*(1), 53–61. doi:10.1016/j.tourman.2009.11.004

Chiou, C. (2008). The effect of concept mapping on students' learning achievements and interests. *Innovations in Education and Teaching International*, *45*(4), 375–387. doi:10.1080/14703290802377240

Cohen, E., & Cohen, S. A. (2015). Beyond Eurocentrism in tourism: A paradigm shift to mobilities. *Tourism Recreation Research*, *40*(2), 157–168. doi:10.1080/02508281.2015.1039331

Cole, S. (2012). Synergy and congestion in the tourist destination life cycle. *Tourism Management*, *33*(5), 1128–1140. doi:10.1016/j.tourman.2011.11.018

Corneo, G., & Jeanne, O. (1999). Segmented communication and fashionable behaviour. *Journal of Economic Behavior & Organization*, *39*(4), 371–385. doi:10.1016/S0167-2681(99)00046-3

Costa, J., Montenegro, M., & Gomes, J. (2018). What challenges and opportunities will lead to success? *Worldwide Hospitality and Tourism Themes*, *10*(6), 631–634. doi:10.1108/WHATT-08-2018-0053

Ditton, R. B., Fedler, A. J., & Graefe, A. R. (1983). Factors contributing to perceptions in recreational crowding. *Leisure Sciences*, *5*(4), 273–288. doi:10.1080/01490408309513009

Du Cros, H. (2012). *Tourism and intangible cultural heritage*. Madrid, Spain: UNWTO.

Fink, A. (2005). *Conducting Research Literature Reviews: From the Internet to Paper* (2nd ed.). Thousand Oaks, CA: Sage.

Fistola, R., & La Rocca, R. A. (2017). Driving Functions for Urban Sustainability: The Double-Edged Nature of Urban Tourism. *International Journal of Sustainable Development and Planning*, *12*(3), 425–434. doi:10.2495/SDP-V12-N3-425-434

Fotiadis, A. K. (2016). Modifying and applying time and cost blocks: The case of E-Da theme park, Kaohsiung, Taiwan. *Tourism Management*, *54*, 34–42. doi:10.1016/j.tourman.2015.10.013

Freeman, R. E. (1984). *Strategic Management: A Stakeholder Approach*. Boston, MA: Pitman.

Fuller, H., & Michel, B. (2014). Stop being a tourist! New dynamics of urban tourism in Berlin-Kreuzberg. *International Journal of Urban and Regional Research*, *38*(4), 1304–1318. doi:10.1111/1468-2427.12124

Gao, W., Zhang, Q., Lu, Z., Wu, D., & Du, X. (2018). „Modelling and application of fuzzy adaptive minimum spanning tree in tourism agglomeration area division. *Knowledge-Based Systems*, *143*, 317–326. doi:10.1016/j.knosys.2017.06.007

Garrigós Simón, F. J., Narangajavana, Y., & Marqués, D. P. (2004). Carrying capacity in the tourism industry: A case study of Hengistbury Head. *Tourism Management*, *25*(2), 275–283. doi:10.1016/S0261-5177(03)00089-X

Gürsoy, I. (2019). Beauty and the Beast: A Fairy Tale of Tourismphobia. *Tourism Planning & Development*, *16*(4), 434–451. doi:10.1080/21568316.2019.1582086

Hartman, S. (2018). Resilient tourism destinations? Governance implications of bringing theories of resilience and adaptive capacity to tourism practice. In Innerhofer, Fontanari, & Pechlaner (Eds.), Destination Resilience – Challenges and Opportunities for Destination Management and Governance. 66-75. Abingdon, UK: Routledge doi:10.4324/9780203701904-5

International Conference on Responsible Tourism in Destinations. (2002). *The Cape Town Declaration, Cape Town.* Retrieved from http://responsibletourismpartnership.org/cape-town-declaration-on-responsible-tourism/

Jacobsen, J. K. S., Iversen, N. M., & Hem, L. E. (2019). Hotspot crowding and over-tourism: Antecedents of destination attractiveness. *Annals of Tourism Research*, *76*, 53–66. doi:10.1016/j.annals.2019.02.011

Jogaratnam, G., Mccleary, K., Mena, M., & Yoo, J. J. (2005). An Analysis of Hospitality and Tourism Research: Institutional Contributions. *Journal of Hospitality & Tourism Research (Washington, D.C.)*, *29*(3), 356–371. doi:10.1177/1096348005276929

Joppe, M. (2018). Tourism policy and governance: Quo vadis? *Tourism Management Perspectives*, *25*, 201–204. doi:10.1016/j.tmp.2017.11.011

Jordan, E. J., Spencer, D. M., & Prayag, G. (2019). Tourism impacts, emotions and stress. *Annals of Tourism Research*, *75*, 213–226. doi:10.1016/j.annals.2019.01.011

Kandampully, J., Zhang, T. C., & Jaakkola, E. (2018). Customer experience management in hospitality: A literature synthesis, new understanding and research agenda. *International Journal of Contemporary Hospitality Management*, *30*(1), 21–56. doi:10.1108/IJCHM-10-2015-0549

Kaspar, C. (1973). *"Tourism Ecology – New Dimension of Tourism Science". Festschrift zur Vollendung des 65. Lebensjahres von P. Bernecker. Beitraege zur Fremdenverkehrsforschung* (pp. 139–143). Wien: Ender W.

King, W., & He, J. (2005). Understanding the Role and Methods of Meta-Analysis in IS Research. *Communications of the Association for Information Systems*, *16*, 665–686. doi:10.17705/1CAIS.01632

Klein, R. (2009). Keeping the Cruise Tourism Responsible: The Challenge for Ports to Maintain High Self Esteem. Available at www.cruisejunkie.com/icrt2.pdf

Koens, K., Melissen, F., Aall, C., & Mayer, I. (2019). The Smart City Hospitality Framework: A conceptual framework for collaborative reflections on tourism's contribution to a transition towards sustainable urban development. *Journal of Destination Marketing & Management. Sustainability, 11*(15), 4228. doi:10.3390u11154228

Koens, K., Postma, A., & Papp, B. (2018). Is Overtourism Overused? Understanding the Impact of Tourism in a City Context. *Sustainability, 4384*(10). doi:10.3390u10124384

Krippendorf, J. (1999). *The Holliday Makers Understanding the Impact of Leisure and Travel*. London, UK: Routledge.

Larsen, S., Wolff, K., Marnburg, E., & Øgaard, T. (2013). Belly full, purse closed: Cruise line passengers' expenditures. *Tourism Management Perspectives, 6*, 142–148. doi:10.1016/j.tmp.2013.02.002

Lerario, A., & Di Turi, S. (2018). Sustainable Urban Tourism: Reflections on the Need for Building-Related Indicators. *Sustainability, 10*(6), 1981. doi:10.3390u10061981

Lew, A. A., & Cheer, J. M. (Eds.). (2018). *Tourism Resilience and Adaptation to Environmental Change*. London, UK: Routledge.

Lingard, L. (2018). Mapping the gap. *Perspectives on Medical Education, 7*(1), 47–49. doi:10.100740037-017-0401-x PMID:29260402

Litchfield, R. C. (2008). Brainstorming Reconsidered: A goal-based view. *Academy of Management Review, 33*(3), 649–668. doi:10.5465/amr.2008.32465708

Liu, J., Liu, N., Zhang, Y., Qu, Z., & Yu, J. (2019). Evaluation of the non-use value of beach tourism resources: A case study of Qingdao coastal scenic area, China. *Ocean and Coastal Management, 168*, 63–71. doi:10.1016/j.ocecoaman.2018.10.030

Luque-Gil, A. M., Gómez-Moreno, M. L., & Peláez-Fernández, M. A. (2018). Starting to enjoy nature in Mediterranean mountains: Crowding perception and satisfaction. *Tourism Management Perspectives, 25*, 93–103. doi:10.1016/j.tmp.2017.11.006

Maoz, D. (2006). The mutual gaze. *Annals of Tourism Research, 33*(1), 221–239. doi:10.1016/j.annals.2005.10.010

Marrero Rodríguez, J. R. (2006). "El discurso de rechazo al turismo en Canarias: Una aproximación cualitativa". *PASOS Revista de Turismo y Patrimonio Cultural, 4*(3), 327-341. Instituto de Investigación Social y Turismo, 2006.

Martín, J. M., Guaita Martínez, J. M., & Fernández, J. A. (2018). An Analysis of the Factors behind the Citizen's Attitude of Rejection towards Tourism in a Context of Overtourism and Economic Dependence on This Activity. *Sustainability, 10*(8), 2851. doi:10.3390u10082851

Medway, D., Warnaby, G., & Dharni, S. (2010). Demarketing places: Rationales and strategies. *Journal of Marketing Management, 27*(1-2), 124–142. doi:10.1080/02672571003719096

Mihalic, T. (2016). "Sustainable-responsible tourism discourse–Towards 'responsustable' tourism". *Journal of Cleaner Production*, 111(B), 461–470.

Milano, C. (2018). Overtourism, Social Unrest, and Tourismphobia. A controversial debate. *Pasos: Revista de Turismo y Patrimonio Cultural*, *16*(3), 551–564. doi:10.25145/j.pasos.2018.16.041

Milano, C., Novelli, M., & Cheer, J. M. (2019). Overtourism and degrowth: A social movements perspective. *Journal of Sustainable Tourism*, *27*(12), 1857–1875. doi:10.1080/09669582.2019.1650054

Milman, A., & Asli, D. A. (2018). Exploring the experiential and sociodemographic drivers of satisfaction and loyalty in the theme park context. *Journal of Destination Marketing & Management*, *8*, 385–395. doi:10.1016/j.jdmm.2017.06.005

Moher, D., Liberati, A., Tetzlaff, J., & Altman, D. G.The PRISMA Group. (2009). Preferred Reporting Items for Systematic Reviews and Meta-Analyses: The PRISMA Statement. *PLoS Medicine*, *6*(7). doi:10.1371/journal.pmed.1000097

Moreno-Izquierdo, L., Ramón-Rodríguez, A., & Such-Devesa, M. J. (2018). The challenge of long-term tourism competitiveness in the age of innovation: Spain as a case study. *Investigaciones Regionales – Journal of Regional Research*, 42, 13-34.

Morteza, Z., Reza, F., Seddiq, M., Sharareh, P., & Jamal, G. (2016). Selection of the optimal tourism site using the ANP and fuzzy TOPSIS in the framework of Integrated Coastal Zone Management: A case of Qeshm Island. *Ocean and Coastal Management*, *130*, 179–187. doi:10.1016/j.ocecoaman.2016.06.012

Muler, V., Coromina, L., & Gali, N. (2018). Overtourism: Residents' Perceptions of Tourism Impact as an Indicator of Resident Social Carrying Capacity - Case Study of a Spanish Heritage Town. *Tourism Review*, *73*(3), 277–296. doi:10.1108/TR-08-2017-0138

Novy, J. (2019). Urban tourism as a bone of contention: Four explanatory hypotheses and a caveat. *International Journal of Tourism Cities*, *5*(1), 63–74. doi:10.1108/IJTC-01-2018-0011

Ordiz, E. (2017). "Turismofobia en España: Dónde se producen las protestas y cuál es su objetivo". *20 Minutos*. Available at https://www.20minutos.es/noticia/3111276/0/turismofobia-en-espana-claves/#xtor=%20AD-15andxts=467263

Papathanassis, A. (2016). Curing the 'Beach Disease': Corruption and the Potential of Tourism-led Transformation for Developing Countries and Transitional Economies. *Ovidius University Annals Economic Sciences Series*, *16*(1), 75–80.

Papathanassis, A. (2016b). Combating tourism-related corruption: Effective countermeasures derived from analysing tourists' perceptions and experiences. *Ovidius University Annals Economic Sciences Series*, *16*(2), 248–255.

Papathanassis, A. (2017). Cruise tourism management: State of the art. *Tourism Review*, *72*(1), 104–119. doi:10.1108/TR-01-2017-0003

Papathanassis, A. (2017b). Over-Tourism and Anti-Tourist Sentiment: An Exploratory Analysis and Discussion. "Ovidius" University Annals. *Economic Sciences Series.*, *XVII*(2), 288–293.

Papathanassis, A. (2019). The growth and development of the cruise sector: A perspective article. *Tourism Review*, TR-02-2019-0037. doi:10.1108/TR-02-2019-0037

Paré, G., Trudel, M.-C., Jaana, M., & Kitsiou, S. (2015). Synthesizing information systems knowledge: A typology of literature reviews. *Information & Management*, *52*(2), 183–199. doi:10.1016/j.im.2014.08.008

Perkumienė, D., & Pranskūnienė, R. (2019). Overtourism: Between the Right to Travel and Residents' Rights. *Sustainability*, *11*(7), 2138. doi:10.3390u11072138

Pile, T. (2017). Who are the world's worst tourists? Six nations that stand out - you may be surprised. [Online] Available at http://www.sohu.com/a/157741590_644346, Accessed date: October 30, 2019.

Plichta, G. (2018). The Problem of Short-term Rental as a Manifestation of the Overtourism Effect on the Housing Market in the Historical City of Krakow, *Świat Nieruchomości*, *4*(106) 57-64.

Podilchak, W. (1991). Distinctions of fun, enjoyment and leisure. *Leisure Studies*, *10*(2), 133–148. doi:10.1080/02614369100390131

Postma, A., & Schmuecker, D. (2017). Understanding and overcoming negative impacts of tourism in city destinations: Conceptual model and strategic framework. *Journal of Tourism Futures*, *3*(2), 144–156. doi:10.1108/JTF-04-2017-0022

Prakash, S. L., Perera, P., Newsome, D., Kusuminda, T., & Walker, O. (2019). Reasons for visitor dissatisfaction with wildlife tourism experiences at highly visited national parks in Sri Lanka. *Journal of Outdoor Recreation and Tourism*, *25*, 102–112. doi:10.1016/j.jort.2018.07.004

Rees, W. E. (1997). Is sustainable city an oxymoron? *Local Environment*, *2*(3), 303–310. doi:10.1080/13549839708725535

Richards, G. (2018). Cultural tourism: A review of recent research and trends. *Journal of Hospitality and Tourism Management*, *36*, 12–21. doi:10.1016/j.jhtm.2018.03.005

Royo, M., & Ruíz, M. E. (2009). Actitud del residente hacia el turismo y el visitante: Factores determinantes en el turismo y excursionismo rural-cultural. *Cuadernos de Turismo*, *23*, 217–236.

Russo, A. P. (2002). The "vicious circle" of tourism development in heritage cities. *Annals of Tourism Research*, *29*(1), 165–182. doi:10.1016/S0160-7383(01)00029-9

Sanchez, P. M., & Adams, K. M. (2008). The Janus-faced character of tourism in Cuba. *Annals of Tourism Research*, *35*(1), 27–46. doi:10.1016/j.annals.2007.06.004

Sarantakou, E., & Terkenli, T. S. (2019). Non-Institutionalized Forms of Tourism Accommodation and Overtourism Impacts on the Landscape: The Case of Santorini, Greece. *Tourism Planning & Development*, *16*(4), 411–433. doi:10.1080/21568316.2019.1569122

Saveriades, S. (2000). Establishing the social tourism carrying capacity for the tourist resorts of the east coast of the Republic of Cyprus. *Tourism Management*, *21*(2), 147–156. doi:10.1016/S0261-5177(99)00044-8

Séraphin, H., Gowreesunkar, V., Zaman, M., & Bourliataux-Lajoin, S. (2019). Community based festivals as a tool to tackle tourismphobia and antitourism movements. *Journal of Hospitality and Tourism Management*, *39*, 219–223. doi:10.1016/j.jhtm.2018.12.001

Séraphin, H., Sheeran, P., & Pilato, M. (2018b). Over-tourism and the fall of Venice as destination. *Journal of Destination Marketing & Management*, *9*, 374–376. doi:10.1016/j.jdmm.2018.01.011

Séraphin, H., Yallop, A., Capatina, A., & Gowreesunkar, V. (2018). Heritage in tourism organisations' branding strategy: The case of a post-colonial, post-conñict and post disaster destination. *International Journal of Culture, Tourism and Hospitality Research*, *12*(1), 89–105. doi:10.1108/IJCTHR-05-2017-0057

Séraphin, H., Zaman, M., Olver, S., Bourliataux-Lajoinied, S., & Dosquete, F. (2019). Destination branding and overtourism. *Journal of Hospitality and Tourism Management*, *38*, 1–4. doi:10.1016/j.jhtm.2018.11.003

Shafiee, G., & Hasanzadeh, J. (2019). Developing a model for sustainable smart tourism destinations: A systematic review. *Tourism Management Perspectives*, *31*, 287–300. doi:10.1016/j.tmp.2019.06.002

Shea, B. J., Grimshaw, J. M., Wells, G. A., Boers, M., Andersson, N., Hamel, C., ... Bouter, L. M. (2007). Development of AMSTAR: A measurement tool to assess the methodological quality of systematic reviews. *BMC Medical Research Methodology*, *7*(1), 10. doi:10.1186/1471-2288-7-10 PMID:17302989

Shelby, B., Heberlein, T. H., Vaske, J. J., & Alfano, G. (1983). Expectations, preferences, and feeling crowded in recreation activities. *Leisure Sciences*, *11*, 269–291. doi:10.1080/01490408909512227

Shin, H. J., Kim, H. N., & Son, J. Y. (2017). Measuring the economic impact of rural tourism membership on local economy: A Korean case study. *Sustainability*, *9*(4), 639. doi:10.3390u9040639

Simancas Cruz, M., & Zaragoza, M. P. P. (2019). Analysis of the Accommodation Density in Coastal Tourism Areas of Insular Destinations from the Perspective of Overtourism. *Sustainability*, *3031*(11), 3031. doi:10.3390u11113031

Sotiriadou, P., Le, A., & Brouwers, J. (2014). Choosing a qualitative data analysis tool: A comparison of NVivo and Leximancer. *Annals of Leisure Research*, *17*(2), 218–234. doi:10.1080/11745398.2014.902292

Stefanidaki, E., & Lekakou, M. (2014). Cruise carrying capacity: A conceptual approach. *Research in Transportation Business & Management*, *13*, 43–52. doi:10.1016/j.rtbm.2014.11.005

Steven, R., Pickering, C., & Castley, J. G. (2011). A review of the impacts of nature-based recreation on birds. *Journal of Environmental Management*, *92*(10), 2287–2294. doi:10.1016/j.jenvman.2011.05.005 PMID:21640470

Tao, K., Li, X., Zhou, Q., Moher, D., Ling, C., & Yu, W. (2011). From QUOROM to PRISMA: A Survey of High-Impact Medical Journals' Instructions to Authors and a Review of Systematic Reviews in Anesthesia Literature. *PLoS One*, *6*(11), 1–5. doi:10.1371/journal.pone.0027611 PMID:22110690

Templier, M., & Pare, G. (2015). A Framework for Guiding and Evaluating Literature Reviews. *Communications of the Association for Information Systems*, *37*(6), 112–137.

Timothy, D. (2007). Empowerment and stakeholder participation in tourism destination communities. In A. Church, & T. Coles (Eds.), Tourism, Power, and Space, 199-216. London: Routledge.

Tscheu, F., & Buhalis, D. (2016). Augmented Reality at Cultural Heritage sites. In A. Inversini, & R. Schegg (Eds.), *Information and Communication Technologies in Tourism 2016*. Cham, Switzerland: Springer. doi:10.1007/978-3-319-28231-2_44

van Doren, C. S., Koh, Y. K., & McCahill, A. (1994). Tourism research: a state-of-the art citation analysis (1971 -1990). In A. V. Seaton (Ed.), Tourism: The state of the art (308-315). Chichester, UK: John Wiley &Sons.

Van Niekerk, M. (2016). Community perceptions on the impacts of art festivals and its impact on overall quality of life: A case study of the innibos National Art festivals, South Africa. In C. Newbold, & J. Jordan (Eds.). Focus of world festivals: Contemporary case studies and perspectives, 152–163. Oxford: Goodfellow Publishers.

Veiga, C., Santos, M. C., Águas, P., & Santos, J. A. C. (2018). Sustainability as a key driver to address challenges. *Worldwide Hospitality and Tourism Themes*, *10*(6), 662–673. doi:10.1108/WHATT-08-2018-0054

Wall, G. (1997). Is ecotourism sustainable? *Environmental Management*, *21*(4), 483–491. doi:10.1007002679900044 PMID:9175538

Walmsley, A. (2017). *Overtourism and underemployment: a modern labour market dilemma*. Paper presented at Responsible Tourism in Destinations 13 – Tackling Overtourism – Local Responses, September 29-30, 2017 – Hannesarholt, Reykjavik, Iceland.

Wang, D., Li, X., & Li, Y. (2013). China's "smart tourism destination" initiative: A taste of the service-dominant logic. *Journal of Destination Marketing & Management*, *2*(2), 59–61. doi:10.1016/j.jdmm.2013.05.004

World Tourism Organization (UNWTO). *Tourism Towards 2030 / Global Overview* - Advance edition presented at UNWTO 19th General Assembly - October 10, 2011.

World Tourism Organization (UNWTO). *Tourism Highlights*, 2016 Edition. Downloaded on 6/05/2019 from https://www.e-unwto.org/doi/pdf/10.18111/9789284418145

Xiao, H., & Smith, S. (2006). The maturation of tourism research: Evidence from a content analysis. *Tourism Analysis*, *10*(4), 335–348. doi:10.3727/108354206776162796

Zhang, C., Pearce, P., & Chen, G. (2019). Not losing our collective face: Social identity and Chinese tourists' reflections on uncivilised behaviour. *Tourism Management*, *73*, 71–82. doi:10.1016/j.tourman.2019.01.020

Zhang, Y., Xiang, L., & Qin, S. (2017b). Does spatial layout matter to theme park tourism carrying-capacity? *Tourism Management*, *61*, 82–95. doi:10.1016/j.tourman.2017.01.020

Zhang, Y., Xiang, L., Qin, S., & Hu, X. (2017). Exploring a theme park's tourism carrying capacity: A demand-side analysis. *Tourism Management*, *59*, 564–578. doi:10.1016/j.tourman.2016.08.019

Chapter 3
Methods of Published Articles on Overtourism:
Analysis and Reflections of Scopus and Web of Science (1998–2018)

Luciana Aparecida Barbieri da Rosa
https://orcid.org/0000-0001-9240-0236
Federal University of Santa Maria, Brazil

Tais Pentiado Godoy
https://orcid.org/0000-0001-5021-5315
Federal University of Santa Maria, Brazil

Maria Carolina Martins-Rodrigues
https://orcid.org/0000-0003-2575-8611
CIEO, University of Algarve, Portugal

Luana Inês Damke
https://orcid.org/0000-0003-0379-6262
Federal University of Santa Maria, Brazil

Clandia Maffini Gomes
Federal University of Santa Maria, Brazil

ABSTRACT

Studies on overtourism have been growing in the last decade. This article aims to understand the characteristics of the international scientific literature that links the theme Smart Cities in the Web of Science and Scopus databases, from 1998 to 2018. The established methodology was a bibliometric survey. The sample was composed of 24 articles selected in WOS and Scopus. The survey results show that the year with the most publications was 2018. The most relevant article is: "Understanding and overcoming the negative impacts of tourism in city destinations: a conceptual model and strategic framework" by A. Postma, & D. Schmuecker, published in the Journal of Tourism Futures in 2017 with 10 citations. Thus, future studies should expand the search for articles on the topics in other bases as well as national and international scientific events.

DOI: 10.4018/978-1-7998-2224-0.ch003

INTRODUCTION

Tourism provides socio-economic progress and social welfare. In addition, it is mentioned as a significant activity for countries. In the recent years, world tourism has become one of the most important global economic spheres, participating in the gross domestic product (GDP) of different countries, as well as inducing employment (Costa et al., 2014).

According to WTO (2016), the tourism business dimension is as essential as oil exports, food products, or technologies. Tourism is a set of services and activities that converge to the traveler's experience, whether in a group or by itself (Goeldner & Ritchie, 2011). Authors Lohman & Beer (2013) mention that tourism is a cross-industry, greatly influenced by internal and external factors, especially in the areas of the economy, environment, politics, society, and technology.

In this perspective, tourism has been referred to as a critical factor in the progress of cities and regions worldwide. In addition to the economic characteristic, it aims to provide for the perpetuation of natural resources and the cultural valuation of individuals such as history, geology and fauna, and flora. Thus, a new tourist profile emerges, attracted by factors specific to destination locations.

Corroborating Scotolo & Netto (2015), they point out that tourism, apart from fostering the local and regional economy, drives the progress that is connected to the strategies of the public power directed to the needs and peculiarities of the local population. However, Blake (2008); Cleverdon & Kalisch (2000) emphasize that the driving force for socio-economic development needs to have families included in tourism activities.

To improve the performance parameters of tourism, the primary focus of managers and public authorities must be the quality of tourism supply and the skills of individuals working in the value chain, with particular attention being paid to startle and captivate tourists beyond their expectations. Besides, plans must be drawn up to provide long-lasting performance and sustainability to a destination (Costa et al., 2014).

Fernandes & Virgínio (2011), state that Social Responsibility is a factor capable of adding value to the tourism chain, together with government entities, private initiative and organized civil society, to establish citizenship actions. Thinking of social responsibility linked to tourism, one soon becomes an association restricted to environmental preservation, but tourism being multidisciplinary, can encompass various forms of Socioenvironmental Responsibility which a company in this area is capable of performing (Topke & Vidal, 2013).

Several studies focus on the global level of tourism. However, the World Tourism Organization - UNWTO (2017), analyzes the impacts generated by this sector. In this context, a new theme has been researched and debated worldwide in order to preserve and maintain tourist destinations: overtourism in cities such as Lisbon, Barcelona, Venice, Amsterdam, Berlin, Rome, Paris, Fernando de Noronha, etc. (Seraphin et al., 2018, Koens et al. 2018, Pearce, 2018, Pappalepore et al. 2014, Koens & Thomas, 2016, Zanirato & Tomazzoni, 2014).

In the words of World Tourism Organization- UNWTO (2017), overtourism is the destination where frequent visitation occurs and that the quality of life in the area or the quality of the experience has had an unacceptable impact and is to be preserved. For Aramberri (2015), it does not assume that residents are "anti-tourism," but this term seeks to emphasize that there must be control and legislation in tourist locations to maintain the site. Seraphin et al. (2018), emphasizes that awareness of the need to implement ambitious measures and imaginative policies capable of guaranteeing a socially sustainable destination rather than give in to the demands of pressure groups that only demand the reduction of tourist activity.

Starting from such inferences and knowing the importance of tourism for the development of the communities, the research problem was elaborated: "What are the contributions of the international publications on Overtourism?". Therefore, this study has a general objective: To analyze the characteristics of publications related to tourism and overtourism themes, in the Scopus and Web of Science database from 1998 to 2018.

The study is organized into six sections. After the brief introduction, the theoretical-methodological procedures of the research were described and, in the following section, the main results found. In order to complete the study, the final considerations, as well as their limitations and future lines of research, are carried out.

THEORETICAL REFERENCE

Tourism on Different Aspects

Tourism is one of the most globalized sectors of the world's economy. It is one of the most important instruments for creating jobs and income worldwide. In the 1960s and 1970s, tourism was one of the activities that grew most, due to the development of air transport, the generalization of tourism practice to various social classes and tourist services (Straaten, 2000). Over the past 100 years, there was an intense development and improvement, and the twentieth century came to be called the "century of tourism" (Netto, 2017).

For Leiper (1990), the term tourism originates from the Greek, meaning "circular movement" or "tool that makes a circle." Initially, an expression used in French and adapted by the Normans to express a return to the walls of the castle to see the landscape. The first attempt to define tourism, according Ignarra (1999), was in 1910, presented by the Austrian Herman Von Schullard, an economist, who described it as, "(...) addition of transactions, especially those of an economic nature, directly related to the entry, permanence and movement of foreigners into and out of a country, city or region."

In a way, the word gained visibility with the modern trips that the English made through the European continent, being this called the trip of " Grab Tour," from which emanated the term tourists to classify the participants, and tourism to define the activity (Boyer, 2003). Therefore, the tourism activity occurs due to the origin of the tourism phenomenon, which is a process whose occurrence requires the simultaneous exchange of various systems with performances that are added to give the final effect (Beni, 2019).

Thus, it appears that tourism is a diverse and complex social phenomenon, as there are various types of tourism, which can be sorted by different criteria. According to Barreto (2003), tourism is classified by its very nature: the receptive, receive the tourists coming from outside, and the emissive sends tourists out of place. Inbound tourism is active because it aims to move the local economy, since the emissive is economically passive, as it does not bring dividends to the issuing nucleus (Torre, 1992).

To provide tourism with a broader meaning, several stakeholders added new subjects to the definition of tourism. For Weaver & Lawton (2007), tourism can be defined as the sum of the processes, results, activities, that affect the interactions of tourists, between public and private organizations, that is, are organs that deal with the tourism area in the process of transportation, management, and attraction of tourists. It is noted that tourism can be seen as a range of businesses, individuals, places, and organizations, which in any way interact to provide a travel experience. Thus, the definition of tourism is difficult to understand, since some authors, as Cooper et al. (2008) base the definitions on demand and supply.

From the point of view of demand, Cunha (2013) states that tourism results from the movement of people to places outside their usual environment, and tourism based on supply, it's the whole set of organizations, professions, companies, coordinating all, in order to satisfy the necessities arising from the tourist trips. However, Bernardo (2013) describes through a more interactive and integrated approach, defining tourism as a set of interactions, relationships, and temporary activities among residents, mediators, and tourists, formed in response to the wishes and needs of tourists.

Tourism is a phenomenon, that can affect positively or negatively the region, once its ideology is towards the economic side, neglecting the fact that well-developed tourism could also impact socially. According to Körössy (2008), during an extended period, the emphasis of the tourist activity was directed only to the contributions to the Gross Domestic Product (GDP), that is, economic aspects. However, these definitions do not reflect the extent of tourism on social and environmental impact, either in the context of human rights, or cultural dimension, justice, and international citizenship. (Wohlmuther & Wintersteiner, 2013).

Tourism is one of the largest sectors of the world in terms of economic movement and its social impact. As a result, the concern on the impact that tourism could cause has increased, the Manila International Conference sought answers to this concern (Macleod, 2005). For Wohlmuther & Wintersteiner (2013), their concern is about the social impact that tourism would have in promoting fairer tourism with local cultures, providing a balanced division of economic gains from tourism.

Authors argue that tourism can also be seen as an alternative to the decline of agriculture and traditional industries in general, more specifically in regions that are considered backward and peripheral (Weidfeld, 2018). For tourism to have a positive impact, it is necessary coordination in tourism activity, developing strategies able to promote social and environmental development. Thus, even if tourism has a multiplier effect on the development into the visited region, it could create a positive social and environmental impact like any other economic activity (Gogonea et al., 2017).

Thus, there is a concern regarding the social and environmental impact that tourism activity can bring to the region that is being explored. World Tourism Organization (WTO) ranks sustainable tourism as a way to consider the needs of stakeholders such as local communities, tourists, environment, and tourism, taking into account the current and future economic, social and environmental impact (Liu et al., 2018). In such a way, sustainable tourism is idealized as a way to manage resources by satisfying the social, aesthetic, and economic needs and above all respecting the cultural integrity of the place, biological diversity, ecological processes and all the systems that sustain natural life (Brandão et al. 2015).

A study carried out in India shows that sustainable tourism practices are linked to the residents' quality of life, being the path of progress towards the sustainable development of tourism (Mathew & Sreejesh, 2017). In addition, in Malaysia, a survey was conducted with residents, where they believe that the solution to combat the diverse effects of tourism is responsible practices of tourism (Hanafiah et al., 2016).

Therefore, tourism contributes to the most diverse sectors, characterizing itself as an activity that links cultural, economic, social, and environmental spheres (Valencia, 2017). Consequently, to promote the concept of sustainable tourism, it is necessary to analyze practices effectively and existing initiatives, aiming to improve the coordinated and harmonious development of tourism and the environment. In order to turn it into a need, social responsibility must be analyzed in tourism as it can serve as a strategic mechanism for the development of sustainable tourism. The next section describes what social responsibility is, within organizations, and that tourism can develop social responsibility.

Social Responsibility

The theme of social responsibility (SR) has become, increasingly crucial to the performance of companies, as a result of a scenario in which shareholders, customers, suppliers, employees and especially society seek for more reliable and stable relationships.

The first theoretical approaches to social responsibility began in the 1920s, being the first appearance in Sheldon's work (1924), by addressing the existence of social aspect of community-based management and those governing the company (András & Rajcsányi-Molnár, 2015). In the year 1930, Berle (1932) highlighted the responsibility of managers as being alternative support to the community, which is unable to make a living through work or activity. In the following decade, the author Kreps (1940) illustrated that entrepreneurs are part of the community, and their actions are part of the community activity.

In the 1950s, social responsibility emphasized concern about moral issues such as worker welfare, medical care, retirement and safety programs, environmental protection, ethical behavior (Carroll & Buchholtz, 2000; al., 2005). The work that stood out in this decade was one of Eells (1958), which stated that companies also have a philanthropic function, aiming to manage scientific, educational, and charitable activities.

Whereas, in the following decade, the concept of social responsibility was highlighted, in the definition by Davis (1960), being the management of decisions and measures taken by executives due to economic interests. The author also stated that social responsibility was a vague idea, once it could be defined in several ways. In the same decade, the author Frederick (1960) in his work mentioned that the human and economic resources of a company should not be used only for individual interests, but also for great social purposes. As a result, Friedman (1962) discussed the issue by running away from the philanthropic question, saying that the only social responsibility of a business, is to increase profits, provided it complies with the rules of the market.

But it was along with the decade and 1970, that authors describe that the proliferation of concepts related to social responsibility occurred. For Davis (1973), social responsibility arises where the law ends, and Sethi (1975) associates social responsibility with the company's ability to adapt itself to society needs. Fitch (1976), describes that Corporate Social Responsibility is defined as an attempt to solve the social problems caused by companies.

In the 1980s, social responsibility was described as a voluntary effort by business managers to respond to the needs of society (Jones, 1980). In the view of the author Epstein (1989), is discussed that corporate social responsibility has an association with corporate citizenship, especially about the consequences or results of corporate actions on external and internal stakeholders.

Therefore, in the 1990s, it was the decade that companies sought to restructure to increase international competitiveness, thus taking initiatives on social responsibility. Carroll (1991), distinguished himself from the other authors on the subject, by presenting social responsibility, through a pyramid, having its base formed by economic responsibility, followed by legal responsibility, the next stage of the pyramid is formed by ethical responsibility and ultimately, philanthropic responsibility. Thus, in 1998, the Ethos Institute adheres to the discourse on social responsibility.

As from the 2000s, a company is classified socially responsible, when it achieves its liquidity results, honors commitments manufacture correct products, uses clean technology, generates employment, without losing sight of its ethical management function (Melo & Froes 2004). However, Matten and Moon (2008), expose that social responsibility reflects the social needs and social consequences of business success. It should also be noted that although the company can be seen as an enterprise, commercial company, and physical entity, it's a social institution, which ensures the production of goods and services necessary for society (Rocha, 2010).

However, it is not enough to pay taxes, generate jobs and income; it is also necessary to embrace social, environmental, and economic practices. Thus, the company goes from being just a technical and productive system, to be considered as a community of people (D' Arc & Felix, 2009). Therefore, the company's actions reflect directly or indirectly in society, so social responsibility is developed to promote cleanliness, not exploitation of labor, to seek competitiveness (Raman et al., 2012; Sweetin et al., 2013).

In addition, social responsibility has been widely researched, and companies should seek alternatives to act in a balanced way (Alexander, 2015; Armstrong & Green, 2013). Thus, the authors began to assign companies with a social agent role, being responsible for the welfare, quality of life, and for sustainable development (Jesus et al., 2017). The concept of social responsibility can be understood as an open system that interacts with several internal and external publics and stakeholders, defining the ethical, transparent and responsible management of the company (Instituto Ethos, 2012). For Alencastro (2013), corporate social responsibility represents a seal of quality that directs the consumption of services and products, ethical values, and transparency, which involves a good relationship between the community.

The concept of SR is associated with the relationship between internal Stakeholders, corporate responsibility, coordination of relations with the State and other interstate Institutions, and performance among the community (Baumann-Pauly et al., 2013). From Witkowska, (2016), we can mention characteristics of SR, such as the need to integrate social, economic and environmental responsibility in business operations and decision-making, voluntary, focused on the management of external effects, going beyond philanthropy, being focused on operational considerations.

The social responsibility has been signed as a management principle systemic, where the development of value in the social, environmental and economic dimension is assumed, assuming more and more strategies organizations foment balanced enterprise development (NUNES et al., 2014). Thus, SR is a way of rewarding society for an achieved goal, changing customs and habits, or even locations and profile of the individuals who receive the impact of transformation.

In the next section, it will be discussed the contextualization and concepts of Overtourism, is this a set of activities developed by people during their trips, being in places outside their usual environment, for a period that does not exceed one year.

Overtourism

Currently, tourism is a constant and is defined as "an economic activity due to the displacement and permanence of visitors" (Cunha and Abrantes, 2013, p.15). Over the years, the definition of tourism has undergone changes and adaptations, and several were the authors who contributed to the evolution of the definition of tourism.

Initially, the concept of tourist emerges, and only then does the tourism appear, these two concepts being interconnected, since without tourism, there is no tourism. Thus, the first definitions of tourism emerged in the twentieth century, and several authors have considered tourism as a phenomenon (Burkart & Medlik 1981, Mathieson & Wall, 1982). Other authors, however, argue that tourism is not a phenomenon, but a human activity that depends on the interaction between people and the exchange of economic, environmental and socio-cultural resources (Quadros, 2017).

According to Cunha & Abrantes (2013, p.17), the concept of tourism can be the one proposed by the World Tourism Organization, which states that tourism is "the set of activities carried out by people during travels and stayed in places outside their usual environment for a consecutive period not exceeding one year, for reasons of leisure, business, among others." However, this definition is focused on the demand side, excluding the supply side, and forgetting that to carry out activities; it is necessary that they are provided by the destinations, to serve directly or indirectly the visitors.

Still, according to Mathieson and Wall (1982), tourism is the temporary movement of individuals to destinations outside their usual places of work and residence, their activities during the stay at the destinations and facilities created to meet their needs.

Tourism development, results from changes and transformations in various sectors of society, reverting to an evolutionary process that must be structured at an international, national, regional and local level to be successful, as in addition of being an international movement of people, impacts on destination (Quadros, 2017).

Quadros (2017) and Postma & Schmuecker (2017), stresses that tourism development is influenced by the development of visitor's regions of origin, economically, socially, culturally and environmentally, affecting mainly residents, as the main stakeholders, since they are the only ones exposed to tourism development.

For Andereck & Vogt (2000), it is fundamental that the acceptance by the local community is taken into account since tourism development is not to be considered simply by the combination of offering tourist products with the tourist demand; furthermore, tourism involves a set of interactions between tourists and residents. If the community is not involved in the planning process, it may turn into a negative perception of tourism. In this regard, Quadros (2017) states that community involvement in the planning of tourism strategies is an important factor since it depends on the success of tourism development.

Tourism is a growing activity in the world and is increasingly capital-intensive, and as an important source of employment and income, it has been highly valued in the development strategies of several regions, states, and countries (Rezende et al. Rezende, 2005; Ram & Hall, 2018; Joppe, 2018; Butler, 2018). However, the use of tourism for this purpose must be carefully analyzed because of the serious problems brought about mass and exploratory tourism.

In contrast to the models adopted until the 1990s, when investments in infrastructure and facilities stimulated exploratory tourism (Rezende and Rezende, 2005), the emerging ecotourism term presents itself as an alternative form of activity in which innovation, preservation, and sustainability are key factors. For some, the sustainability of tourism is a major concern of planners, and it is necessary to define a development model allowing the growth of activity without quality loss.

According to the World Tourism Organization (WTO, 2003), this is also a concern of an increasing number of tourists who are interested in visiting places well planned, organized and that does not favor the emergence of economic, environmental and social problems.

Traditionally, planning of tourism activities has been more geared to meet tourist demands by investing in the facilities and services they need, without taking into account the well-being of residents (WTO, 2003). According to this organization, such an approach cannot be maintained if the long-term success of tourism interventions is to be ensured, and quality experiments are carried out to improve the destination's retention capacity.

According to Oliveira (2008), the difficulties in integrating the communities in the planning of tourism activity do not justify the abandonment of this practice, as they do not eliminate the communities' desire to be development subjects. According to the author, this position becomes clear when the target population of his research states that his views on community development were never heard - and that they were excluded from all stages of the planning process, showing some discontent.

From this, it is understood that it is necessary and primordial to seek more effective ways of making this participation feasible. One possible path is the empowerment of local people, which ultimately allows them to decide on the main development guidelines they want for their region, and how they want the costs and benefits to be shared among the different stakeholders (Oliveira, 2008).

Studies have argued that it is essential for the development of tourism that local people are informed and involved in strategic planning (Scalabrini et al., 2014). This becomes fundamental for the development of tourism in these communities. Involving residents in planning makes their perceptions and attitudes more positive about tourism and engages with visitors. For Quadros (2017), when the visitor interacts with the resident, a certain type of empathy with the destination is developed, contributing to increasing the quality of the experience and the loyalty to the destination.

In addition to everything that has already been stated, it is understood that the participation of residents in the whole process is fundamental, so that tourism is effectively sustainable, environmentally, economically and socially, for they are the cultural agents and the social group in which tourism is delivered. It is not a question of eliminating tourism, but of striving for balance, that is, avoiding those excesses affect the preservation of cities and the quality of life of their residents, generating inevitable chaos, which is being approached in current studies as overtourism.

Overturism is a novel concept that is being discussed mainly in the major capitals of world tourism. The term means over-tourism and has the effect of damaging the ecosystem. It is considered as something negative, which goes against sustainable tourism practices. Currently, this impacts on the tourism industry, which is facing various difficulties, such as operating with sustainability without deteriorating the environment or creating social and cultural problems. Such a concept, overtourism has gained prominence as one of the most discussed issues concerning tourism in the popular media and, increasingly, in academia. Despite its popularity, the term is still not delineated and remains open to multiple interpretations (Koens et al. 2018; Costa, Montenegro,Gomes, 2018; Moreno-Izquierdo, Ramon-Rodriguez, Jesus Such-Devesa, 2018).

For Koens et al. (2018), the term overtouism arose largely from media discourses without much theoretical foundation. The issue that describes an excessively negative impact of tourism on host communities and/or the natural environment has been a critical concern within the academy for many years. However, when dealing with the topic of overtourism, such authors emphasize the need for government regulation and leadership. This perceived, the need is in stark contrast to the prospects for more autonomy and self-governance that have dominated tourism discourses for several decades (Pearce, 2018).

It is a fact that in recent years, the perception of tourism in cities has changed dramatically. Public transportation, infrastructure, roads, museums, attractions, and other services that were created mainly for local use suffer from the increase in the number of tourists. The growing popularity of online services (Airbnb, HomeAway, Uber) and the desire to see 'real' and 'authentic' everyday life have made tourist activities even more intertwined with local life outside the main tourist areas of cities (Pappalepore et al. 2014).

Such developments have led to an increasing appeal of residents and stakeholders, and protests have been observed in several high-profile destinations. Although the issue is more prominent in European cities, similar feelings have been reported in other destinations as well, for example, tropical islands, backpackers' ghettos or even favelas (Koens & Thomas, 2016). According to Koens et al. (2018), describe these pa ra tourism disturbances, the term "overtourism" rapidly popularized.

Overture or over-tourism can be described as a situation in which many tourists visit a specific destination. However, the term "many" tourists are relative and must be defined within the context of a particular destination. Therefore, overtourism occurs when there are more visitors than facilities at a destination to accommodate them, and when the quality of life or experience in a certain area has been noticeably deteriorated (Misachi, 2019).

According to Goodwin (2017), overtourism describes destinations where hosts, locals or tourists feel that there are many visitors and that the quality of life in the area or the quality of the experience has been unacceptably deteriorated. It is the opposite of responsible tourism, in which tourism is used to provide better places to live in and visit. Often both visitors and guests and locals experience this deterioration at the same time and rebel against it (Goodwin, 2017).

According to the United Nations World Tourism Organization (UNWTO), the number of tourists visiting other countries is expected to grow by about 3.3 percent a year from 2010 to 2030. Given these statistics, the outbreak is inevitable in some places, although contributing factors vary from place to place.

Long-Term Study on Tourism and Overtourism

Popular vacation destinations are suffering increasingly with overloads and overcrowding. Crowded beaches, long queues at attractions, constant traffic jams, and crowded urban centers are just a few of the effects that both guests and locals have faced in various destinations around the world. More than ever, developments in tourism are influenced by global events and trends. These include changes in relations with mobility, general demographic change, and digitization, as well as increased urbanization and consequent migration to urban centers (Erschbamer et al. 2018).

Several low-cost airlines have allowed them to travel long distances in a short period at cheaper prices, thus making destinations around the world accessible to more people. For Erschbamer et al. (2018), the result of all this is that not only the most popular places have suffered from overtourism, but even the second-level destinations end up being crowded.

Overtourism may have several causes and may differ depending on the destination. On the one hand, there are global developments - for example, social changes such as increased mobility. On the other hand, location-specific factors come into play and must be taken into account. Thus, for example, better connections and accessibility to a destination, professionalism in the service sector, or aspects of the local monetary policy can have the effect of increasing the number of visitors and tourists.

Several authors have concentrated their studies on the spheres of tourism growing worldwide, however, it is also fundamental to analyze the impacts generated by this sector, and this has been presented in studies such as Pinke-Sziva et al. (2019), in which the objective was to analyze the phenomenon of overtourism with specific reference to the night economy in Budapest. The survey took place between September and December 2017 in the so-called "party block" of Budapest - District VII. Methods chosen included mapping, observation, interviews, and questionnaires with residents, visitors, and tourists. The results show that tourists highly value holiday opportunities, and they frequent most of the bars in the region.

Already Koens et al. (2018), carried out a qualitative investigation among 80 interested parties in 13 European cities. The results highlight that overtourism describes a multidimensional and complex question. Not only problems caused by tourism and non-tourism stakeholders, but they must also be seen in the context of the wider development of society and the city. The article concludes by arguing that while the over-news debate has once again drawn attention to the old problem of managing the negative impacts of tourism, it is not well-regarded. Seven myths about tourism are identified and can inhibit a complete understanding of the concept.

Thus, to deepen and strengthen the theme and have a more complete and contextualized understanding it is necessary that researchers from the most diverse areas get involved with the topic to reach new perceptions since the term is rather recent and there is still little research published on it.

METHODOLOGY

Description of Research

In the present article, the technique of bibliometry, quantitative and statistical type, according to Tague-Sutckiffe (1992) and Garousi (2015), is a set of mathematical and statistical techniques and procedures that allows among other to measure the diffusion of the production and dissemination of the scientific knowledge in a certain subject to evaluate the growth of this production, in books, articles, documents, newspapers, authors.

In March 2019, the data of this bibliometric analysis were collected, using Boolean operators to expand the research, so that one can have a term of comparison with reality, in two large multidisciplinary bases, Thomson Reuters's Web of Science (WOS) and Elsevier's Scopus.

The two databases were used to reduce the risk of missing documents due to different coverage of publication and research. The research was refined in such a way that it only included scientific papers in both databases, and the terms of "Overtourism" were searched in all areas of research without differentiation of the terms searched. All publications were surveyed in the period from 1998 to 2018, once the year 2019 is underway.

Thus, in order to compare the articles collected in the Web of Science and in Scopus, the main authors who wrote on the subject were identified, along with the growth of the researches in the terms during the period in question, the leading newspapers, main institutions, central countries and languages, areas of knowledge and the relationship between authors with more publications and those most cited. In Excel, comparison tables were created to analyze the data.

It is noted that there is an index of citations in both the Web of Science and in Scopus, where it is mentioned for each article, its title, authors, year, number of citations, type of documents as well as the year of publication (Figure 1).

Stages of Data Collection

This research was divided into two stages. The terms "Overtourism" were initially typed in the WOS and Scopus research field, delimiting the period between 1998 and 2018. The general characteristics of the publications were then surveyed. In the second stage, the most cited publications were compared with the authors who published the most in the same period. Figure 2 shows the steps of the research.

It was performed bibliometric analysis of this study, according to the steps outlined in Figure 2 and are shown below.

Figure 1. Distribution of articles by type of publication

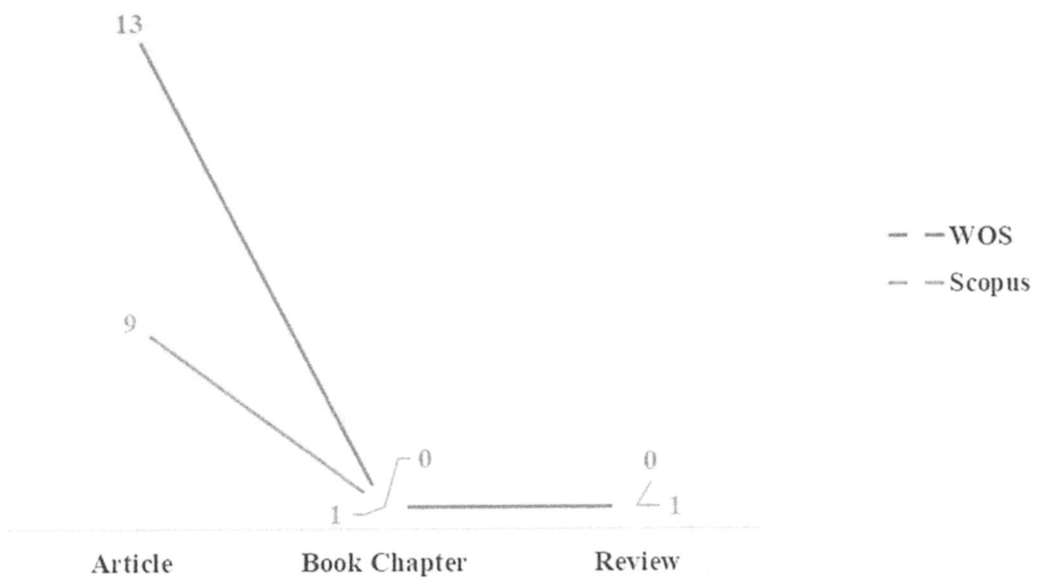

Figure 2. Stages of research
Source: Authors' elaboration.

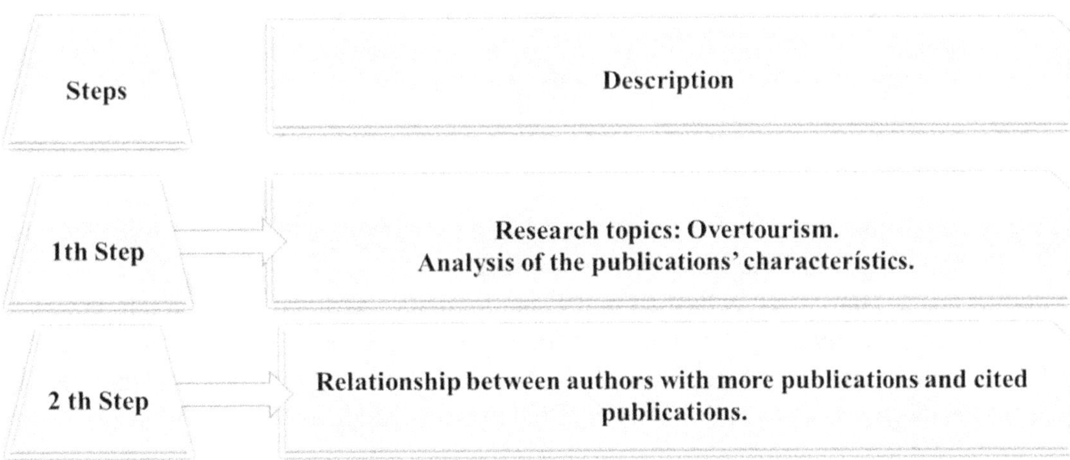

ANALYSIS AND DISCUSSION OF RESULTS

The search results are displayed at this point. To expand the scope of research, and to encompass as many results as possible, it was performed based on the appropriate Boolean operators for refinement. The authors found 24 scientific articles distributed by Web of Science and Scopus, as indicated in Table 1.

Table 1. Search terms

Terms	Web of Science	Scopus
"Overtourism"	13	11

Source: Authors' elaboration.

General Characteristics of Publications

The following are the general characteristics of the publications related to the theme "Overtourism": 1. Main authors, 2. Title of sources, 3. Main institutions, 4. Year of publications, 5. Main countries, 6. Languages, 7. Research areas, and 8. Keywords.

Then, the number of publications per author and the number of citations will be displayed.

Main Authors

The first analysis was concerning the most prolific authors (54 articles). Figure 3 presents the 20 authors and the corresponding number of publications, most published on the topic "Overtourism," in the two databases, during the analyzed period.

As for the representativeness of the articles by author, it is verified that eight authors are common to both bases: Butler, R., Cardoso, C., Coromina, L., Costa, J., Fernández, JAS, Galí, N., Gomes, J. and Postma, A.

The most published author has two publications in Scopus, and the WOS was Albert Postma.

Main Title of the Sources

Most of the scientific journals with the greatest impact on the subject in the Web of Science and Scopus are the Worldwide Hospitality and Tourism Themes (Figure 4).

Most of the articles have been published in the area journals which are not yet a reference as we are facing an emerging issue (International Journal of Tourism Cities, Regional Research Journal of Regional Research, Journal of Tourism Futures, Steps Magazine and Tourism, and Cultural Heritage, Tourism Management Perspectives, Tourism Review, Worldwide Hospitality, and Tourism Themes).

Main Institutions

The institutions that most published works related to the theme "Overtourism" are presented in Figure 5.

As we are faced with an emerging theme and the publications on the subject began in 2017, we do not have yet an institution that stands out in the two bases, Web of Science and Scopus. In the portfolio of the Organizations are two Portuguese universities: The University Algarve and the University Catolica Portuguesa.

Articles Published Per Year

In Figure 6, we can follow the growth of the publications in the study period.

Figure 3. Distribution of articles by the author
Source: Authors' elaboration.

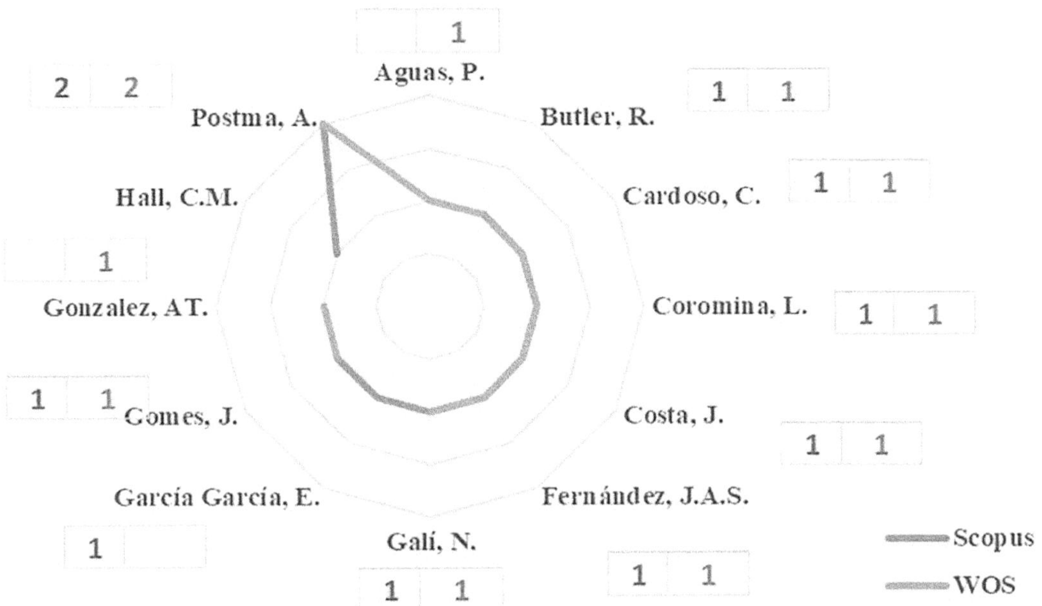

Figure 4. Distribution of articles by type of source
Source: Authors' elaboration.

Figure 5. Distribution of articles by the main institutions
Source: Authors' elaboration.

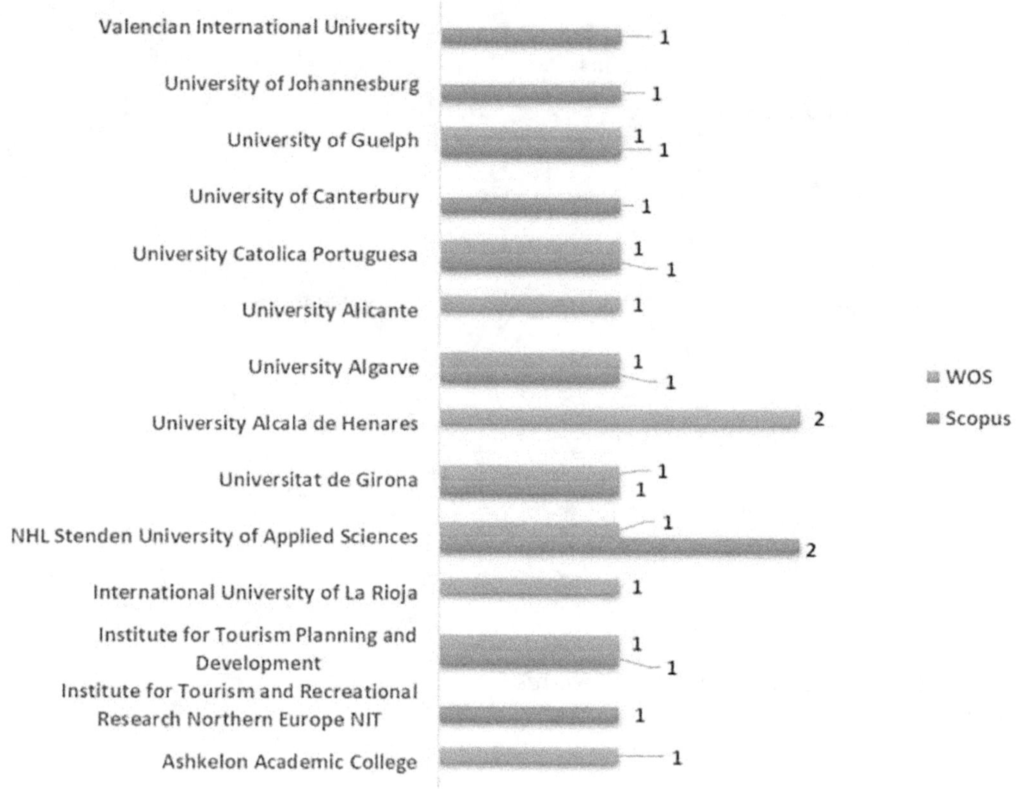

Figure 6. Distribution of publications per year
Source: Authors' elaboration.

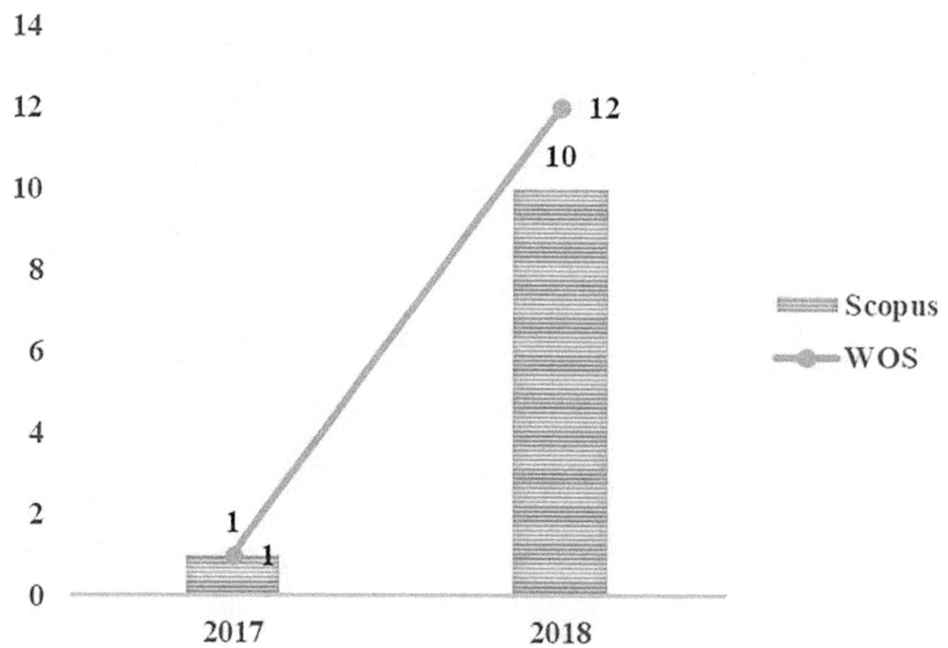

Main Countries

The number of publications per country and database is illustrated in Figure 7.

It is highlighted that it is a very topical issue and the academy plays a decisive role by bringing new ideas and projects that, in the existing relationship and interaction with the Government and Industry, apply in practice the model of the Triplex Helix proposed by Etzkowitz and Leydesdorff (1997). This interaction between the three sectors has to be dynamic.

Main Languages

In this variable, the languages of the papers published in the study theme are presented. The privileged language is English with 33 articles (Web of Science 12 and 11 Scopus), as shown in Figure 8.

Research Areas

Concerning the main areas of research in publications on the subject, presented a convergence in most of the published articles, in Figure 9, the research areas are shown.

In the next section, the keywords in the articles will be analyzed.

Figure 7. Distribution of publications by country
Source: Authors' elaboration.

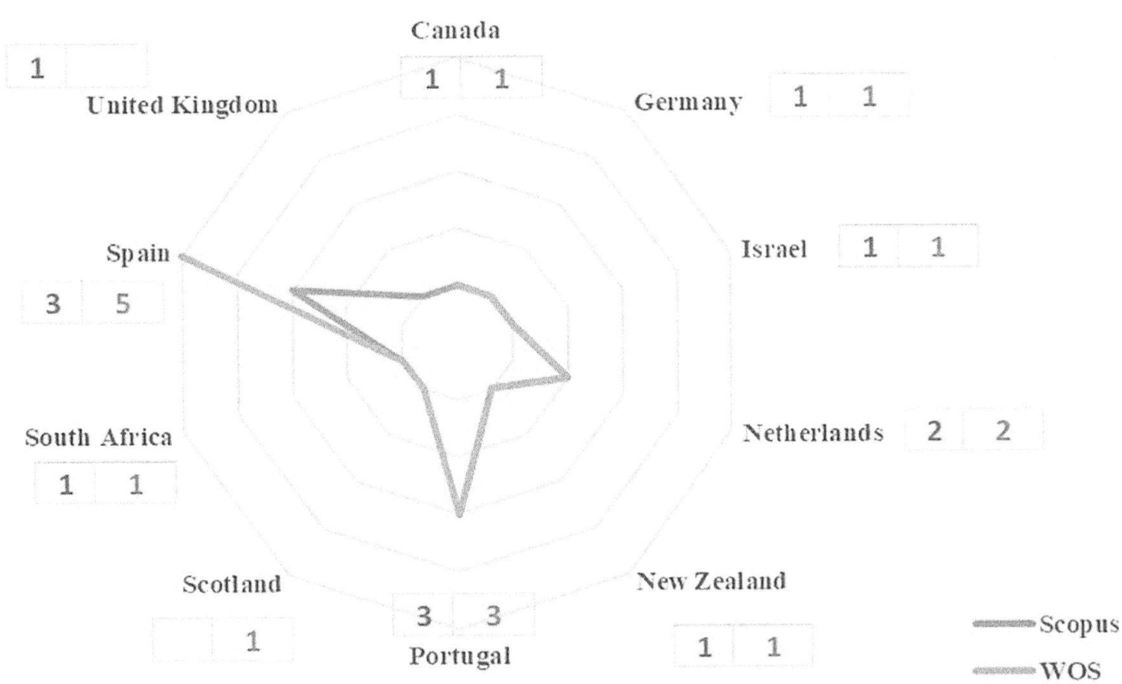

Figure 8. Distribution of articles by language
Source: Authors' elaboration.

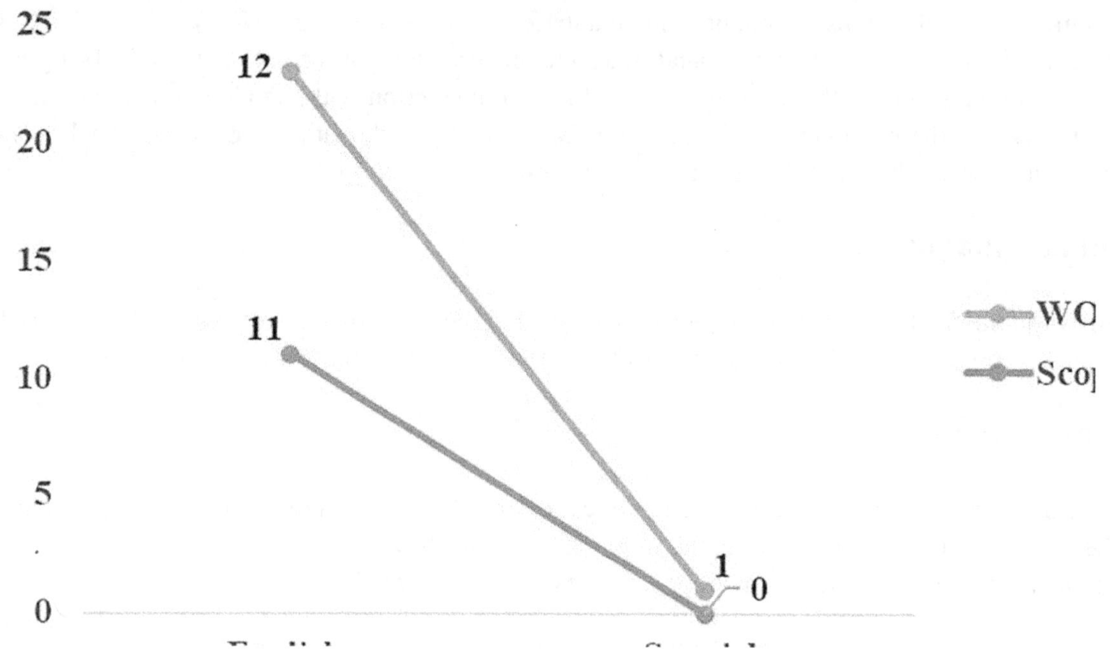

Most Used Keywords

It was verified that in the articles of the study were found 50 keywords in the text related to the theme (Figure 10). The keywords represent a source of access to scientific articles, briefly present the content used in the studies.

The most used keyword was "Overtourism," is cohesive with the purpose of the research, which strengthens the selection of the Bibliographic Portfolio.

Most Cited Articles from 1998 to 2018

Among the data from the Web of Science and Scopus survey on "Overtourism" from 1998 to 2018, it is essential to note that we only found two articles published in 2017 and twenty-two in 2018. Table 2 shows the leading publications in the number of citations, shows a slight juxtaposition between the two bases.

Of the seven most cited articles, six appear in both databases. The most relevant article is: "Understanding and overcoming the negative impacts of tourism in city destinations: a conceptual model and strategic framework" by Postma, Albert; Schmuecker, Dirk, published in the Journal of Tourism Futures, in 2017 with the total of ten citations.

Figure 9. Research areas of the articles
Source: Authors' elaboration.

Figure 10. Keywords of the research articles
Source: Authors' elaboration.

FINAL CONSIDERATIONS

This article aimed to verify the characteristics of the national and international scientific production that relate the theme "Overtourism." A bibliometric survey was conducted on the Web of Science and Scopus databases, obtaining thirteen scientific articles in the Web of Science and eleven in Scopus to achieve the goal.

Table 2. Distribution of the most cited articles in the period from 1998 to 2018

Number of Quotations		Title	Author	Journal	Year
SCOPUS	WOS				
6	4	Understanding and overcoming the negative impacts of tourism in city destinations: a conceptual model and strategic framework	Postma, Albert; Schmuecker, Dirk	Journal of Tourism Futures	2017
4	3	Walk Score and tourist accommodation.	Ram, Yael; Hall, C. Michael	International Journal of Tourism Cities	2018
3	2	Overtourism, Social Unrest, and Tourismphobia. A controversial debate.	Milano, Claudio	Pasos-Revista de Turismo y Patrimonio Cultural	2018
	2	Tourism policy and governance: Quo Vadis?	Joppe, Marion	Tourism Management Perspectives	2018
2	1	Is Overtourism Overused? Understanding the Impact of Tourism in a City Context.	Koens, Ko; Postma, Albert; Papp, Bernadett	Sustainability	2018
2	1	An Analysis of the Factors behind the Citizen's Attitude of Rejection towards Tourism in a Context of Overtourism and Economic Dependence on This Activity.	Martin Martin, Jose Maria; Guaita Martinez, Jose Manuel; Salinas Fernandez, Jose Antonio	Sustainability	2018
1	1	Overtourism: residents' perceptions of tourism impact as an indicator of resident social carrying capacity - case study of a Spanish heritage town.	Muler Gonzalez, Vanessa; Coromina, Lluis; Gali, Nuria	Tourism Review	2018

Source: Authors' elaboration.

Thus, in order to compare the articles collected at WOS and at Scopus, we identified the main authors who wrote on the subject, the growth of the publications in the survey period, the leading newspapers, institutions, countries, languages, areas of knowledge and the relationship between the authors that more publicated and those most cited.

The research results showed that it is an emerging topic in the literature, with the number of publications increased from 2017 (2) to 2018 (22). In this context, it is important to highlight that this topic has been growing. As can be seen, even the longitudinal period in the last twenty years, studies were only found from 2017 onwards being intensified in 2018. Concerning the interest for the academy, it has been found that both exist on the European and international scene.

The literature focuses on articles published in journals specialized in Tourism. It was also verified that, in both bases, Spain, Portugal and Holland, lead the ranking of the countries that published most on the subject, and the English language is predominant in the publications. It is noteworthy that Portugal ranks second in the ranking of publications, and Brazil is not on the list. The analysis of the distribution of publications on "Overtourism" in the Scopus and WOS databases shows that of the seven most cited articles, six appear in both databases. The co-author Albert Postma of the article "Understanding and overcoming the negative impacts of tourism in city destinations: a conceptual model and strategic framework" is one of the ten most published authors on the subject during the survey period. Other authors point out Coromina, Lluis; Gali, Nuria Hall, C. Michael; Salinas Fernandez, Jose Antonio, on the Top of those who wrote the most, are among the most cited. In addition, the articles published by researchers Costa, Jorge; Gomes, Joao; Butler, Richard; Aguas, Paulo and Cardoso, Carla, who are also on the same list, have not yet been mentioned, because the publications are recent.

Thus, it is possible to verify that there is a relation between the publications with the largest publications in "Overtourism," in the two databases.

The present study found the following limitations: 1) the main limitation of this study refers to the fact that only the search term "Overtourism" was used; 2) the information presented is limited to articles selected only on international bases, WOS and Scopus; 3) limitation bibliometric analysis limited to authors, articles, sources, institutions, countries, language and period; 4) Conduct research by summary and keywords; 5) perform the bibliometric analysis of the abstracts of the selected articles.

The contribution of this article to the studies in Management is due to the resulting indicators regarding the research institutions and the newspapers that stand out most in the production of the knowledge on the subject at international level. Considering that this is an emerging theme, it is suggested that for future studies, articles should be selected in others national and international bases in order to compare results; to undertake other bibliometric analyzes, such as the formation of networks of authors and institutions; to carry out a systemic analysis with the aim of analyzing other content different from those analyzed in the present study, such as: which theories have been worked together and in which countries. There is no research to understand how governments are taking steps to address the Overtourism problem.

ACKNOWLEDGMENT

This paper is financed by National Funds provided by FCT-Foundation for Science and Technology through project UIDB/04020/2020.

REFERENCES

Andereck, K., & Vogt, C. (2000). The Relationship between Residents' Attitudes toward Tourism and Tourism Development Options. *Journal of Travel Research*, *39*(1), 27–36.

Barreto, M. (2013). *Manual de iniciação ao estudo do turismo* (13th ed.). Campinas, SP: Editora Papirus.

Beni, M. C. (2019). *Análise estrutural do turismo* (14th ed., p. 558). São Paulo, Brazil: Editora Senac.

Bernardo, E. (2013). *Uma introdução ao turismo: conceitos, classificações e tipologias*. CIES E-Working Paper no. 164, Instituto Universitário de Lisboa, Lisboa.

Boyer, M. (2003). *História do turismo de massa*. Bauru, SP: EDUSC.

Brandão, C. DO N.; Barbieri, J C., & Reyes Junior, E. (2015). Análise da sustentabilidade do turismo: um estudo em comunidades indígenas no Estado de Roraima, Brasil. *Revista Brasileira de Pesquisa em Turismo. São Paulo, 9*(2), 500-518.

Burkart, A., & Medlik, S. (1981). *Tourism, Past Present and Future*. Oxford, UK: Heinemann.

Butler, R. (2018). Challenges and opportunities. *Worldwide Hospitality and Tourism Themes*, *10*(6), 635–641. PMID:30591573

Cardoso, C., & Silva, M. (2018). Residents' perceptions and attitudes towards future tourism development A challenge for tourism planners. *Worldwide Hospitality and Tourism Themes, 10*(6), 688–697.

Cooper, C., Fletcher, J., Fyall, A., Gilbert, D., & Wanltill, S. (2008). *Turismo: princípios e práticas* (3rd ed.). Bookman, São Paulo: Editora.

Costa, J., Montenegro, M., & Gomes, J. (2018). What challenges and opportunities will lead to success? *Worldwide Hospitality and Tourism Themes, 10*(6), 631–634.

Cunha, L. (2013). *Economia e política do Turismo.* Lisboa, Portugal: Lidel Edições Técnica.

Cunha, L., & Abrantes, A. (2013). *Introdução ao Turismo, 5ª Edição.* Lidel.

Erschbamer, G., Innerhofer, E., & Pechlaner, H. (2018). *Overtourism: How Much Tourism is Too Much.* Bolzano, Italy: Eurac Research.

Garousi, V. (2015). A bibliometric analysis of the Turkish software engineering research community. *Scientometrics, 105*(1), 23–49.

Goeldner, C. R., & Ritchie, J. R. B. (2011). *Tourism: Principles, Practices, Philosophies* (12th ed.). New York: John Wiley.

Gogonea, R. M., Baltalunga, A. A., Nedelcu, A., & Dumitrescu, D. (2017). Tourism Pressure at the Regional Level in the Context of Sustainable Development in Romania. *Sustainability, 9*(5), 698.

Goodwin, H. (2017). *The challenge of overtourism. Responsible Tourism Partnership* Working Paper 4. October.

Hanafiah, M.H.; Azman, I.; Jamaluddin, M.R., & Aminuddin, N. (2016). Responsible Tourism Practices and Quality of Life: Perspective of Langkawi Island communities. *Procedia - Social 128 and Behavioral Sciences*, 222, 406–413.

Ignarra, L. R. (2000). *Fundamentos do Turismo.* São Paulo: Pioneira.

Joppe, M. (2018). Tourism policy and governance: Quo vadis? *Tourism Management Perspectives, 25*, 201–204.

Koens, K., Postma, A., & Papp, B. (2018). Is overtourism overused? Understanding the impact of tourism in a city context. *Sustainability, 10*(12), 4384.

Koens, K., Postma, A., & Papp, B. (2018). Is Overtourism Overused? Understanding the Impact of Tourism in a City Context, *Sustainability*, 10(12), 4384.

Körössy, N. (2008). Do "Turismo predatório" ao "Turismo sustentável": Uma revisão sobre a origem e a consolidação do discurso da sustentabilidade na atividade turística. *Caderno Virtual de Turismo, 8*(2), 56–68.

Leiper, N. (1990). Partial industrialization of tourism systems. *Annals of Tourism Research, 17*, 600–605.

Liu, C., Zhang, R., Wang, M., & Xu, J. (2018). Measurement and Prediction of Regional Tourism Sustainability: An Analysis of the Yangtze River Economic Zone, China. *Sustainability, 10*(5), 1321.

Lohman, M., & Beer, H. (2013). Fundamentals of tourism: What makes a person a potential tourist and a region a potential tourism destination? Poznan University of Economic Review, 13(4), 83–96.

Macleod, D. V. L. (2005). *Alternative Tourism: A comparative analysis of meaning and impact*. Oxford, UK: Butterworth-Heinemann.

Martin, J. M. M., Guaita, J. M. M., & Salinas, J. A. F. (2018). An Analysis of the Factors behind the Citizen's Attitude of Rejection towards Tourism in a Context of Overtourism and Economic Dependence on This Activity. *Sustainability*, *10*(8), 2851.

Mathew, P. V., & Sreejesh, S. (2017). Impact of responsible tourism on destination sustainability and quality of life of the community in tourism destinations. *Journal of Hospitality and Tourism Management*, *31*, 83–89.

Mathieson, A., & Wall, G. (1982). *Tourism, economic, physical, and social impacts*. Harlow: Pearson Education.

Milano, C. (2018). Overtourism, Social Unrest and Tourismphobia. A controversial debate, PASOS-REVISTA DE TURISMO Y PATRIMONIO CULTURAL, 16(3), 551-564.

Misachi, J. (2019). *What Is Overtourism? Overtourism occurs when a destination receives too many visitors*. Retrieved from https://www.worldatlas.com/articles/what-is-overtourism.html

Moreno-Izquierdo, L., Ramon-Rodriguez, A., & Jesus, M. S. D. (2018). The challenge of long-term tourism competitiveness in the age of innovation: Spain as a case study, Investigaciones Regionales-. *Journal of Regional Research*, *42*, 13–34.

Muler, V. G., Coromina, L., & Gali, N. U. (2018). Overtourism: Residents' perceptions of tourism impact as an indicator of resident social carrying capacity - case study of a Spanish heritage town. *Tourism Review*, *73*(3), 277–296.

Netto, L. G. P. (2017). *O que é turismo*. São Paulo, Brazil: Editora Brasileiense.

Oliveira, E. S. (2008). Impactos socioambientais e econômicos do turismo e suas repercussões no desenvolvimento local: o caso de Itacare – Bahia / Elton Silva Oliveira. – Ilheus, BA: UESC, 153 fls.

OMT – Organização Mundial de Turismo. (1999). *Conta Satélite do Turismo*. Madrid, Spain: Quadro Conceptual.

OMT - Organização Mundial de Turismo. (2003). *Guia de desenvolvimento do turismo sustentável*. Porto Alegre, Brazil: Bookman.

Pappalepore, I., Maitland, R., & Smith, A. (2014). Presuming creative urban areas. Evidence from East London. *Annals of Tourism Research*, *44*, 227–240.

Pearce, P. L. (2018). Limiting overtourism; the desirable new behaviours of the smart tourist. In *Proceedings of the Tourism Intelligence Forum (t-Forum) Global Exchange Conference 2018,* Palma de Mallorca, Spain, March 11-13.

Pinke-Sziva, I., Smith, M., Olt, G., & Berezvai, Z. (2019). Overtourism and the night-time economy: A case study of Budapest. *International Journal of Tourism Cities*, *5*(1), 1–16.

Postma, A., & Schmuecker, D. (2017). Understanding and overcoming negative impacts of tourism in city destinations: Conceptual model and strategic framework. *Journal of Tourism Futures*, *3*(2), 144–156.

Quadros, M. (2017). *Percepções dos residentes sobre os impactos do turismo na comunidade local. 77f. Dissertação (Mestrado em Gestão do turismo internacional). Universidade dos Açores faculdade de economia*. Ponta Delgada, Portugal.

Ram, Y., Hall, C., & Michael, W. (2018). Score and tourist accommodation. *International Journal of Tourism Cities*, *4*(3), 369–375.

Rezende, C. F., & Rezende, D. D. (2005). *Impactos do turismo: uma análise sob a ótica da população receptora*. Anais do Encontro Nacional de Pós-Graduação e Pesquisa em Administração-ENANPAD.

Scalabrini, E., Remoaldo, P., & Lourenço, J. (2014). Percepções de residentes a respeito dos impactos da atividade turística: Uma análise das publicações brasileiras sobre o tema. *Tourism and Hospitality International Journal*, *2*(2), 12–31.

Seraphin, H., Sheeran, P., & Pilato, M. (2018). Over-tourism and the fall of Venice as a destination. *Journal of Destination Marketing & Management*, *9*, 374–376.

Straaten, J. V. (2000). Can sustainable tourism positively influence rural regions? In G. Richards, & D. Hall (Eds.), *Tourism and Sustainable Community Development* (pp. 221–232). London, UK: Routledge.

Tague-Sutcliffe, J. (1992). An introduction to infometrics. *Information Processing & Management, Oxford*, *28*(1), 1–3.

Torre, O. (1992). *El Turismo, Fenómeno Social* (1st ed.). Mexico: Fondo de Cultura Económica.

Trancoso, A. V. G. (2018). The problem of overtourism and the impact of cruises, Investigaciones Regionales. *Journal of Regional Research*, *42*, 35–51.

United Nations. World Tourism Organization. (1994). Recommendations on Tourisme Statistics, Serie M, nº 83, United Nations, New York.

United Nations, World Tourism Organization, (UNWTO). (2017). *Yearbook of Tourism Statistics. Madrid, UNWTO* (p. 15). Madrid, Spain: Calle Poeta Joan Maragall.

Valencia, A. G. (2017). *El eje económico de la sostenibilidad turística*. México.

Veiga, C., Santos, M. C., Aguas, P., & Santos, J. A. C. (2018). Sustainability as a key driver to address challenges. *Worldwide Hospitality and Tourism Themes*, *10*(6), 662–673.

Weaver, D. B., & Lawton, L. J. (2007). Twenty Years On: The State of Contemporary Ecotourism Research. *Tourism Management*, *28*(5), 1168–1179.

Weidenfeld, A. (2018). Tourism Diversification and Its Implications for Smart Specialisation. *Sustainability*, *10*(2), 319.

Wohlmuther, C., & Wintersteiner, W. (2013). Peace Sensitive Tourism: How Tourism Can Contribute to Peace. In C. Wohlmuther, & W. Wintersteiner (Eds.), International Handbook on Tourism and Peace. Klagenfurt: Drava, 1-379.

Zanirato, S. H., & Tomazzoni, E. L. (2014). A sustentabilidade do turismo em Fernando de Noronha (PE-Brasil). *Turydes. Malaga*, *7*(17), 1–24.

Chapter 4
Beyond Tourismphobia:
Conceptualizing a New Framework to Analyze Attitudes Towards Tourism

Alejandro Mantecón
https://orcid.org/0000-0002-3208-2114
University of Alicante, Spain

María Velasco
https://orcid.org/0000-0001-8590-5869
Complutense University of Madrid, Spain

ABSTRACT

This chapter discusses the factors that have led to the emergence of expressions of criticism toward tourism. This review serves to frame the original contribution of this text: a theoretical model that clarifies the defining features of the main attitudes towards tourism. Merton's model is here adjusted for the analysis of a new relationship between social ends and economic means. In this case, the end is economic progress. The way is the tourism, conceived as a massive social phenomenon. The relation between goals and means generates tensions. Its management derives in strategies of adaptation that include different ways of identification or discussion. The five types of adaptation of the new model are useful for addressing subject positions, political discourses, or attitudinal dispositions towards tourism. To illustrate this typology a purposive sampling of news on the tourismphobia has been selected, with no statistical generalization reflecting the constituent elements of each of the types: legitimization, innovative criticism, resignation, radical criticism, and subversive utopia.

INTRODUCTION

The crises suffered by capitalism during the decade of the 1970s were used by different elite groups to promote the substitution of Keynesian policies, which were influential in the thirty years following the Second World War, by policies adapted to new times, more commonly known as neoliberal and flexibly

DOI: 10.4018/978-1-7998-2224-0.ch004

policies (Schoenberger, 1988; Harvey, 2005). Changes to the principles that oriented the management of the most powerful capitalist states affected millions of people, both in periphery countries (Chossudovsky, 2003) and in regions that had been more central (Bauman, 1998).

At the same time, the digitalization of the economy and the application of technological innovations to movement of people and commodities served to extend and intensify historical processes of dislocation of production and transnational consumption. Among these latter processes, we highlight the growth of international tourism, which was limited up to the middle of the 20th Century but has now become a massive industry. In 2018 the World Tourism Organization (UNWTO) registered 1.4 billion international tourist movements, such that a great number of cities that were important industrial centers in the Fordist Age (as well as regions of the Global South that aim to better position themselves in the world system) now have opportunities for development thanks to the flow of tourism and leisure-oriented mobility around the planet (Burns & Novelli, 2008). Under this scenario, urban areas thousands of kilometers apart compete to position themselves as consumption spaces capable of attracting the tourist gaze (Urry, 1990). The success of tourism is also accompanied by a very positive public image, that of being a universal social right (UNWTO, 2001).

In the face of this scenario public institutions must restrict their basic functions as defined by the era of welfare capitalism (Esping-Andersen, 1990), by which they had to guarantee to their citizenry (and the cultural and natural patrimony) some level of independence and protection with respect to market ups and downs. Now, they must act as specialized promoters capable of attracting private investment and managing physical and human resources they are in charge of according to efficiency criteria (Cochrane, 2007; Harvey, 1989). In this way, a new order is being constructed which is characterized by the commoditization of urban space. It is conceived above all in terms of a set of opportunities for business.

WHY DO THEY CALL IT OVERTOURISM WHEN THEY MEAN CAPITALISM?

The concept of overtourism has become popular recently and refers to a phenomenon that has been studied for decades: the influence of tourism in the massification of spaces. This term has been defined as "the situation in which the impact of tourism, at certain times and in certain locations, exceeds physical, ecological, social, economic, psychological, and/or political capacity thresholds" (Peeters et al., 2018, p. 15).

However, as explained in detail by Habermas (1973), the saturation limits of a cultural system are much more difficult to determine than the limits that permit survival of a natural system. The increase in the number of people that circulate in the cities that are the primary receivers of tourist flow requires a constant increase in available lodging. It requires the availability of those attracted by new work opportunities, and it requires a reorganization of urban space and restructuring of the business fabric at the same time that it implies resettlement of historical residents and redefines the functions carried out by places. As in many other capitalist processes, the logic of growth gives way to a shock in terms of the ecological capacity of the environment and in terms of existing ways of life which, however, rarely results in unequivocal criticism of the reasons for these changes. On the contrary, space becomes a battle ground in which old and new social actors participate with different interests and unequal strengths (Milano & Mansilla, 2018).

The use of tourism by neoliberal capitalism along with the confluence of the real estate economy has promoted the apparition of social problems that require profound analysis: more precarious labor conditions of workers in the sector, the irresponsible consumption of public resources, noise pollution and the degradation of the urban environment, the increase in uncivil behaviors, congestion problems around tourist attractions, the expulsion of residents of tourist zones due to increasing rents, occupation of public spaces by businesses serving tourists, changes to the identities of cities, and integration of housing into an informal economy sustained by illegal tourist housing (promoted by P2P platforms). This brings about impacts including the loss of quality of life for permanent residents and proliferation of business activates linked to fiscal fraud, and therefore, not appropriate for modern societies (Álvarez-Sousa, 2018; Briggs, 2013; Cabrerizo, 2016; Cañada, 2018; Milano, Cheer, & Novelli, 2019; Milano, Novelli, & Cheer, 2019).

Great social changes manifest themselves in different ways and provoke very diverse reactions, especially in a world in which tourism penetrates both small rural communities and global cities. In this sense, tourism becomes an excellent means through which international networks of power and money recompose the spaces in which millions of human beings live their daily lives. The protests of the most affected citizens are often diluted, because their demands do not reach those responsible for the changes that transform their lives. After all, some of these changes do not depend as much on specific localized individuals in local political and economic institutions as on, perhaps, an investment fund registered in Qatar (Sennett, 2018).

THE USUAL SOCIAL EXCHANGE THEORY FOCUS

The studies that analyze the effects of tourism on the host society have been articulated from various theoretical approaches, although the Social Exchange Theory (SET) is the main approach that has guided most of the research (Harrill, 2004; Sharpley, 2014; Rasoolimanesh, Jaafar, Kock, & Ramayah, 2015). The central idea of this theory suggests that residents will support tourism development as long as they perceive that the expected benefits will outweigh the costs. A review of the studies inspired more or less by the SET allows us to recognize a wide variety of contradictory and little connected explanations:

a) Pizam (1978) believed that workers in the tourism sector express more positive opinions about tourism development than those who do not. His research gave way to a series of studies that confirmed his finding. However, the need to enrich the interpretation of the data has also been suggested when discrepant results are observed (Andereck, Valentine, Knopf, & Vogt, 2005, Marrero & Huete, 2013). Teye, Sönmez and Sirakaya (2002) suggest that people employed in jobs related to tourism have a more negative attitude.

b) In terms of the degree of interaction with tourists, some researchers point out that a smaller distance between the homes of residents and the places most frequented by tourists implies more positive attitudes, because these locations tend to be inhabited by those more economically dependent on tourism (Belisle & Hoy, 1980; Mansfeld, 1992). However, other research seems to demonstrate the opposite, and argues that a negative attitude is derived from greater proximity: residents would see tourists as a source of competition for the consumption of scarce resources. In addition, they would consider the influx of tourists to be the cause of the discomforts that they tolerate in their daily lives (Jurowski & Gursoy, 2004).

c) Some experts conclude that the longer a person lives in a tourist location the more critical their attitude towards tourism and, therefore, the less time has passed the more favorable their attitude would be (Brunt & Courtney, 1999; Cavus & Tanrisevdi, 2003; Ryan & Montgomery, 1994). However, other studies explain that newly arrived residents are the most critical (Brougham & Butler, 1981) towards tourism, perhaps because they feel it endangers the tranquility they sought when changing their residence (Faulkner & Tideswell, 1997).

d) The researchers who study the relationship between effects of tourism according to the residents and the usual sociodemographic and socioeconomic variables (gender, age, income, occupation, educational level, rural or urban area) have not reached conclusive results (Liu & Var, 1986 Williams & Lawson, 2001), or the findings are discordant (Harrill & Potts, 2003; Jackson & Inbakaran, 2006; Rasoolimanesh et al., 2015).

e) Those who have tried to elucidate this issue by looking at the stage of development of the tourist destination also have not reached an agreement (Vargas-Sánchez, Oom-do-Valle, da-Costa-Mendes, & Silva, 2015). It might seem reasonable to think that a moderate development of tourism would tend to be perceived positively and that as this development intensifies public opinion would become more negative (Harrill, 2004, Diedrich & García-Buades, 2009), but this argument has also been questioned (Liu & Var, 1986, Faulkner & Tideswall, 1997, Royo & Ruiz, 2009).

In fact, this line of research has not produced relevant empirical generalizations, apart from confirming the evident fact that in opinion studies on the effects of tourism, there is a greater concentration of positive attitudes around the economic effects and more critical attitudes related to tourism's environmental and socio-cultural impacts. Beyond the accumulation of case studies carried out according to the predominant (and not very innovative) logic of gap-spotting (Alvesson & Sandberg, 2013), the truth is that the application of SET to understanding social perceptions of tourism shows symptoms of exhaustion. The inertia observed in its use has long promoted a sort of "theoretical isomorphism" (Pfeffer, 2007, p.1341) that under-problematizes the existing literature, reduces risk and facilitates the existence of those who live under the law of "publish or perish".

The overconfidence that SET places in individual rationality, its underestimation of the emotional aspects, and the scant attention paid to social change are problematic. Also, its difficulty in orienting qualitative methodological designs and its disinterest in explaining the influence that power relations exert on the formation of public opinion invites us to explore alternative ways to help us understand the ways people experience tourism (Mantecón, 2016).

A NEW FRAMEWORK BASED ON MERTON'S CLASSIC STRAIN THEORY

The recent increase in public rejection of tourism is modifying the social perception of a phenomenon that, until recently, had managed to avoid major controversies. Criticism of certain effects of tourism is being incorporated into the demands of different social movements and neighborhood associations affected by the impacts of excessive tourism (Milano & Mansilla, 2018). Tourism has become problematized and today is part of the political and social agenda of dozens of cities around the world (UNWTO, 2019). The new conflictive image that surrounds tourism originated in the summer of 2017, when the media spread the use of the expression "tourismphobia" (coined in 2008 by the geographer José Antonio Donaire), resulting in a media earthquake whose epicenter was in Spain and had repercussions in different countries. However, rather than helping to better understand the origin and characteristics of anti-tourism attitudes, the media dimension acquired by this term resulted in politicization, ultimately hindering the success of any comprehensive effort to address the problem (Huete & Mantecón, 2018). In any case, the media pathologization of public expressions of criticism or rejection of certain types of tourism seems a more appropriate way to stoke political confrontation than the analysis of social problems.

Here we present a theoretical model that aims to clarify the defining features of different attitudes towards tourism. It consists of a novel adaptation of the classic typology developed by the sociologist Robert K. Merton to study the modes of adaptation by individuals within the culture-bearing society. Merton's proposal distinguishes between five distinctive patterns of relations between cultural goals and institutionalized means: conformity, innovation, ritualism, retreatism and rebellion. The author's intention was to explain the ways in which people manage the strain and the anomic risk produced by social demands and pressures (Merton, 1938).

Here Merton's model is adapted to analyze a new relationship between ends and means. In this case, the end is economic growth under the logic of capitalism, understood as a system of division of labor based on the incessant accumulation of capital through the commoditization of more and more elements in more and more places. This objective is inserted into a neoliberal cultural, political and economic context characterized by the enthusiastic exaltation of the consumer society and a scarcely regulated free market economy. The medium is tourism, conceived as a massive social phenomenon that causes very different effects in different but related areas. The relationship between goals and means generates tensions. Its management is based on adaptation strategies that include different modes of identification or questioning, both with respect to the majority tourist behavior and to the values of the capitalist culture. These strategies are presented schematically in Table 1, where, in a generic sense, (+) represents acceptance, (-) represents non-identification and (- / +) represents non-identification with existing reality and substitution of reality with new proposals.

This typology is useful for studying subject positions, social discourses or attitudinal dispositions towards tourism. It can guide the collection and analysis of data in quantitative, qualitative or mixed methodological designs, and be integrated into exploratory, descriptive, and causal approaches, etc. It also allows for carrying out critical and diachronic approaches to investigate the historical configuration of the different positions and the power relations established between the social actors that represent them. With the intention of illustrating this typology, and without any pretense of statistical generalization, we carried out a purposive sampling of news that reflects the constitutive elements of each of the types. The origin of the news are newspapers, radio broadcasts and Spanish digital publications, because, as previously noted, Spain is the epicenter of the earthquake that popularized the term tourismphobia. That is why, in media discourses, there are evaluations and diagnoses laden with emotional connotations. In a practical sense, this situation helps to establish relationships in our theoretical model:

Table 1. Patterns of Relationships Between Capitalism and Tourism

	Culture Goal: Capitalist Development	Institutionalized Mean: Known Tourist Dynamics
I. Legitimization	+	+
II. Innovative criticism	+	-
III. Resignation	-	+
IV. Radical criticism	-	-
V. Subversive utopia	-/+	-/+

Source: own construction based on Merton (1938).

Legitimization

This type of adaptation supposes high acceptance of means and goals. It is the most common strategy and also the most conservative one: it guarantees the reproduction of the social order and the stability of the existing system of relationships and favors the development and intensification of already established guidelines of action. Here, it is represented by news in which arguments are developed that extol the value of traditional tourism (along the lines of Fordist or mass tourism) as a legitimate means to guarantee the production of wealth and the progress of society. This type seeks to persist with known dynamics because they are perceived as the best option. It denies the existence of significant problems caused by tourism. The moral commitment acquired by those who most identify with legitimization implies vehement action aimed at discrediting any questioning. They are the guardians of the status quo. The following excerpts exemplify this position:

A substantial reduction in tourist flows from abroad or from among Spanish citizens would have a very negative economic and social effect [...] The hypothesis known as Venice Syndrome, according to which excessive tourism, for example in urban or historical centers of certain cities, it is harmful to those who live in those habitats, is incorrect. On the contrary, the tourist flows in these areas have resulted in an improvement of their general endowment -infrastructure, security, etc.- in order to satisfy and adapt to the demands of a greater demand. (Turismofobia, *El Mundo*, September 3, 2017)

Tourismphobia does not exist, and in fact it is an insulting term for societies that have lived and worked successfully with tourism for decades, and now do not deserve in any way be considered suspicious of anything because of a few idiots" (Ja ens han colat la turismofòbia, *ara.cat*, 2 August 2017)

Remember that Spain has more than fifty years of perfect coexistence with tourism and that the income earned from the sector exceeds what Saudi Arabia can get for the exploitation of oil. It's something that we cannot let go of, nor can we damage it. (Abrazos y claveles a visitantes para combatir la turismofobia, *Cadena SER*, 15 August 2017)

Innovative Criticism

This corresponds with a high identification and acceptance of the goals and a questioning of the means. This includes news in which traditional tourism strategies are questioned or criticized as legitimate means to achieve the desired goal: wealth generation and progress. This position can be bifurcated in two:

1) On one hand, this includes those frequent speeches of social democratic inspiration, which, despite defending the economic activities that produce material wealth and generate social progress within the framework of a free market economy, doubt the effectiveness of the practices of dominant tourist destinations as a means to achieve that end. This implies a proposal to adopt other types of tourism practices that are more sustainable and typical of a post-Fordist capitalist context. Post-Fordism is reflected in the area of tourism in terms of the proliferation of new attitudes and typologies that overlap and coexist with mass tourism. Post-Fordist tourism is characterized by the diversification of products, which are integrated into more segmented markets, with more demanding and experienced consumers who are less prone to repeat experiences and interested in alternative typologies to mass tourism. The context of restructuring in which today hundreds of territories with a significant industrial past are inserted, together with the incorporation of new spaces in the global exchange system, motivates many distant cities and regions to compete for future development by relying on the tourist economy, far from deep crises. This competition is determined by the possibilities that all regions have in facing the challenges

posed by tourism markets in a post-Fordist scenario. These include the difficulties and challenges faced by specialized destinations for decades in the management of mass tourism. Some illustrative fragments include the following:

We must move from a model of mass tourism towards a model of qualified tourism. A tourism model that allows its own economic and social reproduction, without devastating territories socially and environmentally. Emerging tourism markets do not demand anonymous destinations and infinite skyscrapers. They seek differentiation, distinction, traditions, authenticity. The new tourist is willing to pay more in order to enjoy a respectful natural environment [...] That is why it is necessary to govern tourism. Direct it to ethical and community objectives that allow us to redistribute the benefits of industry without renouncing competitiveness margins. (Gobernar el turismo, *Diario El Salto*, 5 December 2017)

The debate is on the street [...] over the issue of the damage that illegal lodging is doing to historic districts, which causes inconveniences to the neighbors. But the answer is to persecute them, not to end tourism [...] You do not have to criminalize, you have to regulate, and control, which is not the same thing. (Granada es turismoadicta, *Ideal*, 10 September 2017)

The problem is not to say yes or no to tourism but to rethink the way in which the city offers itself to tourists to avoid the dynamics of 'copy and paste' between cities that simplify and banalize the local culture and make it such that, in the long run, the tourist loses interest. (El desafío del turismo masivo, *El País*, 9 August 2015)

2) On the other hand, there are also those who, from less moderate approaches, rely on the capitalist end of production of material wealth and progress but consider that the loss of legitimacy of hegemonic tourist practices in the past must give step to a new era. This new era considers the path of development through tourism to be exhausted and encourages its replacement by another economic dynamic. This type of reasoning has very low public projection, with an almost anecdotal presence in the media. The following news is echoed by the publication of a report where the issue is raised in this way:

Attitudes contrary to tourism in Barcelona could have to do with a current that is being promoted from the higher-income echelon and biased political orientations, which would support the hypothesis that true economic and social development is only possible with high rates of expansion in the industrial sectors and not through tourism. (Turismofobia se debe a la falta de ordenación del sector, según un estudio, *eldiario.es*, 18 June 2018)

Resignation

In this case, a relationship is established between tourism and the production of wealth that is not based on convictions about the value of the proposed purposes, although this does not imply abandonment of known tourism practices. This type is illustrated by news that recognizes the importance that traditional tourism activities have had in the generation of wealth and progress, which justifies the convenience of continuing to bet on them. A blind faith in these practices is not argued here, but the need to persist in these practices is justified, because there is no other remedy. It is not the best option; it is simply the only viable one. This profile is not based on trust but on resignation. Therefore, there are doubts and skepticism about the repercussions of hegemonic patterns, but they are insisted upon because it is not possible to imagine other reliable alternatives. This is a very common position, and together with Type I (Legitimization), it contributes effectively to the reproduction of established models, drowning at the same time the formulation of options different from those already known:

Surely, it would have been better if jobs had been created with high technological qualifications, but with the de-industrialization that we have suffered, tourism has been a lifeline for the economy. Those who protest tourism will have to say what they plan to replace it with. Do they have any alternatives? (Corregir la turismofòbia, *La Vanguardia*, 20 May 2017)

To the group of friends that comes five days to play golf in Benidorm, or to the English bachelorette party that gets off the plane drunk in Malaga, we will not be able to teach anything of Cervantes, of Galician music or of the different grape varieties while they are drinking beers on the beach. The offer is what it is, the Spanish brand sells what it sells -and tens of millions of foreigners buy it willingly year after year- and the national value chain is set up to serve that clientele in that way. (Cuatro pintadas anti-turistas no van a cambiar el turismo en España, *Vozpopuli*, 15 August 2017)

If they stopped coming, what would we live off of? [...] Let's try to think with our heads before doing something with our hearts [...] let's stop demonstrating against the hand that feeds us. (Turismofobia en el Mediterráneo, *MallorcaDiario.com*, 26 July 2018)

Radical Criticism

Here the absence of identification acquires a sense of rejection that comes to question the validity of both predominant tourism practices and the capitalist purposes of wealth production. This profile is expressed in attitudes of direct criticism towards tourism and capitalism that emerged in the journalistic discourse as of 2017. Tourism is recognized as another instrument of the capitalist system directed toward economic exploitation, to the accumulation of wealth, to the decline of social justice and the destruction of the environment. This type includes positions that are not identified neither for the purpose (development of the capitalist society) nor with the means destined to the conquest of that end (the economic activities that are configured around the business of tourism and real estate). The practical consequences derived from this type of diagnosis do not necessarily lead to significant changes, and they sometimes even contribute to the perpetuation of existing realities. It usually happens when, after criticism, all that is observed is a mixture of discomfort and passivity that, at most, crystallizes in a kind of retreatism at touristic places, that is, in daily behaviors that try to avoid contact with tourists. The following fragment is a good example of this type of approach:

The 2008 post-bubble cycle makes it even more evident that the accumulation of benefits of the tourism industry is due to the degradation of life in the city and the dispossession of the neighbors of their communities, which are fundamental for maintaining reproduction [...] The most touristic cities are those that suffer greatest real estate pressure; more evictions and mobbing, making the right to housing of their neighbors inaccessible [...] The beast of tourism, as we understand it, is a false solution that makes work precarious, revives urban speculative economies and destroys the natural and social environments needed for the reproduction of life, especially in neighborhoods where the activity of this industry is concentrated [...] the real estate and holiday rental lobbies at both the state and European level to stop any public initiative that aims to act in the direction of de-commodifying housing, land, or protect the rights of workers [...] It is not viable to build marketable urban environments, or adapt the territory to the consumption needs of those who do not live in the city. (El municipalismo contra la falsa solución turística: hacia una sociedad urbana vivible, *Diario El Salto*, 27 February 2019).

Subversive Utopia

This strategy is very similar to the previous one. In fact, it is normal for the arguments to overlap. However, there is a key element that gives it its own identity: in this case the criticism is linked to the formulation of concrete actions (not explicit in type IV). The proposals for action reveal more or less latent social conflicts and produce political controversies that drive social changes, even if only by incorporating new issues and arguments into public discussion. The rejection of both the means and the known goals is associated with proposing substitution by new ones. The objective is to subvert the social order. Following the classic terminology of Mannheim (1929), we would speak of a utopian attitude when the analysis of reality is oriented towards the explanation of situations that do not exist and whose hypothetical existence is incompatible with the perpetuation of social order; the resulting diagnoses and forecasts are linked to therapies that require a total or partial elimination of the existing order. Thus, power structures often use the utopian adjective as a mechanism of social control. They do it every time they label as ridiculous, naive, unrealistic, absurd or disproportionate the proposals that challenge the stability of their positions of power, thus discrediting those who propose them. However, history teaches us that the definition of what is absurd and what is reasonable often has more to do with the power that some social actors have over others, and not so much with the soundness of the arguments put forward (Berger & Luckmann, 1966). Here the lack of identification with the means (tourism) and the deep questioning of the goals (capitalist) is accompanied by a proposal for social transformation: the substitution of the political-economic order that regulates the cities affected by overtourism by another type of order where social justice replaces economic interest as a vital leitmotiv. In the new order, tourism would be managed according to decreasing growth principles or would be replaced by other types of economic activities that do not generate such aggressive social and environmental impacts. A social actor that produces a discourse along this line of argument is the Network of Southern European Cities against Touristification, known as SetNet:

SetNet warns that local populations are organizing themselves to defend their social rights, especially the right to decent and affordable housing and the right to the city. To achieve a change of model and fight against the problems derived from tourism, they propose the establishment of limits to the tourist industry, the adoption of different fiscal policies for housing and tourist accommodation, the delinking of tourism from the economy of the city, or the decrease in tourism, accompanied by policies to promote just social and environmental economics. (Ciudades del sur de Europa se unen ante la turistización, *Diario El Salto*, 25 April 2018)

More social support is needed to put a brake on this model and also to move in another direction in which we disengage the development of the increase in consumption of materials and energy, the increase in the number of tourists, the increase in urbanization and cement [...] There is no lack of proposals or pathways to move towards greater territorial sovereignty, such as renewable energy, agroecology and responsible consumption, to move towards a new energy model. (¿Quién le pone puertas al turismo?, *eldiario.es*, 22 April 2018)

CONCLUSION

This chapter has proposed a theoretical and analytical framework from which to investigate attitudes towards tourism. More specifically, this work (supported by a selection of news reports) is not intended to develop a detailed analysis of discursive positions, but to explain how a classic theory model of Sociology can be valuable to understand attitudes towards tourism. The adaptation of the model of Merton is compatible with a plurality of methodological options and allows taking into account the political and conflictive dimensions that surround tourist activity. It also offers the possibility of tracing the genesis of public opinion, the actors that produce it, the links that are established and the evolution of discourses about the repercussions of tourism. This perspective surpasses the usual approaches influenced by the Social Exchange Theory.

The five basic positions or attitudes shown here (legitimization, innovative criticism, resignation, radical criticism and subversive utopia) are only a simplified representation of reality. They should not be understood as closed concepts, but as dynamic and connected categories that capture a part of the complexity of tourism within the framework of capitalist societies. Beyond its technical operation, this argument has sought, on one hand, to better understand the ways in which people think about tourism and organize their opinions regarding its effects, and, second, to transcend both media approaches that affect the antagonistic nature of discourses (situation already criticized and illustrated in detail by Huete and Mantecón, 2018) as well as some research perspectives articulated from dichotomous approaches (see, for example, Zerva, Palou, Blasco, & Donaire, 2019) .

The recent increase in expressions of opposition to tourism has loaded the image of tourism with certain negative connotations. Critical evaluations of tourism's impacts have a long history in research (especially in anthropological and sociological studies), but the social visibility of disagreement have not gone beyond exceptional and restricted situations. The perceived economic benefits have tended to eclipse, or at least compensate, the costs suffered. This situation would have favored the propagation of attitudes similar to legitimation (position I) and resignation (position III), with a marginal presence of radical criticism (position IV). Being coherent with the characterization of the types explained above, the result would reflect an ideological context that sustains the statu quo (in line with what was pointed out by Mantecón, 2010, Mantecón & Huete, 2011). In turn, this would help explain why the profound changes experienced by the Spanish political system have not triggered significant changes in the evolution of the tourism policy (Velasco, 2016). However, the proliferation of public complaints regarding tourist saturation and the problems it produces has complicated the aforementioned context. As a result, other types of discourses have emerged with force, referred to here as innovative criticism (position II) and subversive utopia (position V), whose most important common denominator is the justification of more or less disruptive processes of change.

The last wave of explanations suggests the existence of a correlation between overtourism and the increase in public expressions of rejection of tourism. But the empirical reality refutes this hypothesis. For a long time, there have been situations of overcrowding caused by the influx of tourists, which are accompanied by problems of very different intensity that have not materialized in collective protests. It could also be that these protests have not been so relevant as to have resonance in the media and be dealt with by political institutions. In fact, there is nothing specifically new in overtourism, beyond being accompanied by an increase in social protest movements. So, what is really new is not so much the overcoming of tourism capacity (since the thresholds of tolerance have been redefined again and again in each concrete case) as the citizens' actions of response (and their communication through the media). To advance in the understanding of this phenomenon, other factors must be explored (Koens, Postma, & Papp, 2018). It is important to underline that the existence of objective problems motivated by mass tourism is not being denied here. What is argued is that the verification of these problems is a necessary but not sufficient condition for the emergence of rebellious expressions.

In this regard, one of those intervening factors that is usually stressed is the application of different tourism management policies. Thus, the relationship between the number of tourists and the number of residents would not have an important explanatory value, but rather the proper management of tourism in each destination. Undoubtedly, this is an essential line of work and of great practical interest for those who must make decisions regarding planning. However, there are two issues that are often overlooked. In addition, research in this area requires the application of mixed methods, approaches that address historical-social change and perspectives that give decisive weight to the analysis of power relations:

The first issue refers to the role that tourism has played in the political, economic and social history of those cities in which the most intense expressions of criticism and rejection have been identified. Regardless of the objective conditions that explain the current economic situation, it is likely that in those cities where the resident population relies on alternative routes of development besides tourism, they also have a greater predisposition to claim different proposals. The memory of a happy past without tourism (although it is about what Bauman called "retrotopia") makes it easier to imagine an acceptable future without it. Similarly, in those cities where tourism is perceived as the primary (or only) means by which the population has been able to modernize its socio-economic structures and produce material well-being and wealth, protest is perhaps more unlikely. At the same time there is an increase in the degree of tolerance for the inconveniences that the influx of tourists can generate. Thus, the analysis of the processes of social construction of collective memory is an interesting analytical point.

The second question refers to the presence of a previous network of organizations and citizen groups committed to political struggle and social demands. In regions where socio-political activity is concentrated around a dense citizen network, it is easier for the problems linked to overtourism to be incorporated into the civil society's agenda at a given moment. In those cases in which, despite lack of awareness of previous political activity, there have been manifestations of criticism of the impacts of tourism, it is crucial to investigate the configuration of the group dynamics that have articulated these critical expressions (origins, components, objectives, values, proposed actions). The study of the political instrumentalization of social unrest is presented here as a line of relevant research.

The debates about the unexpected effects of tourism are ramifications of major polemics about the benefits and perversions of capitalism, about its exploitative or prosperity-generating character, and about the greater or lesser convenience of subjecting it to regulation. The discussion, then, would have to face the dilemma of whether tourism is configured by a set of activities that sooner or later will produce tensions and conflicts or whether, instead, the problems are more related to inadequate management. Perhaps, reflection on the functioning of tourism cannot and should not be removed from a broader reflection about the type of society in which we live. Ultimately, it is through tourism that conflicts and contradictions inserted in the social, political, cultural, economic and environmental arenas are expressed in our daily lives.

REFERENCES

Álvarez-Sousa, A. (2018). The Problems of Tourist Sustainability in Cultural Cities: Socio-Political Perceptions and Interests Management. *Sustainability*, *10*(2), 503. doi:10.3390u10020503

Alvesson, M., & Sandberg, J. (2013). *Constructing Research Questions: Doing Interesting Research*. London, UK: Sage. doi:10.4135/9781446270035

Andereck, K. L., Valentine, K. M., Knopf, R. C., & Vogt, C. A. (2005). Residents' perceptions of community tourism impacts. *Annals of Tourism Research*, *32*(4), 1056–1076. doi:10.1016/j.annals.2005.03.001

Bauman, Z. (1998). *Work, Consumerism, and the New Poor*. Philadelphia: Open University Press.

Belisle, F., & Hoy, D. (1980). The Perceived Impact of Tourism by Residents. *Annals of Tourism Research*, *7*(1), 83–101. doi:10.1016/S0160-7383(80)80008-9

Berger, P., & Luckmann, T. (1966). *The Social Construction of Reality. A Treatise in the Sociology of Knowledge*. New York: Anchor Books.

Briggs, D. (2013). *Deviance and Risk on Holiday: An Ethnography of British Tourists in Ibiza*. London, UK: Palgrave MacMillan. doi:10.1057/9781137022400

Brougham, J. E., & Butler, R. W. (1981). A Segmentation Analysis of Resident Attitudes to Social Impact of Tourism. *Annals of Tourism Research*, *8*(4), 569–590. doi:10.1016/0160-7383(81)90042-6

Brunt, P., & Courtney, P. (1999). Host Perceptions of Sociocultural Impacts. *Annals of Tourism Research*, *26*(3), 493–515. doi:10.1016/S0160-7383(99)00003-1

Burns, P. M., & Novelli, M. (Eds.). (2008). *Tourism and Mobilities. Local-Global Connections*. Wallingford: CABI. doi:10.1079/9781845934040.0000

Cabrerizo, C. (2016). *La ciudad negocio: turismo y movilización social en pugna*. Madrid, Spain: Cisma.

Cañada, E. (2018). Too Precarious to be Inclusive? Hotel Maid Employment in Spain. *Tourism Geographies*, *20*(4), 653–674. doi:10.1080/14616688.2018.1437765

Cavus, S., & Tanrisevdi, A. (2003). Residents' Attitudes toward Tourism Development: A Case Study in Kusadasi, Turkey. *Tourism Analysis*, *7*(3), 259–269. doi:10.3727/108354203108750102

Chossudovsky, M. (2003). *The Globalization of Poverty and The New World Order*. Ontario, Canada: Global Outlook.

Cochrane, A. (2007). *Understanding Urban Policy*. Oxford, UK: Blackwell.

Diedrich, A., & García-Buades, E. (2009). Local Perceptions of Tourism as Indicators of Destination Decline. *Tourism Management*, *30*(4), 512–521. doi:10.1016/j.tourman.2008.10.009

Esping-Andersen, G. (1990). *The Three Worlds of Welfare Capitalism*. Cambridge, UK: Polity Press.

Faulkner, B., & Tideswell, C. (1997). A Framework for Monitoring Community Impacts of Tourism. *Journal of Sustainable Tourism*, *5*(1), 3–28. doi:10.1080/09669589708667273

Habermas, J. (1973). *Legitimationsprobleme im Spätkapitalismus*. Frankfurt, Germany: Suhrkamp.

Harrill, R. (2004). Residents' Attitudes toward Tourism Development: A Literature Review with Implications for Tourism Planning. *Journal of Planning Literature*, *18*(3), 251–266. doi:10.1177/0885412203260306

Harrill, R., & Potts, T. (2003). Tourism Planning in Historic Districts: Attitudes toward Tourism Development in Charleston. *Journal of the American Planning Association*, *69*(3), 233–244. doi:10.1080/01944360308978017

Harvey, D. (1989). From Managerialism to Entrepreneurialism: The Transformation in Urban Governance in Late Capitalism. *Geografiska Annaler. Series B, Human Geography*, *71*(1), 3–17. doi:10.1080/04353684.1989.11879583

Harvey, D. (2005). *A Brief History of Neoliberalism*. New York: Oxford University Press.

Huete, R., & Mantecón, A. (2018). El auge de la turismofobia ¿hipótesis de investigación o ruido ideológico? *Pasos. Revista de Turismo y Patrimonio Cultural*, *16*(1), 9–19. doi:10.25145/j.pasos.2018.16.001

Jackson, M. S., & Inbakaran, R. J. (2006). Evaluating Residents' Attitudes and Intentions to Act toward Tourism Development in Regional Victoria, Australia. *International Journal of Tourism Research*, *8*(5), 355–366. doi:10.1002/jtr.584

Jurowski, C., & Gursoy, D. (2004). Distance Effects on Residents' Attitudes toward Tourism. *Annals of Tourism Research*, *31*(2), 296–312. doi:10.1016/j.annals.2003.12.005

Koens, K., Postma, A., & Papp, B. (2018). Is Overtourism Overused? Understanding the Impact of Tourism in a City Context. *Sustainability*, *10*(12), 4384. doi:10.3390u10124384

Liu, J., & Var, T. (1986). Residents Attitudes toward Tourism Impacts in Hawaii. *Annals of Tourism Research*, *13*(2), 193–214. doi:10.1016/0160-7383(86)90037-X

Mannheim, K. (1929). *Ideologie und Utopie*. Bonn, Germany: Cohen.

Mansfeld, Y. (1992). Group-differentiated Perceptions of Social Impacts Related to Tourism Development. *The Professional Geographer*, *44*(4), 377–392. doi:10.1111/j.0033-0124.1992.00377.x

Mantecón, A. (2010). Tourist Modernisation and Social Legitimation in Spain. *International Journal of Tourism Research*, *12*(5), 617–626.

Mantecón, A. (2016). La opinión pública sobre el turismo. Crítica y propuesta para una nueva orientación. *Estudios y Perspectivas en Turismo*, *25*(2), 110–123.

Mantecón, A., & Huete, R. (2011). Sociological Insights on Residential Tourism: Host Society Attitudes in a Mature Destination. *European Journal of Tourism Research*, *4*(2), 109–122.

Marrero, J. R., & Huete, R. (2013). La opinión pública sobre el empleo turístico en la Comunidad Valenciana. *Cuadernos de Turismo*, *32*, 189–206.

Merton, R. K. (1938). Social Structure and Anomie. *American Sociological Review*, *3*(5), 672–682. doi:10.2307/2084686

Milano, C., & Mansilla, J. A. (Eds.). (2018). Ciudad de vacaciones. Conflictos urbanos en espacios turísticos. Barcelona, Spain: Pol·len Edicions.

Milano, C., Cheer, J., & Novelli, M. (Eds.). (2019). *Overtourism: Excesses, Discontents and Measures in Travel & Tourism*. Wallingford: CABI. doi:10.1079/9781786399823.0000

Milano, C., Novelli, M., & Cheer, J. (2019). Overtourism and Tourismphobia: A Journey through Four Decades of Tourism Development, Planning and Local Concerns. *Tourism Planning & Development*, *16*(4), 353–357.

Peeters, P., Gössling, S., Klijs, J., Milano, C., Novelli, M., Dijkmans, C., & Postma, A. (2018). *Research for TRAN Committee - Overtourism: Impact and Possible Policy Responses*. Brussels, Belgium: European Parliament, Policy Department for Structural and Cohesion Policies.

Pfeffer, J. (2007). A Modest Proposal: How We Might Change the Process and Product of Managerial Research. *Academy of Management Journal*, *50*(6), 1334–1345. doi:10.5465/amj.2007.28166117

Pizam, A. (1978). Tourism's Impacts: The Social Costs to the Destination Community as Perceived by Its Residents. *Journal of Travel Research*, *16*(4), 8–12. doi:10.1177/004728757801600402

Rasoolimanesh, S. M., Jaafar, M., Kock, N., & Ramayah, T. (2015). A Revised Framework of Social Exchange Theory to Investigate the Factors Influencing Residents' Perceptions. *Tourism Management Perspectives*, *16*, 335–345. doi:10.1016/j.tmp.2015.10.001

Royo, M., & Ruiz, M. E. (2009). Actitud del residente hacia el turismo y el visitante: Factores determinantes en el turismo y excursionismo rural-cultural. *Cuadernos de Turismo*, *23*, 217–236.

Ryan, C., & Montgomery, D. (1994). The Attitude of Bakewell Residents to Tourism and Issues in Community Response Tourism. *Tourism Management*, *15*(5), 358–369. doi:10.1016/0261-5177(94)90090-6

Schoenberger, E. (1988). From Fordism to Flexible Accumulation: Technology, Competitive Strategies and International Location. *Environment and Planning. D, Society, & Space*, *6*(3), 245–262. doi:10.1068/d060245

Sennett, R. (2018). *Building and Dwelling: Ethics for the City*. New York: Farrar, Straus, and Giroux.

Sharpley, R. (2014). Host Perceptions of Tourism: A Review of the Research. *Tourism Management*, *42*, 37–49. doi:10.1016/j.tourman.2013.10.007

Teye, V., Sönmez, S. F., & Sirakaya, E. (2002). Residents' attitudes toward tourism development. *Annals of Tourism Research*, *29*(3), 668–688. doi:10.1016/S0160-7383(01)00074-3

UNWTO - World Tourism Organization. (2001). *Global Code of Ethics for Tourism*. Madrid, Spain: UNWTO.

UNWTO - World Tourism Organization. (2019). Overtourism? – Understanding and Managing Urban Tourism Growth beyond Perceptions: Vol. 2. *Case Studies*. Madrid, Spain: UNWTO.

Urry, J. (1990). *The Tourist Gaze. Leisure and Travel in Contemporary Societies*. London, UK: Sage.

Vargas-Sánchez, A., Oom-do-Valle, P., da-Costa-Mendes, J., & Silva, J. A. (2015). Residents' Attitude and Level of Destination Development: An International Comparison. *Tourism Management*, *48*, 199–210. doi:10.1016/j.tourman.2014.11.005

Velasco, M. (2016). Dynamics of Spanish Tourism Policy: The Political System as a Driver and Policy Instruments as Indicators of Change (1952–2015). *International Journal of Tourism Policy*, *6*(3/4), 256–272. doi:10.1504/IJTP.2016.081526

Williams, J., & Lawson, R. (2001). Community Issues and Resident Opinions of Tourism. *Annals of Tourism Research*, *28*(2), 269–290. doi:10.1016/S0160-7383(00)00030-X

Zerva, K., Palou, S., Blasco, D., & Donaire, J. A. (2019). Tourism-philia versus Tourism-phobia: Residents and Destination Management Organization's Publicly Expressed Tourism Perceptions in Barcelona. *Tourism Geographies*, *21*(2), 306–329. doi:10.1080/14616688.2018.1522510

KEY TERMS AND DEFINITIONS

Capitalism: A system of organization of social and economic relations based on the commoditization of the greatest number of elements in the greatest number of places.

Commoditization: Process by which something or someone is converted into a commodity, into an object available in a market.

Fordist Tourism: Forms of organization of the tourist industry that emerged after the Second World War characterized by the large-scale commercialization of standardized products, based on a basic offer of "sun, sand and sea", aimed at homogeneous and undemanding markets.

Overtourism: A buzzword that refers to issues similar to the established concept of "tourism carrying capacity". This neologism is often used to allude to problems caused by tourist saturation with regard to the emergence of public expressions of criticism or rejection of tourism or any of its effects.

Post-Fordist Tourism: Form of organization of the tourism industry that emerged in the 1990s, characterized by the diversification of products and their integration into more segmented markets, with more demanding, experienced consumers who are less likely to repeat experiences and interested in alternative typologies to mass or Fordist tourism.

Strain Theory: A theory proposed by R.K. Merton in 1938 in order to explain the types of adaptation of individuals to a society that defines success in terms of accumulation of wealth and that typifies the acceptable and unacceptable means of achieving it.

Tourismphobia: An expression used mainly by the media with the aim of stigmatizing public expressions of rejection of tourism or any of its effects.

Chapter 5
Overtourism:
Impacts on Residents' Quality of Life – Evidence From Macau

Hio Kuan Lai
Faculty of Economics, University of Algarve, Portugal

Patrícia Pinto
https://orcid.org/0000-0002-3153-2830
Faculty of Economics, Research Centre for Tourism, Sustainability, and Well-Being (CinTurs), University of Algarve, Portugal

Pedro Pintassilgo
https://orcid.org/0000-0001-9115-4301
Faculty of Economics, CEFAGE, University of Algarve, Portugal

ABSTRACT

The concept of overtourism has been popularly discussed in recent years, upon local residents starting to recognize the tourism disturbances in some high-profile destinations. For a sustainable tourism development, it is crucial to investigate potential impacts of overtourism towards the local residents living in the destination. Thus, this chapter employs the existing academic work on quality of life (QoL) concepts, to examine the correlations between residents' perception of tourism's impact to them in material, community, emotional, and health & safety life domains, and their satisfaction with QoL in corresponding life domains, and their overall life satisfaction. This chapter involves a survey taken place in Macau, which represents a case of plausible overtourism. The research reveals that residents who perceived more negative impact from tourism on QoL, were more unsatisfied with QoL. Thus, tourism planners and policy makers should be aware of such threat on sustainable tourism development and search out a solution balancing different stakeholders in overtourism destinations.

DOI: 10.4018/978-1-7998-2224-0.ch005

INTRODUCTION

Sustainability has been a prevalent topic of interest in regard of tourism destination development (Buckley, 2012; Kuščer & Mihalič, 2019; Lu & Nepal, 2009). Additionally, in response to the growing importance of community involvement in the tourism sector, the research on residents' perception of tourism's impact has also proliferate recently (Nunkoo, 2016; Nunkoo & Ramkissoon, 2011; Uysal, Sirgy, Woo, & Kim, 2016; Woosnam, 2012). However, while the host community's participation is being recognized in sustaining tourism development, it is quite discomfiting that the term "overtourism" has cropped up lately. Overtourism can be explained as the tourism impact that immensely and negatively affects residents' perception of quality of life (QoL) and/or the quality of tourists' experience (UNWTO, 2018). The term overtourism has just been popularly used for around two years yet the problematic phenomenon has been a research topic for more than 30 years (Perkumiene & Pranskuniene, 2019). Among the host community of a destination, the residents are the stakeholders largely and directly affected by tourism development (Li & Wan, 2017). Their imbalance perceptions between the positive and negative impacts derived from tourism paves the way for the rise of overtourism (Gonzalez, Coromina, & Galí, 2018), which is a sign of failure on sustainable tourism development.

Most definitions of overtourism concern about the worsening of residents' QoL (e.g. Oxford Dictionary, 2019; Responsible Tourism, 2019; UNWTO, 2018). Hence, the authors aim to contribute to the literature by testing the existing QoL construct in an overtourism destination. This study examines the correlations among residents' perception of tourism's impact and satisfaction in material, community, emotional, and health & safety life domains, together with their overall life satisfaction. Considering that case studies of overtourism were mostly done in Europe (e.g. Gonzalez, Coromina, & Galí, 2018; Kuščer & Mihalič, 2019; Seraphin, Sheeran, & Pilato, 2018), this study makes an attempt to specify the examination of the phenomena in an Asian city - Macau (China), where overtourism has been a topic of interest for government officials, lawmakers and the wider public in recent years (Macau Daily Times, 2019).

OVERTOURISM: A PHENOMENON THAT DEPENDS ON RESIDENTS' PERCEPTIONS

According to UNWTO (2018, p.4), overtourism can be defined as "the impact of tourism on a destination, or parts thereof, that excessively influences perceived quality of life of citizens and/or quality of visitors experiences in a negative way". Responsible Tourism (2019, p.1) describes overtourism as "destinations where locals or visitors feel that there are too many visitors and that the quality of life in the area or the quality of the experience has deteriorated unacceptably". As the concept of overtourism and related issues have been popularly discussed in both tourism media and academic research in recent years, there are different interpretations of overtourism from different perspectives, and the usage of which has been doubted. CNN Travel carried out interviews with three tourism experts (The head of Barcelona's city council's Strategic Plan for Tourism, the representative of Iceland's Department of Tourism, and the general manager for South America at the travel company G Adventures) asking for their opinion of overtourism (Street, 2018). All of them agreed that the term "overtourism" is an overused buzzword, notwithstanding, it is a real issue derived from the pressure that has been established for a long time. Concurrently, one of the recent academic studies was also dubious of the usage of the term "overtourism" and included in its title the question: "Is Overtourism Overused?" (Koens, Postma

& Papp, 2018). Based on their qualitative investigations among 80 stakeholders in 13 European cities, Koen, Postma and Papp (2018) concluded that overtourism should not be solely considered as a tourism problem, but a wider issue in societal development. They recommended that more research from other disciplines are needed to generate new insights. The term "overtourism" is newly introduced to academia and further research is needed for a holistic understanding. However, the phenomenon of overtourism itself is definitely not a new concept (Pinke-Sziva, Smith, Olt, & Berezvai, 2019). Kuščer and Mihalič (2019) emphasized that there is actually extensive academic work available to serve as the foundation for future overtourism study. Thus, it is suggested to establish further research based on the existing work rather than re-establish a new overtourism debate.

Although the continuously growing number of tourists is usually seen as the stimulus of overtourism, interestingly, some prominent destinations could withstand the soaring tourist arrivals while the others would suffer from notable negative impacts by a slight increase in the number of tourist arrivals (Koens, Postma, & Papp, 2018). Thereupon, overtourism is determined differently in different tourism settings (Kuščer & Mihalič, 2019). The problem lies to a great extent on residents' view on how tourism encounters with their lives (Postma, 2013). As a result, subjective measurement in multidimensional values maybe an appropriate approach for studying overtourism. Many scholars have already recognized the importance to take residents' view into account for the tourism industry to be sustainable in long-term (Doxey, 1975; Nunkoo & Ramkissoon, 2011; Uysal, Sirgy, Woo & Kim, 2016; William & Lawson, 2001; Woosnam, 2012). Thus, not surprisingly, academic research on the topic of tourism's impacts on local communities is heretofore massive and customary (Dioko, 2017). Likewise, the problem of residents' dissatisfaction with tourism is also far from being a new topic (Pinke-Sziva, Smith, Olt, & Berezvai, 2019). If the phenomenon of overtourism is closely related to the worsening of residents' QoL, then it makes sense to carry out an in-depth research about the residents' perceptions of QoL, in a destination where the residents' QoL is likely being influenced negatively.

OVERVIEW OF QOL

The Concept of QoL in Tourism Research

The term 'quality of life' and 'well-being' are usually interchangeably used in the academic literature. Researchers have been continuously debating about the definition of quality of life (QoL) since 1960s (Woo, Kim, & Uysal, 2015). Although there are over a hundred definitions of the term quality of life (QoL), it is considered as a universal value that describes a person's life satisfaction and feeling of fulfilment appertaining to his or her experiences (Andereck & Nyaupane, 2011). Specifically for tourism studies, researchers started to specify the research by examining QoL and its interrelationship with residents' perception of tourism's impacts from 1980 (Uysal, Sirgy, Woo, & Kim, 2016). To access residents' perception of QoL in the tourism context, QoL can be specified as how residents view their living conditions in a tourism destination and how those conditions exert an influence over their satisfaction in diverse life domains and overall life.

When a community develops into a tourism destination, the residents' way of living undergoes changes in multiple ways (Liang & Hui, 2016). Concerning this, Uysal, Sirgy, Woo and Kim (2016) have conducted a qualitative work to review 36 existing studies and revealed three conclusions. First, the impact of tourism development plays an important role in residents' QoL, in both positive and negative ways (e.g. Kim, Uysal, & Sirgy, 2013; Nichols, Stitt, & Giacopassi, 2002). For instance, tourism development can generate more employment opportunities, meanwhile can cause striking inflation in the destination. Second, residents' perception of QoL may differ along different personal variables. This is supported by the recent work of Woo, Uysal and Sirgy (2018), which indicated that residents not affiliated with the tourism industry have less-positive perceptions of both tourism's impact in material life and the related satisfaction level than those who were affiliated. Moreover, Nichols, Stitt and Giacopassi (2002) also revealed that residents' perceptions of QoL are different depending on their different characteristics (e.g. demographic). Third, residents' perceptions of QoL differ along the stage of tourism development, with less benefits being recognized in the later stage. This is concurrent with the Irridex model developed by Doxey (1975), suggesting that residents' perceptions of tourism may encounter an array of stages from 'euphoria', 'apathy', 'annoyance' to 'antagonism'.

From the literature review, it is empirically supported that tourism's impacts and some personal factors have a significant influence on residents' perceptions of QoL. Jennings and Nickerson (2006) underscored that the quality of tourists' experiences depends on a receptive host population. So, the continued degradation of residents' QoL may threaten the long-term development of a destination. Although scholars have examined QoL at different perspectives, exploring QoL in the current study in an overtourism destination (i.e. residents experienced excessive tourism impact in a negative way) may lead to insight into retrieving sustainable tourism development.

The QoL Measurement in Tourism Research

QoL can usually be measured in two dimensions — subjective and objective. Subjective dimension refers to an individual's feeling and perception towards different aspects of QoL, while objective dimension make reference to the externality to an individual (e.g. income) (Andereck & Nyaupane, 2011; Jennings & Nickerson, 2006). Considering that different people can have different perceptions of QoL even living in similar conditions, QoL is best studied from an individual's own perception (Andereck & Nyaupane, 2011). Subjective indicators are commonly used in tourism studies, for instance, subjective well-being measured by an array of life domains (Liang & Hui, 2016; Nawijn & Mitas, 2012). As a result, the authors adopted subjective indicators in the current study. To access a more holistic understanding of residents' QoL, Woo, Kim, and Uysal (2015) suggested a multidimensional instead of an uni-dimensional approach, indicating that the overall life satisfaction is determined and measured by various life domains.

In the tourism context, academic works have adopted various life domains, comprising different subjective indicators to measure residents' QoL (e.g. Andereck & Nyaupane, 2011; Jennings & Nickerson, 2006; Kim, Uysal, & Sirgy, 2013; Nawijn & Mitas, 2012; Uysal, Sirgy, Woo, & Kim, 2016; Woo, Kim, & Uysal, 2015; Woo, Uysal, & Sirgy, 2018; Yamada, Heo, King & Fu, 2009). Specifically, the bottom-up spillover theory, explains that the overall life satisfaction is at the top of a hierarchy and is influenced by the satisfaction with various life domains (Sirgy & Lee, 2006). All these researches provided evidence that measurement in different life domains could ease a more comprehensive representative of residents' QoL. In this context, this study adopts indicators under four different life dimensions (i.e. material, community, emotional, and health & safety life domains). The material life domain refers to an individual's economic and living situation, income level, standard of living, housing and socio-economic status (Cummins, 1996). The material life domain is recognized as an uncontested criterion in measuring QoL (Dolnicar, Yanamandram & Cliff, 2012). The community life domain associates with the community's public space, facilities and services (Puczkó & Smith, 2011). The emotional life domain reveals the fulfilment of an individual's spiritual needs, consisting of the activities that could impel such fulfilment (Woo, Kim, & Uysal, 2015). Lastly, health and safety life domain is also seen as an uncontestable element connecting to QoL (Dolnicar, Yanamandram, & Cliff, 2012; Woo, Kim, & Uysal, 2015).

Scholars have provided evidence that tourism's impact has significant influence on residents' satisfaction with all these four domains (Kim, Uysal, & Sirgy, 2013). Subsequently, Woo, Kim, and Uysal (2015) and Woo, Uysal, and Sirgy (2018) adopted these four domains in their studies in the United States. According to their studies, residents' perception of tourism development has a predictable effect on the satisfaction with material and non-material life (derived from community, emotional, and health & safety domains) (Woo, Kim, & Uysal, 2015). Moreover, residents' overall life satisfaction are positively influenced by their satisfaction with these life domains (Woo, Kim, & Uysal, 2015; Woo, Uysal, & Sirgy, 2018).

From previous research carried out in western destinations, interrelationships were found among perceived value of tourism's impact in different life domains, satisfaction with different life domains and overall satisfaction with QoL. As a result, the following hypotheses are proposed:

Hypothesis One: Residents' perception of tourism's impact on QoL is positively correlated with their satisfaction with QoL in various life domains (i.e. material, community, emotional, health & safety).

Hypothesis Two: Residents' satisfaction with QoL in various life domains (i.e. material, community, emotional, health & safety) is positively correlated with their overall satisfaction with QoL.

MACAU: THE DESTINATION CONTEXT

The city of Macau, is one of the two Special Administrative Regions (SAR) in China. Macau is located on the southeastern coast of China, to the west of Hong Kong SAR. Macau is a former Portuguese colony that reverted to China in 1999. Since the liberalization of the gaming market in 2002, the city has experienced an exponential growth in visitor arrivals. The recorded visitor arrivals have increased

Figure 1. Conceptual Model of Residents' perceptions and satisfaction with QoL
Source: Own elaboration based on literature review

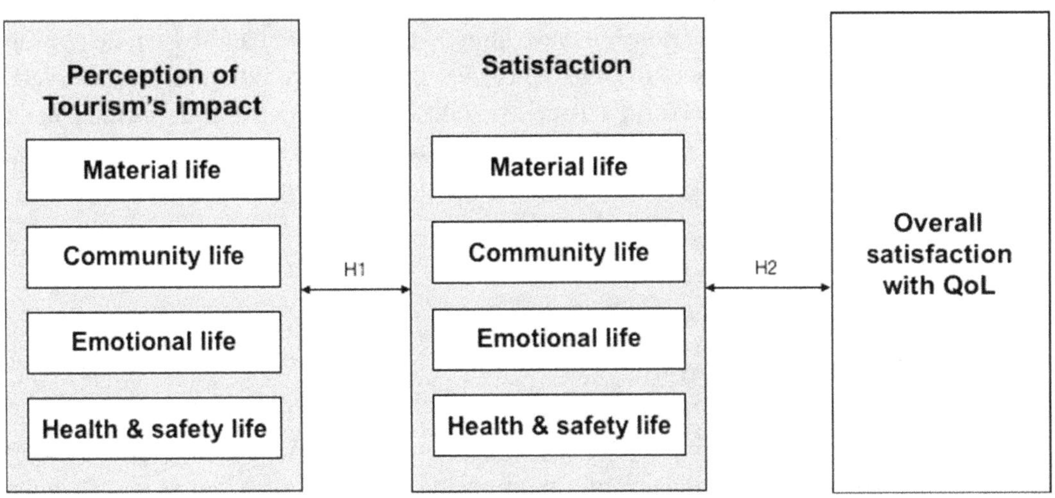

from 11.5 millions in 2002 to more than 32.6 millions in 2017, which has increased further to a peak of 35.8 millions in 2018 (DSEC, 2019). The city is famous for its casino gambling industry, and its gross revenue (US$33.02 billion in 2017 (DICJ, 2017)) has recently surpassed Las Vegas. Meanwhile, the city is also known for its cultural heritage by virtue of the inscription on UNESCO World Heritage List in 2005 (UNESCO, 2005). Specifically, according to Macau Statistical Department, the main motives for the visits to Macau in 2018 were: vacation (54.5%), transit (14.9%), shopping (8.9%), visiting relatives and friends (5.5%), gaming (2.6%) (DSEC, 2018a). According to WTTC (2018), 61.3% of the city's GDP was derived from travel and tourism, which supported 213,000 jobs. The economy of Macau is based enormously on tourism.

Overtourism in Macau

With a land area of just 30.5 km^2 (DSEC, 2017) and a population of 653,100 residents (DSEC, 2018b), the continued growth in the number of visitors has resulted in significant negative impacts to the community. Long before the attention on "overtourism" across the globe, Wan and Li (2013) revealed various negative tourism impacts occurring in Macau, such as, increasing crime rate, traffic congestion, pollution, energy consumption, and land use conflicts. Another research published by the city's Institute for Tourism Studies (IFT) also mentioned that there were possible negative impacts on residents' perceived QoL, as they were beginning to get tiresome of the side effects of the tourism growth (IFT, 2015). Lately, more and more government officials and lawmakers in Macau have been raising concerns on overtourism and the possible life quality disruptions it might have brought to the local community. The tag of overtourism has already been put on the city. One of Macau's lawmakers, Ka Hou Sou, urged the Macau government to consider introducing a cap on the number of visitor arrivals (GGR Asia, 2019). He also expressed his concern on the idea to divert tourists to outlying local neighbourhoods, as the businesses targeting tourists are capable to pay higher rentals, thus further threatening the small-sized local businesses' survival in the neighbourhoods. Another lawmaker, Pereira Coutinho, alarmed that Macau would soon meet the breaking point like Barcelona, Venice, or Hanoi (Macau Daily Times, 2019). He addressed that Macau residents has been losing their QoL and a balanced solution is in need as soon as possible. More importantly, Macau tourism office director, Fernandes, also admitted in the Pacific-Asia Travel Association (PATA) annual summit that Macau faced capacity issues and "the sheer numbers are causing problems – site overcrowding, transport congestion and a deteriorating quality of life for local residents" (Taylor, 2018, p.1). Although the threats of overtourism have already been publicly addressed, no consensus has been reached by stakeholders on how to resolve the issues. The alarm of potential overtourism is quite abash as the city positions itself as a World's Tourism and Leisure Center in long-term. Thus, the current study examining from residents' standpoint is expected to shed light for the government of Macau, particularly on finding a solution before anti-tourism activities are catalysed.

METHODOLOGY

Sampling and Data Collection

The target population of this study is the permanent residents living in Macau, who were aged above 18 years old. Permanent residents indicate a minimum residency length of seven years. They were considered as the potential participants to ensure that he/she would be familiar with the changes resulted from tourism development.

In western countries, it is common to apply probability sampling method by carrying out a systematic mail survey. However, in Macau residents tend not to return a mail survey, as they sense of risk by providing their personal information linked with their residential addresses (Lai & Hitchcock, 2017). As a result, considering the lack of an adequate sampling frame, a convenience sampling method was adopted. This is consistent with the previous residents' studies in Macau (Lai & Hitchcock, 2017; Li & Wan, 2017). In particular, an online survey tool (www.google.com/forms/about/) was employed to conduct the survey. During the period of April and May in 2018, the survey link was placed to high-traffic online forums and social media groups in Macau for data collection. A total number of 415 questionnaires were received. Out of these, eight were excluded because their length of residency was less than seven years and/or they were at the moment living abroad. Eventually, 407 valid questionnaires were employed for data analysis, accounting for 98.1% of the total sample size.

Survey Instrument

A questionnaire was developed in order to test the research hypotheses, the contents of which were based on the literature review. It was originally designed in English before translated into Traditional Chinese. To ensure the comparability of two versions, two bilingual individuals assisted in verifying the translation. Subsequently, the Traditional Chinese version was pilot tested with five Macau residents holding a Bachelor degree on tourism related studies, by using the convenience sampling method. After improving the validity of the questionnaire, the final version was then established online.

The survey instrument consisted of three constructs: 1) residents' perception of tourism's impact in four life domains; 2) residents' level of satisfaction with four life domains and overall satisfaction with QoL; 3) sociodemographic information.

Residents' perception of tourism's impact on QoL and the related satisfaction level were measured with the same items under material, community, emotional, and health and safety life domains (Woo, Kim, & Uysal, 2015). However, different question settings have been used to measure the two dimensions, by asking "How does tourism affect your life in Macau?" and "How satisfied are you with your life?" in an array of 26 indicators borrowed from Woo, Uysal and Sirgy's (2018), a recent study which measured perceived QoL using the four domains. A few modifications have been conducted for better adaptation to the study site. The last construct contains the sociodemographic information. Some specific questions were also placed to address how much the respondents are involved with tourism, which allowed the authors to better understand the characteristics of the sample.

Data Analysis

The SPSS 24 statistical software was used to analyse the data. Descriptive analysis was initially undertaken to evaluate the sample's characteristics, followed by the reliability test to measure the internal consistency of the QoL construct. In examining the research hypotheses, Pearson correlation analysis was undertaken to test the relationships among the grand means of the QoL construct, considering its four domains (material, community, emotional, health & safety). Lastly, the differences in sociodemographic factors and tourism related characteristics were examined by using the independent sample t-test or the independent sample Kruskal-Wallis test, depending on the number of variables and the result of normality test. As the sample is not random, the tests' results should be interpreted with care.

The independent sample t-test is a parametric test used to compare the means of two independent groups, so as to determine whether the difference between the groups is statistically significant (Laerd Statistics, 2018a). As the independent sample t-test can only compare the means for two groups, the one-way analysis of variance (ANOVA) is usually used to determine the differences between three or more groups (Kent State University, 2018). However, if the ANOVA's assumptions are not met, the Kruskal-Wallis test is recommended as an alternative (STHDA, 2018). The Kruskal-Wallis test is a nonparametric test based on ranks, which is also used to determine if there are statistically significant differences among groups (Laerd Statistics, 2018b). The test's results should be interpreted carefully as the Kruskal-Wallis test does not show which particular groups are significantly different from each other. All the test results were considered at a significance level of 5%.

RESULTS

Samples' Characteristics

The sample comprises slightly more females (62.9%) than males; 52.3% of the respondents were in the age group of 25-34, followed by the age group of 35-44 (27.3%) and the age group of 18-24 (16.2%). Around half of the respondents were single. Most of the respondents were born in Macau (83%). The majority (79.1%) lived in the area of Macau Peninsula, which is similar to the residential statistics in Macau (DSEC, 2016). Overall, respondents were generally well-educated with at least a bachelor's degree (81.3%). The majority (65.8%) worked in the private sector. Around 72% of respondents earned a monthly net income of at least MOP 20,000 (72%), which is the median earnings of the employed residents in Macau (DSEC, 2018c). In comparison with the demographic statistics in Macau, the sample showed a similar distribution in gender and residential area, while it tended to be younger and better educated, which resembles Li and Wan's (2017) study on residents' attitude in Macau. In terms of tourism related characteristics, 34.9% of the respondents were working in the tourism sector, 20.4% had direct contact with tourists in their current jobs, and 45.7% had a family member(s) working in tourism sector. In regard to residents' perception of tourists' contribution, on average, respondents perceived that 39.2% of their individual income was directly or indirectly derived from tourists' spending; while 38% of them perceived 0-20% and 16.5% perceived that 80-100% of their income derived from tourists. In respect of traveling experiences, on average the respondents had travelled 7 times outside Macau in the previous two years.

Reliability Test

In statistics, reliability refers to the overall consistency of the measure. This study measures reliability through the Cronbach's alpha, a measure of internal consistency. The values of Cronbach's alpha range between 0 and 1, interpreting from "no error" to "much error", respectively. According to Nunnally (1978), Cronbach's alpha with value greater than 0.70 indicates an acceptable internal consistency in the measured dimension and hence is considered as reliable. The reliability test was conducted for the QoL construct, including tourism's impact on QoL and residents' satisfaction with QoL in material, community, emotional, and health & safety life domains, and their overall QoL satisfaction. The alpha values lie in the ranges 0.737 - 0.926, which indicates that the measurement under each group of variables is internally consistent and has a satisfactory reliability level.

Descriptive Analyses of QoL Construct

Respondents' perception of tourism's impact on QoL was measured on the 5-point impact scale (1 = Very negatively affected and 5 = Very positively affected). As shown in Table 1, the factor grand mean of tourism's impact to material life ($M = 2.92$) is the highest, followed by emotional life ($M = 2.62$), health and safety life ($M = 2.52$) and community life ($M = 2.40$), all between neutral and negatively affected. Specifically, more positively rated items were those under the material life, such as the security of their current job ($M = 3.43$), the income at their current job ($M = 3.42$), the family income ($M = 3.41$) and the benefits they get from the government ($M = 3.18$). The more negatively rated items were the public transportation ($M = 1.82$), the cost of housing ($M = 2.17$), the cost of basic necessities ($M = 2.23$) and the air quality in their community ($M = 2.23$).

In terms of respondents' level of satisfaction under the same items (Table 1), which was measured on the 5-point satisfactory scale (1 = Very unsatisfied and 5 = Very satisfied). The factor grand mean ranked exactly as the tourism's impact score, material life satisfaction ($M = 2.75$) is the highest, followed by the emotional life ($M = 2.68$), health and safety life ($M = 2.60$) and community life ($M = 2.57$). All between neutral and unsatisfied. Particularly, except the security of their current job ($M = 3.28$) and the family income ($M = 3.10$), all other items were rated between neutral and very unsatisfied ($M < 3$). The lowest rated items were the cost of housing ($M = 1.92$), the public transportation ($M = 2.02$), the air quality in their community ($M = 2.37$) and the cost of basic necessities ($M = 2.41$). For the overall life satisfaction, the last section of Table 1, the grand factor mean ($M = 2.76$) is derived from six items, measured on a the 5-point agreement scale (1 = Strongly disagreed and 5 = Strongly agreed).

Correlation among the QoL Construct

For testing the hypotheses, Pearson's r was employed to measure the correlations among the QoL construct and ESS factors. Theoretically, the higher the absolute value of the correlation value (range from -1 to 1) between the pair of variables, the more associated they are with each other. A positive value denotes a positive association between the variables, and vice versa. The result (Table 2) reveals that the variables are all positively correlated at the significant level at 0.01 or 0.05. It implies that residents who perceived relatively more positive tourism's impact on their QoL in four life domains, were more satisfied with the corresponding domains and overall QoL. The findings suggested that the two aforementioned hypotheses were supported (Table 3).

Differences in Sociodemographic Factors and Tourism Related Characteristics

In order to explore if there were other factors affecting perceptions of tourism impacts on QoL, satisfaction with the corresponding domains, and overall QoL, statistical tests were conducted to examine if there were significant differences across sociodemographic factors. Independent sample t-tests were performed based on respondents' gender, marital status and residential area; and independent sample Kruskal-Wallis tests were performed based on their age, birthplace, length of residency, education, occu-

Table 1. Descriptive analysis of QoL construct

	Tourism's Impact		Satisfaction	
	Mean	**SD**	**Mean**	**SD**
Material life domain[ab]	**2.92[d]**	**0.50**	**2.75[d]**	**0.61**
- Overall cost of living	2.60	0.87	2.66	0.88
- Cost of housing	2.17	0.90	1.92	0.91
- Cost of basic necessities (food, clothing, transportation)	2.23	0.81	2.41	0.82
- Income at your current job	3.42	0.77	2.95	0.88
- Security of your current job	3.43	0.72	3.28	0.83
- Family income	3.41	0.69	3.10	0.81
- Benefits you get from the government	3.18	0.85	2.94	0.90
Community life domain[ab]	**2.40[d]**	**0.64**	**2.57[d]**	**0.69**
- Conditions of community environment (air, water, green space)	2.37	0.87	2.59	0.86
- People who live in your community	2.52	0.79	2.74	0.80
- Services and facilities you get in your community	2.69	0.80	2.71	0.84
- Overall community life	2.62	0.77	2.78	0.84
- Public transportation	1.82	0.89	2.02	0.91
Emotional life domain[ab]	**2.62[d]**	**0.67**	**2.68[d]**	**0.67**
- Spare time	2.59	0.86	2.70	0.84
- Leisure life	2.48	0.88	2.66	0.85
- Leisure activity in your community	2.62	0.79	2.62	0.80
- The way culture is preserved in your community	2.64	0.88	2.62	0.83
- Spiritual/religious life in the community	2.79	0.68	2.79	0.67
Health & Safety life domain[ab]	**2.52[d]**	**0.58**	**2.60[d]**	**0.68**
- Health facilities in your community	2.76	0.75	2.71	0.81
- Health service quality in your community	2.70	0.76	2.53	0.85
- Air quality in your community	2.23	0.81	2.37	0.86
- Water quality in your community	2.70	0.69	2.69	0.80
- Environmental quality in your community	2.46	0.81	2.56	0.87
- Environmental cleanliness in your community	2.32	0.82	2.52	0.90
- Safety and security in your community	2.47	0.83	2.73	0.89
- Traffic accident rate in your community	2.48	0.80	2.58	0.85
- Crime rate in your community	2.54	0.76	2.72	0.84
Overall satisfaction with QoL[c]	-	-	**2.76[d]**	**0.74**
- I am satisfied with my life as a whole			3.00	0.86
- The conditions of my life are excellent			2.79	0.82
- In most ways my life is close to ideal			2.65	0.91
- So far I have gotten the important things I want in life	-	-	2.61	0.96
- If I could live my life over, I would change almost nothing			2.41	0.95
- In general, I am a happy person			3.14	0.88

[a] Each item for tourism's impact was asked on a 5-point Likert scale: 1=Very negatively affected and 5=Very positive affected.
[b] Each item for satisfaction was asked on a 5-point Likert scale: 1=Very unsatisfied and 5=Very satisfied.
[c] Each item was asked on a 5-point Likert scale: 1=Strongly disagreed and 5=Strongly agreed.
[d] Factor grand means used for correlation analysis.

pation and income. According to the test results, no significant differences were found across categories of sociodemographic factors (p-values in all texts >.10).

Further statistical tests were performed across the tourism related characteristics on residents' perceptions of tourism impacts on QoL, satisfaction with the corresponding domains and overall QoL. Independent sample t-tests were performed based on whether residents' worked in tourism sector or not, their jobs had direct contact with tourists or not, they had a family member(s) working in the tour-

Table 2. Correlations of QoL construct

	(1)	**(2)**	**(3)**	**(4)**	**(5)**	**(6)**	**(7)**	**(8)**	**(9)**
(1)	**1**								
(2)	0.496**	1							
(3)	0.486**	0.662**	1						
(4)	0.472**	0.731**	0.737**	1					
(5)	**0.608****	0.433**	0.474**	0.461**	1				
(6)	0.447**	**0.632****	0.563**	0.629**	0.609**	1			
(7)	0.453**	0.574**	**0.706****	0.627**	0.572**	0.754**	1		
(8)	0.440**	0.585**	0.586**	**0.700****	0.585**	0.802**	0.771**	1	
(9)	0.494**	0.394**	0.485**	0.456**	**0.684****	**0.554****	**0.538****	**0.532****	1

**Correlation is significant at the 0.01 level (2-tailed).
*Correlation is significant at the 0.05 level (2-tailed).
Note: The correlations under the hypotheses of this case study are shown in bold.
(1) Tourism's impact on material life; (2) Tourism's impact on community life; (3) Tourism's impact on emotional life; (4) Tourism's impact on health & safety life; (5) Satisfaction with material life; (6) Satisfaction with community life; (7) Satisfaction with emotional life; (8) Satisfaction with health & safety life; (9) Overall satisfaction.

Table 3. Summary of the hypotheses testing

	Correlation Hypotheses	**Results**
H1	Tourism's impact on QoL in material/community/emotional/health & safety life domains ←(+)→ Satisfaction with QoL in material/community/emotional/health & safety life domains	Support
H2	Satisfaction with QoL in material/community/emotional/health & safety life domains ←(+)→ Overall satisfaction with QoL	Support

ism sector or not, and the perceived percentage of income derived from tourists' spending. According to the results, significant differences were found regarding the material life domain across categories of the variables "working in tourism sector or not" and "perceived percentage of income derived from tourists' spending" (Table 4).

CONCLUSION

In a plausible overtourism destination—Macau, the authors attempted to measure how residents perceived tourism's impact on their QoL and how they were satisfied with QoL in four domains (i.e. material, community, emotional, and health & safety), and to estimate how life satisfaction in the four life domains correlated with the overall life satisfaction.

The findings indicated that the proposed hypotheses were supported and all factors were positively correlated. It indicated that residents who perceived more negative tourism's impact on their QoL were more unsatisfied with QoL in the corresponding life domains and also with overall life. These findings support the conclusion by Uysal, Sirgy, Woo, and Kim's (2016), based on many studies (e.g. Renda,

Table 4. Analysis of difference in tourism related characteristics – t test and Kruskal-Wallis test

Variables (n = 407)	n	Tourism's Impact on Material Life		Satisfaction with Material Life	
		Mean	t / χ^2	Mean	t / χ^2
Working in a tourism related job					
Yes	142	3.08	4.733**	2.70	1.494
No	265	2.83		2.74	
% of income derived from tourists' spending					
0-30%	201	2.82	25.03**	2.70	9.447**
40-70%	139	2.95		2.74	
80-100%	67	3.15		2.92	

** p-value < .01.

Mendes, & Valle, 2011; Urtasun & Gutierrez, 2006), that the impact of tourism development plays an important role in residents' QoL.

On average, respondents only perceived a slightly positive impact from tourism on their QoL in regard to their individual and family income, security of job, and benefits they can get from the government, all of them under the material domain. Conversely, they perceived negative impact on all other indicators across the four domains, especially felt a vigorous negative impact on public transportation, under the community domain, and cost of housing, under the material domain. There were 22 out of 26 indicators (85%) being negatively rated by the residents. As argued by Gonzalez, Coromina, and Galí (2018), imbalanced perceptions of the positive and negative impacts derived from tourism may signify the phenomenon of overtourism (Gonzalez, Coromina, & Galí, 2018). The growing tension may eventually catalyse anti-tourist emotions, thus unsatisfying the tourists and threatening the sustainable development of Macau as a tourism destination. In fact, anti-tourist movements took place in different destinations after the tourism activities crossed the threshold of the residents' tolerance (Colomb & Novy, 2016). The results also supported the conclusion by Koen, Postma and Papp (2018) that overtourism is not simply a tourism problem but issues across the societal development.

Regarding the respondents' satisfaction with QoL, they only showed modest satisfaction with security of job and family income under the material domain. They felt unsatisfied with all other indicators, with strong dissatisfaction with public transportation and the cost of housing, which were also perceived as rather negatively affected by tourism. Ultimately, the value obtained for the overall satisfaction with QoL (2.76) indicated that residents were not satisfied with their life in general. As suggested by Doxey(1975), residents' attitude may become more negative in a later stage of tourism development when they perceived more costs than the anticipated benefits. This is manifestly the situation in Macau, not only given by the immense growth of visitor arrivals since 2002, but also by residents' negative perceptions of their QoL being revealed from this study.

Tourism has been growing and became a vital industry in many cities. In some of these cities, the immense negative impacts from tourism paved the way to the rise of overtourism. Overtourism signify the lack of sustainable tourism development, which is conditional on the perception of the local residents. The findings of this study allow different stakeholders to better understand how residents perceived tourism's impacts, which may degrade their life satisfaction in multidimensional values (i.e. material, community, emotional, and health & safety). Hence, this study has implications for tourism planners and policy makers, especially for the Macau government officials. Tourism planners and policy makers should put efforts to help residents avoid perceiving tourism as leading to the degradation of their QoL. The study sheds some light on how to avoid overtourism and pursue sustainable development in tourism destinations.

Implications also exist specifically for the Macau's governmental authorities and different stakeholders. As residents perceived rather negative impacts from tourism, and showed dissatisfaction with their QoL, there are potential signs of tension between residents and tourism development. Albeit no anti-tourism protest has been recorded in Macau, it is obvious that the current level of tourism development affects residents' QoL negatively.

Although Macau was ranked the third richest territory in the world (Gregson, 2017), the residents perceived negative impacts from tourism and dissatisfaction with their material life. Additionally, the city's population density was reported as the highest in the world (Macau News, 2016), and the surge of visitor arrivals made the community further overstretched. Plausibly, that was the reason why the community life was rated the lowest score. In order to help the residents to avoid perceiving tourism as contributing to the degradation of their QoL, thus to mitigate the overtourism risk, the Macau government may consider:

1) Allocate the public resources (e.g. land resources) carefully among tourism development and public needs. For example, Williams, Penrose, and Hawkes (1998) suggest that governments should incorporate stakeholders into the land use planning process for a shared-decision making. In the case of Macau, this could avoid residents perceiving that the insufficiency of land for building public houses is a result of the construction of casinos and/or hotels.
2) Manage the number of visitor arrivals so as to avoid exceeding the bottom line of the carrying capacity of the city's 30.5km^2. According to UNWTO (1981, p.4), carrying capacity refers to the maximum number of visitors a destination can host at a given time without harming itself. Although there is not a single measurement of carrying capacity applicable to all destinations, the negative perceptions of residents is an alarm to manage the number of visitor arrivals.
3) Encourage residents' involvement in tourism planning. Lalicic and Onder (2018) pointed out that decision makers can make better decisions and prevent failure with residents' participation, for instance, through online voting, online debates, consultation and campaigning.
4) Understand the perception of the local residents and other stakeholders by conducting regular surveys. A long-term success and sustainable tourism development rely on the co-construction of solutions balancing all stakeholders.

More noticeably, a decline of Macau visitor satisfaction was recorded in 2017 (Macau Daily Times, 2018). This may be due to the negative attitude from the residents, and/or by the degradation of QoL in the destination (e.g. poor air quality and public transportation) encountered by the visitors. This evidence may signify the potential threats of overtourism to the city of Macau. Therefore, it is crucial to enhance residents' perception of QoL in order to make the destination more sustainable and competitive in the long term.

FUTURE RESEARCH DIRECTIONS

Despite the findings of this study, several limitations should be addressed, which also provide opportunities for future research. First, the sampling distribution of the residents in Macau is rather uneven due to the use of an online survey instrument. Online survey respondents are usually younger, richer and better educated owing to the different familiarity in using technology across demographic groups (Yetter & Capaccioli, 2010). To avoid this shortcoming, future research may take note of this and put more efforts to have more diversified sample.

Second, the research was conducted at a particular point of time and circumstances, thus the result may not reveal the comprehensive view of residents' perception (García, Vázquez, & Macía, 2015). Residents' perception is dynamic and changing over time due to the change of governmental policies and/or according to different tourism development stages (Li & Wan, 2017). To avoid this shortcoming, future research may explore residents' perception of QoL in a longitudinal approach. Apart from the time factor, the characteristics of the destination should also be considered. This study focused in Macau, a petite tourism destination being developed intensely. Macau residents' perceptions of QoL may differ from destinations in different countries, cultures, tourism types and/or development stages. The findings of this study may only be a reference for small sized destinations with established tourism industry, particularly in a plausible overtourism setting. Wherefore more studies on residents' QoL should be undertaken in different overtourism destinations, in order to allow other destinations alike to benefit from the lessons learnt.

Third, this study exclusively focused on four life domains (i.e. material, community, emotional, and health & safety) that may be affected by overtourism. Other life domains may also be examined in future studies to obtain a more comprehensive understanding.

Lastly, aside from considering residents' perceptions of QoL, future research should consider to examine how the decrease of residents' QoL associates with other indicators, to explore the tension and relationship between residents and tourists. The present study is a preliminary work to adopt available academic work to examine overtourism. As this challenging phenomenon is becoming more and more active across the globe, further research is needed in order to deepen the understanding of overtourism and to find concrete solutions for it.

ACKNOWLEDGMENT

This paper is financed by National Funds provided by FCT-Foundation for Science and Technology through project UIDB/04020/2020.

REFERENCES

Andereck, K. L., & Nyaupane, G. P. (2011). Exploring the nature of tourism and quality of life perceptions among residents. *Journal of Travel Research*, *50*(3), 248–260. doi:10.1177/0047287510362918

Asia, G. G. R. (2019). *Macau's assembly to debate city tourist capacity*, available at https://www.ggrasia.com/macaus-assembly-to-debate-city-tourist-capacity/ (Accessed on May 30, 2019).

Buckley, R. (2012). Sustainable Tourism: Research & Reality. *Annals of Tourism Research*, *39*(2), 528–546. doi:10.1016/j.annals.2012.02.003

Colomb, C., & Novy, J. (2016). *Protest and Resistance in the Tourist City*. London, UK: Routledge. doi:10.4324/9781315719306

Cummins, R. A. (1996). Assessing quality of life. *Quality of life for people with disabilities: model, research and practice*, 116–150.

DICJ (Direcção de Inspecção e Coordenação de Jogos). (2017). *Gaming Statistics*, available at http://www.dicj.gov.mo/web/en/information/DadosEstat/2017/content.html#n1 (Accessed on March 15, 2018).

Dioko, L. (2017). The problem of rapid tourism growth – an overview of the strategic question. *Worldwide Hospitality and Tourism Themes*, *9*(3), 252–259. doi:10.1108/WHATT-02-2017-0005

Dolnicar, S., Yanamandram, V., & Cliff, K. (2012). The contribution of vacations to quality of life. *Annals of Tourism Research*, *39*(1), 59–83. doi:10.1016/j.annals.2011.04.015

Doxey, G. (1975). A causation theory of visitor-resident irritants: Methodology and research inferences. In *Proceedings from Travel and Tourism Research Association's (TTRA) 6th Annual Conference*, 195-198. San Diego, CA: TTRA.

DSEC. (Direcção dos Serviços de Estatística e Censos). (2016). *Intercensos 2016 resultados globais*, available at http://www.dsec.gov.mo/getAttachment/bfa0112a-eaf3-49a9-9168-b5add46e9d65/C_ICEN_PUB_2016_Y.aspx (Accessed on May 8, 2018) (In Portuguese and Chinese).

DSEC. (Direcção dos Serviços de Estatística e Censos). (2017). *Environmental statistics 2016*, available at http://www.dsec.gov.mo/getAttachment/5399842e-51c4-462c-81eb-942cf94778 6f/E_AMB_PUB_2016_Y.aspx?disposition=attachment (Accessed on March 15, 2018).

DSEC. (Direcção dos Serviços de Estatística e Censos). (2018a). *Tourism Statistics 2018*, available at https://www.dsec.gov.mo/Statistic.aspx?NodeGuid=7b23463a-d253-4750-bd12-958030df5ccb#Pba46f0c9-8b5b-46b6-b346-8cd1ded3c1f5 (Accessed on May 6, 2019).

DSEC. (Direcção dos Serviços de Estatística e Censos). (2018b). *Demographic statistic (4th Quarter/2017)*, available at http://www.dsec.gov.mo/getAttachment/9ea0efa4-6dfc-4e55-b829-743acd429c17/E_DEM_FR_2017_Q4.aspx?disposition=attachment (Accessed on March 15, 2018).

DSEC. (Direcção dos Serviços de Estatística e Censos). (2018c). *Employment survey* (January - March/2018), available at http://www.dsec.gov.mo/getAttachment/3579e9e4-5a5c-4717-b fbc-cccc8b9e1365/C_IE_FR_2018_M01.aspx (Accessed on May 8, 2018).

DSEC. (Direcção dos Serviços de Estatística e Censos). (2019). *Visitor arrivals*, available at http://www.dsec.gov.mo/TimeSeriesDatabase.aspx (Accessed on May 15, 2019).

García, F. A., Vázquez, A. B., & Macía, R. C. (2015). Resident's attitudes towards the impacts of tourism. *Tourism Management Perspectives*, *13*, 33–40. doi:10.1016/j.tmp.2014.11.002

Gonzalez, V. M., Coromina, L., & Galí, N. (2018). Overtourism: Residents' perceptions of tourism impact as an indicator of resident social carrying capacity - case study of a Spanish heritage town. *Tourism Review*, *73*(3), 277–296. doi:10.1108/TR-08-2017-0138

Gregson, J. (2017, March 1). *The richest countries in the world*, available at https://www.gfmag.com/global-data/economic-data/richest-countries-in-the-world?page=9 (Accessed on June 1, 2018).

IFT (Institude for Tourism Studies). (2015). *Findings of the annual study of Macao's Tourism Carrying Capacity, 2013 and 2014,* available at https://www.ift.edu.mo/Membership/ Download/TCC]_[2013and2014/2013-2014_Press_Release.pdf (Accessed on June 1, 2019).

Jennings, G., & Nickerson, N. (Eds.). (2006). *Quality tourism experiences*. Burlington, MA: Elsevier Butterworth-Heinemann.

Kent State University. (2018). *SPSS tutorials: independent samples t test*, available at https://libguides.library.kent.edu/SPSS/IndependentTTest (Accessed on Jun 2, 2018).

Kim, K., Uysal, M., & Sirgy, M. J. (2013). How does tourism in a community impact the quality of life of community residents? *Tourism Management*, *36*, 527–540. doi:10.1016/j.tourman.2012.09.005

Koens, K., Postma, A., & Papp, B. (2018). Is overtourism overused? Understanding the impact of tourism in a city context? *Sustainability*, *10*(12), 4384. doi:10.3390u10124384

Kuščer, K., & Mihalič, T. (2019). Residents' Attitudes towards overtourism from the perspective of tourism impacts and cooperation—The case of Ljubljana. *Sustainability*, *11*(6), 1823. doi:10.3390u11061823

Laerd Statistics. (2018a). *Independent t-test using SPSS Statistics*, available at https:// statistics.laerd.com/spss-tutorials/independent-t-test-using-spss-statistics.php (Accessed on June 2, 2018).

Laerd Statistics. (2018b). *Kruskal-Wallis H test using SPSS Statistics*, available at https://statistics.laerd.com/spss-tutorials/kruskal-wallis-h-test-using-spss-statistics.php (Accessed on Jun 2, 2018).

Lai, I. K. W., & Hitchcock, M. (2017). Local reactions to mass tourism and community tourism development in Macau. *Journal of Sustainable Tourism*, *25*(4), 451–470. doi:10.1080/09669582.2016.1221413

Lalicic, L., & Onder, I. (2018). Residents' Involvement in Urban Tourism Planning: Opportunities from a Smart City Perspective. *Sustainability*, *10*(6). doi:10.3390u10061852

Li, X., & Wan, Y. K. P. (2017). Residents' support for festivals: Integration of emotional solidarity. *Journal of Sustainable Tourism*, *25*(4), 517–535. doi:10.1080/09669582.2016.1224889

Liang, Z., & Hui, T. (2016). Residents' quality of life and attitudes toward tourism development in China. *Tourism Management*, *57*, 56–67. doi:10.1016/j.tourman.2016.05.001

Lu, J., & Nepal, S. K. (2009). Sustainable tourism research: An analysis of papers. *Journal of Sustainable Tourism*, *17*(1), 5–16. doi:10.1080/09669580802582480

Macau Daily Times. (2018, February 26). *More tourists, less satisfaction in 2017*, available at https://macaudailytimes.com.mo/tourists-less-satisfaction-2017.html (Accessed on May 17, 2018).

Macau Daily Times. (2019, February 1). *Lawmaker calls for action to avoid 'over-Tourism'*, available at https://macaudailytimes.com.mo/lawmaker-calls-for-action-to-avoid-over-tourism.html (Accessed on May 5, 2019).

Macau News. (2016, May 11). *Macau records the highest population density per sq km*, available at https://macaunews.mo/macau-records-highest-population-density-per-sq-km/ (Accessed on May 17, 2018).

Nawijn, J., & Mitas, O. (2012). Resident attitudes to tourism and their effect on subjective well-being: The case of Palma de Mallorca. *Journal of Travel Research*, *51*(5), 531–541. doi:10.1177/0047287511426482

Nichols, M., Stitt, B. G., & Giacopassi, D. (2002). Community assessment of the effects of casinos on quality of life. *Social Indicators Research*, *57*(3), 229–262. doi:10.1023/A:1014777323534

Nunkoo, R. (2016). Toward a more comprehensive use of social exchange theory to study residents' attitudes to tourism. *Procedia Economics and Finance*, *39*, 588–596. doi:10.1016/S2212-5671(16)30303-3

Nunkoo, R., & Ramkissoon, H. (2011). Residents' satisfaction with community attributes and support for tourism. *Journal of Hospitality & Tourism Research (Washington, D.C.)*, *35*(2), 171–190. doi:10.1177/1096348010384600

Nunnally, J. C. (1978). *Psychometric theory*. New Jersey: McGraw-Hill Book Company.

Oxford Dictionary. (2019). *Word of the Year 2018: shortlist,* available at https://en.oxforddictionaries.com/word-of-the-year/shortlist-2018 (Accessed on June 10, 2019).

Perkumiene, D., & Pranskuniene, R. (2019). Overtourism: Between the right to travel and residents' rights. *Sustainability*, *11*(7), 2138. doi:10.3390u11072138

Pinke-Sziva, I., Smith, M., Olt, G., & Berezvai, Z. (2019). Overtourism and the night-time economy: A case study of Budapest. *International Journal of Tourism Cities*, *5*(1), 1–16. doi:10.1108/IJTC-04-2018-0028

Postma, A. (2013). When the tourists flew in: Critical encounters in the development of tourism. The Netherlands: Groningen.

Puczkó, L., & Smith, M. (2011). Tourism-specific quality-of-life index: the Budapest model. In M. Budruk, & R. Phillips (Eds.), *Quality-of-life community indicators for parks, recreation and tourism management* (Vol. 43, pp. 163–183). The Netherlands: Springer. doi:10.1007/978-90-481-9861-0_9

Renda, A. I., Mendes, J. C., & Valle, P. O. (2011). A structural model approach of residents' perception of tourism impacts in their own quality of life: the municipality of Loule, Algarve. In *Proceedings International Conference on Tourism & Management Studies - Algarve, Vol. II.* October 26-29, 2011, Faro, Algarve, Portugal.

Responsible Tourism. (2019). *Overtourism,* available at http://responsibletourismpartnership .org/ overtourism/ (Accessed on June 10, 2019).

Seraphin, H., Sheeran, P., & Pilato, M. (2018). Over-tourism and the fall of Venice as a destination. *Journal of Destination Marketing & Management*, *9*, 374–376. doi:10.1016/j.jdmm.2018.01.011

Sirgy, M. J., & Lee, D. J. (2006). Macro measures of consumer well-being (CWB): A critical analysis and a research agenda. *Journal of Macromarketing*, *26*(1), 27–44. doi:10.1177/0276146705285669

STHDA. (2018). *Kruskal-Wallis test in R*, available at http://www.sthda.com/english/ wiki/kruskal-wallis-test-in-r (Accessed on Jun 2, 2018).

Street, F. (2018, October 3). *Can the world be saved from overtourism?* Available at https://edition.cnn.com/travel/article/overtourism-solutions/index.html (Accessed on June 10, 2019).

Taylor, I. (2018). *Over tourism 'highest risk' warns Macao chief,* available at http://www.travelweekly.co.uk/articles/304025/over-tourism-highest-risk-warns-macao-chief (Accessed on June 1, 2019).

UNESCO. (2005). *Mostar, Macao and Biblical vestiges in Israel are among the 17 cultural sites inscribed on UNESCO's World Heritage List*, available at http://whc.unesco.org/en/news/135/ (Accessed on March 15, 2018).

UNWTO. (1981). *Saturation of Tourist Destinations. Report of the Secretary General*. Madrid, Spain: World Tourism Organisation.

UNWTO. (2018). *'Overtourism'? Understanding and Managing Urban Tourism Growth beyond Perceptions,* available at https://www.e-unwto.org/doi/pdf/10.18111/9789284420070 (Accessed on May 5, 2019).

Urtasun, A., & Gutierrez, I. (2006). Tourism agglomeration and its impact on social welfare: An empirical approach to the Spanish case. *Tourism Management*, *27*(5), 901–912. doi:10.1016/j.tourman.2005.05.004

Uysal, M., Sirgy, M. J., Woo, E., & Kim, H. (2016). Quality of life (QoL) and well-being research in tourism. *Tourism Management*, *53*, 244–261. doi:10.1016/j.tourman.2015.07.013

Wan, Y. K. P., & Li, X. (2013). Sustainability of tourism development in Macao, China. *International Journal of Tourism Research*, *15*(1), 52–65. doi:10.1002/jtr.873

Williams, J., & Lawson, R. (2001). Community issues and resident opinions of tourism. *Annals of Tourism Research*, *28*(2), 269–290. doi:10.1016/S0160-7383(00)00030-X

Williams, P. W., Penrose, R. W., & Hawkes, S. (1998). Shared decision-making in tourism land use planning. *Annals of Tourism Research*, *25*(4), 860–889. doi:10.1016/S0160-7383(98)00037-1

Woo, E., Kim, H., & Uysal, M. (2015). Life satisfaction and support for tourism development. *Annals of Tourism Research*, *50*, 84–97. doi:10.1016/j.annals.2014.11.001

Woo, E., Uysal, M., & Sirgy, M. J. (2018). Tourism impact and stakeholders' quality of life. *Journal of Hospitality & Tourism Research (Washington, D.C.)*, *42*(2), 260–286. doi:10.1177/1096348016654971

Woosnam, K. M. (2012). Using emotional solidarity to explain residents' attitudes about tourism and tourism development. *Journal of Travel Research*, *51*(3), 315–327. doi:10.1177/0047287511410351

WTTC. (2018). *Travel & tourism economic impact 2018 - Macau*, available at https://www.wttc.org/-/media/files/reports/economic-impact-research/countries-2018/macau2018.pdf (Accessed on May 29, 2018).

Yamada, N., Heo, J., King, C., & Fu, Y. (2009). Life satisfaction of urban residents: do health perception, wealth, safety, community pride and, and cultural tourism matter? In International CHRIE Conference-Refereed Track. Paper 24.

Yetter, G., & Capaccioli, K. (2010). Differences in responses to web and paper surveys among school professionals. *Behavior Research Methods*, *42*(1), 266–272. doi:10.3758/BRM.42.1.266 PMID:20160305

KEY TERMS AND DEFINITIONS

Overtourism: Tourism impact that immensely and negatively affects residents' life and/or tourists' experience.

Quality of Life: Individuals' view on their living conditions and how such conditions affect their satisfaction in different life domains and overall life satisfaction.

Special Administrative Region (SAR): A region in China that maintains separate legal, administrative and judicial systems from the rest of the country. There are two SARs in China – Hong Kong and Macau.

Tourists: People who visit a place other than their usual place of residence for more than 24 hours but less than a consecutive year for business, leisure or other personal purposes.

Travel and Tourism GDP Contribution: The total contribution comprises of the direct, indirect and induced effects derived from of Travel and Tourism economic activities.

Visitor: People who visit a place other than their usual place of residence for less than a consecutive year for business, leisure or other personal purposes.

Chapter 6
Demarketing:
A Marketing Framework for Overtourism

Pinaz Tiwari
Jamia Millia Islamia, India

Snigdha Kainthola
Jamia Millia Islamia, India

Nimit Ranjan Chowdhary
https://orcid.org/0000-0002-7335-2379
Jamia Millia Islamia, India

ABSTRACT

Ensuring sustainability at a tourist destination is a challenge which is faced by many nations. The challenge compounds since short term monetary gains often blind the desire for sustainable development. The term overtourism has gained popularity during the last few years as instances of anti-tourism reactions have been reported from several cities like Venice, Shimla, Barcelona, etc. An acceptable solution for countering overtourism has not yet been found because of the subjectivity and complexity of the situation. This chapter focuses on deconstructing the situation of overtourism in different parts of the world and how de-marketing can be used as one of the strategies for sustainable tourism. It shows the demarketing structure in marketing framework by modifying the 5As of tourism. It also shows the marketing mix framework within the domain of demarketing. It provides an insight into the role of de-marketing in striking a balance between the interests of local communities and stakeholders on one hand and enhancing the tourist experience at a destination on the other.

INTRODUCTION

The year 2018 has proved to be significant in the history of the tourism industry as it has reached up to 1.4 billion international tourist arrivals in the world (UNWTO Communications Department, 2019). The estimated goal set for the year 2020 has been achieved two years earlier. The economic prospects are wider

DOI: 10.4018/978-1-7998-2224-0.ch006

as it creates direct, indirect and induced employment opportunities for people across the globe. There has been an average increase in the levels of disposable income of the world's population. The process of booking travel has also gotten convenient. This convenience has led to an increase in the proportion of first-time travellers (Glaesser, Kester, Paulose, & Abbas, 2017). However, growth in numbers may not necessarily mean a sustainable growth of the destination. The problem is that "too" many travellers are converging in the same cities and visiting the same places in those cities during a particular period. (Poll, 2019). The challenge of sustainable development compounds since short term economic benefits often blind the desire for sustainable growth. The term "overtourism" is used to define the phenomenon where tourism negatively affects the residents' quality of life as well as tourists' experience at a destination. With an increase in tourists' arrivals, especially in Europe agitation among the local community against tourists has also increased. Notions like gentrification, anti-tourist movements, and *tourismphobia* are frequently used to refer to instances of overtourism in famous cities like Amsterdam, Venice, Dubrovnik, Barcelona, Rome and many cities in Thailand and India.

Third world economies promote tourism activities as a greater number of tourists would mean more revenue for them. Thailand is a case in point, where tourism is a major contributor in Gross Domestic Product of the country (WTTC, 2019). No doubt, tourism brings economic viability for any country engaged in tourism activities. An example of which can be seen in Ireland which opted for tourism to recover its debt in the year 2011. In summers of 2017, the gaze of tourism's negative impacts was observed in European cities where residents came out on roads to protest and asking tourists to go back home. Albert Arias Sans noted that overtourism is nothing but an effect of success (Sans & Quaglieri, 2016). The effect of having "too" many tourists at a destination results in damage to the ecological environment, increase in the cost of living, and displacing residents. However, the ascendant standard is to believe that tourism is broadly an unparalleled good where sustainability liberally asserted to encourage adversely affected critics and local communities alike. In several tourist destinations, a point of saturation has been achieved, which can be observed in the way residents' perception is changing for tourists, and tourism has become a tipping point for politics.

There is no single solution that can fit into all the destinations affected by overtourism. The nature, extent and impact of overtourism at each destination are different. Solutions, therefore, need to be unique to each destination. However, every destination struggling with overtourism would like to check the inflow of visitors. Demarketing is one of the plausible solutions. Authors, look forward to demarketing of a tourist destination as one of the strategies to deal with the phenomenon. Examples have been taken from different places which have witnessed an overflow of tourists and compared with those that are not yet affected by overtourism. Thus, this chapter shall focus on deconstructing the situation of overtourism in different parts of the world and how de-marketing can be used by altering the marketing mix.

BACKGROUND

Overtourism

#Overtourism was first used on Twitter in August 2012 and Harold Goodwin observed that the scenario is probable to become a common sight in several destinations soon (Goodwin, 2017). The term Overtourism is coined by media, specifically by Skift Magazine in 2016 and later trademarked in 2017. However, the negative consequences of tourism were discussed earlier way back in the 1960s (Forster;

Wagar, 1964). After these prominent works, two more studies came up, namely Doxey's Irritation Index (Doxey, 1975) and Butler's tourism product life cycle (Butler, 1980). These two models laid the foundation in understanding the harmful effects caused by excessive tourism on the environment and residents. During the same time, the concept of carrying capacity emerged to explore the problem quantitatively. The UNWTO defined carrying capacity as "the maximum number of people that may visit a tourist destination at the same time, without destroying the physical, economic and socio-cultural environment, and an unacceptable decrease in the quality of visitor's satisfaction (UNWTO, 1981). Today, more than 70% of the tourists are concentrated in only 20% of the destinations in the world (McKinsey & Co and WTTC, 2017).

Overtourism lacks a standardized characterization since it is in its early stages (Capocchi, Vallone, Pierotti, & Amaduzzi, 2019). Collins dictionary defined it as a phenomenon wherein a famous destination is overrun by tourists in an unsustainable manner. Harold Goodwin has described overtourism as "a situation where either local people or the tourists feel that the place is over visited, and its character is changing. So, for the tourist, it loses its authenticity, and for the local people, it just causes irritation and annoyance" (Goodwin, 2017). In simple terms, overtourism transpires when the number of tourists is much more than the capacity of a destination to withstand. In other terms, overtourism can be a psychological or cultural imbalance between the local communities and the tourists or the tourism stakeholders or it can be the concerns about the usage of the community's resources and economy of the place.

Major Causes and Impacts

In the present scenario, the success of any industry is justified by looking at the growth in numbers, and tourism is no exception. Any increase in the number of international tourists or cruise passengers or number of bed nights depicts a favourable picture of the destination. Tourism, it is believed, brings economic benefits to the destination as well as to residents but an unplanned growth can lead to a decline in the quality of life of locals and tourists' experience. Overtourism occurs when tourism activities

Figure 1. Aspects of Overtourism

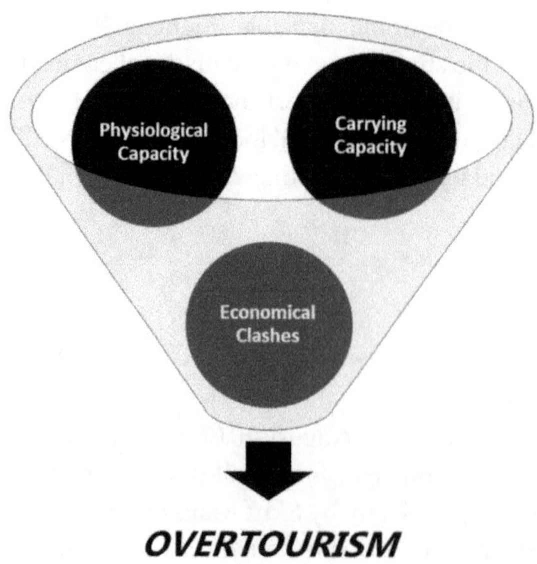

keep on increasing without acknowledging that capacity constraints also exist. There are several reasons which make a destination popular for mass tourists to flock in a particular period. The foremost being the geographical factors, accessibility and costs at the destination. The demography of the destination, their acceptance of diverse cultures of the tourists, the sharing of limited natural resources of the place among different stakeholders along with the administrative strategies employed manoeuvre the consequences of tourism on a destination.

The problem of overtourism is much more complex than just the carrying capacity of a destination. The psychological predicaments faced by the host community due to the actions or even the mere presence of the tourists also is termed as overtourism. The retaliation against tourism is often the consequences of lack of political vision, will, and management. The backlash often comes due to overcrowding or through excessive negative influence on the residents, which can be a result of rising costs, the steadily diminishing rental income, or the abundance of low-quality jobs. For instance, people have rallied in Barcelona and asking the Spanish government to rather allow refugees to stay than visitors (Burgen, 2018).

Major causes of overtourism are listed as (Figure 2):

1. **Population**: Since 1960, the world's population has nearly tripled. More number of people want to travel. The purpose of travelling is different and may have shifted from mass to alternative forms of tourism. The biggest market player in tourism in Asian countries. For instance, only 8% of the Chinese population hold passports and those 8% accounts for nearly 100 million travellers (CNT Editors, 2018).
2. **Income**: The average disposable income of the people is rising. The percentage of the middle class and upper-middle-class is increasing over the decade. As affluence increases, people save money to travel with family. The world's middle-class population is expected to add 160 million people each year up till 2030 (Kharas, 2017)

Figure 2. Causes of Overtourism

3. **Bucket List:** Listing famous places as "must visit" is a trend. People get psychological satisfaction in "ticking off" a destination where they have visited. This indicates that people are travelling but with sheer ignorance. Ignorance about the place, people, their culture and traditions.
4. **Social Media Marketing**: Social media has contributed significantly to influencing people to travel to 'Instagrammable' locations, get perfect selfies at famous tourist spots and travel to come out of the FOMO (Fear of Missing out). Destinations promote themselves by collaborating with social media influencers and companies for branding and campaigning. Travel blogs and articles have started to put catchy headlines to get more clicks, and subsequently, more views mean more money. For example- "Top 10 things to do at Bali" would attract potential travellers to read the article. Such articles may deter travellers from exploring the destination by limiting themselves to these 10 activities.
5. **Films/series Induced Travel**: Films and net series also induce travel to a specific destination, which contributes to the issue of overtourism. The Maya beach is famous for shooting of the firm "The Beach", Disneyland for film Star Wars, Cornwall for Poldark and Dubrovnik for Game of Thrones. When fans are attracted to visit these places, it becomes difficult to manage the crowd and sustainability of the place at danger.
6. **Low-Cost Carriers (LCC):** Airlines are exempted from paying any tax or duty on aviation fuel, the benefit of which is given to travellers. Introduction of low-cost carriers has aggravated the problem. For example, to boost travel, the Government of India has launched a scheme called UDAN. Under this scheme, flights of less than one-hour duration or less than 500 km, the fare for at least 50% seats is capped at INR 2500 (around USD 36)
7. **Budgeted Accommodation**: Sharing accommodation has replaced the conventional idea of staying in a hotel. Peer to peer accommodation, homestays and youth hostels are low budget and preferred by the travellers. Platforms like Airbnb and Booking.com, on the one hand, offer alternate accommodation at a lower tariff, as compared to commercial accommodations like hotels and lodges. On the other hand, this has resulted in high rentals and illegitimate owning of properties in famous cities like Barcelona. Owners prefer to rent out properties to visitors rather than the locals. Locals are gradually pushed out of the tourist districts at the destination.

Major impacts of overtourism are listed below:

1. **Poorer Quality of Life for the Residents**: Movement of tourists in city centres often disrupts the daily routines of locals. They have to stand in long queues at eateries and get stuck in traffic jams due to large tourist vehicles. Night-time economy also disturbs residents' life due to noise (Smith, Sziva, & Olt, 2019) and drunkards who puke in front of their houses.
2. **Displacement of Locals**: The phenomenon of excessive *touristification* at the destinations, residents are forced to leave their homes and shift to different cities (Barata Salguiero, Mendes, & Guimarães, 2017). The case of collective displacement can be witnessed in Venice too. People in Barcelona have left their homes because of a hike in rent prices due to increased demand for rental accommodation by tourists.
3. **Decline in Tourists' Experience**: It is not only the residents who get affected but also many tourists who visit a destination to relax and spend leisure time. Overcrowding by tourists affects tourists themselves. Promotion attracts tourists, but overtourism may create a gap between physical and virtual reality.

4. **Infrastructure Overloading**: A form of carrying capacity, namely infrastructural carrying capacity describes that based on the prevailing infrastructure, only a limited number of tourists can be accommodated (Swarbooke, 1999). One of the main challenges faced in natural parks is that the infrastructure doesn't support excessive tourist footprints (Weber, et al., 2017)
5. **Environmental Pollution**: Overtourism impacts the environment severely. Greater number of tourist vehicles causes air pollution, and beachgoers using sunscreen causes water pollution, which damages coastal flora and fauna (Sánchez-Quiles & Tovar-Sánchez, 2015).

Demarketing

Marketing is indispensable to carrying forward of any product. It has evolved in the past five decades to become a crucial part of the lives of the producers, administrators and the consumers alike. Kotler defines marketing as the function of the business, which is broadly focused on managing the level of demand and its composition for the company. Demarketing is one of the intrinsic aspects of marketing, which was first introduced by Kotler and Levy in 1971. It is defined as a strategy to curtail consumer demand for a product or service for either a limited or permanent basis amongst all the customers or by targeting a certain class of customers (Kotler & Levy, 1971). It is generally adopted in situations of scarce supply or when a product or a service is considered undesirable. Demarketing manoeuvres the market in numbers and consciously aim to influence the demand for a product through various deliberate strategies. The marketers can use demarketing to tactically manage the perceptions of the buyers towards the product to manoeuvre the desired image (Miklós-Thal & Zhang, 2011). Instead of rapidly and thoughtlessly increasing the sales of a product, it the responsibility of the marketer is to engineer the demand in order to align with the long-term objectives (Kotler & Levy,1971).

In 1971, Kotler and Levy explained three types of demarketing. The first one is *General Demarketing,* where the company wishes to reduce the total demand for a product. The second one is *Selective Demarketing* in which the firms aim to discourage the demand of a certain class of consumers through various strategies. This demarketing is often beneficial to safeguard loyal or core consumers. The third strategy is of *Ostensible Demarketing,* where the company creates artificial or perceived shortage of products to stimulate customers' desire to buy the products. Limited supplies of products are created, which consequentially lead to the stocking of these sparse goods by the consumers. Within demarketing, Cullwick (1975) mentioned two categories, namely- *ration demand* and *reduce demand.* He specified that rationing demand is adopted when attempts are made by the marketer to spread the limited supply of product in the market. Whereas reducing demand requires efforts to decrease the levels of demand in the market.

Little, Lee, & Nair (2019) has mentioned that demarketing is reversed marketing, from the drivers to the potential inhibitors of the product. Demarketing is set of strategies that compliments the sustainable ecological and social marketing (Wearing, Schweinsberg, & Tower, 2016). Peattie, Peattie, and Newcombe (2016) elaborates demarketing to be more than just demand reduction strategy. They suggest that it is a highly simple strategy used in a complicated environment. Step 'A' may not always result in behaviour 'B' due to dynamic condition. In the context of the tourism market, reducing demand strategy has been implemented in most of the cities.

APPLICATION OF DEMARKETING IN THE TOURISM INDUSTRY

The application of demarketing strategy in tourism industry is limited, majorly because of its complexity. In the tourism industry, most of the goods and services catered to tourists are furnished by private players. The public sector plays a crucial role in administering, organising and controlling the overall development of tourism at a destination. Goals, strategies, and policies are decided by public sector officials who are fundamentally supposed to work to bring balance between the economic development of the stakeholders and sustainable growth of all. Hence equilibrium between the state and the private parties in tourism is essential to achieve to form a beneficial structure for mutual growth of all the stakeholders along with sustainable development of a destination.

Despite the convolution in governance regarding tourism policies in several places, there are some examples where demarketing has been utilised on a large scale. Blue Mountains Park in Australia is one of the most prominent examples. The Blue Mountains National Park is a popular destination for both domestic and international tourists in Australia. It suffers from a larger number of footfalls, excess demand. Several measures have been taken to organize and manage the demand of the tourists and are applied to particular activities so that tourists' experiences can be improved. Amongst many measures, the strategies include restricting the number of recreational activities, capping the period of the activities, non-publicity of places and closing of certain areas or features. The messages about restrictions in the park were circulated in the promotional material along with the responsibility code of conduct required to show respect to the environment (Armstrong & Kern, 2011)

Demarketing steps have been implemented at heritage properties like Sissinghurst to minimise overcrowding. Further, the measures of demarketing have been used in Napa Valley and the Republic of Cyprus. The Cyprus Tourist Organisation (CTO) conducted an audit on their natural resources, the underlying interest of tourists, spending levels of individual tourists, etc. CTO segmented their market on these grounds, and high and middle-income tourist groups were targeted (Clements, 1989). Modifications in tourism policies to achieve the goals of sustainable destination has also been applied at various places, for instance, Amsterdam has come up with their new tourism strategy which states- "Develop and Discourage" to deal with overcrowding at hotspots. Maya Beach in Thailand is also an example of general demarketing strategy wherein the beach is closed for all tourists for a temporary period. The step has been taken to protect the natural resources which are damaged due to excess number of tourists.

A destination seller can take advantage from the concept of "demarketing," as a practice that consciously restricts marketing efforts to discourage demand image (Miklós-Thal & Zhang, 2011). However, the question arises as to how can quality tourism be justified? Does an increase in museum fees or charging tourist taxes shall imply a responsible attitude amongst tourists? These questions need to be analysed before going for implementation of demarketing at a destination.

Scope of Demarketing

Overtourism is a problem which arises due to the lack of vision and the will of the tourism stakeholders to manage a destination with responsibility. The adequate management of tourism will not only benefit the local population but also ensure easy movement of tourists. Demarketing may prove to help deal with overtourism. The aim of demarketing, as Kotler and Armstrong (Kotler, Armstrong, Saunders, & Wong, 2001) points out "is not to destroy demand but only to reduce or shift it" purposively. Through appropriate marketing, which includes demarketing as one of its facets, the level, timing, and character of the demand for a product or service can be managed according to the objectives of the organisation (Kotler, 2000).

Demarketing as strategy to curb overtourism will be much more fruitful if it is the driving force also spreads awareness to the inhibitors. The marketers can utilize all the tools of Integrated marketing in order to customize the flow of demand for their product (Sakshi & Pradeep, 2017). For example, through the network of social media the marketers have an enormous power to influence the demands towards a destination. Social media is celebrating a great extent of success. In today's global world, larger number of people can be reached on a click. Technology can play a pivotal role in effectively implementing the strategies of demarketing for a destination. Social media has become a marketing platform of its own which can affect the tourist behaviour to a great extent (Sakshi & Pradeep, 2017). This can sensitize the meticulous tourists to behave towards the destination with respect (Kumar & Srivastav, 2019).

Significance of Demarketing

Demarketing is not just a facet of marketing but also is a strong tool. It is a strategy that can help in the reformation of the destination. Demarketing can help in maintaining equilibrium between the number of visitations to a destination, revenue and sustainability. In several examples mentioned above in the chapter, it can be said that demarketing can help in hovering the demand for a tourist destination to suit the providers and tourism administrators. Tourism is likely to be sustainable if injected with community engagement, visitor management, planning, and profiling of visitors. Demarketing may help to deal with overtourism. Appropriate marketing can monitor and control the extent, timing and quality of demand for a product or service.

If marketing focuses on reaching out to the right market segment, then demarketing is about controlling and influencing this demand. Demarketing helps the marketers to further target, position, and segment the demand as per the supply. Often the marketer cannot meet all the demands, and filtering of customers becomes essential to not disrupt the profitable growth as well. The customer mix through demarketing can be influenced by putting the right strategies at the right time. Demarketing plays an optimum role in formulating the right customer mix required by marketers. It can further help in countering the seasonal demand for a product. The marketers can regulate the quality of the customers they receive and make suitable changes to cater to the segment they prefer.

Role of Different Stakeholders

A holistic demarketing strategy is implausible unless implemented centrally. Several industries together form the tourism industry, and hence, coordination among the service providers is important to achieve desirable outcomes. If different producers follow different ideologies, then a sustainable growth of the destination is not likely. The rise in the incidence of backlash against tourism is partly because of the lack of will to administer and manage the destination by a central authority. 'Overtourism' and 'over-crowding' have become hot issues in the past few years in touristic cities like Venice, Dubrovnik, Berlin, and Barcelona where the negative impact of tourism leads to compromises in both the quality of life of locals and the experiences of the visitors.

Government

A centralised strategy which includes demarketing of destination is the need of the hour. Several governments have begun to take up policies to curb the negative impact of overtourism on the destinations and the people of those places (Joppe, 2019) However, if centralised governance was imposed since the beginning of the development of tourism, the negative impact of tourism could have been avoided. The increasing number of anti-tourist movement or the tourismphobia amongst the local community becomes a trigger for the government to take corrective measures.

On the contrary, if all the stakeholders had from the beginning worked on the sustainable development of the destination, then the negative consequences could have been controlled. Instead of short-term goals of profit and sales, the businesses too should look for long term sustainability. To maintain a balance between the aspirations of businesses, locals and the tourists is the function of the destination leadership, which can only come via thoughtful planning. The government should take the responsibility, determine the goals, chalk out the methods required to achieve them and constitute policies that will not only yield profit but also work towards a positive outcome and reduce negative impact due to tourism.

The government can further put a cap on the number of visas issued for a particular destination at a specified time. This shall regulate the footfall of the destinations. Destination leadership may also resort to selective promotion. The brochures or other official promotion from the government's end can exclude the destinations already hit by the overabundance of tourists and can promote sectors that need the requisite push for tourism. The government policy can also target to balance the economic growth for the locals and for the businessmen that came from outside by legislating on investment, land use, owning property, adaptive reuse of heritage, and other policies including the monetary policies.

Service Providers

The tour operators and the service providers are the ones in close to contact to the tourists. They can channelize the demand of the tourists to destinations which are not affected by overtourism. TUI stopped marketing Spain and encouraged the customers to other destinations by educating them of the situation of overtourism in Spain. The Euro Rail increases the prices in the peak season discouraging tourists on an economic basis and slashes the prices in the lean season to boost tourism. Many service providers have promotional activities to increase the sale during the lean season. The tour operators can further remove a particular destination from the itinerary struck with overtourism and offer guests a similar destination on the itinerary. This manoeuvring will also ensure that there are no revenue losses. The internet has a far-reaching impact on the tourism purchase decision making of the visitor. It acts as a dominant push or pulls factor for decision making, especially among the youth. Online portals can disseminate information about parts of the destination that may be facing overcrowding. Such information may help channelize tourists to alternative destinations.

Tourists

There is also a need for the tourists to be aware of the role that they can play to safeguard the destination from the negative impact of overtourism. The foremost step that the tourist can take is research. The tourist must educate themselves before opting for any destination which will enhance their experience. While choosing a destination, the tourist can opt for the places that are not already hit by overtourism.

This step is only possible when the knowledge is disseminated on a global level about destinations suffering from an overabundance of tourist. The most pivotal step is sensitisation, especially in areas of cultural diversity. Further, tourists should be aware of the appropriate behaviour to follow once they visit a destination.

Internal Demarketing and Dispersed Marketing

Here are two marketing concepts that can help tackle the challenge of overtourism.

Internal Demarketing

Internal de-marketing is using internal marketing for demarketing the product. Internal marketing is a strategy wherein employer/organisation motivates its employees to perform well by providing a healthy work atmosphere (Farias, 2010). It is about educating and enabling the service providers to deliver the promise that the organisation has made to its customers. In the context of tourism, internal marketing would mean developing the stakeholders at a destination so that they can deliver better tourists' experience. The term internal demarketing would mean to internally sell the idea of demarketing of destination to the stakeholders at a destination. Internal demarketing is a long-term strategy to balance the supply and demand in the tourist market. It includes creating awareness about the ills and impacts of tourism. It is also about advocating conduct and behaviour which will render consumption of tourism at a destination more sustainable. It may include lobbying for an appropriate hotel policy, renting and rental policy, visa policy, tourism taxation policy, etc.

Dispersed Marketing

It begins where internal demarketing ends. Once the stakeholders are adequately advised, the demand can be directed and channelized to another destination. The stakeholders would appreciate not to seek inordinate demand to maximise their operational efficiency. Effectiveness will take priority over efficiency. Here, travel agencies, tour operators and government play an important role. Promotional strategies should include off-beaten destinations or places that suites the taste and preferences of a traveller. The term literally means dispersing the demand of tourists and shifting them to other destinations. It can be within the same country or from one country to another.

An example is the state government of Uttar Pradesh (India) dropped Taj Mahal from their brochures and introduced new circuits which included less visited places in India. Technology has been an integral part of marketing, and it can be effectively used in dispersed marketing technique. Mobile applications can be developed that list the overcrowded places and alternative destinations within a country based on the preference set by the user. Another strategy would be to not flood the tourist districts. There may be an efficient public transport that can swiftly bring in and take out visitors from a tourist district to off-beat locations for add-on touristic engagements. Thus, the crowding at anchor tenants at the destination will be curtailed. These activities in the remote and outskirts will induce an economic and tourism development of these places.

Demarketing is used differently as a strategy in different situation. In order to reduce excess demand general demarketing is effective, to restrict the demand among undesirable or unprofitable customers selective demarketing is used ostensible marketing is most beneficial to form a tactical image of product shortage to attract additional demand (Chaudhry, Cesareo, & Pastore, 2019). The objective of demarketing is not only restrict the demand for a product but also to increase the level of awareness in some of the instances (Kumar & Srivastav, 2019). One of the classic examples of managing tourism through demarketing is of Cyprus. In 1980s demarketing was used in Cyprus to filter the tourists and discourage the drunken and noisy, tourists. Marketing was limited only to the desirable segments as per age and income, the promotions of certain activities was eliminated along with other measures (Soliman, 2010).

Demarketing for 5As

The 5 A's of a destination elaborate tourism at a place. Addressing these 5As can point at several measures to curb the overtourism at the destinations. The essential requirement for a successful destination lies in the 5 A's of tourism, namely accessibility, accommodation, attraction, amenities and activities. These components are combined to provide a wholesome experience for the tourist. The effective management in all these five areas can minimise the negative effect of tourism for a destination. Demarketing within the marketing framework can be applied to different levels of production of tourism.

Accessibility

Accessibility relates to the infrastructure and technical advancement in modes of transport. Depending upon the cost efficiency and time utilised, a person can use any transport (railways, airlines, roadways or sea) to reach a destination. In the past few years, the aviation and cruise industry has expanded owing to the increase in the average income of people and simpler booking procedures. There has been a steady increase in the use of low-cost carriers and as per reports, LCCs accounts for almost 31% of the total scheduled passengers in the world (Source: CAPA). Cheap flight tickets are a result of subsidised air fuels and large-scale competition in the market. Increase in the air travel demand has out ripped the industry's capacity to expand. Today, all the destinations are easily accessible and implementing demarketing measures will primarily include: discouraging customers to use airlines for domestic sectors by charging higher prices; encourage them to use a surface mode of transportation; increasing fares on busy routes during peak seasons; introducing promotional fares during low seasons. By charging more prices on short-haul flights, people would tend to choose railways or roadways, which will help promote slow tourism.

Accommodation

Accommodation is an important component of a destination as any tourist would need a place to rest or stay overnight. It includes hotels, resorts, bed & breakfast establishments, youth hostels, apartments etc. the Accommodations through peer to peer social networks which aim to provide a local-like experience to the tourists are innovations in the lodging sector. Online platforms such as Airbnb and Booking.com are providing stays at affordable prices and earning huge profits. Entire Sweden is listed on Airbnb. This move is an extension of the Swedish principle of "Allemansrätten' which is freedom to roam across the country as one big home. These portals have helped destination with the surge in tourist demand which at times became unmanageable by the destinations. Cities like Paris and Barcelona have come down heavily on Airbnb and have strict rules in place. Other cities like Berlin, Amsterdam, London and San

Francisco have somewhat lenient restrictions on Airbnb. They hold Airbnb responsible for illegal rentals, which have reduced the housing affordability for the local population. Many real estate companies and individuals understood the tourists' demand and purchased properties in the tourist districts of the city. These properties are given on rent to tourists, which ultimately reduces the supply of rental properties for the locals. There is also a hike in rentals for locals. In this way, residents are displaced from their native places. The destination must deal with issues like licensing, renting of entire homes, the number of nights to be rented, and availability of the host during tourists' stay. Destinations like Tokyo, Venice and Barcelona have decided to keep a check on these rental accommodations by ways of licensing or introducing new regulations. Demarketing in the accommodation sector requires strict monitoring on the construction of hotels in the city centre, licensing of rental accommodations and tourist tax for a stay in luxury hotels.

Attraction

Attraction refers to any element within a destination that motivates people to travel. It includes monument, national parks, beaches, environment, heritage, culture etc. Attractions are most affected by overtourism, specifically the beaches and national parks. Overcrowding has resulted in an obstructed view of the attraction, no parking spaces, long-standing queues and littering. Videos of overcrowding on Mount Everest and littering in Ladakh have recently been doing rounds on social media. Government of different countries has put their step forward in preserving the attractions. For example, the government in Thailand decided to temporarily shut down the famous Maya beach to revive the place by introducing control mechanisms for balancing tourism and natural resources. Similarly, in Machu Pichu, the government has decided to limit the tourists' group to 16 and for a limited period. In Cinque Terre (a village in Italy), authorities have decided to introduce the online ticket booking system and online app which will help tourists to map the places less affected by crowd. In most of the cases, the government is intervening and coming up with measures to save the tourist attractions. Increasing the fee and cutting down on promotional activities can help in sustaining the attractions at a destination. Demarketing, along with technological innovation, can do wonders such as introducing the advance online booking of tickets being properly monitored by the government. This intervention will ensure tapping on the flow of tourists at that attraction and will help in balancing the demand. Adoption of Public-Private partnerships can also be adopted as in case of "Adopt a Heritage" initiative by the Indian government.

Activity

Anything that keeps a tourist engaged at a destination can be considered tourist activity. The set of activities are in support of the attractions. For example, a hill station may offer varied activities such as rock climbing, parachute ride, trekking, camping etc. However, some events and activities are attractions by themselves- for example, participation in the Tomatino festival or the Rio Carnival. During the 49-day Kumbh Fair, the city received 241 million visitors. Activities may not contribute directly to overtourism, but the management of visitors is required. Demarketing measures shall include channelizing the visitors' preference for activities in such a way that it shifts the tourists from honeypots. The Indian government has introduced new tourist circuits to shift the visitors' preference from the traditional travel routes. In Italy, Cinque Terre has put restrictions on the number of hikers visiting the place. For the same purpose, a hiking permit needs to be obtained by a visitor. Sites like AirTreks.com de-market a destination by advising travellers to avoid dates that coincide with activities like regional events and festivals. It suggests that monetary and nonmonetary prices go up during such events. Creative tourism products like theme walk also help disperse tourists.

Amenities

It refers to the basic services which are provided at a destination such as sanitation, public toilets, water, entertainment, public Wi-Fi, signage etc. These are ancillary services which smoothen a tourist's stay at a destination. In the situation of overtourism, the infrastructure and local resources get affected due to capacity constraint. Tourists need to understand that every destination has its own culture, and they should be sensitive towards the locals. Cities affected by overtourism finds it difficult to deal with tourists who litter, change clothes in public, drunkards who puke on streets, bargain and drink and drive. Fines should be imposed on irresponsible behaviours by visitors. Recently, Milan banned selfie sticks, and Hvar (Croatia) charges fine on drunken tourists who create a nuisance.

Using the Marketing Mix to De-Market Applied to Overtourism

Since demarketing is one of the strategies of marketing, the most optimum utilisation of its function is achieved when it is put together with a marketing framework. Demarketing measures can be employed in the marketing mix to manoeuvre the demand for a product or service strategically. The task of a marketer is to shape the demand in such a way that it conforms to the long-term objective of the organisation. The broadened concept of marketing in which the objective is not just to engineer an increase in sales but to adhere to certain goals is especially important while considering natural, scarce and perishable resources. The retention of natural resources which play a crucial role in an attraction of a destination, the sentiments of the local communities and balanced economic development are only the tip of the iceberg when the objectives are underlined for a tourist place.

Demarketing in marketing framework includes an alteration in the variables of the marketing mix to bring a shift in the demand by the following ways:

Product

The first component of the marketing mix is tourism product, which can be defined as any destination, attraction or service that satisfies the need for a tourist. The tourism products are both tangible and intangible and are often perishable, making it complex for the managers to formulate a uniform plan for the entire destination. Tourism has seasonality. If a place is hit with overtourism during the peak season, it returns dissatisfied visitors while alienating the locals. On the other hand, revenues of the destination dip during the lean period causing further losses for the stakeholders of the same destination. Therefore, to mitigate the ill effects of overtourism through demarketing, destination leadership may try to spread demand throughout the year. Demarketing of a tourism product includes measures through which the marketers can architect the product in a way that manipulates the demands. This includes discouraging the product or facilities for a specific group of customers which are undesirable, a substantial change in the product or by changing the quality of the product. For example, at Taj Mahal, a visitor is allowed a stay only for 3 hours. For a longer stay, an additional fee is charged. The tourism planners are also developing other spill-over destinations and accommodations outside the city, which they believe will lower the tourist burden of the city. For tourism, the attractions and experiences formulate the major part of the product line.

Price

Monetary sacrifice plays an important part while choosing any product. The monetary factors make up for a larger part in a customer's decision-making process. The motivation or desire to travel to a certain destination rendered baseless unless it is backed by the ability to purchase. The price of the product can be effectively manipulated, and a substantial control is in the hand of the supplier. The introduction of new prices or increasing and decreasing the prices for an existing product or the ancillary services related to it can alter the demand for a product. The change in the price of the substitute or complimentary service and good can further bring a change in demand for a product. The price in tourism can further be affected by discouraging or discontinuing of the discounts, introduction of differential pricing, and by devising a queuing system that increases the duration and opportunity costs. Destinations also bundle the entry tickets for multiple attractions, thereby pushing the guests to visit lesser-known attractions simultaneously reduce the pressure on popular sites. Implementation of demarketing includes increasing the prices of tickets to overcrowded monuments (E.g. Taj Mahal in India has increased the fee). This may also include discouraging the discounting and introducing differential pricing. Uffizi Gallery in Florence has introduced a system of charging higher prices during peak seasons as compared to the lean season. Similarly, Baleric Island in Spain has introduced a tariff (5 euros) on tourists staying in 5-star hotels during peak season. It has been mentioned under the category of tourists' taxes for infrastructure and maintenance.

Place

The component of 'Place' in the marketing mix can be understood in tourism as accessibility and availability of a destination or attraction to the tourists. Several steps in demarketing can be applied with the component of 'Place'. Introducing the booking system which helps in monitoring footsteps in advance, issuance of time-bounded entry tickets, in adherence to the carrying capacity of the location, especially for national parks an allocation system can be devised. Such allocation can be based on 'first come first served' with a cap on the entries, limiting the total number of accommodations in the destination, or restricting the total number of visitors could be some of the demarketing measures related to 'place'. The limitation on the number of visitors to a destination would help in discouraging demand and shift them to a lesser-known destination for the time being. For instance, Dubrovnik government has decided to put a limit on the number of visitors to the historic city and, police in Venice is ordered to close the pedestrian gates to the city centre in the situation of overcrowding. It is also important to have the necessary infrastructure in place to bring in visitors to a tourist district or an attraction. It is equally worthwhile to enable visitors to quickly get out of the crowded tourist places. If access to a tourist place is efficient, coupled with lower prices for well-connected accommodation, tourist is likely to opt for it. This tactic will help reduce crowding around the attraction. Many touristic destinations have made this arrangement. For example, New Delhi has a *Heritage Metro Line*, which connects many important heritage sites in Delhi. Similarly, in Rome and many other European cities, the metro connects important touristic sites.

Promotion

Promotion in the marketing mix refers to the function aimed at increasing awareness about the product, developing interest of the customers and generating sales while creating brand loyalty. Promotion is a tool catering not just to the profit or sales for an organisation, but it works on the long-term objectives. The target of promotions can be to increase sales, lift the brand image, spread the name, and change in numbers of the buyers or further to select the market segment. Promotions can differ according to the different stages in which an organisation is. While expanding, promotions can be vigorously used. In tourism, promotion can be characterised as the dissemination of information about the destination to the target audience. It generally takes place through roadshows, campaigns, social media, brochures or word of mouth. Implementation of demarketing at a destination may vary from a decrease in the promotional activities of the products in a market segment, stressing the restrictions related to a destination, informing about the difficulties inaccessibility of the place, promoting sustainable practices and disseminating information regarding the appropriate behaviour for the destination are few measures that can help control the flow of the demand. It is also important as to how a destination is positioned (e.g. you get cool European vibes when in Amsterdam and thus alcoholism, and bar culture is common and attracts young people) and promoted to the tourists. Authorities, using dispersed marketing, can exercise strict control and monitoring. For example, in Amsterdam, where a large number of visitors were flocking, the local authorities have decided to do away with the famous "I love Amsterdam" signature from the city centre.

People

People or participants are an important component of the service industry. Tourism is a people's industry. It is contact-intensive on many occasions along the visitor's value chain. Another important component of the 'people' element is the host community. Service providers and the local community are an important component of a visitor's experience. Without the involvement of people at origin and the hosting destinations, tourism cannot exist. Since a tourism experience is created by coming together of multiple service providers and stakeholders, even a single supplier not living up to the expectations would adversely affect a visitor's tourism experience. Overtourism is reported because "participants" (both residents and tourists) were getting affected with excessive tourism. Thus, dealing with the phenomenon will require participation from these people in the decision-making process and implementation of policies. In the time of crisis in Shimla due to water shortages, the local people of Shimla stood against the entry of tourists and the hotels utilising the scares resources. In a situation like this with informed service providers, better management and sustainable approach could have avoided a situation of such sorts. In the first place, the tourism service providers may be educated about the scarce resources at the destination and may be trained for efficient utilisation of the same. This is internal marketing. Subsequently, they may educate the visitors about the challenges that the destination faces. They may manipulate the supply of such scare resources.

Physical Evidence

The environment in which the services are consumed is known as the servicescape, which is composed of ambient characteristics like the architectural layout and its functionality. Tourism, like any other service industry, is largely intangible. The customers, however, respond to physical cues to assess the product before buying. Physical evidence helps in tangibilizing the intangible. This requires extending objectivity to the experience or services of the tourists through tangible pieces of evidence. Therefore, physical evidences formulate the material part of an experience or services. The composition of the service environment also acts as an influencer. This includes how well a place is equipped, the ambiance and the welcoming attitude of the people. The brochures, signage, uniforms, websites, signage, souvenirs all are efforts of tangibilizing the experience. For instance, demarketing in the Blue Mountains National Park in Australia also involved of steering of the tourists by the manipulation of the signage and symbols. The fragile area of the park was left out from the pamphlet, and the signage towards the zone was removed. There was proper signage indicating to the areas that the destination leadership wanted the tourists to go.

Process

A process refers to the varied steps involved in the final delivering of a product to a customer. Processes essentially have inputs, throughputs and outputs (or outcomes). In tourism, the product has several layers to form the final product for the customer. Since a web of industries come together to form the final experience for the guests, therefore, a process is the key component for effective management of visitor's experience at a destination. A process can make or break a destination. It is the process that can optimise the experiences of the tourists. Thus, the 'process' as an aspect of demarketing is a constituent mix of all the 8 Ps of the marketing mix. For example, India trained immigration officers to make the arrival process of the tourists smooth and swift. One of the most prominent examples is the Kumbh Mela organized in Prayagraj, India in 2019, which has been internally applauded for its management. It has been recorded in the Guinness World Records for three achievements. It includes the organisation of the biggest public sites painting exercise, the management plan of the largest traffic and crowd and for developing the biggest sanitation and waste disposal mechanism.

Partnership

Combating overtourism is a shared responsibility, and thus, the last "P" of marketing plays an important role. Partnership refers to the association between individuals or organisation for a common objective. Tourism comprises of several sub-sectors, wherein both the private sector and the government sector play an equally important role in developing and promoting a destination. Therefore, a partnership is required amongst government, residents, Destination Management Organisations (DMOs) and Destination Management Companies (DMCs) not only to cater to the tourists but also to make strategies in sustaining the destination. Any de-marketing strategy can only be implemented through partnering and cooperation among the stakeholders at the destination. Demarketing may require the government to partner with industry stakeholders for discussions and sharing database, which will help in finding a suitable solution for overtourism. For example, Europe has multiple low-cost airlines catering throughout the year. The increase in the number of airlines to Europe has further made accessibility easier in the region. A partnership with the airlines can lead to a spread in the footfall throughout the year. The issue of the seasonality of a destination can be countered through promotional activities provided by the airlines, railways, tour operators and other stakeholders of tourism to gain a mutual benefit from tourism. The prices of Euro Rails are slashed during the lean season to promote people to travel to Europe in those months.

Challenges of Demarketing on Overtourism

In the context of tourism, the focus of implementing demarketing is to get sustainable outcomes. It can also be applied to enhance the perceived quality of a destination. It can be effective if the objectives are clear, and a proper analysis has been conducted about the target market. Owing to its complex nature, demarketing strategy possesses certain limitations:

- **Doesn't Promise Sustainability**: In many cases, as discussed above, cities affected by overtourism have increased the prices to limit the demand. This implies that the economies are shifting from quantity-based to quality-based tourism. Based on "High-Quality Low-Impact" policy, Bhutan tourism for quite some time now, offer higher prices during peak season. But affluence doesn't guarantee sustainability (David Roberts, 2017). People who can afford the increase in costs can pay without giving a thought. They continue to believe their contribution is backing the sustenance of that destination no matter what their patterns of consumption are. Demarketing in such cases would then lead to discouraging the responsible travellers with limited income.
- **Loss of Employment for Locals**: An intentional reduction in the demand by controlling supply would encourage tourists to go to other places. Dispersing tourists to other places would mean dispersing the source of income of residents. This reduction in demand would result in loss of business for residents who are engaged directly or indirectly in any tourism or related activity at a destination. This perceived loss is often the reason while sections of the local community often tend to ignore overtourism.
- **Short Term Strategy**: Demarketing is often a short-term strategy. It is effective in the early stages of overtourism to shape the demand for a destination. Once the balance is achieved, destination leadership should have a suitable long-term plan for managing the destination and flow of visitors. Ensuring a responsible attitude amongst tourists and promoting responsible tourism is what can sustain a destination for a long time.
- **Resistance to Change**: Demarketing is likely to cause inconvenience in the initial stages of its implementation. Tourists would resist paying extra or might get upset being not able to visit a popular destination in their bucket list. Local service providers would resist a reduction in demand. Handling resistance to change amongst tourists, tour operators, DMCs, etc. is a major and cumbersome task.

CONCLUSION

The negative impacts of tourism can be seen when the host community fears and socially rejects tourist activities. The increase in anti-tourist movements in places like Venice, Shimla, Berlin, Dubrovnik, and Amsterdam have brought the adverse aspect of tourism to the forefront. Tourism is a major economic contributor to several economies in the world and is growing exponentially. The challenge rises as short term monetary gains often blind the desire for sustainable development. Overtourism refers to a state where "too many" tourists visit a particular destination and consequentially affect the quality of life of the local communities and the tourist experiences. In simple terms, overtourism transpires when the number of tourists is much more than the capacity of a destination to withstand. Overtourism has its roots in the short-term approach of the policymakers. All the stakeholders need to carry a vision of sustainability and overall development of a destination to foster balanced growth.

As discussed in the chapter, marketing is indispensable to carrying forward of any product. Marketing is the function of the business, which is broadly focused on managing the level of demand and its composition for the company. Demarketing, an intrinsic facet of marketing can be used as a tool to curb the issue of overtourism. The framework of demarketing when used in the 5 A's of tourism namely Accessibility, Accommodation, Attractions, Activities and Ancillary services can provide with strategies of managing the flow of tourist movement and to organize the demand of a place.

Since demarketing is one of the components of marketing its optimum utilisation is when it is put together in a marketing framework. The measures of demarketing can be used in the marketing mix to manage the demand of a tourist destination. One of the most crucial and alarming concerns for the destination managers is a sustainable growth of a destination that is important for all the stakeholders of tourism. Demarketing tool when used under the framework of marketing mix can yield efficient operational measures to cure or prevent overcrowding of a destination. The use of demarketing in the 8 P's of marketing mix can impact the flow of demand and market number of a destination which in turn can bring the desired changes in the class of tourism of a destination. However, there are several challenges in implementing a single strategy in tourism since it is an amalgamation of several industries and a uniform approach is difficult to achieve. Each destination needs to adopt strategies which suit their dynamic trends of tourism.

REFERENCES

Armstrong, E. K., & Kern, C. L. (2011). Demarketing manages visitor demand in the Blue Mountains National Park. *Journal of Ecotourism*, *10*(1), 21–37. doi:10.1080/14724040903427393

Barata Salguiero, T., Mendes, L., & Guimarães, P. (2017). Tourism and Urban changes: Lessons from Lisbon. In *M. Gravari-Barbas, & S. Guinand, Tourism and Gentrification in contemporary monopolises* (pp. 255–275). London, UK: Routledge. doi:10.4324/9781315629759-11

Burgen, S. (2018, June 25). *Tourists go home, refugees welcome': why Barcelona chose migrants over visitors*. Retrieved from The Guardian: https://www.theguardian.com/cities/2018/jun/25/tourists-go-home-refugees-welcome-why-barcelona-chose-migrants-over-visitors

Butler, R. (1980). The concept of a tourist area cycle of evolution: Implications for management of resources. *The Canadian Geographer. Geographe Canadien*, *24*(1), 5–12. doi:10.1111/j.1541-0064.1980.tb00970.x

Capocchi, A., Vallone, C., Pierotti, M., Amaduzzi, A., & Capocchi. (2019). Overtourism: A literature review to assess implications and future perspectives. *Sustainability*, *11*(12), 3303. doi:10.3390u11123303

Chaudhry, P., Cesareo, L., & Pastore, A. (2019). Resolving the jeopardies of consumer demand: Revisiting demarketing concepts. *Business Horizons*, *65*(5), 663–677. doi:10.1016/j.bushor.2019.05.002

Cullwick, D. (1975). Positioning Demarketing Strategy: Marketers must integrate the changes in the business environment into effective new product, pricing, distribution, and promotion strategies. *Journal of Marketing*, *39*(2), 51–57.

Doxey, G. (1975). A causation theory of visitor-resident irritants, methodology and research inferences. The Impact of Tourism. In *Sixth Annual Conference Proceeding of the Travel Research Association*, (pp. 195-198). San Diego, CA.

Editors, C. N. T. (2018, October 30). *Parts of the world we love are straining under too much travel. Some aren't handling it well. But among others, there are signs of hope.* Retrieved from Condé Nast Traveler: https://www.cntraveler.com/story/loving-the-world-to-death-the-good-bad-and-ugly-of-overtourism

Farias, S. A. (2010). Internal Marketing (IM): A literature review and research propositions for service excellence. *Brazilian Business Review*, *7*(2), 99–115. doi:10.15728/bbr.2010.7.2.6

Forster, J. (1964). The Sociological Consequences of Tourism. *International Journal of Comparative Sociology*, *5*(2), 217–227. doi:10.1177/002071526400500208

Glaesser, D., Kester, J., Paulose, H., & Abbas, A. (2017, April). Global Travel Patterns: An overview. *Journal of Travel Medicine*, *24*(4), 1–5. doi:10.1093/jtm/tax007 PMID:28637267

Goodwin, H. (2017, October). The challenge of overtourism. *Responsible Tourism Partnership*. Retrieved from https://haroldgoodwin.info/pubs/RTP'WP4Overtourism01'2017.pdf

Joppe, M. (2019). The roles of policy, planning and governance in preventing and managing overtourism. In R. Dodds, & R. Butler (Eds.), *Overtourism: Issues, Realities, and Solutions* (pp. 250–261). Berlin, Germany: De Gruyter. doi:10.1515/9783110607369-017

Kharas, H. (2017, February). *The Unprecedented Expansion of the Global Middle Class: An Update.* Global Economy and Development. Washington, DC: The Brookings Institution. Retrieved from https://think-asia.org/handle/11540/7251

Kotler, P. (2000). *Marketing Management: The Millennium Edition*. Upper Saddle River, NJ: Prentice-Hall.

Kotler, P., Armstrong, G., Saunders, J., & Wong, V. (2001). *Principles of Marketing*. Upper Saddle River, NJ: Prentice Hall.

Kotler, P., & Levy, S. (1971, November-December). Demarketing- Yes, Demarketing. *Harvard Business Review*, pp. 74-80.

Kumar, G., & Srivastav, A. (2019). Sustainable Tourism Marketing in Andaman and Nicobar Islands. *International Journal of Scientific Research and Review*, *7*(3), 663–671.

Little, V., Lee, C., & Nair, S. (2019). Macro-demarketing: The Key to unlocking unsustainable production and consumption systems? *Journal of Macromarketing*, *39*(2), 166–187. doi:10.1177/0276146718823885

McKinsey & Co., & WTTC. (2017, December). *Coping with success: Managing Overcrowding in tourism destinations.* London, UK: McKinsey & Company, & World Travel & Tourism Council. Retrieved from World's Travel and Tourism Council website.

Miklós-Thal, J., & Zhang, J. (2011, June 29). Strategic Demarketing. *MIT Sloan School Working Paper 4814-10*, 1-52.

Peattie, K., Peattie, S., & Newcombe, R. (2016). Unintended consequences in demarketing antisocial behaviour: Project Bernie. *Journal of Marketing Management, 32*(17-18), 1588–1618. doi:10.1080/0267257X.2016.1244556

Poll, B. (2019, July 25). *What we want to do to help combat overtourism*. Retrieved from Spotted by Locals: www.spottedbylocals.com/blog/overtourism

Roberts, D. (2017, December 1). *Wealthier people produce more carbon pollution — even the "green" ones*. (Vox, Producer, & Vox Media, Inc.) Retrieved from Vox: https://www.vox.com/energy-and-environment/2017/12/1/16718844/green-consumers-climate-change

Rosen, E. (2018, September 6). *New Rankings of The World's Fastest-Growing Tourism Destinations*. (Forbes Media LLC.) Retrieved from Forbes: https://www.forbes.com/sites/ericrosen/2018/09/06/new-rankings-of-the-worlds-fastest-growing-tourism-destinations/#725d17f457ea

Sakshi, P. (2017). Do social media effect tourists' behaviour? *International Journal of Advance Research and Innovative Ideas in Education, 3*(4), 38-45.

Sánchez-Quiles, D., & Tovar-Sánchez, A. (2015, October). Are sunscreens a new environmental risk associated to coastal tourism? *Environment International, 83*, 158–170. doi:10.1016/j.envint.2015.06.007 PMID:26142925

Sans, A. A., & Quaglieri, A. (2016). Unravelling Airbnb: Urban Perspectives from Barcelona. In *G. R. A. P. Russo (Ed.), Reinventing the local in tourism: Producing, consuming and negotiating place* (pp. 209–228). New York: Channel View Publications.

Smith, M., Sziva, I., & Olt, G. (2019). Overtourism and Resident Resistance in Budapest. *Tourism Planning & Development, 16*(4), 376–392. doi:10.1080/21568316.2019.1595705

Soliman, D. (2010). Managing Visitors Via Demarketing in The Egyptian World Heritage Site: Giza Pyramids. *Journal of Association of Arab Universities for Tourism and Hospitality, 7*(1), 15–20.

Swarbooke, J. (1999). *Sustainable Tourism Management*. London, UK: CABI. doi:10.1079/9780851993140.0000

UNWTO. (1981). *Saturation of Tourist Destinations*. Madrid, Spain: General Secretary, World Tourism Organisation.

UNWTO Communications Department. (2019, January 21). *International Tourist Arrivals Reach 1.4 billion Two Years Ahead of Forecasts*. Retrieved from UNWTO: https://www2.unwto.org/press-release

Wagar, J. A. (1964). The Carrying Capacity of wild lands for recreation. *Forest Science*, 1–24.

Wearing, S., Schweinsberg, S., & Tower, J. (2016). *Marketing national parks for sustainable tourism*. Channel View Publications. doi:10.21832/9781845415594

Weber, F., Stettler, J., Priskin, J., Rosenberg-Taufer, B., Ponnapureddy, S., Camp, M.-A., & Bath, M. (2017). *Tourism destinations under pressure: Challenges and Innovative solutions*. Institute of Tourism ITW, Lucerne University of Applied Sciences and Arts, Roesslimatte.

WTTC. (2019, March 20). *One in five Thai baht spent is in Travel & Tourism, according to new WTTC research.* Retrieved from World Travel & Tourism Council: https://www.wttc.org/about/media-centre/press-releases/press-releases/2019/one-in-five-thai-baht-spent-is-in-travel-and-tourism-according-to-new-wttc-research/

KEY TERMS AND DEFINITIONS

Carrying Capacity: The maximum number of footfall a destination can accommodate at a certain period of time, without causing ecological, economical, socio-cultural and physical damage.

Demand: It refers to the requirement of a product. In tourism, demand is the total number of people who desire to travel to or travels to a particular tourist place.

Demarketing: The use of marketing techniques to decrease the demand for a product to meet the long-term objectives of an organization.

Marketing Mix: A coalescence of factors that can be manoeuvred to influence consumer demand and purchase patterns.

Overcrowding: It refers to a permanent or temporary state where there are a larger number of people present in a location than the perceived tolerable level from a safety and security purview.

Stakeholders: Any entity that influences or is influenced by tourism. They can be tour operators, government, media and others.

Sustainability: The ability to live without damaging the natural resources and maintaining the ecological balance. Sustainability means maintain of quality life.

Tourismphobia: The term "tourismphobia" has its inception from mass unsustainable tourism practices which generated the feelings of fear, antipathy, and social rejection amongst the local citizens of the destination towards the tourists whose exploitation is believed to cause the decrease in quality of life.

Section 2

Case Studies: Urban Areas

Chapter 7
Overtourism in Megalopolises:
The Case of St. Petersburg

Valery Gordin
https://orcid.org/0000-0002-6079-4960
National Research University Higher School of Economics, Russia

Irina Borovskaia
https://orcid.org/0000-0002-7869-0684
National Research University Higher School of Economics, Russia

Ekaterina Fedorova
National Research University Higher School of Economics, Russia

ABSTRACT

This chapter discusses manifestations of overtourism in the megapolis of St. Petersburg, which is a major cultural and tourist center. The idiosyncratic nature of overtourism in St. Petersburg is due to the fact that serving as its basis are spatial and temporal factors, which cause the city center to be overcrowded during high season. The chapter reviews the major methods of preventing and overcoming overtourism that are being employed in 18 megapolises that are largest tourist centers, and, based on results of the expert questionnaire, the authors identify methods that would be most suitable for St. Petersburg.

INTRODUCTION

Overtourism is a phenomenon that has in recent years affected not only small and average size cities but also megapolises where the consequences of overtourism are becoming palpable more and more frequently, including the overcrowding of transportation systems, city infrastructure, hotels and fast food places, as well as the overcrowding of cultural attractions, the growth of rent prices, the increase in the noise levels and the increase in ecological pollution levels in cities caused by large numbers of tourists.

DOI: 10.4018/978-1-7998-2224-0.ch007

St. Petersburg is one of the largest European cities: it holds 4th place in terms of the population number (5.4 million of people) and third place in terms of the area (1.4 km^2). At the same time, the city is a major tourist center, which more than 8 million tourists visited in 2018. Despite the city's significant size, at present, certain symptoms of overtourism have already surfaced, which within the next few years may generate acute social and economic conflicts. However, St. Petersburg authorities continue to strive to increase tourist numbers and do not fully appreciate the potential risks associated with newly arising tourism issues in the city.

In the present chapter, the authors explore various reasons why overtourism arises as well as its manifestations in a megapolis such as St. Petersburg. The chapter provides supporting arguments for the claim that, despite significant territorial, human and technological resources, even currently St. Petersburg is facing the problem of overtourism as a result of spatial and temporal idiosyncrasies of tourism development in the city. The spatial idiosyncrasies are due to the fact that most of the tourist infrastructure is located in the city center, which is characterized by a high density of residential buildings and a road network. What exacerbates the problem of overtourism is that, while the dominant type of tourism in St. Petersburg is cultural tourism, the main objects of tourist visitations are 5-7 of the most famous museums and palace ensembles. The second group of factors engendering overtourism issues in St. Petersburg includes temporal ones. The issue is that, due to the city's northern geographical location, city tourism exhibits acute seasonality. The number of tourists who visit the city during high season between April and September approximately threefold exceeds that during low season. As a result, during the summer months, the city is overcrowded by tourists, who create problems for local residents not only in terms of visiting the leading museums, but, most importantly, in terms of overburdening the city infrastructure, overcrowding city cafés and restaurants, and often by making noise during nighttime.

Furthermore, in the present chapter, the authors review and systematize the experience of other megapolises which are the large tourism centers and have high number of museums in a city, which enabled the authors to discuss the level of development of cultural tourism. In this way, 18 megapolises were selected. Based on analyzing data from scientific sources, media publications as well as information on official and unofficial city sites, 25 methods used for containing and regulating excessive tourist flows in megapolises were identified. In the next section of the chapter, the level of acuteness of various aspects of overtourism in St. Petersburg is determined based on the expert questionnaire; the value of using various methods of preventing and overcoming overtourism in the city is assessed.

In the concluding section of the chapter, the authors accomplish the main goal of the current research study: to propose a system of methods aimed at preventing and overcoming overtourism in St. Petersburg that takes into account the goals of the city authorities, tourist business, and the interests of local residents. The tourism planning based on these methods is a way to sustainable tourism development.

BACKGROUND

As of today, the overtourism phenomenon is being tackled in official documents as well as in research studies. Thus UNWTO provides the following definition of overtourism, "the impact of tourism on a destination, or parts thereof, that excessively influences perceived quality of life of citizens and/or quality of visitors experiences in a negative way" (World Tourism Organization [UNWTO], Centre of Expertise Leisure, Tourism & Hospitality, NHTV Breda University of Applied Sciences & NHL Stenden University of Applied Sciences, 2018, p.4). In today's social context, overtourism is defined, first and foremost, by protests by the local population at the destination against the uncontrollable growth of the number of tourists (Goodwin, 2017).

Due to the problematic nature of the overtourism phenomenon, it has been studied extensively by scientists, including the notion of overtourism itself as well as its sources and criteria for identifying the presence of the phenomenon in question in a given destination and its impact on the city, natural and social environments; methods and tools for reining in the enormous tourist flow have been identified as well.

The problem of tourism development bringing about negative consequences for the destination has become an object of research over fifty years ago (Foster, 1964; Wagar, 1964). Based on this approach, Doxey developed a so-called "irritant index" that reflects the locals' attitude toward tourists and toward developing tourism during different stages of the destination's life cycle (Doxey, 1975). Doxey's research was further developed by Butler, who proposed a model of the destination life cycle according to which a destination undergoes five stages of development, provided a free market and a stable demand: exploration, involvement, development, consolidation, and stagnation (Butler, 1980). According to this classification, during the consolidation stage, the number of tourists exceeds that of the local population; the ecological, social and economic pressure on the destination causes the destination's tourism carrying capacity to be exceeded, and in local social circles opponents of tourism gain momentum, blaming all the problems of the destination on tourism. A number of research studies is devoted to exploring the residents' reaction to the growing numbers of tourists during pique seasons and to the development of the "tourist-phobia" phenomenon (Alvarez-Sousa, 2018; Muler Gonzalez, Coromina & Gali, 2018; Cheer, Milano, Novelli, 2018).

Serving as the continuation of the discussion on the negative consequences of developing tourism was a number of research studies that justified the notion of the ultimate tourist carrying capacity of the destination (Mathieson & Wall, 1983; McIntyre, 1993; Inskeep, 1991; Pigram & Wahab, 1997). This concept does not work for megapolises due to their infrastructural features, but becomes a base for some regulation instruments.

It needs to be noted that in recent years researchers' interest in overtourism issues has sharply increased, since many of the most popular tourist destinations have reached the stage of the life cycle during which tourists' interests start to conflict with those of local residents, and, in general, tourists' interests start to affect the stable development of the destination. According to the ITB World Travel Trends Report (2018), every tenth trip has been characterized by the presence of tourist hordes during the course of travel, which testifies to the pervasive nature of overtourism. In Research for TRAN Committee (2018), visually comprehensive maps indicating to what degree overtourism is present as well as forty-one cases of tourist destinations of various types are reviewed in which local residents and the administration have encountered the phenomenon in question.

Since St. Petersburg is a major cultural center, from the current study's vantage point, the problem of cultural overtourism is of most interest. Cultural tourism nowadays constitutes about 40% of tourism worldwide and is concentrated, for the most part, in historic city centers. It is the authentic culture of cities around the world that has become a major tourist attraction, and, as a result, has become a factor in city tourism development (Richards, 2011). Using an example of five cultural destinations, Richards and Marques (2018) have demonstrated the connection between culture development and tourism, given dynamic city development.

It needs to be noted that overtourism affects cities with a great variety of territorial scales and local population numbers. Thus, in central Berlin neighborhoods, there is an ongoing debate concerning instituting a ban on creating new housing for tourists in residential neighborhoods (Novy, 2011). An unprecedented growth of housing rental prices in Amsterdam caused by the increase in the tourist flow in the city has created many issues for the locals (Dredge, Gyimóthy, Birkbak, Jensen & Madsen, 2016). And tourism development in London was placed under the purview of municipal authorities, which, in close cooperation with the local population and the tourist business, promote the policy of sustainable tourism development (Maxim, 2017). The city of Prague presents an interesting case where, in the locals' opinion, the process of touristification has stripped the historical city center of its former authentic character and has transformed it into an entertainment center (Dumbrovská, 2017). Using the case of Rome, Agostinelli (2018) demonstrated the possibility of diversifying the tourist appeal of a destination by employing the marketing tool of storytelling about Rome and the cinema. Nowadays many publications are devoted to the topic of overtourism in Barcelona, where the general situation is becoming more acute with every passing year. The constant growth of tourist arrivals, the "massification" and "standardization" of tourism had caused the local residents' quality of life to decrease, which, in turn, caused Barcelona residents' level of tolerance toward tourists to lower dramatically (Goodwin, 2018). A research study probing tourists' and local residents' opinions on the status of tourism in the megapolis in question and researching media coverage of the issue have shown that tourism in Barcelona has engendered a two-sided social conflict (Alvarez-Sousa, 2018). For this reason, Barcelona authorities have sounded an alarm and currently advocate instituting a number of measures that would rein in the tourist flow (Milano, 2017). In certain residential neighborhoods of Budapest, the presence of large numbers of bars, clubs and restaurants has created a night-time economy; these street establishments are a source of noise and are overcrowded at night, which inconveniences local residents (Pinke-Sziva, Smith, Olt & Berezvai, 2019).

The interweaving of temporary tourists with everyday urban life takes place under the pressure of urbanization. On the one hand, this symbiosis creates the conditions for satisfying the tourists needs in terms of the "real" and "authentic" everyday city life (Pappalepore, Maitland, & Smith, 2014), and on the other, it infringes the boundaries of the residents personal space (Freytag & Bauder, 2018). The result of the conflict of interests takes the form of mass protests as in the case with Barcelona (Koens, Postma & Papp, 2018).

As a conflict solution, researchers offer to use the sustainable urban planning mechanisms that are based on the principles of communication between key stakeholders (Maxim, 2019). When planning a tourist area, it is necessary to take into account the flexibility aspect: the flexibility of culture in the area of study should ergonomically fit into the urban context (De Frantz, 2018). In the modern realities of planning tourist cities well-development infrastructure and mature urban governance are the main tools for adapting the incoming mass flow of tourists (Sheng, Li, & Wang, 2017). Sustainable development requires examination, special policies and planning for all types of tourism at all levels so as to reap the economic, social, and environmental beneðts of tourism and enhance visitor satisfaction (Gunn & Var, 2002).

A special place among scientific publications on overtourism belongs to research on employing effective measures to regulate overtourism in various specific tourist destinations, including city cultural heritage destinations. In their research that employed European tourist cities as case studies, Koens, Postma and Papp (2018), have demonstrated that major manifestations of overtourism are overcrowding, physical touristification, pressure on the local environment, and residents being pushed out of residential areas.

Based on analyzing the generalized experience of eleven tourist destination cases from all over the world, the research team from Lucerne has formulated principles that would promote successful governing of a territory that is subject to overtourism (Weber et al., 2017). The research team has identified the following effective methods of overcoming overtourism: stakeholder involvement in managing tourism, utilizing a variety of regulation tools at different destinations, ensuring a constant dialogue between representatives of the tourism industry and authorities with the goal of reining in overtourism, and taking an active stance in coming up with creative solutions for managing the tourist flow. European Cities Marketing have suggested ten tools aimed at managing the growing tourist flows in European cities, which may be used to a certain degree by DMO or city marketing agencies: strategy creation and territory planning; forming partnerships; smart marketing; on-the-ground visitor management; technological solutions; public education; managing the collaborative economy; taxes and limitations; measurement and monitoring; dialogue and consultation (European Cities Marketing, 2018). Postma and Schmuecker (2017) suggested a framework for developing strategies in order to decrease the negative impact of tourism. They determined "cultural distance" and spatial and temporal distribution as the residents' irritation drivers and, based on potential areas of conflict between locals and tourists, proposed ten measures that can be undertaken in order to prevent a negative impact of overtourism. Joppe (2019) argued that solving overtourism challenges can be accomplished through creating a strategy that would be based on taking into account interests of the local population and of all the stakeholders and using a horizontal and vertical governance model that is socially and environmentally oriented. However, an equally important stage in developing the strategy in question is favoring niche types of tourism that prove to be useful and promising for the city over mass tourism as well as developing niche tourism based on sustainability principles (Benner, 2019). It is the segmentation of the tourist demand that has the potential to divert tourists' attention and free up the most popular attractions. The implementation of smart technologies in the age of innovation among other areas takes place in the field of regulating the tourist flow (Skeli & Schmid, 2019). For example, in European destinations, formerly popular city visitor maps are being replaced by mobile applications that can guide tourists to less overcrowded routes.

MAIN FOCUS OF THE CHAPTER

For several decades now a stable growth of the numbers of tourists in St. Petersburg having been observed. In 2018, in St. Petersburg, the city was visited by 8.2 million people, which includes 3.9 million of foreign tourists and 4.3 million of Russian tourists. In terms of the tourist to local ratio, St. Petersburg has approached Singapore and Rome ("Which City's Populations", 2019). At the same time, in fairly large cities, such as Paris and Barcelona, the ratio in question is even higher. In recent years the tourist flow has been increasingly growing in St. Petersburg; during this time period, the city has been plagued with a number of issues stemming from the pressure of the tourist flow on the transport infrastructure and the fast food system; the tourist flow has negatively affected city center residents' life conditions and the functioning of cultural institutions that service tourists and residents alike. These problems have arisen due to a host of issues with how the tourism industry is organized; these issues are both temporal and spatial in nature.

First and foremost, it is necessary to note that tourism in St. Petersburg is acutely seasonal in nature: during the high season that lasts from April until September, there are 100,000 tourists in the city at any given time; during the low season, there are up to 20,000 tourists (these data were obtained by the authors based on officially available statistics). Moreover, the bulk of the tourist population resides, purchases food, does shopping, visits museums, theaters and other cultural institutions mostly in the historical downtown. The main issues that St. Petersburg faces are the high concentration of tourist flows in a relatively small historical downtown area, which is about 4,0 hectares (3% the total city's area), and the fact that currently a very limited number of tourist attractions is in high demand.

This state of affairs is something of a conundrum, since St. Petersburg is rightfully dubbed "the cultural capital of Russia." The city boasts about 206 museums and 97 theaters. 36 cultural-historical complexes in the city proper include more than 4000 architectural, cultural, and historical artifacts, which are included in UNESCO's World Heritage List. However, currently, the bulk of the tourist population strives to visit no more than 5-7 major places of interest that are St. Petersburg's business card; first and foremost, there are Peterhof's museums (5,300,000 visitors), State Hermitage (4,220,133 visitors), Tsarskoye Selo's museums (3,900,000 visitors), and St. Isaac's Cathedral and Church of the Savior on Spilled Blood (3,814,000 visitors).

When considering the territorial aspect of overtourism in St. Petersburg, it is necessary to note that the issue is not limited to tourists' "obsession" with visiting a narrow range of museums located either downtown or in remote suburbs. Up to the present day, downtown St. Petersburg is characterized by streets that are narrow by modern city standards and by very densely spaced low-rise buildings. Keeping intact the city blocks that were built a few centuries ago imbues the city of St. Petersburg with an aura of uniqueness and authenticity. However, it is important to make three points about the historic neighborhoods that have practically not been renovated during the past century; firstly, around a million residents permanently live in these neighborhoods; secondly, a significant number of company offices, city government offices, and countless educational institutions are located there; thirdly, many services that are popular with residents (e.g., stores, restaurants, and movie theaters) are located there as well.

As a result, the majority of tourist infrastructure objects are located in residential buildings in historical city neighborhoods. City center neighborhoods face acute challenges, as evidenced by the following data: 55% of all city hotels and hostels are located in two central neighborhoods (Popova, 2016).

St. Petersburg residents rightfully resent the fact that mini-hotels, hostels and other hotel infrastructure offices are located in residential housing. The same attitude to this issue is being voiced by other Russians as well: according to results of a national sociological study, 70% of respondents do not want to see hostels and mini-hotels in residential apartment buildings (All-Russian Center for Researching General Public's Opinion, 2019). It is as a result of significant pressure from residents of multi-apartment buildings that amendments to the Housing Code of the Russian Federation were ratified that limit the housing of companies that offer lodging rental services in residential apartments as of November 1st, 2019. Now hotels and hostels may be located only in buildings that have the status of being non-residential, have a separate entrance, and, most importantly, two thirds of tenants of a given building must provide their consent to housing a hotel in the building (The State Duma of the Russian Federation, 2019). The law provides legal grounds for "evicting" a great many mini-hotels and hostels from residential housing, which has caused concern in the hotel industry of the city as well as nationally. This example illustrates how the local population's interests collide with those of the tourist industry whose expansion may have a significant negative impact on city residents' life conditions. Equally, severe clashes of interests characterize the burgeoning restaurant industry. Thus, whole restaurant clusters have come into being in the city center on Rubinstein St. (7 restaurants have opened in a 100 m^2 space), on Nevsky Prospekt (4 restaurants per 100 m^2 throughout), and on Sadovoja St. (3.5 restaurants per 100 m^2). In recent years, residents of buildings on Rubinstein St. have actively campaigned against housing bars and restaurants on the ground floors of residential buildings; a great deal of noise emanating from these establishments in the evening and at night proves bothersome for the building tenants.

To summarize, in St. Petersburg territorial issues with housing the majority of tourist and hospitality companies engender acute social collisions; a solution needs to be found as soon as possible in order to curtail the increasing social tension between tourists and residents. In addition, this solution requires new approaches to city planning and urban development, such as tourist attractions and accompanying infrastructure location, city territory renovation, non-controversial usage of cultural, catering, recreational facilities by tourists and locals, so that to prevent "tourist ghetto" creation. Sustainable tourism development should include the development of infrastructure facilities generating income for tourism industry, but not complicating residents' daily life, in order to provide positive residential attitudes and to create tourism products which give no negative impressions.

In addition to the impact of the territorial aspects of overtourism on St. Petersburg, the temporal factor has an impact as well. The issue is that, even though, by and large, city tourism is less susceptible to the influence of seasons than other types of tourism, the major nature-related brand of St. Petersburg is white nights (dusk lasts almost the entire night). It is this nature-related brand that attracts the majority of tourists. The city actively capitalizes on this brand: "White Nights" and "Stars of the White Nights" festivals are being held; it is during the white nights season that the most highly attended cultural event takes place: a celebration devoted to graduates of the "Scarlet Sails" schools, which attract over 1,200,000 guests including 100,000 tourists (Kommersant, 2018). During these months the most highly attended event in the field of business tourism takes place, the St. Petersburg International Economic Forum, which had about 20,000 participants in 2019, most of whom were tourists (RIA novosti, 2019). This seasonality negatively affects the functioning of city institutions involved in servicing tourists. Thus in 2017, in St. Petersburg hotels in January through March and in November through December 44%-60% of rooms were booked; however, in April through October 64%-87% (Hotel Advisors, 2018) of rooms were booked (data for 2018 are irrelevant due to FIFA World Cup).

It is necessary to note that a number of tourism types are currently burgeoning in St. Petersburg; first and foremost, it is cruise tourism. In 2018, 620,000 cruise tourists visited St. Petersburg; moreover, the growth rate of cruise tourism is 15% per year (The official site of St. Petersburg Authorities, 2018). Even though cruise tourists do not need to be accommodated in city hotels, their presence creates a significant strain on cultural institutions and many other elements of tourist infrastructure. And this causes the decrease of touristic experience' quality, as they lose their time in queues, visit overcrowded sites, miss theatre tickets, etc. Thus, not only locals but tourists themselves are suffering from overtourism impacts as well.

The analysis of the overtourism situation in the city has enabled the authors to propose a hypothesis to the effect that a megapolis such as St. Petersburg has a great deal of resources that can be employed in order to prevent and overcome overtourism; the city administration needs to develop and enact a goal-oriented policy on tourism. The cornerstone of the policy in question should be methods for overcoming overtourism that are relevant, including those that have been tested in other tourist destinations that are similar to St. Petersburg in terms of their population numbers, territory, and numbers of visiting tourists.

METHODS AND RESULTS

The goal of the first stage of research study is to identify various practices employed in megapolises geared toward developing cultural tourism and controlling exceedingly large tourist flows; these experiences of dealing with overtourism can be used by St. Petersburg authorities with the goal of fostering robust tourism development. In the second stage of this study, an expert survey was conducted in order to identify methods of containing and overcoming overtourism that would be most appropriate for St. Petersburg among those used in other millionaire cities.

During the first stage, a selection of urban cultural tourist destinations that currently face overtourism problems was conducted based on data from the Euromonitor International Top 100 City Destinations 2018. For the analysis were chosen millionaire-cities (population over 1,000,000 people) with tourist arrivals exceeding 1,000,000 per annum, with the ratio of tourists to locals as 150% and higher, and with developed cultural tourism, indicated with 50 or more museums existed. 18 city destinations were included in the final selection.

As sources of information on methods of regulating tourist flows, both scientific publications and a wide spectrum of media publications that discussed employing such measures were used; also, information from official and unofficial destination sites was used. A total of 25 methods of containing and regulating excessive tourist flows in megapolises were identified, which can be classified into eight groups.

As it follows from the data in the table, each method group contains methods employed in the majority of 18 megapolises that have been researched. This testifies to the fact that each method group represents a given direction of administrative action aimed at preventing and overcoming overtourism. Moreover, methods that are widely used constitute a universal toolkit that is suitable for the vast majority of megapolises confronting overtourism. Included in this toolkit are pricing regulation methods of controlling tourist flows, instituting a tax for housing lessors, large-scale development of informational technologies, creating and promoting new tourist attractions, especially by means of creating new tourist hubs in areas that are being renovated, and also by actively engaging tourists into non-museum related types of cultural consumption. At the same time, each method group contains a number of methods, which are employed for the following reasons: they may be employed due to a given megapolis being at a certain stage of the tourist destination life cycle that causes an acute overtourism problem in the megapolis; these methods may also be employed due to idiosyncrasies of city development: first and foremost, territorial, and, in part, social city development, which necessitate using specific methods of preventing and overcoming overtourism. These methods include regulating the functioning of the sharing economy system in the real estate field, bans on erecting various types of tourist infrastructure in the city center, and severe restrictions on tourists' inappropriate behavior.

Some methods aim to defuse tensions between locals and tourists, while others tend to increase the quality of the touristic experience. Renovation and expansion of public spaces, tax and price regulation and other planning and specific policy measures are based on sustainable tourism development approach.

Table 1. The final list of surveyed cities

N°	City	N°	City	N°	City
1.	London	2.	Paris	3.	New York
4.	Rome	5.	Prague	6.	Barcelona
7.	Milan	8.	Vienna	9.	Berlin
10.	Madrid	11.	Dublin	12.	Moscow
13.	Athens	14.	Munich	15.	Budapest
16.	St. Petersburg	17.	Kyoto	18.	Warsaw

Table 2. Methods for preventing and overcoming overtourism

Method Groups	Applied Methods	Percentage of megapolises that employ a given method
G1. Pricing regulation	1. Price increases for visits to tourist destinations 2. A differentiated pricing policy determining museum and cultural institution ticket prices based on various factors (seasonality, visitation time, citizenship, etc.)	94% 100%
G 2. Transport infrastructure regulation	3. Limiting transportation access (private and tourist bus) into downtown 4. Reducing the number of free parking spaces / creating a large number of paid parking spaces downtown	28% 61%
G 3. Regulating the sharing economy system	5. Creating a registration system for tourists who find housing through Airbnb and other housing booking services 6. Reducing the duration of the long-term rent time period	11% 33%
G 4. Legal and tax regulation	7. A ban / limitations on new hotel construction downtown 8. A ban / limitations on opening new souvenir and tourist shops 9. Instituting fines for unlawful behavior and public order disturbances 10. A tax for housing lessors (including those who rent housing through Airbnb). 11. A tourist tax	11% 6% 28% 100% 61%
G 5. Development and use of informational technologies	12. Instituting a booking system through tourist services 13. Creating local apps for tourists 14. Creating ways to track the tourist flow and waiting times online / in real time 15. Instituting a mobile registration system	100% 94% 22% 83%
G 6. Creating a comfortable atmosphere and keeping things manageable in the tourist downtown	16. Having longer operating hours during the tourist season 17. Instituting a limit on the number of daily arrivals / museum visitations 18. Renovation and expansion of public spaces 19. Recruiting volunteers for consulting tourists	50% 17% 100% 17%
G 7. Diverting tourists' attention to alternative destinations and attractions	20. Branding suburban areas and territories under renovation 21. Creating new cultural hubs in non-tourist neighborhoods 22. Developing new routes and promoting new places of interest 23. Engaging tourists into non-museum related types of cultural consumption (theaters, circuses and other cultural institutions)	78% 89% 94% 100%
G 8. Regulating and diminishing seasonality	24. Instituting free admission at certain museums during low season 25. Organizing and leading large-scale events during low tourist season	11% 72%

Tacking stock of the overall results at this research stage, the authors conclude that the majority of megapolises actively participating in the tourist business have created a system of general methods and approaches to preventing and overcoming overtourism. At the same time, unique methods are fairly broadly used; these methods are employed due to various factors affecting a given megapolis' development. Therefore, the aim of the second stage of the research study was to identify those established and unique methods for preventing and overcoming overtourism that mesh well with the current stage of development of St. Petersburg as a tourist destination; it has been also aimed to identify factors that encourage the stable growth of tourism in a megapolis.

The second stage of the research study was based on obtaining tourism specialists' expert opinions. The qualitative survey included in-depth interviews with St. Petersburg tourism administrators (the A-group), representatives of the largest tour operator companies (the B-group), managers of cultural institutions, creative industries, and creative spaces (the C-group), city planning researchers, and culture, tourism, and hospitality researchers (the D-group), and with members of protest groups (the E-group). A total of 18 interviews were conducted in April through June of 2019. The experts determined which methods of overtourism regulation currently used in tourist destinations will be most suitable for St. Petersburg in order to combat overtourism and decrease its negative consequences.

As the survey was being conducted, it was immediately apparent that tourism specialists are very concerned about the problem of overtourism in St. Petersburg. However, all the representatives of the tourism governing body, tourist agency executives, as well as some of the city development researchers and those in the field of tourism believe that the overtourism problem has not yet become as acute in St. Petersburg as in many other tourist destinations. Moreover, some of these respondents are of the opinion that much of St. Petersburg's tourism potential has yet to be exploited.

Serving as the basis for these negative views is the fact that many fascinating historical, architectural, and cultural objects located in urban neighborhoods and city suburbs are not being sufficiently involved in the tourist enterprise.

We hold all the events in the city center. There are proposals to do activities at other venues, in other districts, not only in the Central one. Thus, the budgeting will change, and it will be interesting for the district, the district will develop. Here this redistribution will have its effects (A-2).

At the same time, other groups of experts, first and foremost, executives of certain museums, festival organizing committees, representatives of local residents' protest groups, a number of city development researchers, and those in the field of tourism believe that the city has already approached the threshold of overtourism, and that in further development of tourism this fact must be taken into account.

Of course, we have a pandemonium in the city center. Indeed, the city center is overcrowded. We have a lot of interesting things happening on the outskirts (A-1).

They present the following arguments supporting this view: the state of affairs during high season in the largest city museums, the city center being congested with tourist transportation, the unacceptability of servicing tourists during nighttime by the service industry, conflicts between clients of hotels and hostels located in multi-apartment houses and local residents, as well as a host of other problems.

We have queues, and they develop on holidays, but this is primarily due to the way in which the museum layout is organized. Because the museum is small in size, the ticket office and the coat check create queues (C-1).

Protest groups have an issue with a high concentration of tourists in residential areas:

The main problem is an excessive number of catering establishments... It is necessary to carry out the declustering of restaurant businesses in residential areas with the assistance of the city and to create restaurant clusters in depressed areas instead (E-1).

Evaluating the most problematic areas in the field of tourism development, experts have drawn attention to the fact that it is necessary to create a comfortable environment for tourists.

The city administration must provide transportation to Pulkovo, which can't be reached by subway or monorail. Passengers are forced to deal with traffic jams while traveling there by bus. (C-5).

Street directions are not always up to snuff. One can still encounter errors in English translations of city street signs and places of interest. In addition, these are the most popular attractions. The less popular ones can sometimes only be found with the help of local residents (D-3).

Therefore, the experts have identified the measures described below as being high-priority for preventing overtourism in the city: creating an environment that would contribute additional comforts for tourists and, at the same time, would not negatively impact residents' living conditions. Among such high-priority measures are, first and foremost, finding a solution to transportation problems, improving visual navigational services, creating public spaces that are attractive for tourists either in the city center or in neighboring areas, extending the hours of operation of museums and museum spaces during high season as well as those of the creative spaces located outside of residential building neighborhoods that offer cultural, entertainment, and restaurant services.

It is necessary to develop the 'gray belt'. Make it more attractive for tourists. A good example of this are the creative spaces 'New Holland' and 'Port Sevkabel'. While these are burgeoning creative spaces, the majority of visitors are locals. It is necessary to hold events that would interest tourists as well (D-4).

In the experts' opinion, one of the key directions in developing a system of measures for preventing overtourism is correctly steering the urban politics in St. Petersburg, which should be geared toward generating spatial solutions that would divert a significant number of tourist flows from the city center. At the same time, the experts have underscored not only the importance of housing creative spaces, hostels, and restaurants in the "gray belt" area, but also that of creating areas with large-scale complexes in order to attract tourists. Examples of such locales are the Lakhta area that is currently under construction, which boasts the Lakhta Center, the tallest skyscraper in Europe; the Gazprom Arena stadium; as well as a number of museums, and concert and exhibition halls, etc. Without a doubt, this cultural-business area will divert a significant chunk of tourists from the city center.

It is necessary to link the founding of new museums and museum spaces with locations of last subway stops in order to develop those neighborhoods. This will divert folks from the city center and promote budget replenishment but without reducing the tourist numbers. (A-2).

The creation of this type of new large-scale attractions fits into the framework of measures employed for rebranding tourist products, whose goal is to draw attention to cultural-entertainment objects that are located outside the city center and are not subject to overtourism issues. We are referring to small museums and museum spaces that have the potential to attract niche tourism and individual tourists who want to avoid the beaten paths. These sites will also interest tourists who are paying their second visit to St. Petersburg; according to experts, this will become a standard practice once electronic visas have been instituted.

It is possible to reduce the seasonality through promoting small and average-sized museums, through using CityCard, and through supporting projects by providing necessary information. It's necessary because if we have overtourism for real, they will push the Hermitage into the Neva. (B-4).

It's important to work with those who are not visiting the city for the first time, with individual tourists who are not interested in checking in at the "must visit" places. They seek unusual attractions. (D-3).

Among measures employed for rebranding tourist products, experts list promoting small royal and grand duke palaces located in St. Petersburg suburbs. As was previously noted, currently, large tourist flows aim to visit only Peterhof and Tsarskoye Selo (Pushkin). At the same time, palace complexes of Pavlovsk, Gatchina, and Oranienbaum would have been attractive for tourists. Experts note that there are no infrastructure issues in these suburbs, and it is viable to offer more holistic cultural-entertainment services that would be not only museum-based but would also rely on cultural-ethnographic services provided by creative and gastronomic tourism.

I think that the Leningrad Oblast has a great potential in terms of its cultural heritage and based on the fact that the transportation infrastructure is being developed. Tourists who visit the city for a given period of time will deliberate on what they ought to see outside the city proper. This will free up space in the city (A-1).

Among crucially important measures aimed at overcoming overtourism in St. Petersburg, experts have listed those aimed at improving the functioning of leading city museums. It is these museums and their visitors that suffer from overcrowding in exposition halls and building facilities during high season. Although, currently, leading museums are taking a number of steps in order to regulate mass tourism flows (separate entrances for different types of visitors, extending the hours of operation during high season, providing opportunities for individual excursions during hours when the museum is closed and a number of other measures), the experts believe that it is necessary to pursue more effective regulation tools, including employing those based on a flexible pricing policy, varying excursion routes, instituting separate visitation times for groups vs. individual tourists, etc.

We need distributional systems, so that a tourist would buy a ticket to a museum as if he were buying a theater ticket with a clearly indicated visitation time. A more clear visitation time limit can help to somehow solve the issue of excessive visitor flow (B-4).

Extended business hours must be instituted, even though many museum folks are against this. Just like they are against raising prices: museums, first and foremost, carry out a social mission, and any innovations may interfere with this (C-4).

The experts believe that, all things considered, pricing and taxation methods may be more vigorously employed for preventing overtourism. However, there are diverging attitudes to these methods among different groups of experts. Although a tourist tax of about $2.00 on foreigners in St. Petersburg that is in the works will clearly not diminish the city's attractiveness to tourists, instituting the tax is a cause for concern among tourist business representatives because it will increase expenditures related to collecting the tax and transferring the collected money into the budget.

On the whole, the pricing policy may be an effective tool for regulating tourist flows because a significant portion of Russian tourists are youth and families with children, who are fairly sensitive to differences in prices charged for a given service.

Charging every tourist one hundred rubles will create a negative emotional reaction. Non-essential fees can be a psychological turnoff (A-2).

Reviewing measures for overcoming the seasonality of Petersburg tourism, the experts have noted that event tourism is of critical importance. However, the majority of cultural festivals, including those sponsored by the city, are held during high season. The experts have emphasized the importance of coordinating the event activity between the cultural governing body and the tourist industry.

We should hold more events geared not only toward locals but also toward tourists. And not only during high season. If events and concerts are meaningful, people will attend them even in the winter, and not just downtown (D-3).

As far as the most mass type of event tourism is concerned -- sport tourism -- the experts have been fairly cautious in terms of evaluating its potential as a tool to be used for preventing and overcoming overtourism. In recent years, such major football events as the World Cup championship and the Confederations Cup took place in St. Petersburg; some of the European Football championship matches are scheduled to take place in St. Petersburg in 2020. However, despite the arrival of high numbers of fans, sportsmen, football officials, and journalists, on the whole, the number of tourists in the city decreases during such events. The issue is that regular tourists do not want to be in the city that is overflowing with football fans. On the other hand, successfully carrying out large-scale sporting events has a positive impact on St. Petersburg's image as an attractive tourist destination, but does not directly translate into overcoming overtourism.

The experts have drawn attention to the fact that it is necessary to undertake measures to regulate mass tourism and to stimulate niche tourism, which would enable us to successfully address the issue of the seasonality of tourist flows.

With respect to subtypes of niche tourism, which it is advisable to develop in order to overcome seasonality, there is a wide range of expert opinions. The experts have been unanimous in recognizing only MICE tourism as a promising niche subtype of tourism.

After the Expoforum has been built, the city has devoted a great deal of attention to attracting various large exhibitions and congresses. St. Petersburg has very good prospects in this respect because nowadays bleisure is becoming more and more popular, that is, combing business and leisure during a tourist trip. (A-1).

As far as other types of niche tourism are concerned, among those that are supported by the city authorities, the experts have identified those that may realistically be instrumental in decreasing the seasonality of tourism (medical tourism, cinema tourism, ballet tourism), but these tourism types have not been widely developed. Other types of niche tourism whose support is advertised by the city authorities (yacht tourism and wedding tourism), in the experts' opinion, may have the opposite effect of increasing the seasonality of St. Petersburg tourism.

Medical tourism is developing, and the situation with manor tourism is very interesting. That is, in our Leningrad region, there are some manors with accommodations, but it is the ones without accommodations that are most interesting ones. This combination of attractions will free up the city and showcase the city's other side (A-2).

In the experts' opinion, it is expedient to develop other types of niche tourism: industrial, scientific-educational, recreational, creative, musical, as well as a number of other types. Serving as the criterion for prioritizing niche types of tourism must be a high level of collaboration between tourists and local professionals who are interested in communicating with colleagues and like-minded individuals from abroad and from other regions of Russia.

I actively promote industrial tourism. Each of our large enterprises has its own museums that have military and patriotic, revolutionary, industrial, and professional themes. In St. Petersburg, there are more than 750 large and medium-sized enterprises, and 20%-25% of them are among the largest ones in Russia (A-2).

Researching the use of informational technologies as a tool for overcoming overtourism, the experts have noted that advanced IT solutions are currently being used. These tools are being actively supported by the city authorities as part of the program "St. Petersburg the Smart." A large number of tourist services are offered in application format, and there are sharing economy platform services, various navigators and other IT services.

The experts have noted that there is an issue pertaining to providing IT solutions stemming from the requirement that government companies comply with financial accountability norms, whereby many of these companies are not prepared to adopt IT technologies.

Of course, ones and zeros are the way to go, we should go with apps. We have an app ready, but we can't use it because the majority of attractions have not been automatized. You've got a ticket, you've got a stub, you've got a ticket, you've got a stub. Even if there are turnstiles, this makes no difference because a ticket is a document of strict financial accountability (B-4).

As promising information-based methods that would aid in regulating overtourism, experts have identified those for observing tourist flows and constantly informing all the participants of the process about exceedingly large numbers of tourists at tourist sites, about the availability of parking spaces for tourist buses, about that of free spaces in fast food restaurants, as well as about other information important for tourists.

Summarizing the experts' viewpoints on the methods of overcoming overtourism, it is necessary to note that there are differences of opinion on some issues and a consensus on others; these findings demonstrate that it is necessary to continue researching the issue in question.

1. Since St. Petersburg authorities and most members of the tourist industry aim to increase the number of tourists who arrive in the city, the major measures geared toward preventing and overcoming overtourism should not be restrictive but should rather be targeted toward redistributing tourist flows both temporally and spatially, having many opportunities for that.

2. At the present time, there is no unified strategy being used for preventing overtourism in the city. Measures undertaken by the city authorities are unrelated to one another and fall short of defending city dwellers' interests to a sufficient degree. In the near future, the city may enter the consolidation stage (according to R. W. Butler's classification), which implies the existence of an active standoff between tourists and local residents.
3. On the whole, experts gave high ratings to novel urban solutions for preventing and overcoming overtourism in St. Petersburg, pointing out that in the "gray belt" as well as in central city neighborhoods there are a great many spaces, such as former factories, storage facilities, and transportation factories, which could function as creative spaces for local residents and tourists after being properly reconstructed and renovated. However, these solutions are not being supported by efforts to promote new cultural hubs as tourist attractions.
4. A more stringent taxation policy toward the tourist business is not being viewed by the city as a tool for overcoming overtourism, but rather as a way of replenishing the city budget by using the burgeoning tourism industry to this end. Thus experts have suggested undertaking more aggressive measures aimed at monitoring housing booking systems, which are instrumental in housing significant numbers of tourists; however, lessors do not pay the required taxes. Moreover experts called attention to the fact that Chinese tourism ought to be drawn out of the "shadows," given that its scope is steadily increasing; however, as a result of various factors, and first and foremost, due to the corrupt nature of Chinese tourism, it fails to make necessary contributions to the city budget, while applying significant pressure on the tourist infrastructure.
5. Using certain methods aimed at overcoming overtourism, according to experts, is hindered by the conservative stance of executives of the city cultural institutions, who refuse to institute a flexible pricing policy (changing admission prices depending on the tourist season, day of the week, individual on-line ticket purchases, etc.); these executives use as an argument against instituting such policies the idea that it is incumbent upon their institutions to fulfill a certain social mission.
6. As seasonal tourism is exhibited in St. Petersburg to a far bigger degree than in other tourist megapolises, possible methods aimed at overcoming the seasonal nature of tourism are discussed. Experts have emphasized the importance of developing various forms of event-based tourism during low season: cultural, congress, and sport tourism, and a variety of niche tourism types: scientific-educational, ballroom, recreational, creative, musical, as well as some other types of tourism, whose distinguishing characteristic is a high level of collaboration between tourists and local experts. The nature of the majority of these types of tourism is non-seasonal, and they may aid in overcoming the seasonal nature of overtourism in St. Petersburg. Among branches of niche tourism that are currently being actively supported by the city, there are types of tourism characterized by relatively low numbers of tourists whose development will not counteract seasonality (e.g., yacht tourism) or will not generate stable and mass tourist flows during low season (e.g., medical and wedding tourism); at the same time, there are certain types of tourism that may be instrumental in overcoming overtourism (e.g., event and MICE tourism).

Therefore, the overall conclusion is that idiosyncrasies associated with employing methods aimed at overcoming overtourism in St. Petersburg are, firstly, due to the tourist destination being at a given life cycle stage and, secondly, they are due to a pronounced and recalcitrant seasonality of tourism.

FUTURE RESEARCH DIRECTIONS

The first promising overtourism research direction is related to exploring opportunities for controlling individual tourists' impulsive behavior through utilizing digital technologies. The authors are referring to the fact that the ever-increasing tourist population has two major types of consumption behaviors: (1) tourist attraction, cultural event, and restaurant visitations that have been planned in advance in terms of their time and place and (2) unplanned, impulsive visitations resulting from tourists' unrelenting inquiries concerning the hours of operation, the intensity of tourist flows, the potential waiting times, the number of passengers taking public transportation, and other comfort level indicators of tourist attraction visitations. Creating a city-based and institution-based user-friendly system for informing tourists about the current state of affairs at given tourist objects would enable to reroute tourists to different cogent tourist destinations. There are some digital solutions being already tested mainly on cultural attractions of Amsterdam (*'Reis Je Rijk'* app.), Vatican (*Group Tour™ System*), London (*PassageWay*), Boston (*Natural History Museum Explorer* app), etc. What is special about a megapolis such as St. Petersburg is that it affords opportunities for rerouting tourist flows between various visitation objects while emphasizing new tourist attractions. Utilizing these methods of controlling tourist flows will necessitate the use of artificial intelligence technologies that would enable us to identify the user and determine which tourist routes would interest him or her the most. Using these technologies would enable us to significantly lower the attendance of the most attractive visitation objects during the high season.

A second research direction that the authors propose here is exploring the role of international housing booking services such as Airbnb.com, homeaway.com, and a number of others in exacerbating overtourism. The issue is that, as it was noted in the current chapter, the business activity of such services is out of sync with the Russian jurisprudence, as landlords do not register their rental properties properly and oftentimes do not pay the required taxes, which affords them with undeserved competitive advantages over hotels and hostels. It is necessary to undertake large-scale international comparative studies of accumulated organizational and legal experience of interactions between local and state-level authorities and housing booking services in order to determine the most effective methods of preventing overtourism, including methods such as using prices and taxes as tools to attain this end.

CONCLUSION

The experts' opinions as well as results of researching best practices of overcoming overtourism in other megapolises and learning from the experience of employing a few methods of regulating tourism in St. Petersburg have served as the foundation for creating the recommendations. The proposals in question may be classified into two groups: altering the structure of the tourist flow and changing the urban planning policy.

Measures aimed at altering the structure of the tourist flow include the following ones:

- developing niche tourism that is not subject to seasonality: medical, ballroom, recreational, industrial, and cinema tourism;
- developing tourism with high collaboration levels: scientific-educational, MICE, and business tourism, which would effect the growth of economical, social, and cultural capital of the destination through collaborations between tourists and certain groups of residents;

- a purposeful support of mass cultural events by the city authorities during low season;
- regulating cruise tourist flows;
- regulating the functioning of housing search platforms (for example, Airbnb) in order to overcome the excessive pressure on the real estate; ensuring that housing rentals are not part of the shady economy;
- broadening the scope of purposefully collected tourist taxes and collected moneys; a seasonally-differentiated pricing policy;

Measures pertaining to urban planning in areas that attract tourist activity:

- stimulating the creation of tourist clusters isolated from residential areas in the "industrial belt" territories and in suburban areas;
- curtailing the building of new hotels and souvenir shops in the city center in order to reduce overcrowding;
- regulating transport infrastructure (limiting the entry of transportation into the city center and increasing the number of parking spaces).

The city government committees and, to a lesser extent, business community should be the entities responsible for the implementation and monitoring of these measures. They are interested in substantial growth of tourist flows, though sometimes they do not pay attention to its quality. Carrying out this array of measures would enable us to curtail the increase in social unrest between tourists and residents, to improve tourists experience and to boost the effectiveness of managing the tourist activity in the destination.

ACKNOWLEDGMENT

This research was supported by the Basic research program of the National Research University Higher School of Economics.

REFERENCES

Advisors, H. (2018, February 7). *Operatsionnye pokazateliotelej Sankt Peterburga za 2017 god [Performance indicators of St. Petersburg hotels for the year of 2017]*. Retrieved from https://hoteladvisors.ru/blog-2/Operatsionnye-pokazateli-oteley-Sankt-Peterburga-za-2017-god/

Agostinelli, A. (2018). Culture & Tourism in Rome: What is the Future of the Eternal City? A New Storytelling to Save the Old History [PDF document]. Retrieved from https://www.researchgate.net/publication/323855568_Culture_Tourism_in_Rome_What_is_the_Future_of_the_Eternal_City_A_New_Storytelling_to_Save_the_Old_History

All-Russian Center for Researching General Public's Opinion. (2019, March 13). *Hostely v zhilyhdomah: zapretit' iliostavit' [Hostels in residential buildings: ban them or keep them]*. Retrieved from https://wciom.ru/index.php?id=236&uid=9592

Alvarez-Sousa, A. (2018). The problems of tourist sustainability in cultural cities: Socio-political perceptions and interests management. *Sustainability, 10*(2), 503. doi:10.3390u10020503

Benner, M. (2019). *From overtourism to sustainability: A research agenda for qualitative tourism development in the Adriatic*. [PDF document]. Retrieved from https://mpra.ub.uni-muenchen.de/92213/

Butler, R. W. (1980). The concept of a tourist area cycle of evolution: Implications for management of resources. *The Canadian Geographer. Geographe Canadien, 1*(24), 5–12. doi:10.1111/j.1541-0064.1980.tb00970.x

Cheer, J. M., Milano, C., & Novelli, M. (2018, Jul 18). Overtourism: a growing global problem. *The Conversation.*

De Frantz, M. (2018). Tourism marketing and urban politics: Cultural planning in a European capital. *Tourism Geographies, 20*(3), 481–503. doi:10.1080/14616688.2017.1354392

Doxey, G. V. (1975). A causation theory of visitor-resident irritants: Methodology and research inferences. In *Proceedings of Sixth annual conference of the Travel Research Association*, 195-198. San Diego, CA: Travel Research Association Conference.

Dredge, D., Gyimóthy, S., Birkbak, A., Elgaard Jensen, T., & Madsen, A. (2016). *The impact of regulatory approaches targeting collaborative economy in the tourism accommodation sector: Barcelona, Berlin, Amsterdam and Paris*. Copenhagen, Denmark: European Commission DG GROWTH, Aalborg University.

Dumbrovská, V. (2017). Urban Tourism Development in Prague: From Tourist Mecca to Tourist Ghetto. *Tourism in the City: Towards an integrative agenda on urban tourism*, 275-283.

Euromonitor International. (2018). *Top 100 City Destinations 2018*. Retrieved from https://go.euromonitor.com/white-paper-travel-2018-100-cities#download-link

European Cities Marketing. (2018). *Managing tourism growth in Europe: the ECM toolbox*. Dijon, France.

European Parliament's Committee on Transport and Tourism (TRAN Committee). (2018). Overtourism: impact and possible policy responses. Brussels, Belgium. European Parliament, Policy Department for Structural and Cohesion Policies.

Forster, J. (1964). The sociological consequences of tourism. *International Journal of Comparative Sociology, 5*(2), 217–227. doi:10.1177/002071526400500208

Freytag, T., & Bauder, M. (2018). Bottom-up touristification and urban transformations in Paris. *Tourism Geographies, 20*(3), 443–460. doi:10.1080/14616688.2018.1454504

Goodwin, H. (2017, October). *The challenge of overtourism.* Retrieved from https://haroldgoodwin.info/pubs/RTP'WP4Overtourism01'2017.pdf

Goodwin, H. (2018, October). *Managing Tourism in Barcelona.* Retrieved from https://haroldgoodwin.info/RTPWP/012ManagingTourisminBarcelona.pdf

Gunn, C. A., & Var, T. (2002). *Tourism planning: Basics, concepts, cases.* Psychology Press.

Inskeep, E. (1991). *Tourism Planning: An Integrated and Sustainable Development Approach.* New York: Van Nostrand Reinhold.

International, I. P. K. (2018). *ITB World Travel Trends Report.* Berlin, Germany: Messe Berlin GmbH.

Joppe, M. (2019).The roles of policy, planning and governance in preventing and managing overtourism. In R. W. Butler, & R. Dodds (Eds.), Overtourism: Issues, Realities and Solutions (250-261). Berlin, Germany: De Gruyter Oldenbourg. doi:10.1515/9783110607369-017

Koens, K., Postma, A., & Papp, B. (2018). Is overtourism overused? Understanding the impact of tourism in a city context. *Sustainability, 10*(12), 4384. doi:10.3390u10124384

Kommersant. (2018, June 25). *Alye parusa 2018 posetili 1,2 mln chelovek [Scarlet Sails 2018 were visited by 1,2 million of people].* Retrieved from https://www.kommersant.ru

Mathieson, A., & Wall, G. (1982). Tourism, economic, physical and social impacts. *Journal of Travel.*

Maxim, C. (2017). Challenges Faced by World Tourism Cities–London's Perspective. *Current Issues in Tourism,* 1006–1024.

McIntyre, G. (1993). Sustainable tourism development: guide for local planners. Madrid, Spain: World Tourism Organization (UNWTO).

Milano, C. (2017). *Overtourism and Tourismphobia: Global trends and local context.* [PDF document]. Retrieved from https://www.academia.edu/35911428/Overtourism_and_Tourismphobia_Global_trends_and_local_contexts?auto=download

Muler Gonzalez, V., Coromina, L., & Gali, N. (2018). Overtourism: Residents' perceptions of tourism impact as an indicator of resident social carrying capacity-case study of a Spanish heritage town. *Tourism Review, 73*(3), 277–296. doi:10.1108/TR-08-2017-0138

Novy, J. (2011). Berlin does not love you. *The Berlin Reader,* 223-233.

Pappalepore, I., Maitland, R., & Smith, A. (2014). Prosuming creative urban areas. Evidence from East London. *Annals of Tourism Research, 44,* 227–240. doi:10.1016/j.annals.2013.11.001

Pigram, J., & Wahab, S. (1997). *Tourism, Development, and Growth.* Abingdon, UK: Routledge.

Pinke-Sziva, I., Smith, M., Olt, G., & Berezvai, Z. (2019). Overtourism and the night-time economy: A case study of Budapest. *International Journal of Tourism Cities, 5*(1), 1–16. doi:10.1108/IJTC-04-2018-0028

Popova, T. (2016, December 10). *Knight Frank: Rynok gostinichnoj nedvizhimosti. Sankt-Peterburg - 2016 [Knight Frank: The hotel real estate marketplace. St. Petersburg - 2016].* Retrieved from http://www.hotelscatalog.ru/novosti-gostinits-sankt-peterburga/knight-frank-rynok-gostinichnoj-nedvizhimosti-sankt-peterburg-2016.html

Postma, A., & Schmuecker, D. (2017). Understanding and overcoming negative impacts of tourism in city destinations: Conceptual model and strategic framework. *Journal of Tourism Futures, 3*(2), 144–156. doi:10.1108/JTF-04-2017-0022

RIA novosti. (2019, June 8). *PMEF v etom godu posetili bolee 19 tysjach uchastnikov iz 125 stran mira [PMEF was visited this year by more than 19 participants from 125 countries of the world].* Retrieved from https://ria.ru

Richards, G. (2011). *Cultural tourism: global and local perspectives.* Abingdon, UK: Routledge.

Richards, G., & Marques, L. (2018). *Creating synergies between cultural policy and tourism for permanent and temporary citizens.* Retrieved from http://www.agenda21culture.net/our-cities/barcelona

Sheng, L., Li, T., & Wang, J. (2017). Tourism and externalities in an urban context: Theoretical model and empirical evidence. *Cities (London, UK), 70*, 40–45. doi:10.1016/j.cities.2017.06.012

Skeli, S., & Schmid, M. (2019). Mitigating overtourism with the help of smart technology solutions–a situation analysis of European city destinations. In *ISCONTOUR 2019 Tourism Research Perspectives: Proceedings of the International Student Conference in Tourism Research*, 13. Norderstedt, Germany: BoD–Books on Demand.

The official site of St. Petersburg Authorities. (2018, October 25). *V Peterburge podveli itogi navigatsii 2018 goda [Marine navigation in St. Petersburg in 2018 was assessed].* Retrieved from https://www.gov.spb.ru/gov/otrasl/c_tourism/news/149312/

The State Duma of the Russian Federation. (2019, April 4). *Zakon o hostelah prinjat v okonchatel'noj redaktsii [The law on hostels has been ratified in its most recent iteration].* Retrieved from http://duma.gov.ru/news/30244/

Wagar, J. A. (1964). The carrying capacity of wild lands for recreation. *Forest Science, 10*, a0001–a0024.

Weber, F., Stettler, J., Priskin, J., Rosenberg-Taufer, B., Ponnapureddy, S., Fux, S., & Barth, M. (2017). *Tourism destinations under pressure.* Luzern, Switzerland: Lucerne University of Applied Sciences and Arts, Institute of Tourism ITW.

Which city's populations grow the most due to tourism? (2019). [Graph illustration the relationship between total native population and annual visitor numbers in cities]. *Data Access from World Economic Forum and Mastercard Global Destination Index.* Retrieved from https://www.onthegotours.com/city-populations-that-grow-most-from-tourism/

World Tourism Organization (UNWTO). Centre of Expertise Leisure, Tourism, & Hospitality, NHTV Breda University of Applied Sciences & NHL Stenden University of Applied Sciences. (2018). 'Overtourism'? – Understanding and Managing Urban Tourism Growth beyond Perceptions, Executive Summary, Madrid, Spain: UNWTO.

Chapter 8
Bye Bye Lisbon:
Tourism Gentrification Impacts on Lisbon´s Inner-City Housing Market

Luís Mendes
University of Lisbon, Portugal

ABSTRACT

Urban tourism in its various forms has seen a great expansion at the beginning of the 21st century, but the overtourism as a mass phenomenon, hegemonic and dominant in the large Portuguese cities of Lisbon and Porto, is a recent phenomenon. Lisbon is experiencing a peak of international projection as a tourist destination, registering considerable uninterrupted increases in tourist arrivals, overnight stays, and daily revenues, as well as number of hotels, hostels, and short-rental tourist accommodation, with drastic impacts on the social and economic fabric in the historical districts of central Lisbon is recent. This chapter focuses on the driving forces of austerity urbanism and the neoliberal turn in urban politics that fuels the process of tourism gentrification, in the last decade, in Lisbon's historic neighborhoods.

INTRODUCTION

This paper will focus on the driving forces of austerity urbanism and the neoliberal turn in urban politics that fuels the process of tourism gentrification, in the last decade, in Lisbon's historic neighborhoods. Urban tourism in its various forms has seen a great expansion at the beginning of the 21st century, but the overtourism as a mass phenomenon, hegemonic and dominant in the large Portuguese cities of Lisbon and Porto is a recent phenomenon. Lisbon, specifically, is experiencing a peak of international projection as a tourist destination, registering considerable uninterrupted increases in tourist arrivals, overnight stays and daily revenues, as well as number of hotels, hostels and short-rental tourist accommodation.

This evolution has a direct and indirect impact on the national and urban economy, both in terms of wealth created and assured jobs, in addition to the urban revitalization process, evident in the rehabilitation of the building and the housing stock, fundamental to the country's and city´s economic recovery, during and after the capitalist crisis of 2008-2009.

DOI: 10.4018/978-1-7998-2224-0.ch008

However, the apparent absence of any tourism planning and impact assessment strategy coupled with the almost non-existent regulatory process has fueled tourism gentrification. A neoliberal turn on fiscal and urban policies emerged, driven by post-crisis capitalist international austerity intervention 2011-14 in Portugal. Both national as urban government has discovered the potential of touristification and overtourism in regenerating inner-city traditional housing areas, in order to increase the competitiveness of the city and certain neighbourhoods in the global context of urban competition (Mendes, 2013, 2014, 2016a, 2016b, 2017, 2018; Tulumello, 2015; Barata Salgueiro et al., 2017; Barata Salgueiro, 2017; Krähmer, 2017; Alves, 2017; Pavel, 2017; Nofre et al., 2018; Montezuma and McGarrigle, 2018; Lestegás, 2019). The creation of aggressive programs to attract foreign investment (as the Golden Visa and the Non Habitual Residents), the new urban rent law, the new tax regime for Property Investment Funds, the new law of tourist lodging (short-rental), along with an intense rent gap in the inner-city, as well as a strong growth in tourism in the city of Lisbon, introduced significant changes in the residential market, which went from an abrupt pause to a high level of demand, with supply now beginning to fall short. This situation led to a very quick take-up of the new and good quality residential stock that was available, located mostly in the city's historic centre. The rapid take-up of apartments, the new tourism drivers – with an "alternative" demand for stays in apartments – and tax incentives to boost urban regeneration, have created renewed interest among many national and international developers, leading to a great rise in the refurbishment of buildings in Lisbon's historic district.

We have witnessed the transformation of the popular and historical districts of the center city into places of consumption and tourism, by expanding the function of recreation, leisure or tourist accommodation / short-rental lodging that begins to gradually replace the traditional functions of housing for permanent use, long-term lease and traditional local proximity retail, aggravating tendencies of displacement and residential segregation, emptying the neighborhoods of its poor population or preventing low socio-economic population and even middle class from accessing housing in these areas. This is how we define tourism gentrification.

As a methodological line, we would like to emphasize that the text we present is only a mere theoretical essay, problematizing and exploratory of this theme, recognizing that it lacks a more exhaustive reference to not only secondary data, but also primary data to consolidate some of the hypotheses the case of tourism gentrification in Lisbon presents. Following a hypothetical-deductive methodology, the construction of this essay starts from postulates or concepts already established in the consulted literature, through a logical work of relation of explanatory hypotheses, that configures, in our view, a possible perspective of interpretation of the phenomena under study. In this chapter, the theory and empirical approach will be developed in an integrated way in order to facilitate the reading of the phenomenon. Therefore, the theoretical questions will be constantly interspersed and illustrated with reference to small empirical aspects that evidence the deepening of the tendencies of privatization, commodification, financialization of the housing market of Lisbon, in favor of a growing tourism led gentrification. In addition to a review of the international literature on overtourism, tourism gentrification and neoliberal urbanism, the methodological line is composed of both quantitative and qualitative methods and techniques (critical content analysis, collection of secondary data in various specialized reports produced by public and private entities) that allow a macro analysis of the various laws that produced austerity urbanism, crossed with a meso analysis of the expansion trends of short-rental tourist accommodation and its impacts on the local housing market.

The chapter is organized as follows. The first point analyzes the theoretical problem of integration between tourism and gentrification, equating the concept of tourism gentrification and how this process evolves in the city of Lisbon, producing today a new stage of urban regeneration. It addresses the theoretical framework on austerity urbanism and how touristification becomes an urban and economic relevant strategy, legitimized by the post-crisis-capitalist austerity climate of 2008-2009. In the second section and starting from the concepts of overtourism, we will try to understand in which stage Lisbon is in this process. We will review some indicators of this touristification in Lisbon. The third point explores the relationship between expansion of short-rental tourist lodging and financialisation of the housing market, identification of the main drivers of the neoliberal turn on urban policy. The chapter concludes with a summary of main findings.

OVERTOURISM PANACEA OR... TOWARDS A DEFINITION OF TOURISM GENTRIFICATION

The relationship between tourism and the city is by no means recent (Ashworth and Tunbridge, 1990; Mullins, 1991; Law, 2002). Today, however, urban tourism is booming due to market urbanism, an increasingly profligate consumer society and global internationalization of financial systems that recognize the gains from the new economic urban regeneration, supported by urban policies and local governance. This accompanies the changes in the motivations for the demand for urban tourism and the important investments made in the preservation of the built heritage and the promotion of cosmopolitan environments, especially in the historical districts of the city centers, which have triggered several urban changes, namely in the housing market, producing gentrification (Ashworth and Page, 2011; Wilson and Tallon, 2012; Hiernaux and González, 2014; Cocola Gant, 2015, 2018; Delgadillo, 2015; Mendes, 2016a; Gravary-Barbas and Guinand, 2017; Barata Salgueiro *et. al.,* 2017; Barata Salgueiro, 2017).

For a number of decades, the development and management of urban tourism has become a major challenge for sustainable development in many cities in Europe, home to history, culture and heritage, as well as clusters of social and cultural life, important tourist resources, but also as tourist destinations, aggregators of social and environmental externalities less positive due to increasing tourism. The challenge now is the integration of an urban tourism development that rebalances the urban economy by generating new dynamics and activities, promoting urban regeneration in various dimensions of urban social life, but mitigating the negative impacts introduced in the target communities. This challenge implies not only a holistic understanding of tourism and its integration / articulation with other sectors of the urban economy and society in general (housing, transport, energy, waste, employment, etc.) but above all that policies for urban tourism are designed and integrated into urban development policies (Burns and Holden, 1995; Law, 2002; Page, 2003; Hoffman *et al.*, 2003; Ashworth and Page, 2011; Jamal and Camargo, 2017).

Tourism, in addition to the issues associated with the economy, raises the discussion in areas such as mobility and transportation, housing, city uses by the population (local and external), cultural and social resources, and also makes us question the sustainability of this activity, the resources available and the positive and negative impacts, in the form of externalities, that it generates in the destiny communities. Thus, due to a weak mitigation of the effects of tourism massification and the increasing visibility that tourism has gained in the fields of public policies and regional and urban development, innovative studies have emerged advancing the need to better understand the process through more effective new concepts of overtourism (Goodwin, 2017; Milano, 2017, 2018a) resilience of destiny communities (Munro et al., 2010; Luthe e Wyss, 2014; Bec et al., 2016; Sheppard and Williams, 2016; Agyeiwaah et al., 2017;

Prayag, 2017; Espiner et al., 2017; Kato, 2018; Lew and Cheer, 2018; Cheer and Lew, 2018) and tourism gentrification (Gotham, 2005; Wilson and Tallon, 2012; Hiernaux and González, 2014; Cocola Gant, 2015, 2018; Xian Liang and Gang Bao, 2015; Pinkster and Boterman, 2017; Mendes, 2016b; Gravary-Barbas and Guinand, 2017), seeking to overcome the limitations of already classic studies on the impacts and externalities of tourism in the environments where it occurs.

In terms of theoretical paradigms and methods of study, tourism and gentrification, although both studied in the field of urban studies, have only recently begun to be understood theoretically and empirically in an integrated way. The important point to stress here is that the literature notes that both processes feed each other and they overlap in time and space. While in some cases the proliferation of gentrified spaces become tourist destinations, in other cases tourism-oriented urban promotion strategies produce a new built environment that, in turn, attracts new residents with higher incomes and, therefore, encourages processes of gentrification. In fact, the two processes coexist and are not distinguished in the empirical fact being generally interconnected and feeding each other. Regardless of which process encourages the other, the conclusion is that both tend to coexist in the same urban environment, resulting in what is called tourism gentrification (Gotham, 2005; Cocola Gant, 2015, 2018).

But why can we speak on tourism gentrification? In Lisbon, for instance, the flows of capital in the real estate market combined with the shift to tourism explain gentrification more fully than do alternative accounts that focus on consumer demand or cultural preferences of a new middle class for upscale neighbourhoods. Specially in the context of the financial post-crisis, the social and economic urban structure of Lisbon was profoundly transformed to embrace the growing demand of international tourism. Tourism in inner city of Lisbon is understood by the neoliberal subjectivity of austerity as a kind of "panacea" that can cure all the diseases of the urban crisis. There has been some discussion lately about whether touristification is a kind of gentrification, since processes often share common traits with one another. So, first, we can´t understand this tourism fever without seeing the increasing mesh of gentrification into global systems of real estate and banking finance, as well the ideology of free market, austerity urban regime, financialisation of society and economy, which curiously has been advanced as a cure for the problems created by the capitalist crisis 2008-2010, when in fact they are the very cause of the whole problem of financial deregulation created.

One of the early seminal works on the subject is Gotham's study (2005) of New Orleans' *Vieux Carre*. In his work Gotham coins the concept of tourism gentrification "to highlight the role of state policy in encouraging both gentrification and tourism development, and the actions of large entertainment corporations" (p.1100). This concept also helps to explain the "transformation of a middle-class neighbourhood into a relatively affluent and exclusive enclave marked by a proliferation of corporate entertainment and tourism venues" (p.1102).

Expanding the definition of Gotham (2005) and opening the doors of the conceptual framework for the empirical reality of Lisbon, in this essay, I understand "tourism gentrification" as the transformation of popular and working class neighborhoods of the inner city in consumption places and tourism sites, so that the recreation, leisure or lodging function begin to gradually replace the residential and retail traditional functions, aggravating tendencies of retail and housing displacement and residential segregation, emptying the neighborhoods of its original population or preventing low socio-economic population from accessing housing in these areas (Mendes, 2016a, 2016b). Historically, the traditional center of Lisbon has been home to diverse groups of people. Over the past ten years especially, however, the property values has increased. This, coupled with the economic crisis, the financial austerity and the new law of urban rental (bulwark of a neoliberal turn in legal and urban policy frame) legitimized the "tourism panacea" in the city center. Escalating rents have pushed out the poor people and immigrants, so that the tourist attractions, restaurants, entertainment bars and shops for visitors and tourists now dominate much of the central districts. We argue that this shift in capital flows to the real estate market of tourist accommodation combined with the growth of tourism and, lately, with a neoliberal trend increasing in the urban regeneration policies, all together reinforce the importance of activities aimed for tourism consumption and encourage displacement and gentrification.

As in Berlin, Barcelona, Venice or Amsterdam, so it was also in Lisbon. In the last decade the frontier of gentrification has been advancing and much, thanks to an accelerated touristification. Not only has it expanded in scale but it has changed contours, forms and protagonists. We know that until the beginning of the XXI the process was marginal and embryonic in the two great Portuguese cities of Lisbon and Porto. I called it "pocket gentrification" or "embryonic gentrification" (Mendes, 2006, 2008, 2014), because it was a kind of soft gentrification, its stage was primary, so much that its growth was slow and sporadic, manifesting itself in the urban space in a punctual and fragmented form, in a small circumscribed scale and limited to only some apartments or, at most, a few blocks from the neighborhood. Displacement was small or even non-existent. This happened because all the urban rehabilitation policies undertaken from the 1970s to the beginning of the 21st century were very protective of the tenants and the most vulnerable people who lived in the historical center for decades, trying to fix them in countercurrent with the intense process of depopulation suffered, as the suburban expansion consolidated the metropolitan area. To this end also contributed the law of the frozen rents in 1948 and that maintained the value of the rents paid to a very low level and also contributed to the advance of the state of degradation of the building.

At the moment, Lisbon is experiencing a new stage of gentrification in everything different from the previous one, due to the explosion of several forms of tourist accommodation, promoted mainly by the great foreign investment injected by big owners and economic groups of real estate promotion. While the marginal gentrifiers continue to influence the area, gentrification is often accompanied by larger real estate agents and urban rehabilitation with tourism interests begins to appear as a political and economic priority strategy for revitalizing the historic center. As a result of the increasing volume of real estate interventions, physical and architectural improvements become increasingly visible at this stage. As a result, house prices in historic neighborhoods begin to climb very rapidly. Without regulation or moderate control over rent increases, the eviction process expands to more aggressive forms as neighborhood real estate values also increase and the state approves legislation that facilitates private initiative and evicts local residents and retailers from local and traditional commerce. The better maintained housing and commercial properties become part of the upper- and upper-middle class market as homeowners seek to take advantage of the area's enhanced notoriety, which in turn leads to further displacement.

BUT IS LISBON A CITY REALLY AFFECTED BY OVERTOURISM?

In social and economic studies dedicated to tourism, the concept of overtourism has emerged to designate «destinations where hosts or guests, locals or visitors, feel that there are too many visitors and that the quality of life in the area or the quality of the experience has deteriorated unacceptably. It is the opposite of Responsible Tourism which is about using tourism to make better places to live in and better places to visit» (Goodwin, 2017, p.1). The team leaded by Dichter and Manzo (2017) in the McKinsey & Company and World Travel & Tourism Council report on how to manage overcrowded tourism destinations focus on five types of problems associated with tourist overcrowding: alienated local residents, a degraded tourist experience, overloaded infrastructure, damage to nature, and threats to culture and heritage. Some places endure two or more of these problems at the same time.

Peeters et al. (2018, p.24) defines overtourism as «a complex phenomenon that strongly affects the livability of a place, as well as the experiences of residents, visitors and different stakeholders who are either directly or indirectly involved with or affected by tourism». This EU study refer that overtourism causes might be driven by factors such as: the accessibility and affordability of travel, the traditional policy focused on promoting volume, an increases in international arrivals, the urbanisation pressure, the gentrification and increasing prices in city centres and new neighbourhoods, the proliferation of unregulated tourist accommodations, and the concentration of large groups of tourists.

The idea constantly reiterated by the Secretary of State for Tourism that Portugal and Lisbon are far from overtourism also sums up what has also been the position of other governors, city officials and entrepreneurs in the sector: Lisbon has experienced rapid and explosive tourism growth in recent years, but it is still far from becoming a Barcelona, Venice, Berlin or Amsterdam, all cities where the concept of "overtourism" is not a taboo, but a reality treated with legislation. Where is Lisbon, after all, when compared to these four cities? Looking at the evolution of tourism in Lisbon over a decade is looking at a history of growth with few parallels in other sectors of the Portuguese economy. Of course, it is important to distinguish metropolitan area of city. In fact, with regard to the metropolitan area, which includes peripheral municipalities such as Vila Franca de Xira, Alcochete or Mafra, it can be said that Lisbon is far from tourist saturation. On the other hand, when the analysis is limited to the city of Lisbon, comparing it to the cities of Barcelona, Berlin, Amsterdam and Venice, the case is different and the indicators of both are not so different. Disaggregating the city of the Metropolitan Area, there is no indicator that does not increase to two digits, or even three, during the analyzed period, from 2005 to 2016. Already in 2017, the Lisbon Metropolitan Area registered more than six million guests and 14 million overnight stays, but the National Statistics Institute (INE) has not yet made available these data, for the whole of last year, disaggregated for the city of Lisbon. In recent years, the city has responded, on average, to 72% of guests and overnight stays in the metropolitan area. Assuming this proportion, the city of Lisbon will have registered 4.4 million guests and 10.3 million overnight stays in 2017. This year it had 204 hotels and pensions totaling more than 21,000 rooms and 36,000 beds, according to the Lisbon Tourism Observatory. At the same time, there were 10,700 local accommodation with a capacity of 24,000 rooms and 52,000 beds, according to data provided by the National Local Lodging Registry. Although the percentage evolution is higher (also because the base of departure was lower), the absolute numbers of Lisbon fall short of the other cities analyzed.

It is true that the absolute numbers of the Portuguese capital are almost neither comparable to those of these cities, but it is also important to place them in their proper place: at scale. Here are some of them: the city of Lisbon receives eight thousand tourists per thousand inhabitants, a ratio higher than Barcelona or Berlin, and has more than 19,000 nights per capita, being above Amsterdam. It has two hotels per square kilometer, number only lower than Barcelona. It has 107 local accommodation per square kilometer, five times more than in Berlin and twice as many as in Amsterdam[1]. Focusing on the proportion of guests and nights spent in relation to the number of inhabitants. Contrary to what happens in Barcelona, Berlin and Amsterdam, where the population has increased in the last decade, the city of Lisbon has lost inhabitants, going from a population of 555 thousand people in 2005 to about 505 thousand in 2016. Thus, in that year, the capital recorded eight thousand tourists and 19 thousand overnight stays per thousand inhabitants. In the European Union, according to Eurostat data, an average of 5,292 overnight stays per thousand inhabitants was recorded in 2015. The European statistics office considers that, as far as overnight stays, there is an overload from 20,000 nights per thousand inhabitants. As for the ratio of guests per inhabitant, Lisbon is only behind Amsterdam (although very close) and Venice (Figure 1).

When analyzing the proportion between the inhabitants of the cities and the annual number of tourists, to conclude that the visitors that flock to Lisbon (4.4 million) are nine times higher than the number of inhabitants (505 thousand), while in Oporto it is almost eight times higher (1.6 million tourists to 214 thousand residents). In Prague and Barcelona, it is five times more, in London, only four. In some cases, the number of guests and overnight stays is not enough to assess the tourist overload, since the number of tourists who visit it but who do not stay are significant. In Venice, for example, an estimated 30 million visitors are expected each year. However, the city only registers 4.6 million guests and 10.5 million overnight stays, similar to those in Lisbon.

In a study on tourism gentrification in the city of Lisbon, Barata Salgueiro et. al. (2017) calculated for the neighborhood of Alfama a tourist saturation index reaching 2.2 (there´re 2 residents for 1 tourist), considering only the supply of tourist beds on the booking.com platform in February 2016 and the population residing in the old parishes of São Miguel e Santo Estevão (that now belong to Santa Maria Maior), with reference to the 2011 census, without considering the massive offer on the Airbnb platform, nor the direct evictions provoked by the New Urban Rental Law from 2012. Therefore, the value, which is clearly underestimated, means that in practice, there are two residents for each tourist in the Bairro de Alfama, with a future tendency to approach dangerously the number of a tourist for each resident, which would amount to a situation of tourist saturation. In Alfama, Mouraria or Castelo, for example, tourists already occupy, on average, 34% of the houses. In the case of the Alfama neighborhood, specifically, the number of tourist short-rental reaches 55% of all the houses existing.

The distribution of tourist establishments throughout the city can also be counted. Last year, Lisbon had two hotels and 107 local accommodations per square kilometer, numbers that are only surpassed in Barcelona. In the Catalan city, there are 6.8 hotels and 186 local accommodations per square kilometer. Berlin does not have one hotel per square kilometer and only has 20 local lodges in that area. Amsterdam has, like Lisbon, two hotels per square kilometer, but only 58 local lodgings. And Venice has one hotel and 16 local lodgings (Figure 2).

Figure 1. Number of guests and overnight stays per thousand inhabitants, in 2017
Source: INE, Eurostat, World Travel and Tourism Council.

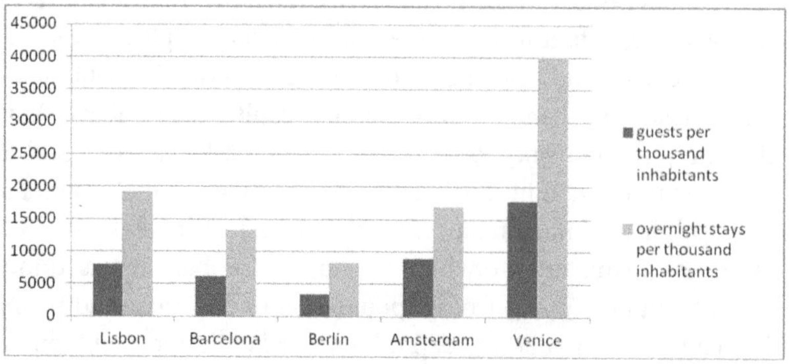

Figure 2. Number of airbnb per square kilometer, in 2017
Source: www.airdna.co

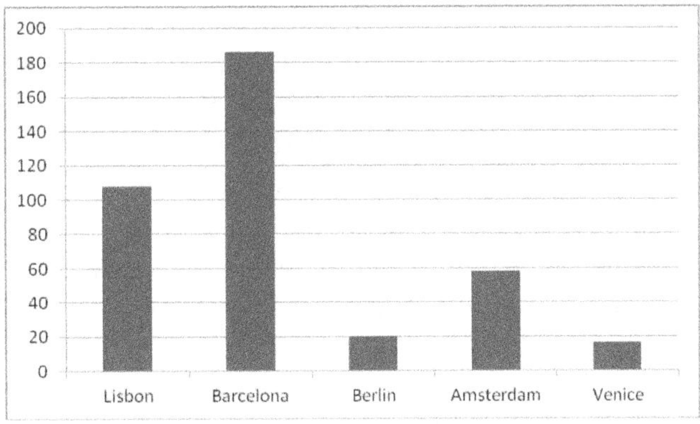

The tourist boom that gives new life and new businesses to Lisbon and Oporto is also enhancing latent tensions and generate new problems and urban and fiscal challenges. In the name of tourism, rehabilitation of vacant buildings is done, but the rents skyrocket, multiplying the evictions of vulnerable inhabitants and closures of historic shops, ie, residential and commercial displacement. On the other side, thanks to tourism, many Portuguese are launched in the business of rental rooms, many unemployed young people take the first steps in the labour market, but this does not always mean jobs with rights and income above average. In fact, it is undeniable that Lisbon is experiencing a peak of international projection and that tourism revenue can contribute to the economic recovery of the country and the city, during and after the time of capitalist crisis. Urban tourism may also be an urban regeneration engine for preservation of the architectural heritage, rehabilitation of buildings in an advanced state of degradation and contribute to job creation. But the apparent absence of either planning strategy and the evaluation of its impact, together with the almost non-existent process of regulation, have overwhelming consequences.

BEYOND TOURISTIFICATION AND THE EXPANSION OF SHORT-RENTAL TOURIST LODGING: FINANCIALISATION OF THE HOUSING MARKET AND THE NEOLIBERAL TURN ON URBAN POLICY

Fieldwork reveals that some causes behind the process are: the rise of the internet platforms peer to peer that allow clients to book directly their accommodation, such as Booking.com, Airbnb, Windu, Homeaway, etc.; the lack of hotel supply in the historic centre, where at the same time, there was a huge amount of vacant properties; the attraction of the "authenticity" of historic centre of the city to a more critical and emancipator post-tourist demand that prefers heritage and patrimony, instead of other forms of modern and massif urban tourism; the perception that the profit is higher than the traditional residential long-term leasing; the lack of legislation that allowed an easy entry for everyone interested in touristic lodging; the continued cheapening of international mobility, including through low-cost airlines. These factors triggered the entrance of new operators, lots of them individual and informal players, creating a deregulated market with under pressure margins. Also, the unexpected increase of tourism in Portugal derives largely from the decline in tourism in several Arab countries – Egypt, Tunisia, Morocco. Lisbon has a crescent notoriety worldwide – it´s an International Brand recognized as result in massive Marketing and Promotion in recent years, since Expo'98 campaign. However, these causes are just the tip of the iceberg and we need to adopt a more structural view of the process, to discover the forces behind it (Mendes, 2016a, 2018).

Since 2013, the Portuguese real estate market has been recovering the confidence of international and national investors. With the investment incentive and attraction programs – Golden Visa (GV) and the tax regime of Non-Habitual Residents (NHR) – Portugal has become one of the most attractive real estate investment destinations in Europe. At the same time, tourism has been growing and Lisbon has become one of the most important European capitals for city-break. These two factors boosted and revitalized the real estate market and forced developers to bet on previously poorly developed concepts. The strong demand compelled the developers to develop projects designed for pure real estate investment, becoming common the tourist apartments and the concept of Hotel Apartments, all with guaranteed yields and positioned for the short term rent, in the form of the so-called Local Accommodation (LA). In fact, the growth in the tourist market, through the demand for tourist apartments, has had a re-invigorating impact on the residential market. Tourist apartments, serviced apartments and local accommodation are relatively recent products in the tourist accommodation segment in Portugal. This type of accommodation has gained popularity especially in recent years and has become an alternative offer compared to the conventional and massive hotel offer. Especially for holidays and city breaks, the LA often offers a more central location - next to the historic center where there is a lack of hotel offer - and a more authentic and informal stay, also giving the possibility to accommodate the whole family or groups of friends at very competitive prices (JLL, 2015). These apartments have appeared in the market individually or in buildings totally dedicated to this end, being able to be managed by professional operators or by their owners, as it happens with the platform Airbnb, being not restricted already to the mere economy of sharing, with rooms in houses shared with the owner himself, but functioning in the form of micro-capitalist rentier, with LA practiced in whole houses, being this the regime of occupation that dominates already more than 90% now (against 75% in 2016) of the market in Lisbon.

A query to the website www.airdna.com (platform that monitors the evolution of Airbnb in several cities of the world) allows to conclude that in July 2018[2], in Lisbon, there are registered more than 14 thousand ads in the platform Airbnb and that this number has grown exponentially at an average annual rate of change of 42% since at least 2010 (Figure 3). The format of AL has been expanding, hence it is natural that the number of offers of this type of accommodation has more than quadrupled since 2014, with a very intense and concentrated territorial expression in the historical center of the city (Figure 4). This means that the offer has gone from about 13 thousand establishments in 2014 to more than 55 thousand spaces, nationally in 2017. This type of accommodation already concentrates a third of the national total of overnight stays and is close to that in the case of the number of guests. Its massive and disproportionate proliferation has fueled distortions and imbalances in the housing market for home ownership and the provision of permanent or long-term rental housing, jeopardizing the right to housing in the city center.

The players in this market can be divided into 4 large groups: the Promoters, who develop buildings totally allocated to this use and later sell the fractions with the guarantee of the management of the operation; the Operators / Owners who promote and operate their tourist apartment buildings; the Operators, who normally have, under management, third-party units (single apartments or dedicated buildings), such as "Feels Like Home", or "Lisbon Serviced Apartments"; and Individuals who manage their own apartments and who, in a market that tends to become more professional and for reasons of economy of scale, should be increasingly absorbed by large operators (JLL, 2015).

The adverse impacts on the rental market and the eviction of former residents are a reality – supported by the New Urban Rental Law of 2012 – to give rise to various forms of tourist accommodation, often of luxury, although the study of Pereira and Teixeira (2017) evidences that the majority of the dwelling rehabilitated for LA was vacant before the intervention. Numbers provided by the Local Accommodation Association in Portugal indicate that 60% of the properties were vacant and that 64% of the apartments were in very poor state of conservation before being converted into LA. Real estate owners are betting on local tourist accommodation as a more profitable and secure investment because of the general instability of the classic long-term housing. At the moment many homeowners find that the investment in LA is safer and allows for more effective and quick reproduction of real estate capital compared to what happens with long term normal rental housing. From the perspective of tenants, the owners' commitment to LA impairs the supply of homes for rent and increases the rental price to unbearable values for most households. The truth is that the investment is also more profitable even under the existing tax regime until a few months ago. The taxation regime discriminated between classic rent and rent to tourists. Normal and classical rent had an income tax rate of 28%, while tourist short rental, which is practiced by LA, only had a tax rate of 15% of the amount received.

In a study carried out in November 2016 by the Nova School of Business and Economics and by the Faculty of Law of the New University of Lisbon for the Portuguese Hotel Association, it was concluded that, from an economic point of view, LA creates pressure for rising real estate prices if it is not accompanied by an increase in supply of affordable housing. It also has a reallocation effect of properties from the traditional long term rental market to the new short rental housing market. The generated econometric model pointed to an increase in rental prices of 13.2% and 30.5% in real estate sale prices in areas where LA has a higher percentage of fires, as is the case of Misericórdia and Santa Maria Maior (both parishes in the inner city), where the LA registered in the Airbnb platform reaches on average 20% (1/5) of the total of the existing classic accommodations (AHP, 2016).

Figure 3. Number of properties available at airbnb in Lisbon (2009-2018)
Source: www.airdna.co

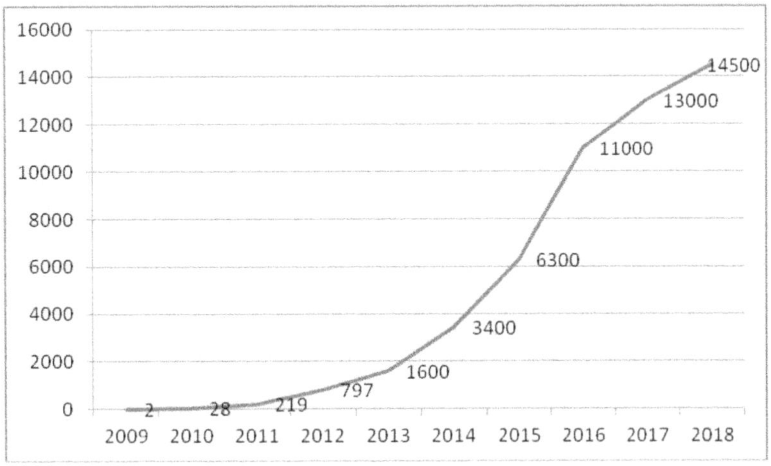

Figure 4. Distribution of properties available at airbnb in Lisbon in July 2018
Source: www.airdna.co

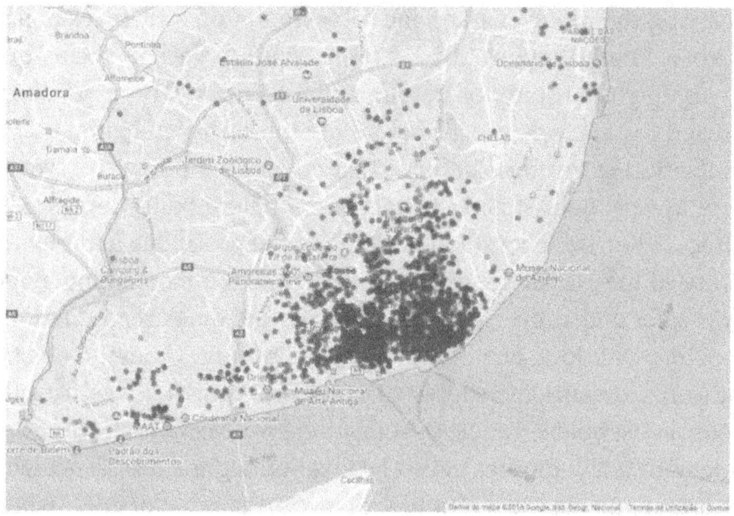

A combination of factors has led to enormous new pressures on the housing market, leveraged by the combination of liberalization of the urban lease law, the gains in tourist accommodation short-rental, and broad benefits the international real estate investment. Such pressures have resulted in a very high increase in real estate and financial speculation on behalf of a supposed urban regeneration that it seems not benefit the local population. Real estate prices skyrocketed and become increasingly difficult to find affordable housing in Lisbon.

Over the past few years we have witnessed a profound change in housing dynamics in the country's metropolitan areas. A drastic rise in housing rental values that has led to the expulsion of the population from the more central areas of the city, coupled with an abrupt drop in supply and an exponential increase in home ownership values have made access to housing in Lisbon privileges of few and almost practically inaccessible to Portuguese families. In the last five years, rental housing prices increased between 13% and 36%, and for acquisition increased to 46%, depending on the areas of the city (Confidencial Imobiliário, 2016), which results, it is estimated, a rate of between 40% and 60% of household income, while common standards recommend an effort rate of up to 30%.

On the other hand, foreign investment in real estate has been intense. In 2017, 152 thousand houses were sold, an increase of 25% to 30% compared to 2016. Brazilians, but mainly French, are the largest foreign home buyers in Portugal and are already worth 25% of the market, that is, a quarter of the houses were bought by foreigners. To these nationalities, also join the British and Chinese who make use of the gold visas for investment in the housing market.

Tourist accommodation can have positive effects on regenerating urban centers, reclassifying wealth, social diversification, supplementing residents' income, job creation and economic recovery, but their excessive supply reduces the supply of houses to rend to those who want to live in the city. The truth is that, today, many investors buy whole buildings (many to rehabilitate) to lease to tourists, mainly in the centers of Lisbon and Porto. Only in the parishes of Santa Maria Maior and Misericórdia, in the center of Lisbon, there are around 2700 local accommodation units, about 8.8% of the national offer. Renting a house in the main Portuguese cities has become an almost impossible task. The problem is not only in rent, but especially in the absence of houses available for rent. According to the Association of Professionals and Real Estate Agents of Portugal (APEMIP), the number of houses for long-term lease has fallen by 30% in the last five years, affecting mainly two cities: Porto and, of course, Lisbon, which housing rental offer reduced by 85 and 75%, respectively. However, in some of the more central neighborhoods, the decrease has exceeded 90% in the last three years.

Regarding the most recent evolution of the housing sales, according to INE, the Portuguese Housing Price Index had an average annual increase of 8.5% in the third quarter of 2017. In an analysis of Lisbon and Porto according to the data available by the end of the 2nd quarter, the average figures per square meter increased by 12.4% in both cities. With regard to the prime values of the Portuguese capital, JLL recorded appreciations throughout the year due to the lack of demand in relation to the offer, with the main areas of Lisbon growing, on average, 10% to 20%, with some areas to increase increases of more than 30% (JLL, 2017). In the first quarter of 2018, the cities of Lisbon and Porto saw the sale price of their homes soar more than 20%. They were the only ones in the country to show such a strong growth, in year-on-year terms with 2017, according to the latest statistics on the price of local housing released by the National Statistics Institute (INE). Lisbon is notably the municipality where the median value of housing prices is highest – reached 2581 euros per square meter, after a growth rate of 20.4%. Among the 24 parishes in Lisbon, Santo António maintained the highest price of housing (4083 euros per square meter) and the highest year-on-year rate of change (+ 39.8%), followed by Campolide (+32,5%), Campo de Ourique (+30,4%), Arroios (+29,5). It curious to see that these last parishes are pericentral to the ones of the inner city like Misericórdia (€ 3667) (+21,9%), which includes the Bairro Alto and Cais do Sodré areas, the parish of Santa Maria Maior (€ 3528) (+14,6%), which includes Alfama, the Castle area and Baixa (Downtown)/Chiado (INE, 2018b).

These last ones, the first to be strongly gentrified, are noble areas that have a great centrality, mainly due to the concentration of historical memory and heritage value, with neighborhoods that reveal a typical, authentic and old identity, therefore, of greater interest to channel investment and reproduction of real estate capital, besides the advantages as tourist destination and accommodation for tourists. As a result, once already gentrified, its real estate valuation has been slowing (even though they concentrate the highest prices per square meter), comparing with the pericentral areas mentioned (first ring to the historical center) that in the last year valued much more, indicating that the housing market in the center of the city is saturated and that real estate capital now seeks the pericentral border to install and reproduce.

These are the districts in Lisbon that concentrate some aspects of patrimonialisation and urban memory which, by itself, reinforce their matrix of tourist attractiveness, in the face of a capitalist and financial valorization of the urban space, but also allow the engagement with the desire for cosmopolitanism and authenticity of the tourist who seeks a tourist experience that allows him "to live like a local", an experience of visual and sensorial consumption with local residents and host communities in historic, heritage, still traditional neighborhoods of the inner city connoted with an urban authenticity, as we saw before. Milano (2018b) argues that there are three motivations (interior and atmosphere, interaction with local inhabitants and immersion in local culture), on the basis of which lies the search for an authentic experience by the tourist. The study highlights that the main characteristics of the first motivation that guests consider authentic atmosphere (the environment of the destination) and the environment of a local accommodation instead of accommodation in traditional structures, like hotels or hostels. With respect to the second, the perception of authenticity of the experience of travel lies in the role of hosts as providers of information. Finally, immersion in the local culture would result from the feeling of having an experience that tourists staying in hotels could not experience due to the standardization of experience. This tourist saturation in the search of these neighborhoods for LA explains in part these significant increases in the values of housing for purchase and lease, which already begin to reach the levels of pre capitalist crisis 2008-2009.

It is true that the local short-rent accommodation for tourists has been one of the responsible for the rise in the prices of rents and housing in the city center, aggravating the conditions of access to housing, however we must not forget that in historic centers, this recent and rampant rise in housing prices, results in a great and greater measure of the huge demand for real estate for acquisition by a transnational, foreign capitalist elite with great financial capacity and attracted by the tax regime of NHR, by the GV Act and encouraged by the tax exemption policy that benefited the Real Estate Investment Funds. These three programs of the Austerity Government (2011-2015) encouraged the accumulation of real estate capital at the expense of the progressive financialisation of the Lisbon housing stock.

One of the important reforms proposed by the Memorandum of Economic and Financial Policies was the new regime of urban rental, also known as New Urban Rental Scheme (the acronym in Portuguese: *NRAU – Novo Regime de Arrendamento Urbano*), promulgated in November 2012[3]. As mentioned earlier, this was subject to vigorous social protest, as it is alleged to have imposed an update mechanism that has caused rents unaffordable for many tenants, because values are established without adequate social support and necessary affecting families of lower socio-economic status. The displacement has been a hallmark of the new law of the urban rental, since it allows facilitating the eviction if the landlord / owner claims to want the house for his own home or descendants, or when he wishes to make claims of more structural works. NRAU it is essentially a diploma that has its primary objective on the termination of lease agreements concluded before the 90s, without the guarantee of rights to the tenants. The growing demand for rental as a result of the market crisis in the construction and real estate and the lack of affordable rental supply determined that the reform of the urban rental was assumed as a priority in housing. In fact, the reform of urban rental 2006 failed to give a sufficient answer to the main problems that face the urban leases, especially those related to contracts with previous budgets to 1990, with the difficulty of carrying out rehabilitation works in rental real estate and with a complex and lengthy eviction procedure (Mendes, 2014, 2016a; Alves, Pereira and Rafeiro, 2015).

The residential market experienced major difficulties during the crisis, causing severe consequences due to the lack of bank financing and the great impact exerted on Portuguese incomes. In an attempt to overcome such difficult times, the government launched two programs to attract foreign investment: Golden Visas and the Non-Habitual Resident Tax Regime.

The "Fiscal Regime for Non-Habitual Residents" and the "Golden Visa Portugal" (Residence Visa for so-called Investors), in turn, contributed significantly to the intensification of real estate speculation. The first is intended for EU citizens with high economic capacity that want to fix their tax address in Portugal. The second is intended for citizens of other countries and allows them residence to carry out investment activities, namely the transfer of capital, creation of employment and purchase of real estate in areas of the historic center of the city. In both cases it benefits foreign citizens with large reductions and even exemptions from taxes, introducing inequality among those benefited by promotions and fiscal balances and permanent residents, Portuguese or foreign, who do not have any tax benefits.

These fiscal programs, by providing unique and extraordinary advantages for the establishment of international financial capital in the housing stock, stimulated the development of various formats and typologies of short-rental tourist apartments generally located in the historical center of Lisbon, where there was a lack of this type of offer, a huge stock of empty and dilapidated buildings, and, at the same time, an ever-increasing tourist demand for a local, welcoming, typical and more "authentic" stay in a historic neighborhood. Simultaneously, investors seeking to obtain Golden Visa have seen in this market an opportunity to earn income from their real estate assets (Mendes, 2016a; Montezuma and McGarrigle, 2018). With all this dynamics, we can now see a great dynamism in the rehabilitation of buildings in the historical center of the city, which contributed to the continuous increase in the supply of this type of real estate product, truly subsidized by a growing financialization of the housing market, same time displacing the poorest population already living in the historic center.

CONCLUSION

Portugal has been regaining confidence of international and national investors in the past three years, while tourism has been growing very fast and Lisbon has become one of the leading European capitals in city break. The combination of these factors, driven by further incentive to foreign investment programs and the growing importance of this segment in urban regeneration, revitalized the residential market and place it in an unbeatable time. The Golden Visa and the NHR programs, the new urban lease law, the new tax regime for Property Investment Funds, the new law of tourist lodging, along with a strong growth in tourism in the city of Lisbon, introduced significant changes in the residential market, which went from an abrupt pause to a high level of demand, with supply now beginning to fall short. This situation led to a very quick take-up of the new and good quality residential stock that was available, located mostly in the city's historic centre. The rapid take-up of apartments, the new tourism drivers – with an "alternative" demand for stays in apartments – and tax incentives to boost urban regeneration; have created renewed interest among many national and international developers, leading to a great rise in the refurbishment of buildings in Lisbon's historic district. In this chapter we present evidence of the deepening of the tendencies of privatization, commodification, financialization of the housing market of Lisbon, in favor of a growing tourism led gentrification.

But it was not always so. In fact, the city of Lisbon has always been very resilient to gentrification, at least until the end of the 20th century. First, the successive legislative packages of urban rehabilitation programs aimed at improving the living conditions of residents until the beginning of the 20th century worked as a brake on the forces of gentrification. All the urban rehabilitation programmes launched by the Government since the mid-1970s, have led to rehabilitating and upgrading existing buildings in the historical city centre of Lisbon by respecting collective public interests (Mendes, 2008, 2014). Second, control of rents established in 1948, with retroactive application to 1942, in Lisbon and Porto, generated a policy of frozen rents in the following decades, which meant that the prices of rents remained very low, which would have justified a weak competitiveness in the rental market, not allowing the eviction and release of leased dwellings. Thus, the immediate consequence was a large housing immobility of those who were already settled in the center city (where the rental market is more representative) and a shift to the periphery of those seeking more modern housing, that is, young couples and immigrants. It also contributed to the acceleration of the degradation of the housing stock, since the rents practiced did not compensate to the landlords the investments necessary to carry out works of conservation and maintenance of the buildings.

However, it is precisely in the capitalist crisis of 2008-2009 that the inflection between these two stages of gentrification takes place and in which tourism is affirmed and legitimized as a panacea for the social and urban crisis. The Portuguese financial crisis (2009-2013) was a major political and economic crisis, related to the European sovereign debt crisis and its strong impact on Portugal. The crisis began to occur in the first weeks of 2010 and only began to disappear with the start of the Portuguese economic recovery at the end of 2013. It was the most severe recession in the Portuguese economy since the 1970s and tourism definitely helped create jobs and to mitigate the consequences and impacts of the crisis and austerity among Portuguese families. In part, this validation by civil society and public opinion legitimized this neoliberal subjectivity. That is why in Lisbon, tourism gentrification has been privileged as a political strategy for urban revitalization promoted by the Capitalist State, based on the neoliberal argument and the ideology of austerity urbanism that legitimize the hegemony of tourism as a panacea in the post-crisis capitalist context.

New trends in urban policies, from the beginning of the 21st century, mainly oriented to a hegemonic touristification in the historic districts of Lisbon, have fostered the emergence of more aggressive forms of tourist and commercial gentrification, which seem to lead to displacement (even that indirect) and worsening of residential segregation and socio-spatial fragmentation. Physical structure and environmental conditions have changed, the poor, immigrant and aged population has been displaced from the inner city area, and the higher income groups had taken the area over. With this transformation, the area has been re-structured according to the expectations of the local and national authority and the increased rent has been facilitated. These conclusions strengthen the idea that gentrification is effective in the regeneration process and this is evolving in Lisbon´s inner city to another level.

Continued growth in the LA market is expected. Investors and national and international promoters continue to buy buildings in the historical center of Lisbon to rehabilitate for this specific use. The conversion of these assets to the use of tourist apartments allows a greater profitability for investors, a tendency increasingly taken into account by the investors GV and NHR that thus derive their income from assets, contributing to the financialization of housing. Due to the still informal process of managing these assets and the lack of regulation, it is difficult to identify new projects in this market and monitor their evolution, but a steady growth rate is expected.

It is true that a significant part of the occupation of the tourist accommodation of residential buildings did not imply eviction and direct eviction of the residents, since the buildings were empty and in an advanced state of bad conservation, presenting itself empty or in urban void condition. On the other hand, the eviction we are witnessing is equally indirect and just as severe. While direct eviction refers to the actual eviction / displacement moment, due to rapid rise in rents and real estate values, which the current residents can no longer afford, what happened in both Lisbon and Porto concerns with indirect eviction, which is a long-term process that results from a set of pressures that makes it progressively difficult for low-income residents to stay longer in the neighborhood or others who want to live and attend the neighborhood can do so. And this is also because traditional retail is disappearing, which was of extreme importance because it served the daily needs of the aging and poorest resident population of the historical center, contributing to the community's resilience and its cohesion and social sustainability, being fragile and at risk now.

Overtourism and tourism gentrification, in specific, and urban tourism, in general, are definitely asserting themselves on the political agenda of the social sciences and urban theory as a question of Right to the City, right to habitat and to inhabit. The right to the city involves collective creation of our niche, our built environments and modified landscapes. And for this it involves the deepening of democracy, de-financialization of housing and the de-commodification of urban space, making room for the common and colective construction of place and housing, in the sense that another city is possible, beyond the correlated processes of financialisation, privatization and appropriation by dispossession of the real estate and the public space of the city, brought by the ideology of austerity and neoliberal urbanism.

ACKNOWLEDGMENT

We acknowledge financial support from the Portuguese Science and Technology Research Council (FCT) within the scope of the SMARTOUR project (Ref: PTDC/GES-URB/30551/2017).

REFERENCES

Aalbers, M. (Ed.). (2012). *Subprime Cities: The Political Economy of Mortgage Markets*. Oxford, UK: Wiley Blackwell. doi:10.1002/9781444347456

Aalbers, M. (2016). *The Financialization of Housing: a Political Economy Approach*. London, UK: Routledge. doi:10.4324/9781315668666

Agyeiwaah, E., McKercher, B., & Suntikul, W. (2017). Identifying core indicators of sustainable tourism: A path forward? *Tourism Management Perspectives*, *24*, 26–33. doi:10.1016/j.tmp.2017.07.005

AHP. (2016). *O Alojamento Local em Portugal: qual o fenómeno?* Lisboa, Portugal: Nova School of Business and Economics e Faculdade de Direito da Universidade Nova de Lisboa.

Alves, R., Pereira, M., & Rafeiro, M. (2015). O Memorando de Entendimento e as reformas no mercado de arrendamento e na reabilitação urbana. In *M. Lurdes Rodrigues, & P. Adão e Silva (Orgs.), Governar com a Troika: Políticas Públicas em Tempo de Austeridade* (pp. 237–259). Coimbra: Almedina.

Alves, S. (2017). Requaliðcação e reabilitação do centro histórico do Porto. *Scripta Nova. Revista Electrónica de Geografía y Ciencias Sociales*, *21*(557), 1–32.

Ashworth, G., & Page, S. (2011). Urban tourism research: Recent progress and current paradoxes. *Tourism Management*, *32*(1), 1–15. doi:10.1016/j.tourman.2010.02.002

Ashworth, G., & Tunbridge, J. (1990). *The tourist-historic city*. London, UK: Belhaven.

Barata Salgueiro, T. (2017). Alojamentos Turísticos em Lisboa. *Scripta Nova. Revista Electrónica de Geografía y Ciencias Sociales*, *21*(578), 1–43.

Barata Salgueiro, T., Mendes, L., & Guimarães, P. (2017). Tourism and urban changes: lessons from Lisbon. In M. Gravary-Barbas, & S. Guinand (Eds.), *Tourism and Gentrification in Contemporary Metropolises: International Perspectives* (pp. 255–275). London, UK: Routledge. doi:10.4324/9781315629759-11

Bec, A., McLennan, C., & Moyle, B. (2016). Community resilience to long-term tourism decline and rejuvenation: A literature review and conceptual model. *Current Issues in Tourism*, *19*(5), 431–457. doi:10.1080/13683500.2015.1083538

Burns, P., & Holden, A. (1995). *Tourism: a New Perspective*. Harlow: Prentice Hall.

Cheer, J. M., & Lew, A. A. (Eds.). (2018). *Tourism, Resilience and Sustainability: Adapting to Social, Political and Economic Change*. London, UK: Routledge.

Cocola-Gant, A. (2015). Tourism and commercial gentrification. *RC21 International Conference on "The Ideal City: between myth and reality. Representations, policies, contradictions, and challenges for tomorrow's urban life"*. Urbino (Italy), August 27-29, 2015.

Cocola-Gant, A. (2018). Tourism gentrification. In L. Lees, & M. Phillips (Eds.), *Handbook of Gentrification Studies* (pp. 281–293). Cheltenham, UK: Edward Elgar Publishing. doi:10.4337/9781785361746.00028

Confidencial Imobiliário. (2016). Portuguese Housing Market Survey. Monthly Report, October. Lisboa.

Delgadillo, V. (2015). Patrimonio urbano, turismo y gentrificación. In V. Delgadillo, I. Díaz, & L. Salinas (Orgs.), Perspectivas del Estudio de la Gentrificacíon en México y América Latina (pp. 113-132). Coyoacán: Instituto de Geografía, UNAM.

Dichter, A., & Manzo, G. (Orgs.) (2017). *Coping with Success. Managing Overcrowding in Tourism Destinations*. London, UK: McKinsey & Company and World Travel & Tourism Council. [Online], [Accessed July 5, 2018], Available from https://www.wttc.org/-/media/files/reports/policy-research/coping-with-success---managing-overcrowding-in-tourism-destinations-2017.pdf

Espiner, S., Orchiston, C., & Higham, J. (2017). Resilience and sustainability: A complementary relationship? Towards a practical conceptual model for the sustainability–resilience nexus in tourism. *Journal of Sustainable Tourism*, *25*(10), 1385–1400. doi:10.1080/09669582.2017.1281929

Goodwin, H. (2017). *The challenge of overtourism*. Responsible Tourism Partnership Working Paper 4. October 2017 [Online], [Accessed 5 July 2018], Available at http://haroldgoodwin.info/pubs/RTP'WP4Overtourism01'2017.pdf

Gotham, K. (2005). Tourism Gentrification: The Case of New Orleans' Vieux Carre (French Quarter). *Urban Studies (Edinburgh, Scotland)*, *42*(7), 1099–1121. doi:10.1080/00420980500120881

Gravary-Barbas, M., & Guinand, S. (2017). *Tourism and Gentrification in Contemporary Metropolises: International Perspectives*. London, UK: Routledge. doi:10.4324/9781315629759

Hiernaux, D., & González, C. (2014). Turismo y gentrificación: Pistas teóricas sobre una articulación. *Revista de Geografía Norte Grande*, *58*(58), 55–70. doi:10.4067/S0718-34022014000200004

Hoffman, L., Fainstein, S., & Judd, D. (2003). *Cities and Visitors: Regulating People, Markets and City Space*. Oxford, UK: Blackwell. doi:10.1002/9780470773673

INE. (2018a). *Resultados Preliminares sobre Actividade Turística 2017*. Lisboa, Portugal: Instituto Nacional de Estatística.

INE. (2018b). *Estatísticas de Preços da Habitação Local*. Lisboa, Portugal: Instituto Nacional de Estatística.

Jamal, T., & Camargo, B. (2017). Tourism governance and policy: Whither justice? *Tourism Management Perspectives*. doi:10.1016/j.tmp.2017.11.009

Kato, K. (2018). Debating Sustainability in Tourism Development: Resilience, Traditional Knowledge and Community: A Post-disaster Perspective. *Tourism Planning & Development*, *15*(1), 55–67. doi:10.1080/21568316.2017.1312508

Krähmer, K. (2017). *Gentrification without gentry? Tourism and Real Estate Investment in Lisbon.* Master's thesis. Torino: Politecnico di Torino, Corso di laurea magistrale in Pianificazione Territoriale, Urbanistica e Paesaggistico-Ambientale.

LaSalle, J. L. (2015). *Apartamentos Turísticos em Lisboa*. Lisboa, Portugal: Um Mercado em Crescimento.

LaSalle, J. L. (2017). Portugal Real State, Market 360º. Lisboa, Portugal.

Law, Ch. (2002). *Urban Tourism. The Visitor Economy and the Growth of Large Cities*. London, UK: Mansell.

Lestegás, I. (2019). Lisbon after the crisis: From credit-fuelled suburbanization to tourist-driven gentrification. *International Journal of Urban and Regional Research*, *43*(4), 705–723. doi:10.1111/1468-2427.12826

Lew, A. A., & Cheer, J. M. (Eds.). (2018). *Tourism resilience and adaptation to environmental change: Definitions and frameworks*. London, UK: Routledge.

Luthe, T., & Wyss, R. (2014). Assessing and planning resilience in tourism. *Tourism Management*, *44*, 161–163. doi:10.1016/j.tourman.2014.03.011

Mansilla, J. (2019). Gentrificaci6n, turistificaci6n y clases socials en las ciudades del Mediterráneo. *Alba Sud*, *9*, 62–65.

Mendes, L. (2006). A nobilitação urbana no Bairro Alto: Análise de um processo de recomposição socio-espacial. *Finisterra*, *41*(81), 57–82.

Mendes, L. (2008). *A Nobilitação Urbana no Bairro Alto: Análise de um Processo de Recomposição Sócio-espacial.* Master's thesis. University of Lisbon, Portugal.

Mendes, L. (2013). Public policies on urban rehabilitation and their effects on gentrification in Lisbon. *AGIR. Revista Interdisciplinar de Ciências Sociais e Humanas*, *1*(5), 200–218.

Mendes, L. (2014). Gentrificação e políticas de reabilitação urbana em Portugal: Uma análise crítica à luz da tese rent gap de Neil Smith. *Cadernos Metrópole*, *16*(32), 487–511. doi:10.1590/2236-9996.2014-3209

Mendes, L. (2016a). Tourism gentrification: touristification as Lisbon´s new urban frontier of gentrification. Master Class *Tourism Gentrification and City Making*, Stadslab and Academia Cidadã, Lisbon, April 16.

Mendes, L. (2016b). What can be done to resist or mitigate tourism gentrification in Lisbon? Some Policy Findings & Recommendations. In M. Glaudemans, & I. Marko (Eds.), *City Making & Tourism Gentrification* (pp. 34–41). Tilburg: Stadslab.

Mendes, L. (2017). Gentrificação turística em Lisboa: Neoliberalismo, financeirização e urbanismo austeritário em tempos de pós-crise capitalista 2008-2009. *Cadernos Metrópole*, *19*(39), 479–512. doi:10.1590/2236-9996.2017-3906

Mendes, L. (2018). Tourism gentrification in Lisbon: The panacea of touristification as a scenario of a post-capitalist crisis. In I. David (Eds.), Crisis, Austerity, and Transformation: How Disciplinary Neoliberalism is Changing Portugal (pp. 25-48). London, UK: Lexington.

Milano, C. (2017). *Overtourism y Turismofobia. Tendencias globales y contextos locales*. Barcelona, Spain: Ostelea School of Tourism & Hospitality.

Milano, C. (2018a). Overtourism, malestar social y turismofobia. Un debate controvertido. *Pasos: Revista de Turismo y Patrimonio Cultural, 16*(3), 551–564. doi:10.25145/j.pasos.2018.16.041

Milano, C. (2018b). *Tendencias en viajes: "live like a local" y saturación turística*. Barcelona, Spain: Ostelea School of Tourism & Hospitality.

Montezuma, J., & McGarrigle, J. (2018). What motivates international homebuyers? Investor to lifestyle 'migrants' in a tourist city. *Tourism Geographies*. doi:10.1080/14616688.2018.1470196

Mullins, P. (1991). Tourism Urbanization. *International Journal of Urban and Regional Research, 15*(3), 326–342. doi:10.1111/j.1468-2427.1991.tb00642.x

Munro, J., Allison, H., & Moore, S. (2010). Using resilience concepts to investigate the impacts of protected area tourism on communities. *Annals of Tourism Research, 37*(2), 499–519. doi:10.1016/j.annals.2009.11.001

Nofre, J., Martins, J. C., Vaz, D., Fina, R., Sequera, J., & Vale, P. (2018). *The 'Pink Street' in Cais do Sodré: urban change and liminal governance in a nightlife district of Lisbon*. Urban Research & Practice; doi:10.1080/17535069.2018.1449010

Page, S. (2003). *Tourism Management: Managing for Change*. Oxford, UK: Elsevier.

Pavel, F. (2017). Turismo e mercantilização da cidade. *Revista espaços vividos e espaços construídos: estudos sobre a cidade, 5*(1), 21-28.

Peeters, P., Gössling, S., Klijs, J., Milano, C., Novelli, M., Dijkmans, C., ... Postma, A. (2018). *Research for TRAN Committee - Overtourism: impact and possible policy responses*. Brussels, Belgium: European Parliament, Policy Department for Structural and Cohesion Policies.

Pereira, H., & Teixeira, N. (2017). *(coord.). O impacto económico do Alojamento Local na Área Metropolitana de Lisboa 2016-2020*. Lisboa, Portugal: AHRESP.

Pinkster, F., & Boterman, W. (2017). When the spell is broken: Gentrification, urban tourism and privileged discontent in the Amsterdam canal district. *Cultural Geographies, 24*(3), 457–472. doi:10.1177/1474474017706176 PMID:29278248

Prayag, G. (2017). Symbiotic relationship or not? Understanding resilience and crisis management in tourism. *Tourism Management Perspectives*. doi:10.1016/j.tmp.2017.11.012

Sheppard, V., & Williams, P. (2016). Factors that strengthen tourism resort resilience. *Journal of Hospitality and Tourism Management, 28*, 20–30. doi:10.1016/j.jhtm.2016.04.006

Tulumello, S. (2015). Reconsidering neoliberal urban planning in times of crisis: Urban regeneration policy in a "dense" space in Lisbon. *Urban Geography*. doi:10.1080/02723638.2015.1056605

Wilson, J., & Tallon, A. (2012). Geographies of gentrification and tourism. In J. Wilson (Ed.), *The Routledge Handbook of Tourism Geographies* (pp. 103–112). London, UK: Routledge. doi:10.4324/9780203859742

WTTC. (2018). *Travel and Tourism Economic Impact 2018 Portugal Report*. London, UK: World Travel & Tourism Council.

Xian Liang, Z., & Gang Bao, J. (2015). Tourism gentrification in Shenzhen, China: Causes and socio-spatial consequences. *Tourism Geographies*, *17*(3), 461–481. doi:10.1080/14616688.2014.1000954

Zwiersa, M., Gideon, B., Van Ham, M., & Van Kempen, R. (2016). The global financial crisis and neighborhood decline. *Urban Geography*, *37*(5), 664–684. Doi:10.1080/02723638.2015.1101251.

ENDNOTES

1. Numbers provided by https://eco.pt/2018/03/18/lisboa-ainda-nao-e-barcelona-nao-esta-assim-tao-longe/, on the basis of INE data and in others used by Eurostat or by organizations such as the World Travel and Tourism Council (accessed 5 June, 2018).
2. According to https://www.airdna.co/market-data/app/pt/lisboa/lisbon/overview (accessed 5 July, 2018).
3. from the Law 31, of 14 August 2012.

Chapter 9
Touristificacion in the Central Market of Valencia:
Fact or Fiction?

Alexandra Joana Bol Esteve
University of Valencia, Spain

Ruben Arnandis-i-Agramunt
https://orcid.org/0000-0002-8928-530X
University of Valencia, Spain

ABSTRACT

This chapter analyses the relationship between cultural heritage and the tourism activity in the ongoing context of globalisation. It establishes the debate among society and tourism considering the process of touristification and its impact in cultural heritage, addressed initially to locals. Concretely, this investigation is focused on Valencia's Central Market, Spain, as a case study. By fieldwork research methods (direct observation and a questionnaire), a diagnosis of the central market's situation and its consumers' perception is presented, which reflects the transformation that it is suffering because of the tourism activity. A process of touristification has been identified in some areas of the Central Market with relation of the stall tipologies and location, related to the main entrances. The saturation of touristic flows during weekends has also been demonstrated in this investigation.

INTRODUCTION

Society changes at a very fast pace. The phenomenon of globalisation prevails in the modern world, mainly in urban habitats, giving rise to homogenization effects and jeopardizing local identity. In this context, a tendency of recovering the past emerges gradually. Studies have shown that a raising awareness of "patrimonialisation" to connect people and identity through heritage flourishes and cultural heritage has a key role in producing a sense of belonging and building local identity.

DOI: 10.4018/978-1-7998-2224-0.ch009

At the same time, another tendency is emerging in tourism demand. After the cultural boom in a globalised environment, tourists look for different and authentic experiences that can only be found in specific places. Travelling is now understood as an elevating and personal growth experience. Furthermore, tourism experiences no longer consist of must-see sights but rather deep experiences as a part of locals every day's life; to feel, listen and behave as if they were so.

In this vein, local markets have gained popularity. On one hand, they represent an opportunity to encourage local values and preserve traditions to create local identity. On the other hand, they meet the needs of the new tourism demand as they reflect the lifestyle of a community; habits and peculiarities allow tourists to be part of this experience interacting with locals in their daily life.

Presenting the previous points in more detail, this chapter aims to introduce the key role that local markets play today in cultural heritage and urban tourism to also understand its potential and opportunities for both local community and tourism demand. This paper analyses, specifically, the span of the touristification process that is undertaking in the Central Market of Valencia, Spain, how it is related to overtourism and the perception by its local consumers.

This local market represents the most outstanding site of Modernism in Valencia, used by local and regional tourism boards as an icon in their marketing campaigns, and it is one of the most visited sites in town. These features contrast to the fact that it is still a much-appreciated market for the community as a place to find a wide variety of fresh local products and even more, as a socializing space. It is not just a market; it is part of their lifestyle identity. As it has become so popular for tourism demand due to all aforementioned, during the last few years it is one of the local customers' concerns. For that reason, the mission of this study is to identify what kind of tourism impacts occur and whether they are transforming both cultural heritage and local customer habits.

GLOBALISATION AND IDENTITY

Today's world is characterized by presenting big changes in short-time periods. The increasingly extensive globalisation and therefore the homogenisation effect alters territories resulting in a sense of uprooting among the local community. At this point, some reflexions appear about whether this development achieved is desired. (Soldino Nicolás, 2004)

Culture in the Globalised Context

Globalisation is settled as an open-ended socio-historical context related to contemporary development. During the last years, this process has been intensified to the extent that modern social life is understood on a global scale with transnational social relations. (Hernández i Martí, Santamarina Campos, Moncusí Ferré, & Albert Rodrigo, 2005). This phenomenon explains that time and space are no longer together, as Giddens (1993) defined it when he talked about "uprooting mechanisms". In this regard, people take distance from their local communities, whose they are spatially linked, to live "the world's time". For this reason, social relations are restructured in new undefined temporary space intervals. (Giddens, 1993)

Set this context, local level affects globality and globalisation, at the same time, affects the local level. Regarding this fact, it can be assumed that the local and the global feed each other. Indeed, Robertson (1995) introduces the concept of "Glocalisation", understood as "the universalisation of the particularisms and the particularisation of the universalisms". Robertson established this concept as the key point to understand culture construction in the modern world. Again, it explains how local cultures are adapted to a global phenomenon and how local creations have relevant global consequences simultaneously. “As to do so, connections along cultural referents are reinforced and they are consumed and interpreted commonly, but diverse”. (Hernández i Martí, 2005)

Then, a process known as the deterritorialization of culture is intensified. Individuals establish social and cultural ties within an external framework outside their territory, unlike traditional cultural relationships that were established at the local level, regional or, at best, national, framed with the near and familiar scope. That is why deterritorialization manifests the loss of the traditional relationship of culture with the geographical and social territories (Hernández i Martí, 2005) and creates a global homogeneous environment that hinders the preservation of local identity.

Looking for Identity Through Heritage

In a globalised world, the surrounding changes are assumed as inexorable. However, these transformations are questioned due to the generation of new impersonal spaces with no relation to the history nor the culture of the place. These variations, affecting especially to big cities, are producing impersonalized spaces and a clear feeling of uprooting in people (Soldino Nicolás, 2004). Therefore, with the emergence of modernity and the conception of the past as a distant and diffused matter, a collective conscience of searching the local identity arises (Hernández i Martí, 2005). That means big changes and cultural homogenisation are translated into an ambition to build an identity story through cultural heritage to generate a sense of belonging among the local society.

Cultural heritage is then presented as the basis of the past and authenticity. According to Prats (1997), the fundamental nature of heritage is the power to express an identity. Therefore, it represents the primary element for the creation of collective identities (Santamarina Campos, 2005), as it gives meaning to the created story and provides a sensation of rooting. In this sense, heritage conservation is the key factor that responds to frustration and impotence in facing threats against cultural identity (Soldino Nicolás, 2004). In other words, heritage conservation enables reterritorialization processes while reaffirming and revitalizing the local specificities. (Hernández i Martí, 2005)

After these previous considerations, it can be confirmed that globalisation contributes to the homogenisation. Concurrently, one of its direct consequences is the heterogeneous process that arises in response to this phenomenon: rebuilding the meaning and the identity of local communities (Robertson, 1995). Besides, facing the threat of the present, the fugacity and the spatial contraction, heritage represents a warning to preserve the past and a call for the construction of this identity. (Hernández i Martí, 2005)

CULTURAL TOURISM AND THE EVOLUTION OF THE TOURISTIC DEMAND

Heritage conservation trend has been accompanied by another growing tendency in tourist demand, cultural tourism. This trend appreciates the intrinsic authenticity of heritage and, specifically, has the objective of experiencing the daily life of each territory by first-hand.

Cultural Tourism

According to Ruíz (1997), cultural tourism has been considered for years as the overcoming of consumerism and evasion tourism, as a way of linking communities and as a lever for economic development for regions. Furthermore, cultural tourism had been characterized by having more purchasing power and more quality than other tourism demands (Ruíz, 1997).

As Richards (2018) pointed out after World War II travelling helped to increase cultural understanding and rebuild shattered economies. From this point, tourism and culture consumption started to rise until the creation of the cultural tourism label in 1980 and it continued growing until the 1990s receiving the identification of a "good type" of tourism. That supposed understanding tourism as a way to stimulate the economy and help culture conservation. (Richards, 2018).

According to Hernández i Martí (2005), cultural tourism helps to promote sustainable development of regions, as it helps personal fulfilment, contributes to education, promotes respect for the identity and dignity of territories and people, affirms diversity and cultural originality and fosters a properly protected environment (Hernández i Martí, 2005). This is also stated in the UNTWO report on synergies between culture and tourism, which explains how tourism and culture can work together to promote cultural development, identity creation, social cohesion and economic growth. (Richards & Marques, 2018)

It can be said then that cultural tourism represents an opportunity for the development of the territories as it allows them to strengthen their cultural identity by enhancing the heritage and fostering the mechanisms of heritage conservation and the creation of identities. "However, the growth of budget airlines, the sharing economy and social media have led to an increase in sheer numbers and a broadening and diversification of the market of cultural tourism" (Pinke-Sziva, Smith, Olt, & Berezvai 2019:1). Richards (2018) also indicates the change of orientation towards a mass-market and its fragmentation into different categories, building a wider concept of cultural tourism which has been redefined by the UNTWO (2019:30): "Cultural tourism is a type of tourism activity in which the visitor's essential motivation is to learn, discover, experience and consume the tangible and intangible cultural attractions/ products in a tourism destination. These attractions/products related to a set of distinctive material, intellectual, spiritual and emotional features of a society that encompasses arts and architecture, historical and cultural heritage, culinary heritage, literature, music, creative industries and the living cultures with their lifestyles, value systems, beliefs and traditions."

Insofar, there is a growing range of cultural phenomena in recent years which is opening the borders of cultural tourism consumption including tangible and intangible, built and mobile assets, and ultimately the daily life of the destination. (Richards & Marques, 2018)

A New Tourism Demand

The end of the 80s supposed an increase in tourism demand interested in culture overturning the hegemony of "sun and beach" (De la Casa Martínez, 2000). The main motivation of the trip was determined by the accomplishment or participation in a cultural activity. At that moment, the importance of the cultural patrimony was based on its physical features and tourists began to consume the main monuments and symbols of the cities, the "must-see sights" (Richards & Marques, 2018).

However, the 90s showed a clear expansion of attractions linked to popular culture, tourists became more interested in the elements that translate the people's way of life history, including intangible elements such as music and gastronomy (Richards, 2007).

As cultural tourism has grown significantly in recent decades (Novy, 2018), the cultural object of tourism has also shifted: the search for the exceptional (Richards, 2018). The latest trends in tourism demand show how "last-or third-generation tourists tend to take advantage of their stay to look for enriching and unique deeper cultural experiences" (Crespi Vallbona & Domínguez Pérez, 2016) Tourists now pursue experiences that make them feel part of the daily life of local people linked to their cultural traditions and giving importance to personal contact. That means, tourists want to experience the authentic nature of different spaces by first-hand and to avoid being labelled as tourists; to live as if they were one local more (Richards, 2018).

Dirksmeier and Helbrecht (2017) identify the concept of New Urban tourism as tourists interested in urban neighbourhoods "off the beaten track". That means, neighbourhoods characterised as authentic and heterogeneous places (edgy, unpolished, creative…) where tourists and locals share the same space (Sommer & Helbrecht, 2017). The 'local' becomes the target of tourism consumption and the producer of the local culture sought by tourists (Richards & Marques, 2018). Maitland (2010) also sustains that every day and mundane activities of residents and "not planned for tourism" areas are seen as an opportunity of discovering the real distinctiveness. As he mentioned, it allows tourists to create their narratives of the city.

Shadows of Cultural Tourism

Cultural tourism growth has also brought some unexpected effects to cities (Soldino Nicolás, 2004) especially historic centers (Yubero, García & de la Calle, 2017) According to Pinke-Sziva, Smith, Olt, & Berezvai (2019) congestion, loss of privacy, gentrification, out-migration and competition over resources and space are some of the effects linked to flawed or absent planning in cultural tourism, which results in developments that are contrary to local communities' interests. When the visit becomes an invasion and when its predatory potential emerges, the right of the visitor hits the inhabitants' right (Ruiz, 1997). At this point, the overuse degrades many places with the consequent destruction of their heritage. Hernández i Martí (2005) declares that the tourist instrumentation of heritage produces unwanted collateral effects such as cultural homogenization, impoverishment and banalization of local traditions that are being transformed into simple and uprooted souvenirs. It is, in fact, a paradox. The same cultural tourism that promotes the creation of collective identity and represents an opportunity for local development is the one which carries out problems of culture banalization and sacralization.

These transformations have damaging consequences, especially for the local community. During the past two years, conflicts with resident population, who can no longer continue living in their areas, have been in the headlines of newspapers. Cities like Barcelona, Madrid, Berlin or Venice have already suffered from many of these problems (Ballester, 2014; Berrueta, 2017; Bocadosdemadriz, 2018; Füller & Michel, 2014; Martín 2017; Navarro, 2018). It reflects the tourism economic logic over humanistic aspects (Hernàndez i Martí, 2005), excluding the two main goals of sustainable development related to society and the environment. At this moment, services, facilities and shops where heritage is located become tourism-oriented because they are being focused on tourists rather than citizens (Canalis, 2017). That is what the literature identifies as touristification.

TOURISTIFICATION: PROCESS DEFINITION, APPROACHES AND RELATION TO OVERTOURISM

In the words of Wöhler (2011), the concept of touristification, even its empirical observation and its description, differs among scholars in tourism studies. This situation implies that a well-established and suitably-defined concept has still not been found (Freytag & Bauder, 2018). At the same time, different conceptual approaches over the past few years have emerged (Sequera & Nofre, 2018). One of these contributions was suggested by Del Romero (2018), who stated that touristification is the result of a process in which places become a tourism commodity itself due to relatively spontaneous, unplanned massive development of tourism. This definition is in line with the works of Belhasen, Uriely & Assor (2014), Fernandes, (2011) and Leite (2008) when touristification is the result of the complete transformation on the urban space into a touristic one.

A more in-deep review of the literature shows that other researches, as Brito (2007), Pérez (2013) and Taleb (2013), define touristification as a systematic removal of uncertainty and randomness from attractions and places, trying to make them more predictable in their smallest details, in the interest of comfort, convenience and efficiency.

In all these cases, which can be defined as a market-oriented approach, touristification focus on top-down initiatives from public authorities (Gotham, 2005), so it has more in common with the term gentrification than with the touristifying process (Cummings, 2015).

Freytag & Bauder (2018) affirmed that touristification is also a transformation of urban spaces into tourist areas that are produced by visitors and their practices. They understand that touristification is a bottom-up process in which tourist actively take part with all different activities in town. What is more, these aforementioned authors give support to the idea that urban transformation (shopping, mobility), tourism growth (accommodations, restaurants) and touristification process are mutually related. That is what Baerenholdt, Haldrup, Larsen & Urry (2004) defined as "performing tourism" approach.

Defining the scope of touristificacion is by no means an easy endeavour. While Gravari-Barvas and Guinand (2017a, 2017b) focus on how touristification, and tourist-based transformation of urban environments and gentrification, can take different forms depending on the place, other scholars put emphasis on the consequences (Table 1): how the number of visitors has exceeded the physical and psychological carrying capacity (Colomb & Novy, 2017), how overcrowded urban spots can provoke feelings of dissatisfaction on tourist (Popp, 2009) or how neighbourhood express their rage against tourism (Füller & Michel, 2014).

Nonetheless, to what extent touristification is provoking these outcomes? Koens, Postma & Papp (2018) highlighted that overtourism cannot be dealt with sufficiently by focusing on tourism alone. In other words: overtourism is not a tourism-only problem. Other issues whether touristification is a positive or a negative process, or if touristification is a positive process when is well-planned or a negative consequence otherwise are still unanswered questions.

What is clear, at least, is that touristification is not a unidimensional, determinist cause-and-effect model (Freitag & Bauder, 2018; Sequera & Nofre, 2018). Policy actions are required to consider wider city usage (leisure, retail trade, transport) and cooperation between multiple city departments and other stakeholders must be into consideration.

Table 1 . Issues attributed to tourism. Extracted from Koens, Postma & Papp (2018)

Issue	Type of impact	Spatial Distribution
Overcrowding in city's public spaces	Overcrowding on streets and pavements, as well as public transport, heavy traffic, loss of local identity	Tourist hotspots and newly developing tourist areas
Pervasiveness of visitor impact due to inappropriate behavior	Noise, disturbance, loss of local identity	Tourist hotspots and newly developing tourist areas
Physical touristification of city centers and other often-visited areas	Loss of amenities for residents due to mono-culture of tourist shops and facilities	Tourist hotspots and city centers
Residents pushed out of residential areas due to AirBnB and similar platforms	Less availability of housing, loss of sense of community and security	Throughout city, mainly near tourist hotspots
Pressure on local environment	Increased waste, water use, air pollution	Throughout the city, near specific sites (harbour, road junctions)

THE ROLE OF LOCAL MARKETS IN THE TOURISTIC OFFER

Local markets embody the history and culture, eating practices, tastes and gastronomic habits of the society in a specific territory. In this way, markets are associated with values such as quality, friendly treatment, trusted products, proximity trade and improvement of the local economy (Crespi Vallbona & Domínguez Pérez, 2016). Thus, both good planning and management, they can be a key point to revitalize a city.

Local Markets as Heritage Attractions

While globalization is affecting and transforming the culture of different places, accompanied by a homogeneous effect, local markets represent an opportunity for the generation of collective images in the representation of cultural specificities.

Local markets are more than shopping places. Throughout history, they have represented the core of cities. Usually, they were erected as the centre of economic development and the venue of great events. On top of that, it was also the space for socializing. Markets became forums of news and events while generating trusted and friendship ties. (Asociación de Vendedores del Mercado Central de Valencia, 2005). Markets signify the affirmation of popular culture, they comprise part of the heritage that value the way of life and the location used for the daily activity (Delgado, 2005)

That is the reason why markets represent a key point when it comes to talking about the generation of images from the local society by appealing to popular culture and heritage and creating a dynamic conception: “a living culture” (Montiel Márquez, 2005).

The new tendency regarding tourist demand is in line with this assertion. According to the latest trends in tourism demand, the last-generation or third-generation tourists tend to take advantage of their stay to fetch enriching and deep cultural experiences, as Crespi Vallbona & Domínguez Pérez (2016) stated. Tourists are now looking for unique and personal development experiences in which they can feel first-hand the local daily life.

In this context, local markets have the necessary characteristics to please this new demand since they are representative spaces of cultural manifestation and the way of life of local people. Markets are conceived as sites where it is possible to feel the essence of a territory; walk around like a local, to be in touch with locals and to eat as one of them (Crespi Vallbona & Domínguez Pérez, 2016). It is a way of understanding local food, cuisine methods and the daily life of communities.

This type of tourism is an opportunity to value and promote local cuisine and local products, it also helps to reaffirm the local identity and it can be a source of income to invest in heritage conservation and local development.

However, whereas this heritage was not originally built to meet tourism requirements, planning is required to make this happen.

Local Markets at the Crossroads Between the Crisis and Tourism

As Crespi and Domínguez (2016) declare local markets are one of the most promoted resources where the local and the touristic go hand to hand. In this vein, Caesares (2003) indicates that commerce symbolises the starting point of social life, tourism, industries, residences... It is necessary to put efforts to achieve a balance between the touristic and the local in a sustainable way to avoid the consequences of inappropriate or no tourism planning. It is high time to undertake a tourism planning approach which cannot be accepted without considering its impact on local development.

Local markets are also suffering from tourism activity, impacts and transformations. These simultaneously affect the markets surroundings and, ultimately, the local population.

The third stage of Cara's proposal (2001), which shows the process of creating images influenced by tourism activity, matches this situation. In the first phase, local society is encouraged by tourism activity to recruit values and specificities in a creative process. This phase is the one that would be considered suitable, which includes local participation and where the process of building a collective identity is reinforced. In the second stage, the stimulus for tourism activity is intensified. A tourist image is defined independently by selection and exaggeration according to other interests. This means that an artificial society image is created parallel to the initial premises. Albeit in this phase, local actors are still participating. In the third stage, a total breakup between the image of local society and the tourist image is produced. At this point, local society is excluded. Consequently, the new imaginary devised has no relation to the essence of the territory, appearing as alienated from society.

The direction that local markets are following in cities such as Madrid and Barcelona, the two biggest cities in Spain, meets this latter phase. During the last years, diverse newspapers have echoed with this issue: La Boqueria market loses its charm (Rufas, 2017); Barcelona restricts tourism to the Boqueria market (Vázquez, 2017); Tourists barely bought in the Market: they merely come and take photos for social networks (Pascual, 2018); The problem is not tourism, it is the management.

These headlines highlight the inadequate planning and management, even non-existent, that has been made in local markets for tourism purposes. These flows lead to the disappointment of the local population and culture sacralisation and banalization.

The Central Market of Valencia

Merchants in the market square had already existed since the Middle Ages. However, at the very beginning of the 20th-century, a new building was constructed in modernist style: The Central Market. Throughout history, the market has been a forum for news, events and the scenario of social relations. It has also been linked to quality products, local commerce and trusted relationships.

The importance of the Central Market in the city of Valencia is reflected in the declaration as a building of Cultural Interest acquired on October 5, 2007. After several reforms, the market stands with an area of 7,972 square meters and it has a total of 231 stalls classified in numerous typologies. It presents an irregular plant divided into two spaces. On one hand, the general market, full of vegetables, fruits, meat, cereals, cheeses, honey, olive oil... and new goods brought worldwide. On the other, the fish market specialized in local product. In terms of access, it has 4 main entries and 4 secondary entries. Besides, it was adapted with a car park and access with an elevator for those who may need it (Ajuntament de Valencia, 2010).

Nowadays, this space located in the historic centre of the city and nearby two other sites of interest (the Silk Market and Santos Juanes Church) represents one of the main tourist attractions in town. The arrival of tourists in a space as unique as the Central Market can offer opportunities for economic diversification and the promotion of local cultural values to avoid getting lost. But, without a proper unified strategy, there will have consequences and touristification may arise, generating a loss of identity, massification, sacralization, population disappointment and transformation of spaces and culture as previously mentioned.

In 2002, for instance, some initial signs of massification forced local authorities to regulate the use of toilets by tourists with a fee (Ros, 2012). Some other attempts to manage the tourists flow and their attitudes were put stickers in the main entrance so as not to sit down on the steps while eating and drinking, not allowing groups from Thursday to Saturday between 11 am and 3 pm, nor using improper clothes to come in (slippers, swimwear, bare-chested)

In light of this situation, to what extent tourism has altered the main use of the Central Market of Valencia? How is this situation perceived by its local customers?

METHODOLOGY

Direct observation at the Central Market was conducted. First, each stall of the Central Market map (more than 200) was classified depending on the products that they were selling to the public (shop typology), being subsequently visited to analyse how they have been adapted to the tourist interests (touristification). The fish market area was dismissed because it is considered that their products are not being affected by the tourism impact. A scale of 1 to 4 reflecting the impact of the tourist activity according to the offered products (Table 2) was applied.

Table 2. Degree of touristification

VARIABLE	DESCRIPTION	VALUE
No Modification	The stall offers the same traditional products to all its clients, with no distinction	1
Innovation	The stall has adapted some of its traditional products to a new demand (tourism)	2
Diversification	The stall offers new products to a new demand (tourism) with no relation to its traditional offer	3
Turistification	The stall is addressed specifically to tourism	4

Source: own elaboration

After recording the degree of transformation, each stall was also classified according to their typology and location in the market. To determine whether the degree of transformation is significative affected by the stall typology and/or location an ANOVA test (alpha = 0.05) was performed using the R program (R Development Core Team, 2008). Last, a local customers' questionnaire was established. The objective was to analyse their perception (Is the central market becoming a touristic and massive site?) and whether this issue is affecting their purchasing decision (are they changing their habits due to overtourism?) A total of 100 were collected from May the 3rd to June the 1st (2019). People were randomly selected among consumers entering and leaving the market through the West Avenue door. Questionnaires were completed from Monday to Saturday to achieve a major diversity of scenarios and adjust as much as possible to reality. Specifically, the distribution of questionnaires was 77 on weekdays and 23 on Saturday.

TOURISTIFICATION DIAGNOSIS OF THE CENTRAL MARKET

After performing direct observation of the market stalls, Table 3 summarises the results regarding the degree of transformation of the Central Market stalls by typology variable.

According to the results, most of the stalls were classified within the degree of transformation 1, specifically 141 out of 231. Therefore, 61% of the stores are found with NO MODIFICATION, that is, they continue to offer their traditional products without client distinction. Next, 46 stalls have been assessed with the degree of transformation 2, so 20% are in INNOVATION stage, that means they have adapted traditional products for new tourism demand. 32 of the stalls (14%) were included in DIVER-

Table 3. Degree of touristification by typology.

TYPOLOGY	1	2	3	4	Total
Poultry and eggs	15		4		19
Snails	1				1
Butchery	25	1			26
Various, non-food supply	4	1	1	7	13
Drugstore and beauty	1		1		2
Spices and herbal products	7	2			9
Fruit and vegetables	39	13	7		59
Nuts	7	1	2		10
Bakery, pastries and sweets	5	8	2		15
To take or to take away	4	4	3	2	13
International products	5	2			7
Salted and pickled products	9	2	2		13
Gourmet shops, drinks and grocery store	9	5	1	2	17
Charcuterie and cheese	10	7	9	1	27
Total general	**141**	**46**	**32**	**12**	**231**

Source: own elaboration

SIFICATION because they have introduced new products which are no related to their initial offer for a new demand (tourism). Finally, 12 of the stalls (5%), were affected by the TOURISTIFICATION, thus they are directly oriented to tourism demand.

A map to illustrate the degree of transformation of their stalls is shown in figure I. As it can be seen, the stalls located in the areas closer to the main and touristic doors (the door of the Llotja and Palafox Street) are the ones that have modified their products to a greater extent. An imaginary line to separate the market in two areas, an area that remains mostly without modifications (stands coloured with light green) and a more tourist zone (stands coloured with dark green) has been drawn up.

The relation between typology and the degree of transformation with the previous results, it arises the hypothesis that typology could be to some extent explanatory of the degree of transformation. To contrast this fact, an ANOVA test using the R statistical program (R Development Core Team, 2008) with an alpha 0.05 was performed. Because of the "Various, no food supply" typology encompasses tourist stops associated with a degree 4 and it could affect the degree of relationship, it was decided to carry out ANOVA test dismissing this typology to verify whether the relationship is representative. The result showed that typology and the degree of transformation were very significant <p-value 0.000163> In the following diagram of boxes, the results are graphically expressed.

Figure 1. Degree of touristification in the Central Market of Valencia
Source: own elaboration

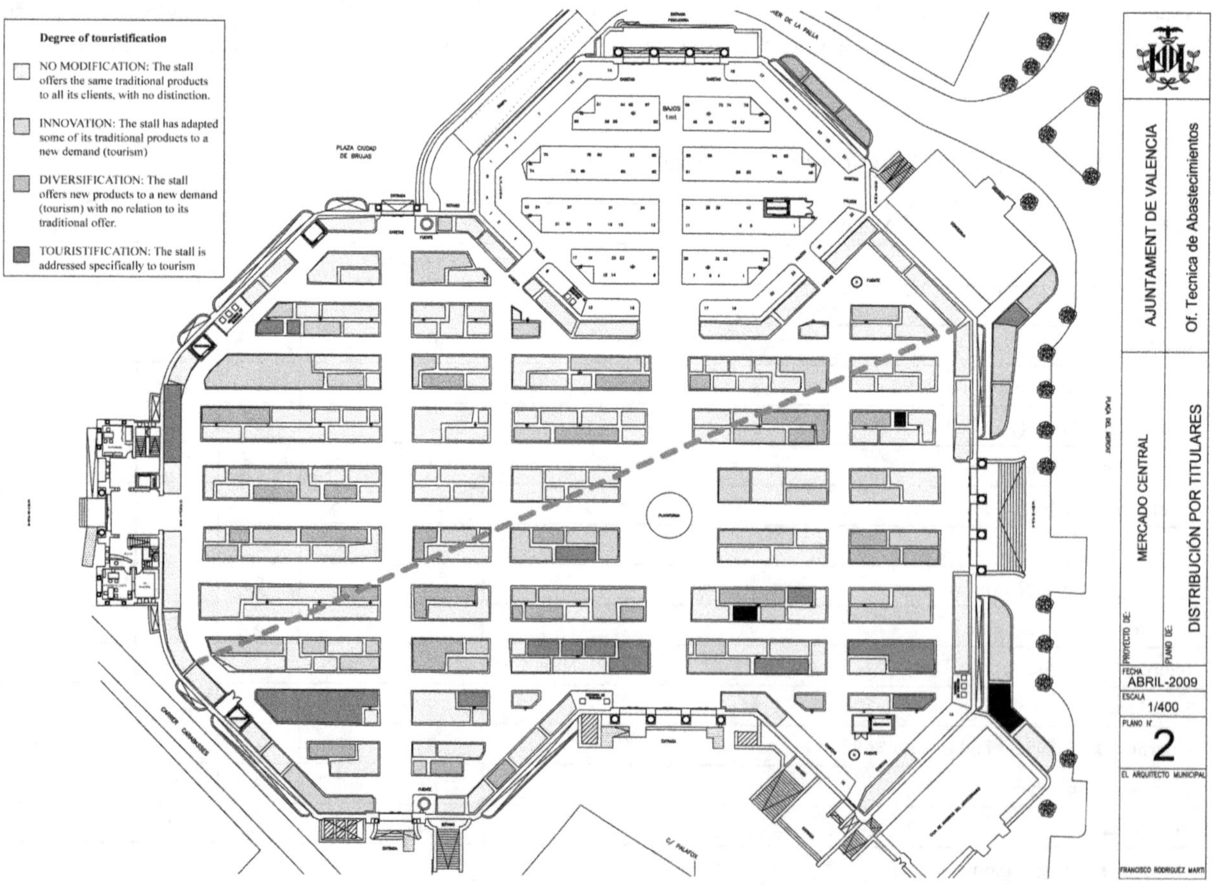

Figure 2. Classification of the stalls according its typology
Source: own elaboration

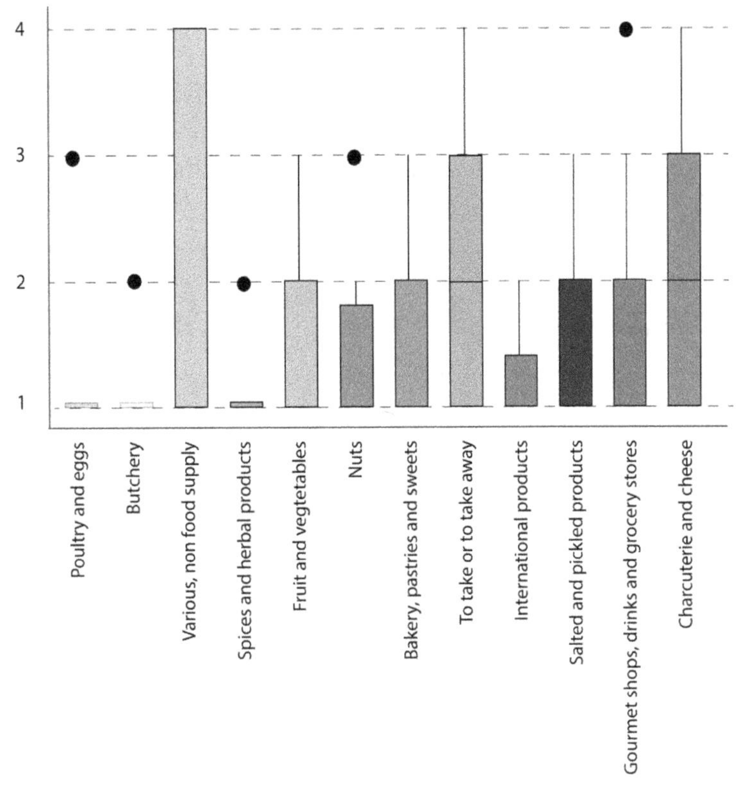

As it is shown in figure 2, the highest median is the one which corresponds to the typology "Various, non-food supply" since it encompasses touristic stores, especially souvenirs stalls. Next, the typologies "Bakery, pastries and sweets", "To take or take away" and "Charcuterie and cheese" have a median of 2, although 3 and 4 values were reached up in the last two types. This fact is due to the incorporation of soft drinks and cold drinks, tasting experiences and the adaptation of traditional products to other formats for tourism demand (ham sandwiches, small pots of ham & cheese...). Also, posters for this type of products and experiences written in English or other foreign languages were detected.

One of the most popular typologies in the Central Market is "Fruits and Vegetables" (n = 59). In this case, it is observed that the median corresponds to 1, that is to say, many of these stores have not yet modified their products, but 50% of the degree of transformation values are between 1 and 2. This is explained due to the sale of natural juices and cut fruit mainly addressed to tourist demand. It is also relevant that some values reached up the degree of transformation 3, as some of the fruit stalls have incorporated cold drinks (water, beer, soft drinks ...) that have no relation to the products of the initial offer.

Regarding the "Salted and pickles products" typology, the incorporation of cold drinks is again noticed in some cases, the median is at grade 1 and 50% of the values are between 1 and 2, which means that some of these stores have adapted products to a new demand (innovation).

The "Gourmet shops, drinks and grocery store" have included tasting experiences ("agua de Valencia", sangria, local wine, drinks, gourmet products) as well as titres written in English. Even so, the median is situated in grade 1.

Finally, "Poultry and eggs", "Butchery", "Spices and herbal products", "Nuts" and "International products", have a median of 1 and 50% of the values are under the grade 2. Those are the typologies that have suffered less transformation.

The Relation Between Location and the Degree of Transformation

On the other hand, a contingency table with the Central Market stalls according to their location and the degree of transformation was created (Table 4).

Table 4. Degree of touristification by location

LOCATION	1	2	3	4	Total general
A4			1		1
B1	7	2		2	11
B2	7	2			9
B3	4	1			5
B4	3				3
B5	5	2			7
C1	10	2	1	1	14
C2	5	1	1		7
C3	9	1	2		12
C4	9	2	3		14
C5	11	2	1		14
D1	11	3	4		18
D2	6	1	1		8
D3	5	3	1	1	10
D4	5	3			8
D5	1	3	3		7
E1	9	4	2		15
E2	1	2	3		6
E3	4	3	3	4	14
E4	7	3	1	2	13
E5	2	3		1	6
F1	5	1	2		8
F2	8	1	3		12
F3	3				3
F4	1	1			2
F5	3			1	4
Total general	**141**	**46**	**32**	**12**	**231**

Source: own elaboration

In the same way, location is likely to be explanatory of the degree of transformation. The relationship between the location and the degree of transformation was also significant <p-value 0.055> by the ANOVA test.

Both location and typology are related to the degree of transformation, as it is true that location explains to a smaller extent this relationship. These results are in line with the map shown in figure 1, where the stalls located in the areas closer to the main and touristic doors (the door of the Llotja and Palafox Street) are the ones that have modified their products to this end, especially, "Fruits and vegetables" and "Salted and pickles products".

Consumers' Perception of Tourism in the Central Market of Valencia

The previous diagnosis indicated that part of the Central Market was undergoing a process of touristification. Despite this result, it was therefore deemed appropriate to know the consumers' perception regarding this fact. Hence, a survey of consumers of the Central Market to know their opinion about tourism impacts was conducted.

Regarding the origin of the people surveyed, 87% came from the municipality of Valencia, while the rest (13%) did from villages of the influence area. About the purchase frequency, most people went to the Central Market once per week (51%), while "occasionally" and "two or more days per week" represented 20% and 23% respectively. A smaller portion, 6% of consumers declared to go to the Central Market daily. One result of the study that is especially worth mentioning was about how emotional was the Central Market for them. 78% of respondents chose 4 or 5, and only 14% rated this affirmation with a 1 or 2. Thus, the questionnaire mostly encompassed local people that appreciate the value of the Market.

About how tourism has affected their decision to go to the Central Market, the average response obtained on weekdays and at the weekend was different (1 and 3 respectively) what it means that people going to the Central Market at the weekend were more aware of how tourism was influencing their choice.

Tourism impacts in the Central Market have changed their purchasing habit (day) because the responses of the consumers surveyed during the week and at the weekend were not the same again. At figure 3, it is seen that the negative response predominant on people surveyed on weekdays since many of them go already during the week. However, among people surveyed at the weekend, the affirmative answer predominated; they preferred going during the week although they had to go on Saturday because it was the only day that their schedule made it possible.

When consumers were asked by new products aimed at tourists in their usual stores, 56% of the sample agreed to this statement. About the perception of reduction of the traditional offer, only 7% admitted it. Almost 50% of respondents declared that tourists did not represent a barrier when choosing the Central Market as a shopping space. Conversely, 35% of people surveyed affirmed that tourists did (figure 4). Focusing on the period that the questionnaire was completed, that is, whether they had answered on weekdays or at the weekend, the results were different depending on the day collected. On weekdays the median reached up 4 while on Saturday 2. This fact shows that problems regarding the saturation of the Central Market are more often at the weekend, especially in the days before the weekend and on Saturday, when tourists are more likely to visit the site.

Only 25% of the consumers surveyed consider that the Central Market is losing its function as a local market, but almost half of the total sample agree that the Central Market is transforming for new tourist demand. This could be the reason why 44% of the sample estimated that the flow of visitors should be

Figure 3. Change of purchase habits regarding the purchasing days
Source: own elaboration

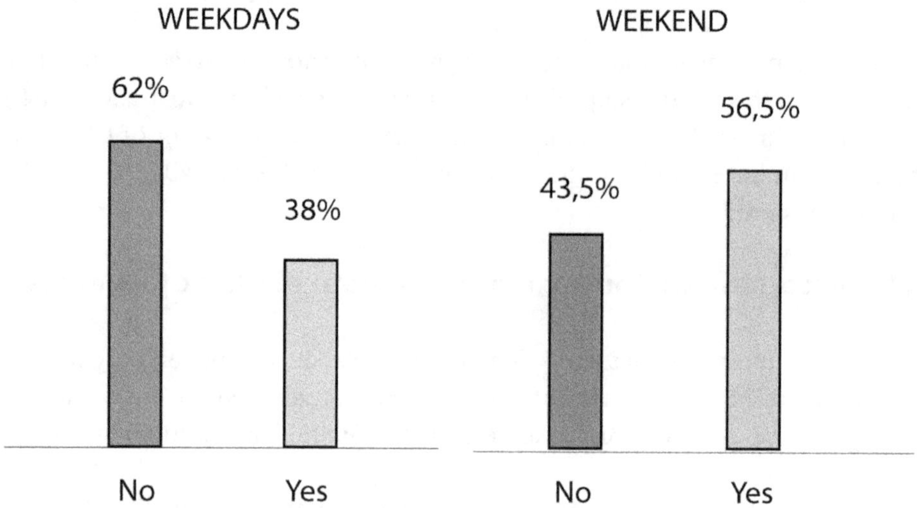

Figure 4. Lost of the function of the Central Market as a Market
Source: own elaboration

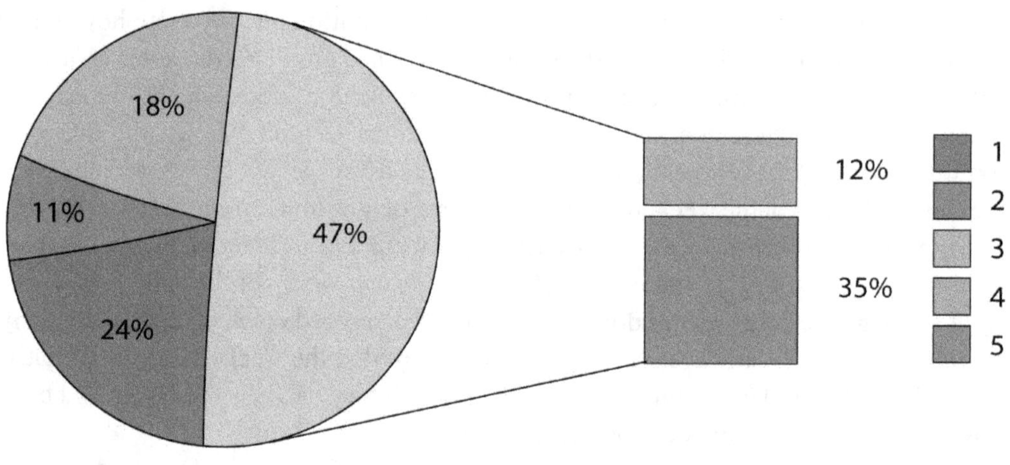

limited or managed in some way so that ordinary residents and consumers can have a better shopping experience.

However, the vast majority of respondents (72%) completely disagrees that they would stop shopping at the Central Market (figure 5), even though the situation would not be controlled. This answer is directly linked to the first question, in which many of them pointed out the emotional value that the Central Market has for them.

Figure 5. Decision of stopping buying at the Central Market
Source: own elaboration

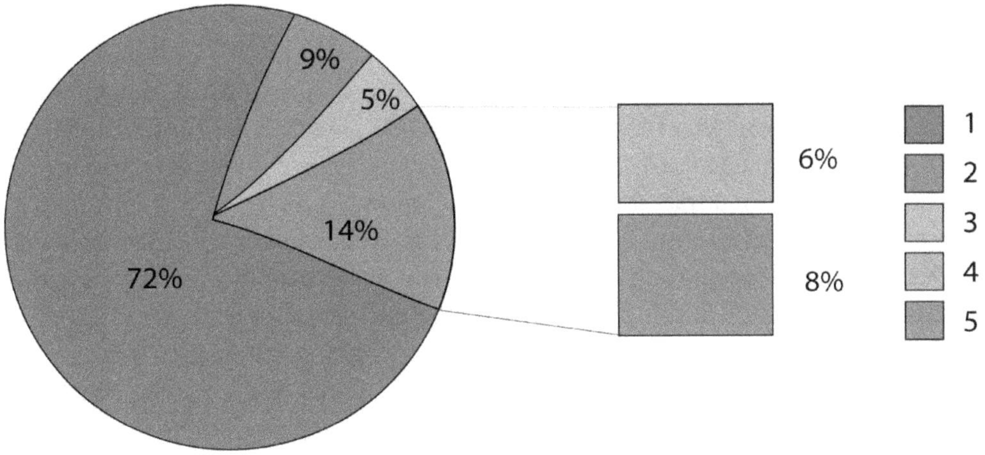

CONCLUSION

In the current context of globalization, heritage is established as the key tool by which generating heterogeneous movements for the creation of the collective local identity at the different territories. Specifically, local markets have become popular for several reasons: they represent values and traditions of the local culture, they satisfy the new tourism demand, personal development experiences more related to the daily life of a territory, that is to say, to be in direct contact with the local community.

Therefore, by increasing the interest in these kinds of markets, some places have suffered from a serious touristification process by visitors and their practices, as Freytag & Bauder (2018) described it. This is the case, for example, of the Boqueria market in Barcelona, which has completely become a tourist spot, isolating itself from the local population and producing cultural disenchantment and cultural banalisation. That is, irrespective of how it had been planned, its main use as a market has been given away to a secondary and predominant use as a tourist space.

Given this context, the Central Market of Valencia is undergoing a touristification process. Although most stalls remain unchanged, both the location where the stores are situated (areas close to the main and touristic entrances) and the typology they belong to (fruits and vegetables, to buy or to take away, charcuterie and cheese), are factors that can explain the transformation of some areas of the Central Market. Soft drinks, souvenirs, tasting experiences and gourmet products that are not related to its traditional products/offer have been introduced for sale. Besides, it is worth highlighting the new stalls in the Central Market completely aimed at tourists.

In the same way, it has been confirmed through the results of the questionnaire that saturation in the Central Market are concentrated at the weekend, especially on Saturdays, when a large number of tourists visit it. Simultaneously, regular consumers have worse shopping experiences.

Many consumers have already perceived that the Central Market is focusing on new touristic demand and, for this reason, they consider managing touristic flows primarily on Saturdays and high seasons. Despite all of this, the market has such a great emotional value for residents that the vast majority would not reject to go, even though the situation continued to deteriorate.

For all those reasons, the challenge of the Central Market is to achieve a balance between society and tourism so that better-planned tourism becomes an opportunity to continue promoting local cultural values associated with proximity and quality commerce, local cuisine, coexistence and sociability, historical past and popular tradition.

The Central Market of Valencia has already begun a process of touristification that, if it is not managed properly, it can convert into another touristic site, completely alien to local society and to the values it represents. If this situation happened, it could provoke new impacts in other local markets (like Russafa or Cabanyal) due to the loss of attractiveness of the Central Market.

It is necessary to put some effort to benefit the community and enhance the development of the city as a tourism destination, preserving and promoting its main use as a market as well, its authenticity and its value as a socializing and identifying space among the population of Valencia.

"Going to the market is a ritual, you will not only supply your kitchen, you go to inspire yourself and let yourself be tempted by the products of the day, but you also go to discuss the problems while you buy. A market is a stimulating source of news, in which you exchange tricks and recipes" a greengrocer of the Central Market of Valencia said.

As a conclusion, some areas of the Central Market of Valencia have set out a touristification process. Not taking them into consideration may bring artificial scenarios and move away from the identity of what represents for residents. Good governance is essential at this stage to avoid these problems before it is too late and, sometimes, overtourism may appear, although overtourism is not a tourism-only problem. Well-Balanced tourism does not depend on the typology, not even in having a tourism plan, but how stakeholders understand the different impacts of the policies in town and work together, how they are affected by tourism and how they should pay attention before they arise.

REFERENCES

Ajuntament de València. (2010). *Revisión simplificada del plan general de Valencia. Catálogo de bienes y espacios protegidos*. Valencia: Ajuntament de València.

Bærenholdt, J. O., Haldrup, M., Larsen, J., & Urry, J. (2004). *Performing tourist places*. London, UK: Ashgate.

Ballester, F. (28 de 08 de 2014). *El turismo urbano de la ciudad de Valencia se enfrenta a un problema de masificación y poco gasto*. Retrieved April 10, 2019 from Valencia plaza: http://epoca1.valenciaplaza.com/ver/137689/el-turismo-urbano-de-la-ciudad-de-valencia-se-enfrenta-a-un-problema-de-masificacion-y-poco-gasto.html

Belhassen, Y., Uriely, N., & Assor, O. (2014). The Touristification of a Conflict Zone: The Case of Bil'in. *Annals of Tourism Research*, *49*, 174–189. doi:10.1016/j.annals.2014.09.007

Berrueta, J. (13 de 08 de 2017). *Venecia se ahoga de turistas*. Retrieved May 27, 2019, from ctxt revista contexto: https://ctxt.es/es/20170809/Politica/14394/venecia-turismo-gentrificacion-ctxt-alquiler.htm

Bocadosdemadriz, E. (01 de 09 de 2018). *La "turistificación" de los barrios de Madrid: Llegaron las chanclas*. Retrieved May 27, 2019 from Huffpost: https://www.huffingtonpost.es/evabocadosdemadriz/la-turistificacion-de-los-barrios-de-madrid-llegaron-las-chanclas_a_23513335/

Brito, M. (2007). Las ciudades históricas como destinos patrimoniales: Potencialidades y requisitos. Fundación Cátedra Iberoamericana (Ed.) Palma de Mallorca.

Canalis, X. (16 de 05 de 2017). *Turistificación: la palabra de moda para demonizar al turismo.* Retrieved May 1, 2019 from Hosteltour: https://www.hosteltur.com/122062_turistificacion-palabra-moda-demonizar-al-turismo.html

Colomb, C., & Novy, J. (Eds.). (2017). *Protest and resistance in the tourist city.* New York, NY: Routledge.

Crespi Vallbona, M., & Domínguez Pérez, M. (2016). Los mercados de abastos y las ciudades turísticas. *Pasos. Revista de Turismo y Patrimonio Cultural*, 401-416.

Cummings, J. (2015). Confronting Favela Chic: The Gentrification of Informal Settlements in Rio de Janeiro, Brazil. In L. Lees, H. B. Shin, & E. Lopez-Morales (Eds.), *Global Gentrifications: Uneven Development and Displacement* (pp. 81–99). Bristol: Policy Press. doi:10.2307/j.ctt1t894bt.11

De la Casa Martínez, C. (2000). El turismo cultural: una alternativa al desarrollo económico. En F. d. León, Turismo cultural: el patrimonio histórico como fuente de riqueza (pp. 97-107). Valladolid: Fundación del patrimonio histórico de Castilla y León.

Del Romero, L. (2018). Tourisitification, sharing economies and the new geography of urban conflicts. *Urban Sci*, *2*(4), 104. doi:10.3390/urbansci2040104

Delgado, M. (s.f.). La cultura de las calles. El espacio público como patrimonio sociocultural. En T. Soldino Nicolás, M. Delgado, & A. Montiel Márquez (Eds.), Patrimoni i societat (pp. 61-77). Valencia, Spain: Universitat de València.

Fernandes, J. R. (2011). Area-Based Initiatives and Urban Dynamics. The Case of the Porto City Centre. *Urban Research & Practice*, *4*(3), 285–307. doi:10.1080/17535069.2011.616747

Freytag, T., & Bauder, M. (2018). Bottom-up touristification and urban transformations in Paris. *Tourism Geographies*, *20*(3), 443–460. doi:10.1080/14616688.2018.1454504

Füller, H., & Michel, B. (2014). Stop being a tourist! New dynamics of urban tourism in Berlin-Kreuzberg. *International Journal of Urban and Regional Research*, *38*(4), 1304–1318. doi:10.1111/1468-2427.12124

García-Hernández, M., de la Calle-Vaquero, M., & Yubero, C. (2017). Cultural heritage and urban tourism: Historic city centres under pressure. *Sustainability*, *9*(8), 1346. doi:10.3390u9081346

Giddens, A. (1993). *Consecuencias de la modernidad.* Madrid, Spain: Alianza.

Gotham, K. F. (2005). Tourism Gentrification: The Case of New Orleans' Vieux Carre (French Quarter). *Urban Studies (Edinburgh, Scotland)*, *42*(7), 1099–1121. doi:10.1080/00420980500120881

Gravari-Barbas, M., & Guinand, S. (Eds.). (2017a). *Tourism and gentrification in contemporary metropolises: International perspectives.* New York, NY: Routledge. doi:10.4324/9781315629759

Gravari-Barbas, M., & Guinand, S. (2017b). Introduction: Addressing tourism-gentrification processes in contemporary metropolises. In M. Gravari-Barbas, & S. Guinand (Eds.), *Tourism and gentrification in contemporary metropolises: International perspectives* (pp. 1–22). New York, NY: Routledge. doi:10.4324/9781315629759-1

Hernández Cordero, A. (2014). Gentrificación comercial y mercados públicos: El mercado de Santa Caterina, Barcelona. *Working paper series. CONTESTED_CITIES*, 2341-2755.

Hernàndez i Martí, G.-M. (2005). La difusión del patrimonio cultural y el turismo. En G.-M. Hernàndez i Martí, B. Santamarina Campos, A. Moncusí Rodrigo, & M. Albert Rodrigo (Eds.), La memoria construida. Patrimonio cultural y modernidad (pp. 159-192). Valencia, Spain: Titant lo Blanch.

Hernàndez i Martí, G.-M. (2005). La globalización y el patrimonio cultural. En G.-M. Hernàndez i Martí, B. Santamarina Campos, A. Moncusí Ferré, & M. Albert Rodrigo (Eds.), La memoria construida. Patrimonio cultural y modernidad (pp. 123-158). Valencia, Spain: Tirant lo Blanch.

Hernàndez i Martí, G.-M., Santamarina Campos, B., Moncusí Ferré, & Albert Rodrigo M. (Eds.), *La memoria construida. Patrimonio cultural y modernidad.* Valencia, Spain: Tirant lo blanch.

Leite, N. K. (2008). Turismo e territorio: um estudo sobre a turistificacao de Portimao (Algarve/Portugal) a partir da geografia do turismo. Master's thesis. Universidade de Lisboa.

Los turistas solo entrarán al Mercado Central con guías oficiales de la Comunitat Valenciana. (02 de 05 de 2018). Retrieved February 2, 2019 from Valencia plaza: https://valenciaplaza.com/los-turistas-solo-entraran-al-mercado-central-con-guias-oficiales-de-la-comunitat-valenciana

Maitland, R. (2010). Everyday life as a creative experience in cities. *Journal of Culture Tourism and Hospitality Research*, 176-185.

Martín Plaza, A. (10 de 08 de 2017). RTVE. La turistificación, una "burbuja" que amenaza no solo a los vecinos sino al propio sector. Barcelona, Spain.

Navarro Castelló, C. (20 de 09 de 2018). *Vecinos del Carme exigen más medidas contra la sobreocupación de las zonas peatonales y la turistificación.* Retrieved May 5, 2019 from El Diario.es: https://www.eldiario.es/cv/Vecinos-Carme-turistificacion-reduccion-terrazas_0_816469343.html

Novy, J. (2018). Urban tourism as a bone of contention: four explanatory hypotheses and a caveat. *International Journal of Tourism Cities.*

Pérez, C. (2013). Patrimonialización, turistificación y autenticidad en exaltación de la Cruz, Argentina. *Estudios y Perspectivas en Turismo*, *22*, 785–804.

Pinke-Sziva, I., Smith, M., Olt, G., & Berezvai, Z. (2019). Overtourism and the night-time economy: A case study of Budapest. *International Journal of Tourism Cities*, *5*(1), 1–16. doi:10.1108/IJTC-04-2018-0028

Popp, M. (2009). Der touristische Blick im St€adtetourismus der Postmoderne. Das Beispiel der italienischen Stadt Florenz [The tourist gaze in postmodern urban tourism. The example of the Italian city of Florence]. *Geographische Rundschau*, *61*(2), 42–48.

Prats, Ll. (1997). *Antropología y patrimonio.* Barcelona, Spain: Ariel.

R Development Core Team. (2008). R: A language and environment for statistical computing. R Foundation for Statistical Computing, Vienna, Austria. Retrieved from http://www.R-project.org

Richards, G. (2007). *Cultural Tourism: Global and local perspectives*. Binghamton: Haworth Hospitality Press.

Richards, G., & Marques, L. (2018). *Creating synergies between cultural policy and tourism for permanent and temporary citizens*. Barcelona, Spain: UCLG Committee.

Robertson, R. (1995). Glocalization: Time - Space and Homogenity - Heterogeneity. En M. Featherstone, S. Lash, & R. Robertson (Eds.), Global Modernities (pp. 25-44). London, UK: Sage.

Ros, M. (11 de 05 de 2012). *El Mercado Central estudia cómo cobrar a los turistas por emplear los aseos*. Retrieved May 05, 2019 from Levante-EMV: https://www.levante-emv.com/valencia/2012/05/11/mercado-central-estudia-cobrar-turistas-aseos/904208.html

Ruiz Baudrihaye, J.-A. (1997). El turismo cultural: Luces y sombras. *Estudios Turísticos*, 43–54.

Santamarina Campos, B. (2005). Una aproximación al patrimonio cultural. En G.-M. Hernànez i Martí, B. Santamarina Campos, A. Moncusí Ferré, & M. Albert Rodrigo, La memoria construida (pp. 21-51). Valencia: Tirant lo Blanch.

Sequera, J., & Nofre, J. (2018). *Shaken, not stirred*. City; doi:10.1080/13604813.2018.1548819

Soldino Nicolás, T. (2004). Patrimonio y Sociedad. Acción ciudadana y compromiso por el patrimonio. En T. Soldino Nicolás, M. Delgado, & A. Montiel Márquez, Patrimoni i Societat (pp. 35-59). Valencia: Universitat de València.

Sommer, C., & Helbrecht, I. (2017). Seeing like a tourist city: how administrative constructions of conflictive urban tourism shape its future. *Journal of Tourism Futures, 3*(2), 157-170.

Troitiño Vinuesa, M. (2005). Retos y oportunidades en la gestión del turismo cultural y de ciudad. En M. Gómez Borja, J. Mondéjar Jiménez, & C. Sevilla Sevilla (Eds.), Gestión del turismo cultural y de ciudad (pp. 15-27). Cuenca: Ediciones de la Universidad de Castilla-La Mancha.

Valencia, A. V. (Dirección). (2005). *Mercado Central. Mil años de historia* [Película].

Williams, S. (1998). *Tourism Geography*. London, UK: Routledge.

Wöhler, K. (2011). *Touristifizierung von Räumen. Kulturwissenschaftliche und soziologische Studien zur Konstruktion von Räumen* [Touristification of spaces. Cultural and social studies research on the construction of spaces]. Wiesbaden: VS.

World Tourism Organization. (2019). *UNWTO Tourism Definitions*. Madrid, Spain: UNWTO; doi:10.18111/9789284420858

Chapter 10
The Impact of Vacation Rentals on Housing:
State of the Art and Proposals

Ricardo Urrestarazu
https://orcid.org/0000-0002-4889-3245
University of Málaga, Spain

Alejandro Garcia Pozo
University of Málaga, Spain

Jose Luis Sanchez-Ollero
University of Málaga, Spain

ABSTRACT

The increase in the price of housing for rent and ownership in the city of Malaga is causing a problem of accessibility to it for the most disadvantaged sectors of the municipality, coinciding at the same time with the expansion of the vacation rentals on housing. This phenomenon is characteristic of the Spanish tourist cities, with growths of this variable greater than the national average; and not only is it depressing family income already reduced in recent years, but they are influencing essential aspects of the urban morphology of the city such as its historical heritage and its commercial fabric. All this is producing a social and academic debate on what public policy proposals might be that could be made to limit the growth of the phenomenon.

INTRODUCTION

This chapter aims to set out the breadth and consequences of housing price increases, as well as the factors that determine it, with particular emphasis on the tourism factor. The author's intention is to establish public policy recommendations that can facilitate access to housing and reduce the effect of factors influencing rising prices, especially those concerning vacation rentals in housing.

DOI: 10.4018/978-1-7998-2224-0.ch010

BACKGROUND

In order to have a sufficiently broad perspective of the topic under study, a methodological approach to literature review was used. To do this, it has been included contributions from social scientists at national and international levels in specialized journal articles, books, and reports, in addition to data available from public bodies and national companies related to the tourism and real estate sector. Being a phenomenon of global significance and precedents in other historical moments, there are multiple studies that have been done in cities around the world where the influence of the tourism sector on access to housing was raised as a factor having influence on rising real estate prices.

MAIN FOCUS OF THE CHAPTER

Access to Housing in Málaga: Price Level

In the city of Málaga, access to decent and affordable housing for the citizens of the metropolitan area of the city is diminishing appreciably both for those who want own homes, and for those who just want to enjoy housing on a rental basis. The impact of this situation is being felt by young people of emancipation age and in the more disadvantaged social sectors that suffer an increase in their prices and a reduction in the available supply of affordable housing, most notably residential rentals. There is a similar phenomenon taking place in 90% of the major cities of the world, wherein they are unable to provide affordable housing to their citizens (World Economic Forum, 2019).

In addition, since 2013, there has been a gradual increase in housing prices in Spain, and although this increase has affected the two main tenure regimes, it has been easier to show this more clearly in the rental housing market that is rented for leisure.

It should be noted, however, that the trend presented in the rental housing market does not present uniformly throughout the country, since, unlike the process of price growth and construction of new housing that occurred in the period 1999 to 2007, on this occasion, there is a great heterogeneity by geographical scope, since the greatest price increases have occurred in the historic centers of the cities and in the areas of greatest economic and tourist dynamism in the country, primarily in the larger cities and tourist attraction areas (Naredo, 2019). In this typology of urban center is found the city of Málaga and its metropolitan area. Being a large municipality with the capacity to attract residents as a consequence of its economic dynamism in recent years related to the main economic activity developed in the province of which it is part, and of which it is currently the protagonist—tourism—Málaga has experienced housing price increases above the national average since 2015, coinciding with the emergence and growth of vacation rentals (See Figure 1).

This prominence has been favored by the last decade's expansion of cultural urban tourism around the world, which responds to a change of tourist habits and tastes in the forms and destinations of new tourists being more likely to increase their leisure time in travel with longer rest periods, to which local tourist offers have been adapted, such as the building and improving of large passenger transport infrastructures, a cheapening of the means of collective transport, as is the case with so-called "low-cost" airline fares (Saladié, 2014), the loss of purchasing power of individuals resulting from falling labor incomes, the emergence of new forms of housing that are cheaper and more accessible than traditional

Figure 1. Annual change in home sale prices per square meter in Spain and Málaga's capital from 2007 to 2019, in the first quarter of the year.
Source: Own elaboration based in Tasaciones Inmobiliarias S.A. (2019)

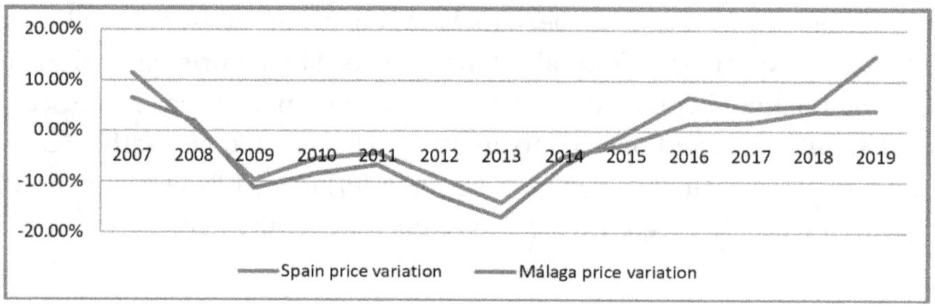

facilities (Bianchini and Parkinson, 1993), regeneration and pedestrianization programs of urban environments (Sans, 2016), and the activities of new Internet marketing agents (Hernandez-Pezzi, 2018).

This expansion of mass tourism, from a cultural and urban orientation around the world, has occurred despite the effects of the economic crisis that began in 2008, thanks to what the tourism sector has allowed, together with real estate, which the global economy would grew during the last years. All its variables have seen the 2008 starting figures increase in recent years, including the number of visitors, overnight stays, visitor expenses, and sector weight in Gross Domestic Product, jobs, receipt of investments, and incorporated companies.

Thus, tourist accommodations, linked as noted to the real estate sector, have seen a type of non-hotel tourist accommodation expansion in recent years, namely holiday tourist housing that is provided in tourist homes, located in areas where families and individuals have their habitual residence or a second residence (Rodríguez-Salmones, 2003). This type of tourist accommodation is marketed through specialized distribution channels, primarily on the Internet (Wegmann & Jiao, 2017), and has led to changes in the way tourists are attracted, in matters that include economic aspects and urban morphology (Brossat, 2019), urban renewal processes (Hernandez-Pezzi, 2018), internal population movements in tourist destinations, and configuration of their commercial fabric (Michael O'Regan & Choe, 2017). The changes in tourist accommodation offerings importance can be seen in the data obtained in recent years relating to the variation of the demand for tourist accommodations in Spain by international customers, being able to appreciate holiday tourist housing and accommodation options that grows the most by the time 2007/2016, which show an increase 89%, well above any other alternative (Statistics of Tourist Movements in Borders (FRONTUR) and National Statistical Institute, as of 2015).

The city of Málaga saw the level of property and rental housing prices grow between 2013 and 2018—although the evolution of the rental price index presents a more stable trend with less oscillations than the housing sales index—showing symptoms of exhaustion in the last year, which could coincide with the first symptoms of the end of the growth phase of the present real estate sector, just as could also happen in other urban centers, such as London, Amsterdam, Vancouver, Toronto, Munich, and Hong Kong, according to an analysis by the UBS Swiss Bank Studies Centre (UBS, 2019).

However, the knowledge of the evolution of the prices and physical and socio-economic characteristics of rental housing in Spain has the serious disadvantage of having a high degree of legal and fiscal opacity of the sector, as well as the absence of official statistics of these characteristics and its price (Naredo, 2013), so that the absence of official data on rental prices has required that in order to determine the extent of the phenomenon, it is necessary to turn to the price index provided by the real estate portal Fotocasa, which bases this on the prices offered for rentals in the market; this information has been gathered by the districts of the city since 2007.

If one compares the sales price increase data per square meter with the price increase per square meter of the rent in the first quarter of each year, it can be seen that these are higher than those that occurred with the former, with a slight uptake in the potted in the last year. So, the rental price in the capital of Costa del Sol grew by 0.61% in the period from 2007 to 2019, while the sale price decreased by 25.32% in the same period, a question that is visible in Figure 2.

Social Consequences of Price Variation

In order to determine the extent and social consequences of these price increases, it is necessary to compare these increases with the evolution of the income of the Málaga citizens during that same period of time; the information has been provided by the State Tax Agency concerning Wage Income Declared per capita from 2007 to 2016, the last published financial year, which collects the income declared by its inhabitants as income from employment and self-employment.

If one compares both magnitudes, evolution of rental prices, and evolution of the average income declared per inhabitant in the city of Málaga from 2007 to 2016, it can be seen that these have decreased by 13.43% compared to the 4.72% decrease in rental rents, that is, the decline in wage income is three times greater than that suffered by rental income. In Figure 3, very moderate developments in the economic levels of families can be seen during this period, with a greater dispersion of the growth of both magnitudes from 2013, confirming a more negative development during these years of household income relative to rental income.

These trends anticipate a greater difficulty in paying monthly rents for Málaga families, which must be confirmed with another study variable necessary to know the degree of accessibility of city housing in Málaga, the ratio that implies the payment of the rent with respect to the total household income, in which the higher percentage, the greater the difficulty of access.

In this case we will detail the degree of accessibility to a rental house in Málaga with data by districts of the year 2016, with the intention of knowing how the phenomenon affects areas of the city with very different socioeconomic situations of its inhabitants. This will require three types of information as reflected in Table 1 the average square meters per house obtained from the consultancy International

Figure 2. Annual change in rental prices and sale per square meter in Málaga's capital from 2007 to 2019, in the first quarter of the year
Source: Own elaboration based in Fotocasa (2019)

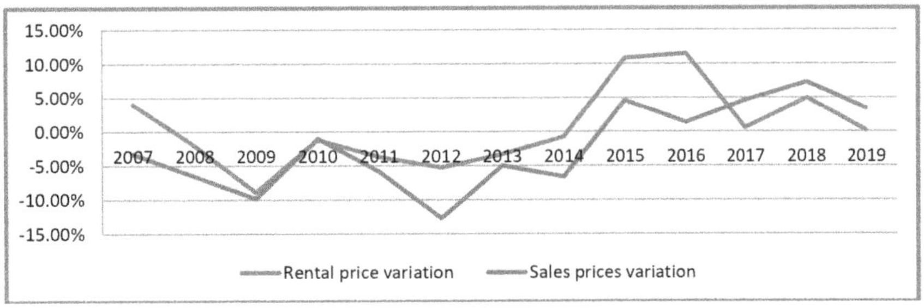

Figure 3. Annual change in rental prices and wage rents declared in Málaga capital from 2007 to 2016
Source: Own elaboration based in Foundation CIEDES (2017), Fotocasa (2019) and Spanish Tax Agency (2019)

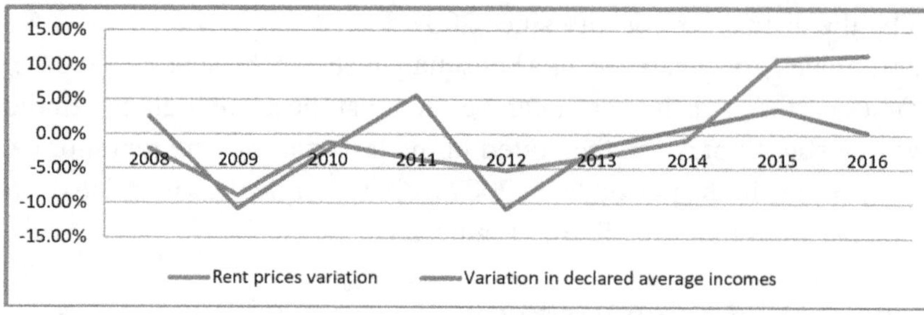

Financial Analysts for the year 2016 with data collected from the Land Registry, the average annual income per household for the year 2016 provided by the Spanish Ministry of Finance, and finally the rental prices per square meter collected on the Fotocasa real estate portal.

The data indicate that, with the exception of the Historic Center, in all areas citizens spent more than 35% of their income to cover residential rental rent, an amount considered as adequate to be able to live with basic needs covered worthily; confirming the existence of housing accessibility problems in the capital. As for the situation of the Historic Center, it should be noted that in this district is where vacation rentals housing has grown the most and where the housing park offered in residential rentals has been reduced the most, with an appreciable trend since 2015, throughout the national territory. Since that time, there has been a gradual decline in the park of housing available for residential rentals, which has increased from 115,000 in that year, to 97,900 in 2017 (Servihabitat, 2018).

Factors Influencing Housing Prices

The situation of average rent increases above those produced in household income, in a working environment with significant percentages of temporarily and precariousness, justify the interest in knowing the reasons for the structure of the housing parks and the consequences they end up producing relating to

Table 1. Comparison of the percentage of annual rent supported on the average annual income of city of Málaga districts in 2016

District	Average square meters per dwelling	Average annual household income	Average annual rental estimate	Estimated rental percentage on average annual income
Centro Histórico	85	27.472	8.854	32,23%
Cerrado de Calderón	145	28.542	14.529	50,90%
Carretera de Cádiz	95	20.521	8.904	43,39%
Bailen Miraflores	86	20.405	7.235	35,46%
Cruz de Humilladero	90	18.079	7.560	41,82%
Teatinos Universidad	130	26.107	11.217	42,97%

Source: Own elaboration based in Observatory of the Urban Environmental Monitoring of Málaga (2016) with data from National Statistical Institute, Fotocasa (2017) and International Financial Analysts (2016)

social welfare. To understand these reasons, it is necessary to start with the knowledge of the elements and factors that influence the supply and demand of the housing market and that determine, among others, its level of prices.

On the supply side, the inelastic nature of the supply cannot be forgotten, as increases in the demand for housing cannot adequately be met by immediately incorporating the production or rehabilitation of new housing into the market (Glaeser, 2008). In this way situations in which increases in the number of homes are followed by an increase in prices, which implies behavior contrary to the logic of the market for goods and services of classical economics (Arrazola, 2015; National Competition Commission, 2013).

This inelastic nature of the housing supply is based on the fact that the price of housing is perceived as a function of land supply and housing regulation (Altuzarra & Esteban, 2011). These elements are the first in which this behavior can be seen. The supply of land has repeatedly suffered sharp increases in its prices, as did those of the housing market (Idoate, 2008), a consequence of not being able to build a real estate project promptly. Additionally, according to Titman (1985), real estate developers who own urban land usually act according to profitability criteria and opportunity cost when starting a new construction project and deciding to do it in the present or in the near future.(Titman, 1985).

Another characteristic element of real estate developers, construction companies, and auxiliary companies is their size. Most small companies with few workers, a low technological level and productivity have a reduced capacity that prevents them from undertaking real estate projects with sufficient agility, even if they have sufficient land. Large companies, in cooperation with national or international financial groups (Whitehead & Williams, 2011), often have the market power to influence the price of land and housing (García Montalvo. 2000). A lot of financial resources frequently allow these large companies to retain soil for future projects in search of greater profits, leading to further price increases (Berges & Ontiveros, 2007).

On the other hand, other elements also influence the supply of housing that prevents the rapid adjustment of the building of new construction or rehabilitation of housing to meet changes in the demand for housing. For example, the availability of skilled labor, fixed capital of materials and soil (López Andión, 1998; Martínez Pagés, 2003), or the permanence of low interest rates, make it easier for developers to acquire and maintain soil to develop with cheap financing and lower levels of risk, as they can contribute less equity (Rodríguez López, 2007).

However, there are other factors that also have a prominent impact on the supply of housing, such as the absence of public housing for rent; the reform of the Urban Rent Act of 2015; and the intervention of the Real Estate Investment Trust (REITs) and real estate investment funds, which are very active in the tourism sector, promoting the construction of new houses and tourist accommodation facilities (Urrestarazu-Capellán, 2019). First, the Spanish public sector should be mentioned in the direct provision of affordable housing to citizens, referring to either of the two main tenure regimes. Despite the intentions and objectives set out in the state housing plans of the last decade, the construction of social housing has focused on social housing in property (Palacios & Vinuesa, 2010), a visible reality in the figures of housing built in recent years, where the majority weight of property housing has not only not decreased, but has otherwise increased from previous periods. Figure 4 shows official protection homes built by the State and the Autonomous Communities during the period of 2005 to 2018. Rent on all protected homes built went from 22.27% in 2005 to 2.95% in 2018.

In the case of the city of Málaga, the work of promoting and building social housing by public administrations has been quite modest as neither City Council nor the Autonomous Community have developed sufficient initiatives able to meet the demands of its citizens in such a way that in 2019, there

Figure 4. Percentage of social protection housing for rent on all protected housing (according to provisional qualification) from 2005 to 2018
Source: Own elaboration based in Spanish Ministry of Development (2019)

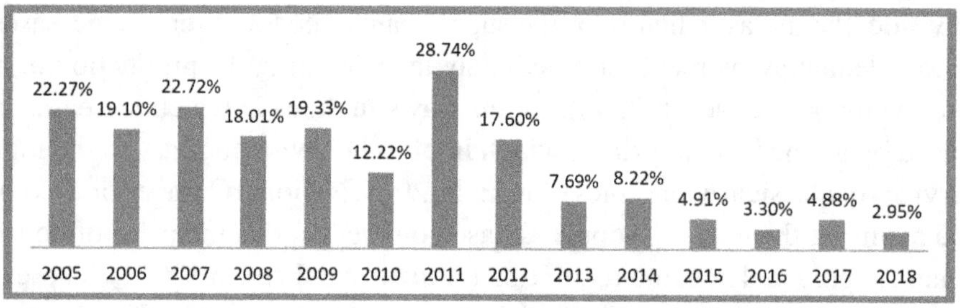

was 18.723 requests for registered housing at the Municipal Málaga Housing Institute, of which 70% were on the way to be rented. The administrations was responded by building 338 social housing units per year from 2010 to 2018, without any future trend change being anticipated as at the time there were 69 homes under construction (Municipal Málaga Housing Institute, 2019). By comparing the approvals figures in this protected scheme against the total number of approvals for housing in any regime during that period, it can be seen that from 2016 to the present year, this has not exceeded 10%, while from 2010 to 2015, the percentages was above 40%, primarily the result of the drastic fall in the construction of free housing, as indicated in Table 2, and not the result of increased construction of social housing, which has maintained throughout the period an appreciable lack of activity.

The second factor cited is the influence of the reform of the Urban Rent Act of 2.013 and 2.015, which revised Act 29/1994, Spain´s Tenancy Act (often abbreviated as LAU) to some degree. The objective of the Act was to boost the rental market and liberalize the sector by promoting freedom of agreements between the parties. This allowed for a reduction in the maximum term of leases from five years to three,

Table 2. Number of social protection and free income approvals per year, and percentage of social protection housing approvals relative to the total number of housing approvals in the city of Málaga from 2010 to 2018

Year	# of social protection approvals housing	# of free rent approvals housing	Percentage of social housing over total
2010	203	374	35,07%
2011	338	156	68,42%
2012	231	227	50,44%
2013	146	157	48,18%
2014	12	154	7,23%
2015	319	393	44,80%
2016	3	852	0,35%
2017	119	1.095	9,80%
2018	224	1.982	10,15%

Source: Own elaboration based in Official College of Architects of Málaga (2019)

which allowed for a quick update of monthly rents by homeowners responding to the price increases of the market, doing the same with their properties in short periods of time (Ardura, 2017).

Another intended objective of legislative reform was to create a stable fabric for landlords, both smallholders and large leasing business corporations, since the market was traditionally in the lead up to the former and recognized the need to incorporate more prominently into the latter. This was intended to create a professional rental market that could permanently offer residential services with greater guarantees for its owners and diversify the rental investment in geographical and risky terms (Inurrieta, 2007).

This objective established by Act 29/1994 became a reality, not only because of the regulatory changes included in it, but also because of other economic and financial factors that led to global financial capital deciding to direct its resources to the Spanish real estate market. For example the liquidity increases in the global financial market and the reduction in official interest rates by the world's major central banks (Stigliz, 2010). This was despite the number of real estate investors in the country remaining national, with almost idle capital on savings products in financial institutions and almost no remuneration or a high level of financial risk (Naredo, 2019).

However, there were other measures promoted by the public sector of a normative nature that shared the objective of creating a professionalized fabric of companies that offered real estate for rent. This included participating in the management and construction of social housing. A specific reference should be made to the rules governing and promoting the Real Estate Investment Trust and in particular, Law 11/2009 of 26 October, and the reform thereof by Law 16/2012, of 27 December. These laws had elements similar to that of other countries around Spain, and these companies were granted tax benefits to avoid double taxation through their exemption to companies, imposing them exclusively on their shareholders (Rabadan, 2009).

These changes, beneficial to their shareholders, coincided in time with the lowest price of the assets of the Spanish real estate market, and with the highest level of unsold real estate stocks of the last decade. These factors, added to excess liquidity and lowered official interest rates mentioned earlier, produced a context conducive to the inflow of international and domestic financial capital (Fields, 2014) that these companies used as an instrument of investment. This facilitated a wide development in recent years, but with the consideration of being a contributing factor in the increase in rental prices, thanks to a traditional relationship between urban growth and finance that in recent years has intensified with the development of urban environments (Sanfelici & Halbert, 2018).

As shown in Figure 5, since 2012, there has been a substantial increase in equity managed by companies that offered real estate for rent compared to previous years.

Also from the demand side, there have been factors conducive to this rise. First, the authors should cite the decline in the personal income of the housing claimants, who suffered from the increase in social inequalities in the last decade. Secondly, in that connection, the difficulty in accessing bank financing necessary for the acquisition of property housing, and thirdly, the need for housing in the most economically dynamic areas of the country and, therefore, with increased population employed. Fourth, and last, the increase in investments in real estate, which unlike the pre-2008 real estate boom, is excelling in housing for rent, either residential or vacation rentals housing (Urrestarazu-Capellán, 2019).

With regard to declining personal income and difficulties in accessing bank financing, household income has worsened over the last decade, leading to a shortage of savings necessary for those wanting to purchase a home. The high price of a house, with an average of cost of 1653 Euros per square meter in Málaga capital (Tasaciones Inmobiliarias S.A., 2019), and the need to obtain bank financing, which will not provide the borrower with more than 80% of the purchase price (Edufinet Financial Education

Figure 5. Equity in thousands of Euros of the real estate investment trust in Spain from 2000 to 2018
Source: Own elaboration based in Inverco (2018), CNMV (2016) and Bank of Spain (2018)

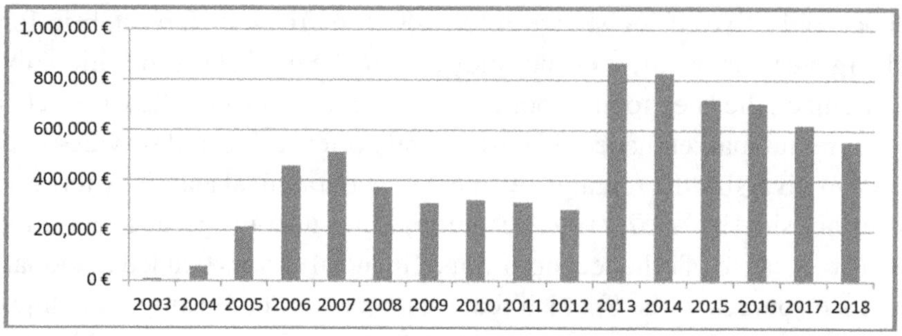

Project, 2019), requires the buyer to have more than 40,000 Euros in cash available to afford the purchase (Sociedad de Tasación, 2018). This is a difficult amount to gather in a country where the average family savings have reached an all-time low rate of 4.9% of the Gross Disposable Income from families destined for it, as is indicated in Figure 6 (Social Barometer, 2019; Laborda, 2018).

The Influence of Vacation Rentals on Housing

The increase in investment in the real estate sector in recent years by Real Estate Investment Trust and local investors of various sizes has also moved to the vacation rentals housing market that has seen an increase in their park offered and the volume of investments received, attracted by levels of profits above other sectors and other types of real estate (Hernández-Pezzi, 2018). Although these benefits are based on the geographical location of the property (historic centers), the degree of fiscal opacity or the number of days to be rented per year (minimum 60% of the days of the year) (Alquiler Seguro, 2019).

This increased expectation of benefits results from a displacement of part of the park offered for residential housing for rent to holiday housing, and a general increase in housing in residential rent as a result of its owners making sure to not spend them on holiday tours (Dredge et al, 2016). This is a reality that can be seen in different cities around the world, such as Madrid, Barcelona (Garcia Lopez et al, 2019), Berlin (Schafer & Hirsch, 2017), Sydney (Gurran, 2017), Los Angeles (Lee, 2016), San Francisco (Brousseau, 2015), New York (BJH Advisors, 2016) and Boston (Horn, 2016).

Figure 6. Annual savings of Spanish households relative to gross disposable income from 2000 to 2018
Source: Own elaboration based in Social Barometer (2019) and National Statistical Institute (2019)

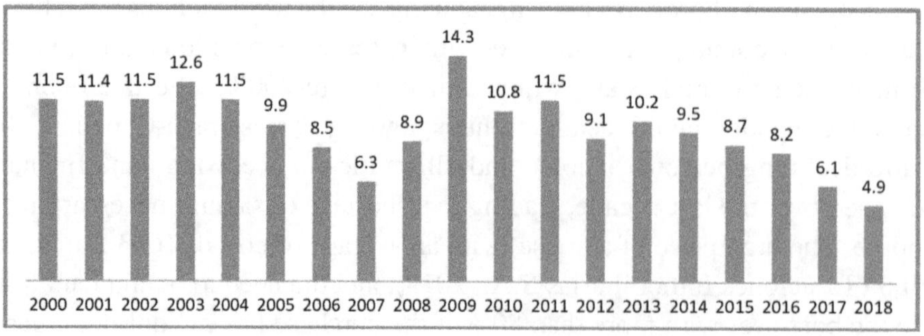

Additionally, these investments have been able to convert the numerous empty houses that existed a few years ago, or secondary houses with little use of the Historic Centre, into vacation rentals, thus available houses have virtually disappeared (García Peña, 2017). Table 3 shows areas of the city that are of greatest interest to tourists, where the largest volume of vacation rental housing is given, representing 27.7% of the total housing park of this district in 2018 (Sánchez & Ordaz, 2018; InsideAirbnb, 2019), while in the rest of districts do not exceed 2.6% of the park in a much more modest contribution. It also shows the high degree of concentration of tourist activity in a single district, since 70.27% of vacation rental homes offered in Málaga are in its Historic Centre.

The reduction of empty housing for exploitation as vacation rentals or residential rentals is a general trend that occurs in all major cities and tourist attraction centers throughout the country, so that the stock accumulated in new housing has risen from 649,000 in 2009, to 432,000 in 2017 (Rodriguez Lopez, 2017 and Ardura, 2016). This is the same circumstance that occurs with the non-main homes, where the secondary ones are located, which have also declined from 7,343,389 homes in 2010, to 6,529,836 at the end of 2016 (Ministry of Development, 2018), (Diaz Hernandez, 2017).

Vacation rentals have become an option of increasing importance in the tourist accommodation sector, partly because of the lower prices and new ways of providing accommodation services, which are more adapted to different types of tourist friendship groups or families with children (Exceltur, 2015). These characteristics have facilitated an increase in tourist demands produced in recent years. The demands that the activity that the vacation rentals contribute to increase; and where other explanatory factors should be included, such as increased expenditure on citizens' tourism, lowered forms of transport, increased urban tourism, longer periods of rest, an increase in the number of retirees with resources, and the activity of new online marketing agents (Malighetti, 2009; Horn & Merante, 2017).

It can be understood that the impact of vacation rentals on housing has ended up producing, in the tourism sector, the same effect that produced the emergence of low-cost air transport companies. Both sectors have adapted to the new framework socio-economic reduction of household income in the middle

Table 3. Number of vacation rentals on housing on total per district in Málaga in 2018

Districts	Housing offered on Airbnb	Percentage of vacation rentals on housing over total
Centro	3.364	27,70%
Este	536	2,60%
Ciudad Jardín	34	1,40%
Bailen Miraflores	132	1,20%
Palma Palmilla	28	1,04%
Cruz Humilladero	174	0,45%
Carretera de Cádiz	370	1,14%
Churriana	50	0,10%
Campanillas	3	0,01%
Puerto de la Torre	29	0,08%
Teatinos Universidad	67	0,65%

Source: Own elaboration based in Project DataHyppo (2018), Inside Airbnb (2019), Eldiario.es (2018) and Population and Housing Census of The National Statistical Institute (2011).

classes of Western countries, which with less income want to continue enjoying affordable holiday periods (Aledo, 2007; Malighetti, 2009).

The vacation rentals on housing need for its development and expansion, the stock of various business instruments that facilitate on the one hand its commercialization and presence in the potential markets to which it wants to address, and on the other, management of the business efficiently (Gyimóthy, 2017). Among the first are the housing reservation platform by peer-to-peer Internet and among the latter, the management of tourist accommodations companies.

The housing reservation platform on the Internet P2P are multinational companies that allow the owners of these homes to access a global market easily and cheaply (Gyimóthy, 2017), since the cost they have to assume is usually between 7% and 12% according to the company. These owners can offer their homes to tourists from all over the world with a guarantee of charging their accommodation service, highlighting amongst these are all the Airbnb and Homeaway, who form a powerful lobby with an oligopoly tendency in their sector (Crommelin & Troy, 2018).

With eleven years of life, these types of companies are part of the phenomenon of technology start-ups (Varma et al, 2016) that are companies founded by entrepreneurs that in a few years, and thanks to the possibilities of reaching large markets offered by the Internet (Oskam & Boswijk, 2016). They have managed to become multinational companies with large sums of capitalization of their stock values and depositaries of the confidence of international financial flows.

These entrepreneurial companies offer faster and cheaper services than the pre-existing companies through so-called Internet search engines, which facilitate the positioning and location of information and services of tourist accommodations. These Internet search engines have knowledge the preferences and tastes of consumers, being able to carry out a more effective work of market research that allows them to estimate the behavior of the demand of tourists (Palos & Correia, 2018).

Tourist accommodation companies make it easier for investors or neophyte homeowners in the business to put their homes on the market without the need to have prior knowledge of the sector, having only to be concerned with putting in capital and getting their real estate up and running. These tourist accommodation companies are responsible for managing the marketing, maintenance, cleaning, decoration, and reception of clients, and they offer additional apartment services in exchange for a percentage of the total amount of the rent that is between 15% and 30% of the total invoiced (Garijo, 2017).

Socioeconomic Effects of Vacation Rentals on Housing Activity

The expansion of vacation rentals on housing has had positive economic effects, since the financial and real estate resources employed have ended up generating an increase in demand for certain necessary goods and services for the commissioning of these properties, as well as the creation of direct jobs for their implementation, and indirect in ancillary sectors. It is estimated that in 2017, the contribution to the Spanish economy of the expenditure on rent and stay of tourists housed in vacation rentals on housing amounted to 24.495 million euros (Homeway, 2018).

However, the vacation rental expansion has also created many social externalities in the localities where they are located. This has motivated questions as to their real value. One of the outstanding externalities is its influence on rising prices of residential rental housing (Biagi et al, 2015). Residential rental housing is usually attended by collectives such as students, families with less economic resources, divorced individuals, and young people who see how the residential housing park is reduced and its price is increased by hindering its accessibility. The expansion also contributes to the displacement of this population from the central areas of the cities to others further away, while their former homes become occupied by wealthier families and tourists (Ardura, 2016). There is a flow of capital from the most disadvantaged social sectors, which have no choice but to pay rising rents to residential rental homeowners, creating a process of inequality between neighborhoods of different income levels, depending on the monthly rental income levels disbursed, in an appreciable worldwide trend (Wachsmuth & Weisler, 2018).

Additionally, the growth of vacation rentals on housing supply has had an impact on the city's historic real estate heritage. If you analyze the total of houses demolished from 2008 to 2016 in the so-called "city almond" formed by the geographical area where the ancient medieval city of Málaga was extended. In this area, one can see that the total number of buildings demolished in the area during those eight years amounted to 76, which represents 6% of the total buildings in the area. These buildings were empty in recent years, and, in most cases, have now become tourist accommodations, which have been set up in a historic center, and which in 2017, had 12 hotels, 16 apartment buildings, and 9 pensions, to which one would have to add more than 2,000 vacation rentals (Ozomek, 2018).

In this way, the urban landscape ended up losing its native elements, including its traditional architectural styles, compared to others standardized according to the conventional and depersonalized tastes of tourists. This happens in such a way that the previous landscape is replaced, altering the identity of resident citizens, in favor of other exogenous and outside local cultural frameworks (Nogués, 2007).

Therefore, there is the paradox that urban cultural tourism has been developed in recent years by changes in the tastes and habits of tourists, who, having a higher level of education, apparently value more leisure facilities and cultural facilities of the cities (Tresserras, 2004), in search of a unique and authentic experience. Although, in the practice, most of these tourists tend to have patterns of behavior similar to mass tourism, demand spaces of rest, relaxation, and good weather, importing less the conservation of the historical and cultural heritage (Santana, 2003).

As for the impact on the commercial fabric, there has been a process of specialization of the activity towards businesses oriented to meet the demands of tourists (Brossat, 2019), and leaving aside that of the residents, motivated by greater possibilities of benefits and by a process of expulsion of traditional businesses. Theses traditionally businesses unable to pay the rental prices of the premises where they were traditionally located, after the completion of the renewals of their leases, having on January 1, 2015, a moment of particular significance, since, on that day, according to the content of Law 29/1994 of November 24, the contracts for "old rent" of rented real estate were terminated before May 9, 1985.

The renewal of prices in tourist areas at the market price resulted in the disappearance of large numbers of businesses that were unable to afford higher rental rents, in an environment already complicated by the reduction of traditional customers that previously had been victims of the same phenomenon of gentrification. Thus, of the 1,332 premises opened in the Historic Centre of Málaga in 2017, 34% were engaged in hotel and restaurant activity. In 2001, this percentage was 23%, with great weight in addition to franchises from multinational companies in the rest of business (Ozomek, 2018), becoming this area of the city in a space dedicated almost exclusively to tourist leisure, which is a limitation of the use of this space for citizens (Gil & Sequera, 2018).

It can, therefore, be seen that tourism activity even that which is motivated by a cultural orientation, can have an impact on historical heritage, a phenomenon which bears similarities to the environmental impact of the activity increase in the number of visitors and the necessary consumption of natural inputs needed to serve them. These include water treatment, water and garbage treatment, energy, which already exceeds the ecological capacity of the tourist destination (Curry & Mounitnho, 1992; Aledo, 2016). The tourism sector in the Mediterranean basin is one of the most vulnerable to the effects that global warming temperatures are producing, as a result of longer periods of drought and extreme climates (Olcina & Vera, 2016).

Finally, some of the tourist activity economic benefits remain to be questioned, as these are being received mainly by domestic and international investors who spend their capital managing vacation rentals, hotel facilities, commercial franchises, and high-profit catering businesses, forming many of the international financial flows that are investing in recent years in the real estate sector (Hernandez-Pezzi, 2018). However, while investors prosper the incomes of work and working conditions that reach residents is characterized by low wages and job precariousness, deepening social inequalities within the city (Gil & Sequera, 2018).

SOLUTIONS AND RECOMMENDATIONS

Having given a general description of the housing market situation in the city of Málaga, it is worth asking what society can do in order to correct its current dynamics, although this would be based on considering the existence of a problem and finding a convenient way to deal with it. This latter statement implicitly carries the idea that there may be another perspective. This accepts that this is not a problem, but that the role of the market as a distributor of goods and services and of rents must be defended. This makes the generation of wealth should not be altered or diverted, in such a way that periods of high price growth are adjusted at a given time automatically by the law of supply and demand without the need for interventionism from the public authorities, and efficiently.

This conception of housing policy can be questioned with the reality experienced in Spain in recent decades, the housing market has shown on several occasions that there is no efficient adjustment, and that its patterns of behavior can be guided by irrational motives, where the individual benefits of various groups bring impoverishment to large segments of the population, as happened in the housing bubble of the previous decade.

It may, therefore, seem more reasonable to face the reality of an inefficient housing market whose dynamics produce price growths and affordable housing shortages that must be resolved, as the European Commission itself acknowledges in 2019 (European Commission, 2019). Also where the tourism sector and, more specifically, vacation rental housing, is helping to amplify, even if it is not the only determining factor in this situation. Tackling this reality is to implement active housing policies with a clear vocation to correct, from the public sector, the dysfunctions and externalities generated by the market, taking measures that have already been implemented in other states of the world, especially on the European continent

These housing policies should consider greater construction and rehabilitation of affordable housing, preferably for rent, including housing built and promoted by the public sector, as well as the promotion of cooperatives and private sector business sectors than through aid, promote and place affordable housing on the market under price controls. Also, conditions by public administrations, giving priority to all of them to environmentally sustainable housing projects; also permanently eliminating deductions for financing the purchase of housing, which maintains a transitional regime until 2013.

Additionally, creation of public rental price indices by city areas should be considered as a measure that would allow municipalities wishing to impose maximum prices on residential rental rents according to objective criteria, such as average rental prices for the area, physical characteristics of housing, and economic levels of the residential population.

Application of moratoriums on the authorization of new vacation rentals in those areas of a city where there is already a high level of these rental homes as happens in the Historic Center of Málaga, including the establishment of fees and taxes on the vacation rentals, which allow cities to compensate for the public and natural resources used to cater for tourism.

FUTURE RESEARCH DIRECTIONS

Regarding future lines of research on the issue under consideration in this chapter, these should incorporate the need to quantitatively measure the impact that the expansion of vacation rentals is having on access to residential housing in relation to other explanatory factors. It is therefore important to analyze precisely what these factors are, paying particular attention to the role of public policies in the development of the real estate sector and the possibilities it has to contribute to a change in the trend of this, as well as its options to combat social inequality.

It would also be appropriate to determine the weight that the construction and promotion of the Real Estate Investment Trust (REITs) has on the current sector, assessing whether the tax benefits they receive and the favorable legislation with which they have social goals set by the state when it promoted its development.

CONCLUSION

The consideration of housing as an economic good incorporated into the market with the intention of making profit remains, as opposed to one that considers it a social good, essential for the coexistence of human beings. It cannot be forgotten that, as an object, which serves to satisfy a fundamental human right recognized in the international legal system, all public policy measures facilitating their access to the most vulnerable social sectors, will be justified.

REFERENCES

Aledo, A. M. (2007). La insostenibilidad del turismo residencial. *Antropología y turismo: claves culturales y disciplinares*, 185-208.

Aledo, A. M. (2016). Turismo residencial y vulnerabilidad en el interior del Levante español. (Ed.) Turismo residencial y gentrificación rural. (pp. 37-60) Alicante. Colección PASOS 16 & Foro de Turismo Responsable.

Alquiler Seguro. (2018). *II Estudio comparativo sobre alquiler turístico vs. residencial*. Madrid, Spain.

Altuzarra, A., & Esteban, M. (2011). Land prices and housing prices: The case of Spain. *Journal of Housing and the Built Environment*, *26*(4), 397–409. doi:10.100710901-011-9235-8

Ardura Urquiaga, A. (2017). La incidencia de la proliferación del alojamiento turístico en viviendas en el centro de Madrid. In *Vivir las ciudades en Europa*. Madrid, Spain: II Encuentro Europeo de Asociaciones de Vecinos.

Ardura Urquiaga, A., & Sorando Ortin, D. (2016). *First we take Manhattan. La destrucción creativa de las ciudades*. Madrid, Spain: Editorial Catarata.

Arias Sans, A., & Dominguez, Q. (2016). Unravelling Airbnb: Urban perspectives from Barcelona. *Reinventing the local in tourism: Producing, consuming, and negotiating place*. 73, 209-228.

Arrazola, M. (2015). Long-run Supply and Demand Elasticities. *Journal of European Real Estate Research,* 37, 371-404.

Association of Collective Investment Institutions and Pension Funds (Inverco). (2018). *Statistics,* Retrieved from http://www.inverco.es/38/0/43

Bank, U. B. S. (2019). *UBS Global Real Estate Bubble Index 2019*. Retrieved from https://www.ubs.com/global/en/ubs-news/r-news-display-ndp/en-20190930-grebi-global.html

Bank of Spain. (2018). *Financial balance of non-bank financial institutions*. Madrid, Spain.

Berges, A., & Ontiveros, E. (2007). La nueva Ley de Suelo desde la perspectiva económica. *Ciudad y Territorios. Estudios Territoriales, 152-153*, 259–275.

Biagi, B., Brandamo, M., & Lambiri, D. (2015). Does Tourism Affect House Prices? Evidence from Italy. *Growth and Change*, *46*(3), 501–528. doi:10.1111/grow.12094

Bianchini, F., & Parkinson, M. (1993). *Cultural Policy and Urban Regeneration: The West European Experience*. Manchester, UK: Manchester University Press.

BJH Advisors L.L.C. (2016). *Short Changing New York City The impact of Airbnb on New York City's housing market. Nueva York: Housing Conservation Coordinators Inc*. MFY Legal Services Inc.

Brossat, I. (2019). *Airbnb. La ciudad Uberizada*. Madrid, Spain: Editorial Katakrak.

Ciedes Foundation. (2017). *Málaga in figures*. Retrieved from http://ciedes.es/gestion-y-seguimiento/malaga-en-cifras.html

Comisión Nacional del Mercado de Valores (CNMV). "Series Estadísticas de las Instituciones de Inversión Colectivas. 2016. www.cnmv.es

Crommelin, L., Troy, L., Martin, C., & Parkinson, S. (2018). *Technological Disruption in Private Housing Markets: The Case of Airbnb. AHURI Final Report 305*. Melbourne, Australia: Australian Housing and Urban Research Institute Limited.

Curry, B., & Mounitnho, L. (1992). Diseño de los aspectos medioambientales del turismo. *Papers de Turisme*, *10*, 7–22.

Díaz Hernández, R. P. (2017). La vivienda secundaria y vacacional en la diversificación de la oferta alojativa de Canarias. *Revista Atlántida: Revista Canaria de Ciencias Sociales*, *8*, 243–264.

Directorate-General of Tributes. (2014). *Recaudación y Estadísticas del Sistema Tributario Español 2006-2016*. Madrid, Spain: Ministry of Finance and Public Administration.

Dredge, D., Gyimothy, S., & Birkbak, A. (2016). The impact of regulatory approaches targeting collaborative economy in the tourism accommodation sector: Barcelona, Berlin, Amsterdam, and Paris. Copenhagen: Paper 9 prepared for the European Commission DG GROWTH. Aalborg University.

Edufinet Financial Education (Ed.). (2018). Guía Financiera. Málaga: Aranzadi S.A. European Commission (2019) Communication from the Commission to the European Parliament, the European Council, the Council, the European Central Bank, the European Economic and Social Committee, the Committee of the Regions and the European Investment Bank. European Semester: country-specific recommendations. Brussels.

Eldiario.es. (2018). *¿Cuántas viviendas de tu barrio están en Airbnb? Descúbrelo en este mapa, manzana a manzana?* Retrieved from https://www.eldiario.es/economia/concentracion-Airbnb-manzana-viviendas-vacacional_0_809119651.html

Exceltur. (2015). *Alojamiento turístico en viviendas de alquiler: Impactos y retos asociados*. Madrid, Spain. Exceltur.

Fields, D., & Uffer, S. (2014). The financialisation of rental housing: A comparative analysis of New York City and Berlin. *Urban Studies (Edinburgh, Scotland)*, *53*(7), 1486–1502. doi:10.1177/0042098014543704

Fotocasa. (2019). *Real estate index*, retrieved from https://www.fotocasa.es/indice/#/filter/eyJ0cmFuc2FjdGlvbiI6ImJ1eSJ9

Garcia-López, M., Jofre-Monseny, A. J., Martínez Mazza, R., & Segú, M. (2019). Do Short-Term Rental Platforms Affect Housing Markets? Evidence From Airbnb in Barcelona. IEB Working Paper, 2019/05.

García Montalvo, J. (2000). El precio del suelo. La polémica interminable. *Instituto Valenciano de Investigaciones Económicas y Universidad de Valencia*, 1-47.

García Peña, M. (2017). *El impacto del turismo sobre la disponibilidad y precio de la vivienda*. Málaga: Fundación CIEDES.

Garijo, M. (in press 2017). Estas son las empresas que explotan el negocio del alquiler turístico al calor de Airbnb. *Eldiario.es*

Gil, J., & Sequera, J. (2018). The expansion of the tourist city and new resistances. The case of Airbnb in Madrid. *Revista de Metodología de Ciencias Sociales*, *41*, 15–32.

Glaser, E., Gyourko, J., & Saiz, A. (2008). Housing supply and housing bubbles. *Journal of Urban Economics*, *64*(2), 198–217. doi:10.1016/j.jue.2008.07.007

Gurran, N. (2017). When Tourists Move In: How Should Urban Planners Respond to Airbnb? Journal of the American Planning Association, 80-92.

Gyimóthy, S. (2017). Business Models of the Collaborative Economy. In D. Dredge, & S. Gyimóthy (Eds.), *Collaborative Economy and Tourism. Tourism on the Verge*. Cham, Switzerland: Springer. doi:10.1007/978-3-319-51799-5_3

Hernandez-Pezzi, C. (2018). *Turismo: ¿truco o trato?: políticas urbanas para el turismo de masas*. Madrid, Spain: Los Libros de la Catarata.

Homeaway. (2018). *V Barómetro del alquiler Vacacional en España*. Salamanca. Universidad de Salamanca.

Horm, K., & Merante, M. (2017). Is home sharing driving up rents? Evidence from Airbnb in Boston. *Journal of Housing Economics*, *38*, 14–24. doi:10.1016/j.jhe.2017.08.002

Idoate, E. Z. (2008). Políticas de vivienda en el Estado español. In Seminario de Economía Crítica Taifa, (pp. 66-80).

InsideAirbnb. (2019). *Sobre Airbnb en Málaga*. Retrieved from http://insideairbnb.com/malaga/

International Financial Analysts. AFI data laboratory. (in press 2016). ¿Cuánto bajaron los precios de la vivienda en tu provincia con la crisis? *Diario El País*.

Inurrieta Berruete, A. (2007). *Mercado de vivienda en alquiler en España: más vivienda social y más mercado profesional*. Madrid, Spain: Fundación Alternativas.

Laborda, J. (in press 2018). Ajuste de cuentas de las élites: pasado, presente y futuro. *Vozpopuli.com*

Lee, D. (2016). How Airbnb short-term rentals exacerbate Los Angeles's affordable housing crisis: Analysis and policy recommendations. *Harvard Law & Policy Review*, *10*, 229–253.

López Andion, M. (1998). Modelos econométricos de la vivienda. *Econometrics Working Paper Series Economic Development de la Universidad de Santiago de Compostela*, 59, 1-46.

Malighetti, P. P., Paleari, S., & Redondi, R. (2009). Pricing strategies of low-cost airlines: The Ryanair case study. *Journal of Air Transport Management*, *15*(4), 195–203. doi:10.1016/j.jairtraman.2008.09.017

Martinez Pagés, J., & Maza, L. (2003). Análisis del precio de la vivienda en España. *Documento de Trabajo. Servicios de Estudios del Banco de España*, *307*, 1–46.

Ministry of development. (2019). *Protected housing and rehabilitation*. Retrieved from https://apps.fomento.gob.es/BoletinOnline2/?nivel=2&orden=31000000

Municipal Málaga Housing Institute. (2019). *Promotions*. Retrieved from http://smv.malaga.eu/es/proyectos/proyectos/#!tab2

Naredo Pérez, J. M. (2013). Patrimonio familiar y crisis inmobiliaria. La apuesta por el alquiler. *Fundación de Estudios Financieros*, 169-186.

Naredo Pérez, J. M. (2019). Diagnóstico del panorama inmobiliario actual. ¿Rebrota una burbuja inmobiliaria comparable a la anterior? *Papeles de relaciones ecosociales y cambio global*, 145, 129-140.

National Competition Commission. (2013). *Problemas de competencia en el mercado del suelo en España*. Madrid, Spain.

National Statistical Institute. (2011) Census of Population and Housing (2011). Retrieved from https://www.ine.es/censos2011_datos/cen11_datos_inicio.htm

National Statistical Institute. (2019). *Rent per household by autonomous communities*. Retrieved from http://www.ine.es/jaxiT3/Tabla.htm?t=9949

National Statistical Institute. (2019) Statistics of tourist movements at the border. Frontur, Arrival of international tourists according to access road. Retrieved from https://www.ine.es/dyngs/INEbase/es/operacion.htm?c=Estadistica_C&cid=1254736176996&menu=ultiDatos&idp=1254735576863

National Stock Market Commission (CNMV). (2019). *Statistical Series of Collective Investment Institutions.* Retrieved from https://www.cnmv.es/portal/Publicaciones/Estadisticas-IIC.aspx

Nogués, A. M. (2007). Lugares y territorios: la segregación social y política en contextos turísticos. (Ed) Antropología y Turismo (pp. 165-184). México DF: Plaza y Valdés.

O'Regan, D. M., & Choe, J. (2017). Airbnb: Turning the Collaborative Economy into a Collaborative Society. In D. Dredge, & S. Gyimóthy (Eds.), Collaborative Economy and Tourism. Tourism on the Verge. Cham, Switzerland: Springer.

Observatory of the Urban Environment of Malaga. (2016). *Annual rent per household in Malaga, 2015.* Retrieved from http://www.omau-malaga.com/2/com1_md3_cd-1585/renta-anual-hogar-malaga

Official College of Architects of Malaga. (2019). *Data concerning the files of new work of housing approvals in the COA Málaga.* Retrieved from http://coamalaga.es/estadisticas/

Olcina Cantos, J., & Vera-Rebollo, J. F. (2016). Adaptación del sector turístico al cambio climático en España. La importancia de las acciones a escala local y en empresas turísticas. *Anales de Geografía de la Universidad Complutense, 36*(2), 321–352.

Oskam, J., & Boswijk, A. (2016). Airbnb: The future of networked hospitality businesses. *Journal of Tourism Futures, 2*(1), 22–42. doi:10.1108/JTF-11-2015-0048

Ozomek, A. (2018). Cronomapa del Expolio en el BIC Centro Histórico de Málaga 1957-2016. Edifeicios: Paisaje Urbano Patrimonio BIC Centro Histórico de Málaga.

Palacios García, A. & Vinuesa Angulo, J. (2010). Un análisis cualitativo sobre la política de vivienda en España. La opinión de los expertos. *Anales de Geografía de la Universidad Complutense,* 101-118.

Palos-Sanchez, P. R., & Correia, M. B. (2018). The Collaborative Economy Based Analysis of Demand: Study of Airbnb Case in Spain and Portugal. *Journal of Theoretical and Applied Electronic Commerce Research, 13*(3), 85–98. doi:10.4067/S0718-18762018000300105

Population and Housing Census of The National Statistical Institute. (2011). Retrieved from https://www.ine.es/censos2011_datos/cen11_datos_inicio.htm

Project DataHippo. (2019). Retrieved from https://datahippo.org/es/

Rabadan Formies, M. (2009). La inversión colectiva: concepto y evolución. Sus funciones. Situación actual y perspectivas. *Fundación de Estudios Financieros,* 27-42.

Rodríguez López, J. (2007). Urbanismo, vivienda y economía en España. *Arquitectura. Ciudad y Entorno, 3,* 80–85.

Rodríguez López, J. (2017). Las viviendas que pudieron hundir la economía española. La caída del mercado de vivienda y sus consecuencias. *Cuadernos de Relaciones Laborales,* 71-99.

Rodríguez-Salmones, N. A. (2003). El alojamiento turístico privado en el contexto del sistema de estadísticas del turismo. *Estudios Turísticos*, n.° 155-156, 7-32.

Saladié, O. A.-J. (2014). La influencia de las rutas de vuelos de bajo coste en la elección del destino turístico. *Cuadernos de Turismo*, *34*, 287–312.

Sánchez, R., & Ordaz, A. (in press 2018). ¿Cuántas viviendas de tu barrio están en Airbnb? *Eldiario.es*

Sanfelici, D., & Halbert, L. (2018). Financial market actors as urban policy-makers: The case of real estate investment trusts in Brazil. *Urban Geography*, *40*(1), 83–103. doi:10.1080/02723638.2018.1500246

Santana Talavera, A. (2003). Turismo cultural, culturas turísticas. *Horizontes Antropológicos*, *20*(9), 31–57. doi:10.1590/S0104-71832003000200003

Schäfer, P., & Hirsch, J. (2017). Do urban tourism hotspots affect Berlin housing rents? *International Journal of Housing Markets and Analysis*, *10*(2), 231–255. doi:10.1108/IJHMA-05-2016-0031

Servihabitat. (2018). *Mercado residencial en España. Segundo semestre 2017*. Barcelona, Spain: Caixabank.

Social Barometer. (2019). *Net disposable household income per capita*. Retrieved from https://barometrosocial.es/renta-y-patrimonio

Sociedad de Tasación. (2018). *El perfil del comprador de vivienda 2018*, Informe planner exhibitions - st sociedad de tasación Octubre 2018.

Spanish Tax Agency. (2019). *Statistics of IRPF declarants by municipalities,* retrieved from https://www.agenciatributaria.es/AEAT.internet/datosabiertos/catalogo/hacienda/Estadistica_de_los_declarantes_del_IRPF_por_municipios.shtml

Stiglitz, J. (2010). *Caída libre*. Madrid, Spain: Taurus.

Tasaciones InmobiliariasS. A. (2019). Retrieved from .https://www.tinsa.es/precio-vivienda

Tinsa. (2019). *Interactive Imie, The price of housing in Spain*. Retrieved from https://www.tinsa.es/precio-vivienda/

Titman, S. (1985). Urban land prices under uncertainty. *The American Economic Review*, *75*, 505–514.

Treserras, J. J. (2004). La tematización cultural de las ciudades como estrategia de desarrollo a través del turismo. *Portal Iberoamericano de Gestión Cultural*, 1-19.

Urrestarazu-Capellán. (2019). *Los factores determinantes de la oferta y precio de la vivienda en alquiler: una visión desde el turismo* (Unpublished doctoral dissertation). University of Málaga.

Varma, A., Jukic, N., Pestek, A., Shultz, C., & Nestorov, S. (2016). Airbnb: Exciting innovation or passing fad? *Tourism Management Perspectives*, *20*, 228–237. doi:10.1016/j.tmp.2016.09.002

Wachsmuth, D., & Weisler, A. (2018). Airbnb and the rent gap: Gentrification through the sharing economy. *Environment and Planning A. Economy and Space*, *50*(6), 1147–1170.

Wegmann, J., & Jiao, J. (2017). Taming Airbnb: Toward guiding principles for local regulation of urban vacation rentals based on empirical results from five US cities. *Land Use Policy*, *69*, 494–501. doi:10.1016/j.landusepol.2017.09.025

Whitehad, C., & Williams, P. (2011). Causes and consequences? Exploring the shape and direction of the housing system in the UK post the financial crisis. *Housing Studies*, *26*(7-8), 1157–1169. doi:10.1080/02673037.2011.618974

World Economic Forum. (2019). *Making Affordable Housing a Reality in Cities.* Cities, Urban Development & Urban Services Platform In Collaboration with PwC. Cologny/Geneva.

ADDITIONAL READING

Barron, K. K. (2017). The Sharing Economy and Housing Affordability: Evidence from Airbnb. *Social Science Research Network*, 1-42.

Cocola Gant, A. (2016). Holiday Rentals: The New Gentrification Battlefront. *Sociological Research Online*, *21*, 1–9.

Pareja Eastaway, M. y. (2015). El sistema de vivienda en España y el papel de las políticas: ¿qué falta por resolver?. *Cuadernos económicos de ICE*, 90, 149-174.

Red2Red consultores. (2017). *Análisis del impacto de las viviendas de uso turístico en el Distrito Centro.* Madrid: Ayuntamiento de Madrid. Distrito Centro.

Stevenson, S., & Young, J. (2014). A Multiple Error-Correction Model of Housing Supply. *Housing Studies*, *29*(3), 362–379. doi:10.1080/02673037.2013.803040

Van Loon, J., Oosterlynck, S., & Aalbers, M. B. (2019). Governing urban development in the Low Countries: From managerialism to entrepreneurialism and financialization. *European Urban and Regional Studies*, *26*(4), 400–418. doi:10.1177/0969776418798673

Vinuesa Angulo, J. (2013). *El festín de la vivienda. Auge y caída del negocio inmobiliario en España.* Madrid: Diaz & Pons Editores S.L.

Vives-Míro, S. (2018). New rent seeking strategies in housing in Spain after the bubble burst. *European Planning Studies*, *26*(10), 1920–1938. doi:10.1080/09654313.2018.1515180

Waldron, R. (2018). Capitalizing on the State: The political economy of Real Estate Investment Trusts and the 'Resolution' of the crisis. *Geoforum*, *90*, 206–218. doi:10.1016/j.geoforum.2018.02.014

KEY TERMS AND DEFINITIONS

Accessibility to Housing: The degree to which citizens of a given place can exercise their right to enjoy decent and adequate housing.

Gentrification: Process by which the resident population of a city with fewer resources is forced to leave their places of residence by being displaced by other residents of greater purchasing power or by the action of a certain economic activity.

Housing Reservation Platform: Companies, mainly multinational, whose activity is carried out on the Internet, and which are responsible for reservation tourist housing worldwide, creating a global market, easy and accessible for owners of tourist housing and a form of cheaper accommodation with a different service to the traditional one for its customers.

Inelastic Housing Supply: A feature of the housing market whereby supply and demand do not adjust to price changes according to the logic of economic theory, so that when supply increases, prices also do.

Management of Tourist Accommodation Company: Companies that are dedicated to managing the operation of homes as tourist accommodations, developing management, reception, maintenance, and cleaning.

Public Protection Housing: Housing built by the public sector of a state with the intention of providing affordable housing to the most disadvantaged sectors of the citizenry and, at the same time, energizing the economy with the activity that is generated around it.

Real Estate Investment Trust: Capital investment companies, which allocate these to the acquisition of all types of real estate with the intention of exploiting them on a rental basis, and which have state tax benefits.

Urban Housing Rental Law: National legislation that includes the main issues related to the rental of real estate.

Vacation Rentals on Housing: Type of tourist accommodations characterized by being offered in houses that share their activity with other properties dedicated to habitual and secondary residence.

Section 3

Case Studies: Coastal Areas

Chapter 11
Collaborative Lodging in Puerto Vallarta:
Who Wins and Who Loses?

María de los Angeles Huízar Sánchez
https://orcid.org/0000-0002-8051-4241
University of Guadalajara, Mexico

Jorge Luis López Ramos
https://orcid.org/0000-0002-2290-7004
University of Guadalajara, Mexico

José Alfonso Baños Francia
https://orcid.org/0000-0003-4734-9368
University of Guadalajara, Mexico

ABSTRACT

The collaborative lodging has expanded globally, driven by companies that through digital platforms enlist the properties of individuals to integrate them into the market of tourist accommodation, a situation that has generated wide socio-economic and territorial problems in the main tourist destinations in the world. This chapter analyzes who receives the benefits of the collaborative lodging, as well as the negative impacts it generates in Puerto Vallarta, through an analysis of the behavior of the offer, as well as in-depth interviews with hosts registered in the Airbnb platform.

INTRODUCTION

The accelerated technological development that post-modern society has experienced in recent years has caused changes in the production and consumption paradigms posed by the traditional economy, moving towards new forms of exchange of products and services through the so-called *collaborative economy.*

DOI: 10.4018/978-1-7998-2224-0.ch011

Under this scheme, a series of companies have emerged acting as intermediaries between producers and consumers through digital platforms, among which Uber and Airbnb stand out and have positioned themselves as icons of this model, the first in the mobility sector and the second in lodging.

Since its creation in 2008, Airbnb has captured international attention due to its rapid growth, placed among the companies best valued by the financial markets. It has grown around the world, taking advantage of privately owned homes to combine the largest number of accommodation units on a global scale.

It is recognized as an innovative and disruptive company that offers unique experiences that contrast with the standardized service of traditional hotel management. However, its operation in cities such as New York, Paris, Barcelona or Amsterdam, among others, has given rise to a myriad of questionings derived from the socio-economic challenges that its proliferation has generated in said cities. After various legal disputes, some of these cities have decided to not allow the operation of the company, or have been seen themselves in the need of generating new regulations for its operation. These aspects call into question the benefits attributed to this model.

The high demand for temporary accommodation puts more pressure on tourist destinations, which experience an accelerated growth in the offer of collaborative options. In Mexico, Puerto Vallarta is the third most visited beach destination in the country, and like others, it has an accelerated growth of lodging units under this modality, which are generating changes in its socioeconomic and territorial structure.

In 2010, the company Airbnb began its operations in this tourist destination and quickly positioned itself as a leader in collaborative lodging, presenting a growth rate of over 60% per year (Airdna.co, 2018a), a figure that shows the interest in this type of lodging among tourists impacting the real estate model in two ways. The first, an increase in the number of registered hosts with more properties to offer. The second, real estate investors who see a good opportunity to increase their profit by building high-rises with a greater number of units.

From the community perspective, residents of neighborhoods that have experienced the real estate boom, such as the Romantic Zone and Marina Vallarta, have expressed their rejection to the accelerated process of urbanization. However, given the lack of attention by the local authorities, they have opted for stronger measures (public demonstrations and closing streets).

Meanwhile, the hotel sector has insisted on the need of regulation, referring to the unfair competition that this type of lodging represents by not paying taxes, not being subject to the same demands as the hotel industry. In their opinion, Government should take actions to ensure equity by all means (fiscal, administrative and urban), establishing compensatory instruments.

The effects of collaborative lodging have shown new forms of accumulation of capital for the intermediary company and a good business opportunity for developers and property managers who are taking advantage of spaces with great touristic value, saturating these with high-rise buildings, in smaller spaces and driving unprecedented gentrification dynamics.

All this take place before the passive eye of local agents, who are more interested in the economic dimension and are overwhelmed by the speed of the process. So far, the community has not taken strong actions but tensions are appearing, paving the way for over-tourism effects, with an ending difficult to predict.

Given this scenario, the question is who wins and who loses? In the current process of expansion of collaborative lodging in Puerto Vallarta.

OBJECTIVES

This research aims to analyze the socioeconomic and territorial effects that the boom and growth of collaborative lodging in Puerto Vallarta is generating. It is considered a factor that affects the loss of competitiveness of the hotel sector, as well as urban saturation and gentrification experienced in touristic neighborhoods.

Collaborative Economy

According to Cañigueral (2014) the origin of the collaborative economy dates back to the famous barter system that was used in most ancient cultures to exchange goods or services, or as a form of support among members of small communities.

With the adoption of currency, the exchange value was replaced by the price that was assigned to each product or service, a fact that gave rise to the market economy. However, with the development of information and communication technologies (IT), the forms of production, consumption and sale are changing, driving the market to an evolutionary process towards the digital level, with individuals and companies operating through computer platforms and transforming economic and social structures, such as the concept of space that becomes ubiquitous.

Under this reality, new ways of exchange, sale, rental and access to products and services that are used temporarily and sporadically developed, so that sharing these resources is considered a viable alternative to reduce the high level of consumption that characterizes current society.

Technological resources facilitate direct contact between suppliers and users of goods and services by the internet network, reducing intermediation costs and strengthening links of trust through a double public assessment in digital media, which serves as a reference for future exchanges.

In 2000, Jeremy Rifkin addressed the changes generated by technological media, motivating markets to yield to the power of digital networks, where sellers and buyers become suppliers and users, and access replaces physical property.

This author pointed out the transition from the property regime to the access regime stating that this new practice "is based on guaranteeing the limited and short-term use of goods controlled by supplier networks" (Rifking, 2000, p. 5). Likewise, he warns that this transition will generate important changes in the structure of economic relations and a broader transformation that takes place in the capitalist system.

Years later, technological advances and the expansion of the use of the Internet and social networks boosted this economic model of access, accelerating the exchange of goods and services between individuals and companies, which gave way to the so-called *collaborative economy*. However, the discourse on the subject adopted a different nuance, pointing out that this model operates as an alternative to the paradigm of capitalist consumption.

In 2010, the book *What´s mine is yours* by Botsman and Rogers (2010), presented a flattering picture about the development of collaborative consumption, as a type of social movement, which highlights the advantages of the exchange of products and services between individuals (*peer to peer*) to reduce the pattern of consumption in modern society.

Under this same approach, Cañigueral presented in 2014 his work *Living better with less* in which he addresses hyper-consumption and the advantages of the economy and collaborative society, where the concepts of sharing, collaborating, reusing and recycling are essential to "the urgent and urgent need to reorient the economy to put it at the service and scale of people. A society with people in the center" (Cañigueral, 2014, p.12).

For this author. "Collaborative consumption proposes to share the goods in front of owning them, and focus on putting into circulation everything that already exists" (Ibid, p. 16). Emphasizing the advantages of access over possession, likewise, he points out that this activity happens on a small scale, when the internet and social networks are added it acquires unimaginable speed and dimension.

According to Alonso and Miranda (2017), the collaborative economy constitutes a new economic paradigm, product of the sum of the economic crisis and the hyper-connected society, which found in technology the exit tool to the problems of classical capitalism. Although these authors warn of the premature announcement of the end of the current capitalist model, they state that society is in a transitional stage, where it will be necessary to identify the roles that the market, the State and the different cultures will take; however, they anticipate that,

The growth and boom of this type of collaborative initiatives demonstrates a widespread desire to find solutions based on collaboration... and the emergence of new social attitudes, increasingly generous, more supportive and less hierarchical. (Alonso and Miranda, 2017, pp. 21-22)

The main attributes of this economic model from the offer, are the ease that individuals have to place in the market under-utilized products that provide them with some income and complementary employment. Likewise, the increase in available products and services favored the demand, accessible by this means.

These aspects give an optimistic stance regarding this collaborative economic model, which is presumed unstoppable, revolutionary or disruptive, since it counteracts the negative effects of capitalism.

Collaborative Company

Over the past few years, a significant number of companies of different spheres have emerged, which are identified within this model, because they operate through digital platforms and function as sellers or as intermediaries in the relationship between suppliers and customers.

The companies that have positioned themselves, as icons of this model are Uber and Airbnb; the first in the mobility sector and the second in collaborative lodging. In the business world these companies are labeled as *unicorns*, based on the metaphor that they are mythological and fabulous animals, because rarely, does a company reaches a value exceeding one billion dollars in a short period of time (Calleja, 2017).

The success of these companies relies on the global expansion they have achieved through the management of supply and demand through a digital platform in which individuals register their assets and increase the volume of supply that is concentrated by the company. At the same time, users register on the same platform to access the service and the company manages the relationship, from which their income is derived.

In this way, the intermediation of companies changed the relationship between individuals (*P2P*) which Bostman & Rogers (2010) presented as collaborative consumption by the business to consumer model (B2C), that companies such as Uber and Airbnb have taken advantage of to expand globally, integrating private property and concentrating most of the lodging offer at a global level, especially in the big cities and in the tourist destinations with the highest demand such as New York, Paris, Venice, Amsterdam, London or Barcelona, triggered a series of socio-economic and territorial changes that affect competition with traditional hosting companies (like hotels), saturation (*over-tourism*) and gentrification, that point to the need for the regulation of this new type of accommodation (Brossat, 2018) where businessmen and local people have raised their voice to denounce problems related to the unfair practices that this represents to formal lodging companies.

In theory, these companies present themselves as benefactors of society by promoting a horizontal distribution of wealth, through the employment they generate and the extra income that improves the living conditions of drivers and hosts who participate in their platforms. In the same way, it offers users of these services the "experience" of owning a car with a driver or a home anywhere in the world.

It has also affected the increase in housing and rental prices that lead to the gentrification of cities, in addition to the conflicts between residents and visitors derived from the presence of tourists in the daily living spaces of the inhabitants. Facts, which have forced the authorities of these cities to intervene in the search for solutions.

Given these tensions, it is difficult to maintain a positive position in the matter, and authors such as Slee (2017, p. 15) notes that "The collaborative economy is introducing a ruthless and deregulated free market in areas of our lives previously protected" when analyzing the way in which this type of business operates.

As well, he states that companies that proclaim the benefits of the collaborative economy for their users "have become corporate monsters and are playing an increasingly intrusive role in the transactions they encourage to earn money and maintain their brand" (Ibid, p. 15-16).

For these companies, the goal is to profit from the assets of third parties that they voluntarily put at their service, and from which they extract the benefits, without taking risks, or any commitment to the individuals that provide them with said goods, or of the employees that contribute to their enrichment. Much less, assume responsibility for the fractures generated in the communities in which they are inserted.

For this author, the effects generated by these companies in the cities are the following:

- Instead of openness and personal trust in interactions, they produce a climate of insecurity and vigilance for employees who fear being reported.
- The extra money that the collaborative economy promises translates into informal employment, without social benefits. The better jobs are for computer engineers and business marketing experts.
- Instead of empowering individuals, it uses them to generate profits for its investors and executives.

The contradictory nature of the positions before the effects of the collaborative economy leads to the question that guides this research and which it tries to discern who wins and who loses in collaborative economics? For this purpose, a study was carried out in one of the sectors that has experienced the strongest impact of the collaborative model: tourist lodging. Puerto Vallarta was chosen as a case study because it is a destination that is experiencing an accelerated growth in this type of lodging.

Method

The research was divided into three stages: The first consisted of a monthly monitoring of the offer of collaborative lodging in the municipality through the main platforms during the year 2018, in order to identify the magnitude, cost, type and territorial distribution.

In the second stage, a study was carried out on the real estate market to register the process of amendments to urban regulations, aspects that contrasted with articles in the main local newspapers to assess the perception of local residents in the face of the rise of buildings and developments. Finally, 23 in-depth interviews were conducted with the three types of hosts registered on the Airbnb platform to learn about the operation and benefits offered by the platform and to identify if there is a relationship between the real estate boom and collaborative lodging.

Collaborative Lodging in Puerto Vallarta

In Mexico, the lease or loan of touristic homes is a common practice in the main destinations, both for domestic and foreign users. However, with the expansion of the internet and the creation of digital platforms, the activity has increased considerably, as Conley (2014) pointed out by saying that "Airbnb is an old idea with a new tech twist". However, this tech twist has taken on large dimensions, changing the global lodging landscape and generating concern in the hotel industry of several countries.

The establishment of Airbnb offices in the country offers proof of the importance that this destination has gained in collaborative lodging in Latin America, positioning itself as the second place site with thirty seven thousand properties, surpassed only by Brazil (a21mx, 2017).

According to data from the Mexican Institute for Competitiveness (IMCO, 2017), Playa del Carmen, Mexico City and Puerto Vallarta, lead the top ten Airbnb destinations in Latin America. The latter is one of the main coastal destinations located on the west coast of the country, which gained popularity in the early 1960´s when it was chosen as the location for the filming of the movie *La noche de la Iguana.* The presence of well-known actors, such as Richard Burton and Ava Gardner attracted the international press, who shared the beauty of the town and its landscape (Munguía, 1997).

With strong support from the government (federal, state and local), the destination was provided with the infrastructure required to cope with the growing number of tourists, arriving to a stage of consolidation in the mid-eighties of the last century. Tourism was positioned as the most important economic activity and the main source of employment in town, especially in the hotel industry, a labor-intensive activity that showed a significant increase from 5,642 rooms in 1983 to 15,400 in 1993 (SECTURJAL, 1994).

In this decade, the accommodation offer through time-share diversified, generating great controversy over the competition it meant to hotels and the small number of jobs it generated, compared to the hotel sector. However, time-share gained acceptance among tourists and its offer reached the figure of 13,584 units in 2016 (RCI, 2017), which increased the hotel offer to 23,797 rooms (SECTURJAL, 2017) that are distributed throughout the touristic zone or strip[1].

According to the study carried out by the Mexican Institute of Competitiveness (IMCO), in 2017, Puerto Vallarta had 4,400 properties registered in Airbnb that year, under the collaborative lodging scheme. A year later, the data of Airdna.co (2018b), showed that the offer increased to 5,268 properties, with an annual growth rate of 63% and an average daily rate of two thousand two hundred pesos[2], however, it should be noted that prices are adjusted based on temporality and type of unit, so a room to be shared could be located outside the touristic area for one hundred and twenty pesos[3] a night, or a luxurious villa in the most exclusive districts of the municipality can be rented for almost sixty thousand pesos per day[4].

Of the properties, considered in the offer of collaborative lodging, 84% correspond to the full house or apartment, something that does not coincide with the objective of sharing and even more so, it does not allow the achievement of that unique experience of immersion in the locality that Airbnb promises.

As for territorial distribution, the offer of collaborative lodging follows the pattern of the hotel model, concentrating along the coastal strip; however, it presents a relevant expansion towards the districts neighboring the tourist area and registers a smaller presence in the interior districts.

Figure 1. The offer of collaborative lodging in Puerto Vallarta

The Romantic Zone and Marina Vallarta are the two spaces of robust interest for tourism and in which exists a wide range of collaborative lodging; they are the ones that experience significant real estate growth, giving rise to a series of neighborhood conflicts, which, according to the hypothesis of this work, are propelled by the rise of collaborative lodging in town.

The Romantic Zone distinguished by its proximity to the city center and for being a popular space of the LGBT community, having a wide range of services aimed at this segment of tourism, with high purchasing capacity (Huízar, Baños and López, 2018).

This area is one of the oldest in the city and around it, an urban image of a traditional Mexican village was configured, expressed with low-rise houses, narrow streets, white painted walls and roofs topped with red ceramic tiles. This attribute was a competitive advantage for Puerto Vallarta over other beach tourist destinations in Mexico, an aspect that has been diminished[5].

It is also the area with the highest concentration of collaborative lodging, with a little more than a thousand active rentals, which are mostly characterized by apartments with one or two rooms, with an average daily rate of 143 dollars, which in the high season reaches 205 dollars (Airdna.co, 2018c).

As of 2010, the year that coincides with the start of Airbnb operations in this destination, an intense urbanization process took place through the construction of 14 real estate developments which translates into 675 new housing units; in addition, currently ten new developments are under construction with more than two hundred new units.

As is shown in Table 1, the urbanization process considers more than one hectare of urbanized land, translated into the construction of almost ten hectares of new developments. That is, densification has increased almost six times in relation to the properties that house them.

What stands out is that of the 675 apartments built, only 107 units comply with urban regulations[6]. For the additional units approved, real estate developers used (in all cases) a bogus tool called *Análisis Técnico Justificativo* (ATJ)[7]. The capital gain stands in the private sector, leaving the opportunity to capture a part of it to strengthen public finances.

The implications for the management of public services are already showing their effects. Thus, according to the Director of the water and sewerage system at the local level, the impacts caused by over urbanization threaten to collapse the drainage system that dates back to more than 40 years and requires an investment of 150 million pesos[8] to replace it (Mora, 2017).

Table 1. Recent real estate developments in the Romantic Zone

Name	Land surface m^2	Built surface (m^2) according to regulation*	Real built surface	Units	Units according to regulation*	Height (stories)	Height according regulation*
Pacífica	641.00	1,346.10	3,846.00	36	5	7	3
The Park	1,240.30	2,604.63	7,982.10	54	10	7	3
Zenith	938.50	1,970.85	9,672.00	46	8	7	3
V399	728.00	728.00		55	11	7	3
Rincón de Almas	798.70	1,677.27	3,511.20	38	7	7	3
Pier 57	1,248.00	2,620.80	9,821.00	69	10	7	3
Oceana	665.80	1,398.18	5,205.85	30	6	7	3
Nayri	1,068.10	2,243.01	6,939.72	54	9	7	3
Loft 268	1,232.90	2,589.09	9,440.99	64	10	7	3
Pavillion	1,240.66	2,605.39	8,209.59	61	10	7	3
Urban	300.00	630.00	2,094.00	12	3	7	3
Pinnacle 1	1,075.90	1,291.08	8,607.20	48	4	7	3
Pinnacle 2	948.28	1,137.94	7,586.24	57	8	7	3
105° Sail View	1,279.50	1,535.10	5,760.00	53	4	10	3
Total	14,024.99	26,479.39	93,771.89	675	107		

* Considering the *Plan Parcial de Plan Parcial de Desarrollo Urbano Los Muertos* approved in 2000 and currently in force.
Source: own elaboration based on empirical research and official information provided by the City Hall of Puerto Vallarta.

However, the strong demand and this type of construction are contributing to an increase in the price of land, since, according to Vidal Meza, president of the LGBT Tourist Trade Association, it is the area "that has the most expensive square meter in the whole city" (García, 2019). This situation generates pressure on homes, public spaces and services such as schools and hospitals and promotes the displacement of the local population, a scenario that has generated multiple protests from neighbors who refuse to move and complain about noise and a decreased quality of life.

In this regard, the *Vallarta Opina* newspaper stated,

The greed of those who build and pay for the work goes hand in hand with the authority that has allowed the creation of these buildings that already show saturation in sight, offering luxury condominiums for a few who will modify the environment of a community that has not known how to defend its identity. (Ocaña, 2017)

As for the Marina Vallarta neighborhood, built in 1985 by taking into consideration a mix of uses with a tourism vocation, with big lots facing the sea for hotels, apartments around a marina and golf course, single-family housing, as well as businesses and shops focused to a segment of high-income population.

The boardwalk that surrounds the marina is surrounded by an important number of restaurants and shops that make it the second point of greatest attraction for tourism and one of the best places to live due to the tranquility, services and communications available.

In 2018, this area had 174 active incomes, with an average daily price of 190 dollars for an apartment with two rooms, but most are villas with prices that exceed five hundred dollars per night.

Figure 2. View of the Romantic Zone in Puerto Vallarta

However, in 2017, a dispute between entrepreneurs and neighbors began with the local government due to the granting of construction licenses for buildings that exceeded eight floors (standardized height in urban planning regulations), causing a saturation of multi-family housing.

According to the *observatoriobahía.com.mx website* "The grievance of the residents is that they were not taken into consideration in the approval of those new developments" (Arenas, 2018).

Due to the lack of response from the local authority, at the end of 2017, the Business Coordinating Council summoned a meeting with the Mayor, business representatives, public and private organizations, researchers and neighbors to address the problem related to the massive and out of control construction dynamics in the Romantic Zone as well as in Marina Vallarta. In this meeting, the grievance in connection with the developments in process was expressed, many of which did not comply with urban regulations, as well as the concern about the effects that these constructions could have on the destination. Given the situation, the Mayor said that everything was carried out in accordance with regulations and offered to carry out meetings to address this problem (Mora, 2017), a situation that never happened.

Months later, before the breach of the agreements reached and particularly after the amendment to the urban regulations of Marina Vallarta[9], the Homeowners Association of Marina Vallarta carried out a public demonstration by blocking the main avenue of the city, demanding that,

No more changes will be made to the use of residential land to mixed or commercial land, maintain the heights of new buildings in the area at 8 levels maximum, and continue to honor the declaration as a Controlled Development Division of the subdivision. (Garda, 2018)

Figure 3. Real Estate Developments in Marina Vallarta, Puerto Vallarta.

To the protests mentioned above, the neighbors of other districts joined in, districts in which similar processes of construction of high rise real estate developments began, a model that seems to be adopted under the discourse of the "compact" city proposed by international organizations such as UN-HABITAT[10].

However, the position of the citizens points towards an effective densification where the capital gain generated by the urbanization is captured for local interest, the public space is dignified and compensatory instruments are fought to contain the voracity of an extractive real estate development that seems to favor the expansion of the temporary rental offer for tourism.

The Host: The Key Piece in the Real Estate Development and Collaborative Hosting Relationship

The collaborative host is the key piece to determine if there is a relationship between the increase in collaborative lodging and the real estate boom experienced in Puerto Vallarta, being the intermediary agent between these two elements.

By March 2018, Puerto Vallarta had 2,257 registered hosts, of which 71% were responsible for several properties (multi-listing) and 21% had only one property in their care (Airdna.co, 2018d).

The high number of hosts with several properties under their care allowed for the verification of the real estate nature of collaborative hosting, if compared with the small number of hosts that have only one property to share as established by the collaborative economy. Likewise, it was possible to identify that the traditional hotel industry is adding units to this model by registering their rooms through the platforms.

Based on this information, it was possible to identify three basic profiles for the hosts registered on the Airbnb website, whose features are described below and allow us to identify the role they play in collaborative hosting.

1. *The collaborative host.* Is named as such because it is the one that best fits the principles of the collaborative economy, because it is the one who has under-used space in their home and decides to offer it for rent through the platform. For this host, the income he obtains from his participation in this activity constitutes extra income, since he retains his main job.
2. *The real estate host.* A host that has several properties registered on the platform, its host job is to manage the properties of third parties in exchange for a payment or commission. For this host, its income depends entirely on the rental of the properties and constantly seeks to increase the number of properties under its management, because these represent its only source of income
3. *The hotel host.* They are owners or administrators of traditional hotels, who made the decision to integrate their hotel rooms to the Airbnb offer with the objective of taking advantage of the positioning of this company to disseminate its facilities and increase its sales.

In order to identify the role played by the host in collaborative lodging, an in-depth interview was conducted with hosts of each of the profiles described above (ten a collaborative host, ten a real estate host and three a hotel host), in which four key points were addressed: Host-guest relationships, collaborative hosting - real estate boom, economic benefits, risk and social responsibility.

The Host-Guest Relationship

The information collected from the Airbnb webpage during 2018 allowed for a more detailed profile of the offer, where it was possible to determine that most of the offer corresponds to full houses and apartments. Aspect that in principle questions the primary objective of the collaborative economy, sharing.

That is why the questions focus on the type of contact that is established between the host and the guest during their stay, that is, if they assume the role of guides and offer support for the development of tourist activities that contribute to enrich the local experience. The response in all cases was negative, limiting the contact to the delivery of the property upon entering and leaving the guest, or even only to share a password for gate admission. However, one of the real estate hosts said that "he offered his guests the transfer from the airport to the property and vice versa, as well as accompaniment to make the first pantry if they wanted to" (R. Cortés, personal interview, March 18, 2018).

To address this issue, the hosts were asked what do you think is the reason that most of the collaborative lodging offer are full houses or apartments for the guest(s) alone and not shared? The answers focused on socio-cultural aspects by pointing out that they were not willing to live with "strangers" and put their family at risk. One collaborative host has even stated in this regard, that he modified his house and the room he rents has an entrance door independent of the house.

The above information allows us to conclude that the offer of full units, the high number of real estate hosts in Puerto Vallarta and the way in which the collaborative hosting operates does not generate the local experience on which the advertising of the leading company in this model of collaborative lodging is based on.

The Collaborative Hosting Relationship: Real Estate Boom

The central aspect was the changes that the urban pattern of Puerto Vallarta is experiencing with respect to the boom in the construction of new real estate developments, for this purpose the hosts were asked if they considered that there was any relationship between the increase in the offer of collaborative lodging and the emergence of new real estate developments?.

The real estate hosts and the hotel hosts agreed that construction, mainly in the downtown of the city, is directly related to this activity, because,

This is the area with the highest demand in the destination, and they buy them in order to spend time here and then rent it out, even those who sell them convince them that the apartment will pay itself of with just the rental income and they even offer to be the ones who manage it. They offer them the full package, and they are selling them so fast that they still have not finished selling one when they are already building another one. (R. Cortés, personal interview, March 18th 2018)

Similarly, a hotel host said that "the local residents who do not want to sell their properties is due to the fact that they have become aware of the potential of the business, and they are building an extension to their house to rent it by Airbnb." In addition, he stated that "Puerto Vallarta is currently on a full honeymoon with Airbnb, but we must take into account what will happen when the excitement passes, because with so much offer prices will go down" (MJ. Zorrilla, personal interview, May 7, 2018).

Figure 4. Property for rent in Puerto Vallarta.

The Economic Benefits of Collaborative Hosting

Another of the pillars on which the benefits of the collaborative economy is based on is the income offered to its hosts. In the case of Puerto Vallarta, the offer of income for its hosts potentially ascends to $ 661 US dollars per month for an entire accommodation with capacity for four guests, $ 517 US dollars for a private room for two guests or $ 176 US dollars for a shared room for one guest (Airbnb.com, 2018).

Figure 5. Airbnb advertisement for hosts in Puerto Vallarta

The offer is tempting compared to an approximate monthly minimum wage of 160 US dollars for eight hours of daily work in Puerto Vallarta, or perhaps the inspiring testimonies of other hosts around the world, who give their account of the economic benefits offered by being a host. Airbnb publishes and spreads these testimonies as part of its lobbying strategy in order to counter the constant negative criticisms of its business model.

Based on the information above, the hosts were questioned about the economic benefits generated by being a host and the conditions in which they carry out their activities. The responses of the real estate hosts and the hotel hosts were similar concerning the time allocated to this activity, which covers 24 hours a day, since they must be available at all times to the guest, otherwise they would be negatively evaluated and that would impact their future income.

For its part, the collaborative host said he looked after the guests, but it was not a job to which he devoted all day to.

Regarding their income, it was inquired if in reality they obtained the amount that the company promised on their website. The answers were different, the real estate hosts commented that it depended on the characteristics of the property and the times it was rented, pointing out as its "star" properties, those that are located in the downtown area, because these are the ones that are rented more frequently during the year, whose profits exceed the Airbnb proposal, however, properties farther from the downtown have lower possibilities and do not regularly generate the indicated income.

The hotel hosts, whose units are located in the Romantic Zone, revealed that the rooms registered on the Airbnb platform had a higher occupancy during the year, compared to those offered in the traditional hotel system, so the income at times was in keeping with what the company offered.

On the other hand, the collaborative host said that he did not obtain the income indicated by the company because his property was not rented too often, but that was not important for him, because it was only extra income.

The Risk and Social Responsibility

The "trust" factor is an essential element in collaborative hosting, since it implies an economic transaction, coexistence and access to private property by unknown people. This has given rise to multiple conflicts around collaborative hosting according to the newspapers of different cities, whose notes tell stories of negative incidents to guests in rented apartments (Infobae, 2018), guests who sue the hosts for video recordings in private spaces (Dixón, 2019; Regueira, 2019) or who complain about fraud with their accommodation (Romera, Brualia and García, 2017). Situations that the host must face directly, but it is unknown if the intermediary company shares the risk and social responsibility.

For this reason, the research explored negative incidents that have occurred on the properties under their care as hosts. In this regard, they pointed out issues related to property damage and theft of objects, as well as minor conflicts over the malfunctioning of property equipment. However, a professional host reported an experience of attempted fraud with a property under his care.

Likewise, the following question was raised ¿Who assumes responsibility for the risks they face in the performance of their role as a host.? They all indicated that the risks were their responsibility, because the company provided them with a guideline of how to proceed in certain situations, but they did not believe the company would respond for property damage or accidents during the performance of their work.

And the Local Authority?

The concerns that have been made known by the business sector, the local population and groups of academics who warn against the strengthening of the extractive version of collaborative hosting, have made little echo on the local authority. Urban regulations were not followed, despite of the obligation to honor them. There have been questions about informal agreements that in many cases look more like corruption.

To mitigate the negative comments, the official discourse focuses on pointing out that they are promoting regulatory measures for this type of accommodation; however, the proposals present a fairly narrow view of the problem, because the regulation focuses solely on the requirement of the payment of taxes by the intermediary digital company, but, it does not address issues such as the regulation of its operations, and the social and economic responsibility that it must assume with the local communities that directly receive the effect of its operations, as well as with the hosts registered on its platforms and with the tourists users of its services, as has happened in other cities such as Amsterdam, Paris or Barcelona.

The situation that prevails in the tourist strip, especially in the Romantic Zone and Marina Vallarta, back up this study because, these spaces require comprehensive measures that contribute to solve the various problems they face, such as the saturation of high-rise buildings to foster foreigners (second homes) and tourist rental, generating a gentrification process by expelling local residents and turning the space to commercial activities to meet the demands of tourists.

However, it seems that local authorities are the least interested in intervening, especially if they are considered responsible for authorizing building licenses and the change in the use of land, where negotiations happened through informal agreements.

CONCLUSION

The assessments of the positions regarding collaborative tourism show two facets. One represents an alternative way to capitalism and excessive consumption through cooperation, collaboration and trust between individuals seeking a mutual benefit. This position has a strong base of enthusiastic followers but still needs to be proven.

The second shows a purely economic behavior, where the goal is to maximize profit, taking benefit of the material, human resources and assets in a myriad of cities through technological means. For this model, the breakdown of local structures is irrelevant to the economic benefits that the model itself represents.

Collaborative lodging in Puerto Vallarta is close to the approach of Rifkin (2000, p. 5) when he stated that tourism, travel and fun through electronic media would be "the center of a new cyber-capitalism that trades with access to cultural experiences" as Airbnb currently does, which has turned the promise of a local experience into a multi-million dollar business.

Something similar happens with the supposed empowerment of individuals, which gives them the chance to improve their living conditions through new employment opportunities. However, the results of the research show unfavorable conditions that induce people to work longer hours to obtain an extra income that will allow them to solve their needs, without other benefits such as social security benefits or responsibility for the company regarding their properties.

The spatial alterations attributed to the presence of Airbnb in cities, are visible in Puerto Vallarta, due to a boost in the construction of new apartment units that are the inner core of the collaborative lodging offer. Other side effects such as speculation increase in the cost of land and rapid urbanization are adding pressure and saturating areas with greater touristic demand. Thus, gentrification is stronger, meaning a fast displacement of local residents through different methods, some persuaded with economic offers for their property, and others, tempted to sell their home when the neighborhood loses the traditional charm and quietness.

As a result, the question *who wins?* Was raised during the research. In our understanding, the way the collaborative lodging model has taken form in the touristic city of Puerto Vallarta presents as an undisputable winner: the company Airbnb, which has achieved a valuation of 25 billion dollars and has expanded to more than 190 countries, bringing together the largest number of rental units around the world, without having a single property.

Investors, builders and real estate developers are taking advantage of the boom in collaborative lodging to strengthen their business, aware that once they sell, they will retire, leaving to the community the negative effects they may generate.

Another winner is the local authority, especially urban planning officers, who frequently agreed to informal agreements with real estate developers. Frequently the only glimpses, with a short-term vision, the economic benefits generated by investment allocated to build and grow, without considering the undesirable effects produced by the saturation of real estate developments for collaborative hosting in town.

In the process, tourists can be considered winners, because they have access to different rental properties and types at a lower price than at a traditional hotel. However, according to the results, the coexistence of the tourist and neighbors is increasing irritation and resentment that can be labeled as *over-tourism* or *turismophobia*.

On the other side of the coin, *who loses?* Can be identified in three dimensions. For the local resident, some want to take advantage of their asset, looking to obtain extra income. Others need to move, give up their neighborhood or share it with the tourists that transform their daily lifestyle.

For real estate hosts, they work as a broker of the property, bringing them to the company and working long workdays. In most cases, collaborative lodging companies offers no social security benefits, nor any guarantee against the risks involved in the activity. In addition, in the process they transfer the economic benefits generated by their property and their workforce to the company, paying a commission for intermediation.

Puerto Vallarta and other coastal destinations, face the challenge of dealing with an increase in the collaborative lodging offer, giving rise to need of addressing expressions of *over-tourism*. The speed, scale and complexity of the trend has surprised everyone, so studying and researching the subject is pertinent in order to preserve the common good of the community.

REFERENCES

A21mx. (2017, November 30). *Airbnb grows 165% in Mexico and goes for national tourists*. A21mx online newspaper. Retrieved from a21.com.mx/innovación/2017/11/30/airbnb-crece-165-en-mexico-y-va-por-turistas-nacionales

Airbnb.com. (2018). *Earn money as a host on Airbnb*. Colorado. Retrieved from https://www.airbnb.mx/host/homes?from_nav=1. https://web.archive.org/web/20120729194252/http://www.airbnb.es/info/ten_million

Airdna.co. (2018a). *Airbnb data & insights to succeed in the sharing economy*. Retrieved from https://airdna.co

Airdna.co. (2018b). *Vacation rental data Puerto Vallarta*. Retrieved from https://www.airdna.co/vacation-rental-data/app/mx/jalisco/puerto-vallarta/overview

Airdna.co. (2018c). *Vacation rental data Puerto Vallarta*. Retrieved from https://www.airdna.co/vacation-rental-data/app/mx/jalisco/puerto-vallarta/overview

Airdna.co. (2018d). *Vacation rental data Puerto Vallarta*. Retrieved from https://www.airdna.co/vacation-rental-data/app/mx/jalisco/puerto-vallarta/overview

Alonso, M., & Miranda, A. (2017). *The collaborative company. The new economic revolution*. Spain: Pearson.

Arenas, S. (2018, July 31). *The residents of Marina Vallarta are worried about building with excess height*. Observatoriobahía.com.mx. Retrieved from http://observatoriobahia.mx/preocupa-a-residentes-marina-vallarta-edificacion-altura-excedente/

Botsman, R., & Rogers, R. (2010). *What´s mine is yours. The rise of collaborative consumption*. USA. Harper Collins e-books.

Brossat, I. (2018). *La ville ubérisée*. France: Éditions La Ville Brule.

Calleja, D. (2017, September 14). *What are unicorn companies?* Entrepreneur Magazine. Retrieved from https://www.entrepreneur.com/article/268414

Cañigueral, A. (2014). *Live better with less. Discover the advantages of the new collaborative economy*. Barcelona, Spain: Editorial Connect.

Conley, Ch. (2014, August 29). *Airbnb is an old idea with a new tech twist*. Hotel News Now. Retrieved from http://www.hotelnewsnow.com/articles/23687/Airbnb-is-an-old-idea-with-a-new-tech-twist

Dixón. E. (2019, April 5). *Family finds hidden camera streaming live from their Airbnb in Ireland*. CNN News in Spanish. Retrieved from https://cnnespanol.cnn.com/2019/04/05/famiia-encuentra-camara-oculta-transmitiendo-en-vivo-desde-en-su-airbnb-en-irlanda/

García, J. G. (2019, May 21). *"Dinkis", the sector that spends more than a family when it comes to Puerto Vallarta*. A. Z News. Retrieved from https://aznoticias.mx/index.php/puerto-vallarta-movil/32365_di...-que-gasta-mas-que-una-familia-cuando-viene-a-puerto-vallarta

Garda, J. C. (2018, September 22). *Demonstration and blockades in Marina Vallarta*. Vallarta Daily Review. Retrieved from http://vallartaopina.net/2018/09/22/local/ciudad/manifestacion-y-bloqueo-en-marina-vallarta/

Huízar, M. A., Baños, J. A., & López, J. L. (2018). The impact of LGBT tourism in Puerto Vallarta. In S. M. Arnaiz, & A. B. Gómez (Eds.), *Tourism against the objectives of Sustainable Development*. Mexico: University of Guadalajara.

Infobae.com. (2018, October 11). *The house of horrors: they rented on Airbnb and have not recovered from the nightmare.* Online publication. Retrieved from https://www.infobae.com/america/eeuu/2018/10/11/la-casa-de-los-horrores-rentaron-en-airbnb-y-no-se-han-recuperado-de-la-pesadilla/

Mexican Institute of Competitiveness (IMCO). (2017). *Airbnb in Mexico City: Extends the economic benefits of tourism and reduces the wage gap. Case study for the capital of the country.* Mexico: IMCO.

Mora, M. (2017). *8 of the 24 constructions of the Romantic Zone were authorized in this administration.* Independent Vallarta, Online Journal. Recovered from https://vallartaindependiente.com/2017/11/20/8-las-24-construc...s-la-zona-romantica-fueron-autorizadas-en-esta-administracion/

Munguía, C. (1997). *Historic panorama of Puerto Vallarta and Banderas Bay.* Mexico: University of Guadalajara.

Ocaña, M. A. (April 10, 2017). *The Romantic Zone loses its identity.* Vallarta Daily Review. Retrieved from http://vallartaopina.net/2017/04/10/s-gente-pv/la-zona-romantica-pierde-identidad/

Regueira, M. (2019). *1 in 10 Airbnb users have found hidden cameras in their accommodations.* The computer group.com. Retrieved from https://www.elgrupoinformatico.com/cada-usuarios-airbnb-encontrado-camaras-ocultas-sus-alojamientos-t73574.html

Resort Company International [RCI]. (2017). Results 2016. EU. Retrieved from https://www.rci.com

Rifkin, J. (2000). *The era of access. The revolution of the new economy.* Spain: Paidós.

Romera, J., Brualia, A., & García, C. S. (2017, May 29). *Airbnb faces a wave of fraud due to false floors.* Online newspaper El Economisa. Retrieved from https://www.eleconomista.es/vivienda/noticias/8390344/05/17/airbnb-afronta-una-oleada-de-fraudes-con-pisos-falsos.html

Secretaría de Turismo del estado de Jalisco [SECTURJAL]. (1994). *Statistical Yearbook 1993.* Mexico. Gobierno de Jalisco.

Secretaría de Turismo del estado de Jalisco [SECTURJAL]. (2017). *Statistical Yearbook 2016.* Mexico. Gobierno de Jalisco.

Slee, T. (2017). *Yours is mine. Against the collaborative economy.* Mexico: Taurus. doi:10.2307/j.ctv62hf03

ENDNOTES

1 The touristic zone or strip in Puerto Vallarta develops in a linear pattern next to the coast with an approximate extension of 25 kilometers and an area close to 1,100 hectares. The urban form privileges private use to the detriment of public use, observing two components: the first one contained on the coast with beachfront properties with attractive visuals; and the second, the mountain area where landscape qualities are used.

2 Equivalent to one hundred and ten United States dollars, according to the exchange rate of the first half of 2019.

3 Around six United States dollars, according to the exchange rate of the first half of 2019.

4 Equivalent to three thousand United States dollars, according to the exchange rate of the first half of 2019.

5 The value of the architectural heritage of the Romantic Zone was recorded when it was declared within the *Perimeter of Protection of Urban Image and Conservation of Cultural Heritage* under the Program for the Conservation of Cultural Heritage of the Ministry of Culture of Jalisco (2014:67).

6 According to the *Plan Parcial de Desarrollo Urbano Los Muertos,* approved by the City Hall of Puerto Vallarta in 2000 and still in force.

7 To support this hypothesis, the files of the fourteen developments in question were obtained and in each of them, the modification procedure was the presentation of an *Análisis Técnico Justificativo* (ATJ), endorsed by a construction professional, as well as by the urban local officers.

8 Around seven and a half million United States dollars, according to the exchange rate of the first half of 2019.

9 Referred to the updated *Plan Parcial de Desarrollo Urbano del sub-distrito 5ª Marina Vallarta,* approved by the City Hall of Puerto Vallarta on September 27, 2018 where the urban control regulations were modified allowing substantial increase in density and height construction in the neighborhood.

10 The *New Urban Agenda* proposed by UN-HABITAT in 2016, establishes in objective 3.5 “Achieve a compact, connected, integrated, safe and inclusive urban spatial form, through key components such as the prevention of urban expansion”.

Chapter 12
The Development of a Scale to Measure Tourism-Phobia:
An Exploratory Case of Residents in Majorca

Francisco Rejón-Guardia
https://orcid.org/0000-0002-5201-8435
University of the Balearic Islands, Spain

Nataša Marković
University of the Balearic Islands, Spain

María Antonia García-Sastre
University of the Balearic Islands, Spain

ABSTRACT

The Balearic Islands in general, and Majorca in particular play a significant role in Spanish tourism but the tremendous growth in the number of tourists has led to the deterioration of the environment, a notable decrease in tourist satisfaction, and is having a negative impact on the welfare of residents. Aversion to and rejection of tourism among residents is popularly called tourism-phobia. This chapter studies the negative aspects which over-tourism is causing in the perception of both native and foreign residents of Majorca, as well as to propose a scale developed to measure tourism-phobia. Data for the study was gathered through a questionnaire, with a total of 149 valid responses. An exploratory factor analysis revealed four main underlying dimensions of tourism-phobia encompassing a number of different variables. The aim is for the scale to be used as part of continuous analysis of sustainable tourism development in Majorca, as well as in other destinations saturated by tourism.

DOI: 10.4018/978-1-7998-2224-0.ch012

INTRODUCTION

Due to the growing importance of travel and tourism in society and the escalating number of international tourists, the World Tourism Organization (UNWTO) declared 2017 as the international year for the development of sustainable tourism (Boulding, 2018). The UNWTO defines sustainable tourism as a type of tourism "that takes full account of its current and future economic, social and environmental impacts, addressing the needs of visitors, the industry, the environment, and host communities" (UNEP & UNWTO, 2005; pg. 11). The United Nations Environment Program (UNEP) and UNWTO (2005) state that guidelines and management principles for the sustainable development of tourism are applicable everywhere, regardless of the type of destination, tourism, number of tourists, and so on. However, due to the high number of visitors to specific tourist destinations and increasing numbers of parties interested in monetizing visitor flow, the sustainable development of the tourism sector is under threat.

Despite this declaration, 2017 was a milestone year concerning aversion towards holidaymakers in many European seaside destinations, and the terms "over-tourism" and "tourism-phobia" started to appear in the press worldwide (Arteaga & Hernández, 2017). Local communities began to draw attention to everyday problems caused by tourist invasions and started protesting against them (Goldstein, 2017). This chapter focuses on this issue and investigates overtourism from the perspective of residents. The study is centered on evaluating residents' fears about sustainability, encompassing different dimensions: environmental, legal, safety, loss of identity, and so forth.

The paper is structured as follows: first, a parallel is drawn between tourism growth in Spain and the Balearics and anti-tourism acts; secondly, a literature review describes the theory of tourism and the development of tourism-phobia; thirdly, the research method, consisting of the generation of scale items and the distribution of the questionnaire is explained. In the next section, data purification, the demography of respondents, and the results are described, followed by a discussion of the main findings from the analysis. Finally, the key findings, limitations, and ideas for future investigation are presented.

The Growth of Tourism-Phobia in Spain and the Balearics

Tourism in Spain reached its maximum level of 82 million international tourists in 2017, an increase of 8,9% on 2016, according to estimates by the Spanish Tourism Board Turespaña (Hosteltur, 2018). This was the first time Spain surpassed the USA in tourist numbers, in second place only to France (Mincotur, 2018). The most significant impact of tourists was to be found in the province of Catalonia, with the capital Barcelona receiving 18,223,294 tourists in 2017 – more than any other Spanish region (Minetad, 2018). In the summer of 2017, many residents of Barcelona started protesting against the tourism industry on the streets, as they believed the city was turning into a theme park, and their interests were being neglected in favor of the needs of tourists (Vidal-Folch, 2018). Some posted slogans on buildings such as "Tourism is killing our neighborhoods" or "Tourists go home." Other anti-tourism activists slashed the tyres of a tour bus, sprayed graffiti onto a tour bus full of tourists, and destroyed tourist bicycles. Since then, new rules have come into force, such as the prohibition of Segway and electric scooters in the Old City and the seafront (Coldwell, 2017).

In the same year (2017), the Balearic Islands were in second place to Catalonia in regard to international tourist numbers, receiving 13,691,618 tourists. Both regions showed an increase of more than 6% in the number of tourists in comparison with the previous year (Minetad, 2018). In a similar vein to Barcelona, more than 3,000 residents protested in the center of Palma chanting phrases such as: "Tourists go home" and "Without limits, there is no future" (Nozari, 2017). Some minor incidents directed towards tourists also occurred on other parts of the island, such as confetti thrown into a café full of tourists, eggs thrown at coaches, as well as the ever-present anti-tourism slogans painted on buildings, or anti-tourism posters displayed on balconies. Those responsible for these expressions of rejection are social movements aimed at raising awareness and making their complaints visible through the Internet and social networks. In turn, repulse actions are condemned by politicians and highlighted by the media that pejoratively call them tourismphobic acts. In this context, several studies show an amplifying effect of the media in the visibility of the resident's discomfort, not coinciding with a real feeling (Huete & Mantecón, 2018). However, the dissatisfaction of residents is mainly directed towards the government, claiming that over-tourism has resulted in crowded roads, increased housing costs, threatened livelihoods, and last but not least the destruction of the environment (Goldstein, 2017). Furthermore, residents' discomfort is increased by the discord between the public policies and the resident's expectations and image of their every-day living space. For example, current policies aimed to extend the tourist season or attempting to improve tourist quality through price increases could aggravate the problem (Blanco-Romero, Blázquez-Salom, & Cànoves, 2018).

This raises the question of how the popularity of tourist destinations can affect the quality of life in the destination and increasingly be the cause of dissatisfaction in local communities. There are a growing number of residents in popular tourist destinations such as Venice, San Sebastian, and Dubrovnik, who have become activists in order to "defend" their cities from tourists (Blanco-Romero, et al., 2018; Egresi, 2018). The primary purpose of this chapter is to understand the main dimensions of the rejection of tourism activity, through the development of a multiple-item scale to measure residents' tourism-phobia. The data gathered for the development of the scale from residents on the island of Majorca reveals the attitude of both native and foreign residents towards (over-) tourism on the island, and to what extent they feel it is affecting their lifestyle. The value of the data gathered in this study may prove useful to all sustainable tourism stakeholders, that is tourist destination marketers, researchers, and destination planners.

BACKGROUND

Tourism Development

According to Butler (1980), tourist areas inevitably go through certain stages of transformation. Starting with the exploration stage, a small number of tourists visit a destination travelling independently, they interact with locals and enjoy the location, which is still pristine and unique. By gradually undergoing the involvement and development stages, tourist destinations reach the consolidation stage. At that point, total visitor numbers exceed the number of locals, and although their economy is highly dependent on tourism, locals start to express dissatisfaction with the development of tourism, which may result in some restrictions. Butler maintains that one of three possible stages may follow: stagnation, decline, or rejuvenation, and this depends mostly on the strategy and decision-making of the local authorities. Butler's model is compatible with Doxey's Irritation Index (1975), which states that there are four levels of irritation caused by the number of tourists and tourism development in the community, namely: euphoria, apathy, annoyance and antagonism. Euphoria is a sense of excitement about tourists and occurs in the initial stages of tourism; this is followed by apathy, as tourists are seen as a source of income and investment. As the consolidation stage approaches, locals will start to express annoyance, which will intensify and transform into antagonism towards tourism activity (Doxey, 1975).

Although local communities in most destinations are not entirely opposed to tourism, many studies show that residents do recognize the adverse effects of tourism. For example, King, Pizam, and Milman(1993) state that residents of the island of Fiji show concern that tourism has led to the commercialization of their culture, and at the same time, has created social problems due to the encounter of overseas tourists with residents who have a lower standard of living. Other studies have also shown that residents do not find tourism entirely beneficial for their community and even find tourism to be a threat to their welfare. Boukas and Ziakas (2016) examined the relationship between the development of tourism and the local community and explored the planning approach of the tourism policy in the development of Cyprus. Since mass tourism was initially seen as a solution to a range of economic problems, locals were mostly treated as a means to achieve this goal. The authors' findings demonstrate that the current tourism policy, which is oriented towards meeting the needs of tourists, must be transformed into the "inside-out" approach, which places the well-being and needs of locals first, to achieve sustainable development.

In the case of Majorca, the beginnings of mass tourism in the 1960s caused considerable impact on the natural environment, and also on the local population with the first inflow of migrants motivated by a large number of employment opportunities in the construction and hospitality sectors. Unlike these early migrants, who arrived from Spanish-speaking countries, the second wave, after the 1990s, brought mostly retired Europeans (Picornell, 2014) and more recently, entrepreneurs and others who have chosen Majorca for their seasonal, temporary or permanent home. Although to some extent in its beginnings, tourism brought economic benefits, its subsequent expansion has caused pollution, overcrowding, a rise in property prices and other living costs, which have all affected residents (Martín, Martínez, & Fernández, 2018). Moreover, the growth of the migrant community has created issues with social and cultural integration within the local community (O'Reilly, 2007). Royle (2009), found that 68% of Germans residing on Majorca did not speak either of the two official languages (Spanish and Catalan), and 40% of these expressed no interest in learning the languages or anything about the local culture. Mass tourism and commercialization of the local culture has led to a feeling of "loss of identity." (De Juan-Vigaray, Sarabia-Sánchez, & Garau-Vadell, 2013).

According to Sastre (2012), the continuous influx of German residents to the island has only served to strengthen the social infrastructure that facilitates the life of the German-speaking population and has eliminated the need for interaction with their hosts. A comparative analysis by Garau-Vadell, Gutierrez-Taño, and Diaz-Armas (2018) investigated perceived personal and general benefits of tourism for locals in Majorca and Tenerife. They found that Majorcans had more pessimistic (subjective) feelings about the impact of tourism, in comparison with their counterparts in Tenerife. This included fear of loss of identity, fear of tourist invasion and harassment, while at the same time locals were suffering from problems of pollution and other adverse effects of tourism on the environment (Garau-Vadell et al., 2018).

Tourism-Phobia

When reference is made to massification and tourism saturation in the literature, the terms "over-tourism" and "tourism-phobia" usually appear together. Boulding (2018, pg. 4) explains that over-tourism "…describes destinations where hosts or guests, locals or visitors, feel that there are too many visitors and that the quality of life in the area or the quality of the experience has deteriorated unacceptably." Therefore, over-tourism is, first and foremost, related to exceeded natural carrying capacity, which affects locals as much as tourists (Boulding, 2018). The phenomenon of over-tourism is directly dependent on and can, therefore, be seen as a direct consequence of the decisions of tourism planners such as travel agencies and local authorities, as well as the decisions made by tourists. The authorities in some destinations affected by over-tourism, such as Venice (Seraphin, Sheeran, & Pilato, 2018), Dubrovnik, or Santorini, are trying to combat this by limiting the number of cruise ships in their ports or controlling the number of tourists entering historical centers. Other measures are directed towards tourists and include encouraging them to book accommodation on the outskirts, to consider visiting in low and off-season, to shop at independent local businesses to support the local economy, and to engage in "off-the-beaten-track experiences" (Lipovčan, Brajša-Žganec, & Poljanec-Borić, 2014).

Tourism-phobia is a relatively new term which researchers and media worldwide have recently started to use to describe the manifestation of residents' rejection of over-tourism. It is expressed in minor acts of vandalism, such as damaging tourist bikes and coaches, assaults on restaurants, and so on. As a result of these incidents, tourism-phobia and anti-tourism movements have recently become a trend in many European cities and have consequently become the focus of research. Díaz-Armas and Gutiérrez-Taño (2017) define tourism-phobia as an attitude of rejection of, or exaggerated aversion to, tourism activities by residents (or an uncontrollable fear of certain situations, ideas, or objects). The authors point out that academic literature is focused on the study of residents' attitudes towards the economic benefits of tourism rather than an examination of their negative attitudes towards and acts of rejection of tourism. The need for this kind of research into these negative attitudes and behaviors by residents towards tourism has become wider with the new disruptive growth of the tourism industry, fueled by new lodging and transport business models.

Most of the studies related to the attitudes of residents towards tourism development have been carried out within the framework of social exchange theory (SET) (Emerson, 1976; Duarte & Nyanjom, 2016). Blau (1964, p. 6) defines SET as: "…limited to actions that are contingent on rewarding reactions from others. It is a two-sided process of mutually rewarding exchanges between smaller informal social groups. The value of trade determines if a person is going to act or not – the more valuable a reward is perceived to be, the more likely the person is to respond." From the perspective of tourism, this would mean that those who are involved in tourism and benefit from it, would continue to support it, while those who benefit little, if at all, from tourism would most probably be opposed to it. This includes not only economic aspects, but also social, cultural, and environmental issues.

In addition to including the findings of SET, this study adds other relevant variables which have not been examined in earlier literature. Considering that academic interest in this topic is relatively recent, the small number of studies undertaken is not surprising. Previous studies linked to the term tourism-phobia include: a revision of literature explaining the negative impacts of tourism (Díaz-Armas & Gutiérrez-Taño, 2017); the identification of discourse in the mass media on this topic (Huete & Mantecón, 2018); how specific issues related to mass tourism affect the attitude of residents of Barcelona (Abril-Sellarés, Azpilicueta, & Sánchez-Fernández,2015); how media coverage, political propaganda and social movements amplify acts of anti-tourism and tourism-phobia; and strategies being promoted as a solution for mass tourism and its rejection (Milano, Novelli, & Cheer, 2019).

Due to the relatively recent strong opposition to and rejection of tourism manifested by residents of popular tourist destinations, it is essential to determine the key drivers influencing the perceptions of local residents about tourism, in order to improve their quality of life and create a friendly and welcome environment for visitors. For this purpose, a scale measuring the level of tourism-phobia among residents was developed, based on feedback from residents of Majorca.

METHODOLOGY

The scale was developed following Churchill's (1979) recommendations of three main steps: the generation of scale items, the refinement of sample item, and the verification of construct reliability and validity.

Generation of Scale Items

After a preliminary literature review, a pool of measurement items was developed from relevant studies (Lankford & Howard, 1994; Jordan, Vogt, & DeShon, 2015; Wen, Meng, Ying, Qi, & Lockyer, 2018). These items were empirically tested in different geographical and cultural areas and demonstrated reliability and validity. They were therefore adopted for this study together with additional items gathered from in-depth interviews.

Semi-structured, individual in-depth interviews with two native and two foreign residents of Majorca were conducted in April and May 2018. For the foreign respondents, a minimum of three years of continuous residence on the island was requisite. Two of the four respondents were working directly in the tourism industry and had daily contact with tourists; the third had employment indirectly related to tourism; the fourth respondent's work was unrelated to tourism. The primary purpose of the interviews was to contextualize the feeling of tourism-phobia on the island, considering the whole island as a single tourist destination. The secondary purpose of the interviews was to verify whether all items from previous scales could be applied to Majorca. For this reason, the scale developed for this study does not have identical items to other scales.

The collection of items included in the questionnaire is represented in Table 1. The items are grouped according to the area of research: demographic characteristics of respondents; their personal background; their involvement in tourism; their opinions about social, environmental and economic issues, and legal questions.

Distribution of the Questionnaire

The questionnaire was distributed from mid-May to mid-June 2018 through various social networks, on forums, and via e-mail (snowball method). Respondents were contacted through social media groups related to residents of Majorca, foreigners on Majorca, or specific nationalities residing on the island. This was an effective way to reach groups of different nationalities from different municipalities on the island in a short period. A small number of questionnaires were distributed face-to-face. The questionnaire was delivered in Catalan, Spanish, English, German, French, Italian, and Portuguese, ensuring a wider coverage of different national groups. To identify eligible respondents, the first question was used to classify the length of residence on the island: those living on the island for less than three years; between three and five years; between five and ten years; more than ten years; or if they were born in Majorca. Those who had lived for less than three years on the island were not eligible, as a period of fewer than three years was considered insufficient, on average, for a foreigner to be able to see, experience and comprehend the issues raised by the tourism industry in the destination.

A five-point scale was used for 25 items divided into four sets, each containing information about one aspect: society, the environment, the economy, and legal matters. Responses were categorized according to the level of agreement with the statements, varying from 1 - "not at all" to 5 - "a great deal." The same scale was used for other items such as satisfaction with the salary in tourism, and overall satisfaction

Table 1. Grouped items

Topic	Variable and items
Demographic characteristics	Gender
	Age
	Nationality
	Place of residence
	Length of residence in Majorca
Personal background information	Psychological profile
	Fluency in either of the official languages of the island
	International life experience
	The size of the network of friends on the island temporarily or permanently
	The extended network of classmates or colleagues on the island temporarily or permanently
	Ancestors, descendants or other relatives on the island
Involvement in tourism	Current or previous employment related to tourism
	Amount of daily contact with tourists
	Amount of positive contact with tourists
	Overall satisfaction with previous experience of tourists
	Satisfaction with the salary from tourism-related work
	Verbal or physical confrontation with holidaymakers
	Membership of community organizations
	Involvement in planning and decision-making in tourism
Social issues (8 items)	Negative attitudes (showing-off, foolishness, carelessness, and so on) caused by tourists
	Negative activities (drunkenness, violence, vandalism, and so on) caused by tourists
	The image of the island damaged by tourists
	An increase in segregation between locals and tourists
	Disturbance in neighborhood communities caused by tourists
	Overcrowding of public spaces by tourists
	Commercialization of Majorcan culture and traditions by the tourism industry
	Quality of life in comparison with other provinces
Environmental threats (9 items)	Quality of sea water (Pollution)
	Damage to protected areas on Majorca
	Disturbance caused to animal species
	Reduction of quality of outdoor recreation
	Traffic congestion
	Tourism architecture and infrastructures
	Waste problems
	Noise problems
	Consumption of natural resources

continues on following page

Table 1. Continued

Topic	Variable and items
Impacts on the economy (4 items)	Economic dependence on tourism
	Economic contribution to community improvements provided by tourism
	Increased cost of living due to tourism
	Level of salaries in tourism
Legal measures (4 items)	Restrictions on the number of tourists
	Management of tourist volumes and movement
	Change in the type of tourism and tourists
	Legal support in the case of dispute

Source: Author's own elaboration

with tourists, using a 5-point smiley-face rating, allowing respondents to express their feelings better. Demographic questions included: the length of residence on the island, the municipality of residence, nationality, gender, and age. Another set of questions inquired about the respondent's background: if they had relatives on Majorca, their international experience, their psychological profile; if they had a network of friends and colleagues living on the island, and whether they were members of any community organization. Respondents were given the opportunity to leave a comment, of which a qualitative analysis was subsequently made. The study was carried out in compliance with European regulations on data protection (Regulation (EU) 2016/679 of the European Parliament and of the Council of 27 April 2016). Respondents were informed that the data would be gathered anonymously and included in one or more files managed by the University of the Balearic Islands (UIB). Furthermore, as all data was collected digitally, it would be stored securely and protected.

RESULTS

Responses

The website platform used for data collection and response purification was Qualtrics. The total number of responses collected was 162. After data purification, this number was reduced to 149. This included cleaning of responses that strongly deviated from the average time taken to finish the questionnaire, which was 13 minutes. Responses completed in more than half an hour or less than three and a half minutes were eliminated from the sample.

The descriptive analysis showed that several variables had an insufficient number of complete responses or questions which were not understood correctly. Once these variables were excluded, an exploratory factor analysis (EFA) was applied using SPSS, to reduce the number of items in the scale, as well as to define its underlying dimensions. This technique allowed examination of the entire variable interdependence relationship and indicated the existence of five primary factors responsible for tourism-phobia. Upon examination of the principal component analysis, these five elements showed significant eigenvalues higher than one, and together explained a large amount of variance, of more than 65%. The Varimax rotation was analyzed, and only factor loadings greater than 0.4 were included. Factor loadings corresponding to more than one factor were retained only with their higher value.

Construct Reliability and Validity

Cronbach's alpha coefficient (α) was carried out to measure internal consistency (reliability) within each group of factors. Items insufficiently related to one factor (< 0.7) were removed to achieve better scale reliability. The items: economic dependence, overcrowding of public spaces, and the commercialization of culture and tradition were removed after the elimination process. The explained variance was still considerably high, at 60.54%. The remaining 21 items were tested in both Kaiser-Meyer-Olkin (KMO) for sample adequacy and Bartlett's test. The KMO result of 0.923 indicated that the data was adequate for further analysis. Bartlett's test was also highly significant, showing the existence of some relationships between variables, therefore, the data was appropriate for factor analysis. Table 2 shows the results of this analysis grouped into four different dimensions, which together explain the underlying relationships between the measured variables. It shows both the mean and standard deviations for each item within a dimension, as well as of the dimensions themselves.

Extraction Method: Principal Component Analysis; Rotation Method: Varimax with Kaiser Normalization; a. Rotation converged in 5 iterations. Higher mean values in bold.

The most significant dimension which explains the highest motivation for tourism-phobia is the physical impact on nature and landscape (F2) (mean = 3.25) which includes waste problems (3.69) and the consumption of natural resources (3.35) as the two most important variables of the factor. The following factor by importance is F4, the impact on employment and costs (mean = 3.07), in which the most important variables were an increase in the cost of living (mean = 3.89) and poorly paid jobs (3.27). The next factor is F3, the impacts of volume and type of tourism on humans and animals (mean = 3.04), consisting of the management of volume and movements of tourists (3.64), changes in the type of tourism (3.18) and other variables of motivation. The least relevant factor concerning motivation is F1, behavioral impacts on communities (mean = 2.95), consisting of traffic congestion (3.81), and other harmful activities (3.06).

Demographic Profile of Respondents

The female population was more dominant in the survey, with more than 60%. The majority of respondents, just over half, (52.3%) were of Spanish nationality, followed by French (14.1%). British and Germans, both with less than 10%, took third and fourth place respectively. There was a diversity of other national groups from Argentina, Bulgaria, Belgium, Catalonia, Croatia, Czech Republic, Chile, Greece, Hungary, Italy, Ireland, Poland, Scotland, Serbia, and the USA, providing a wide variety of cultural backgrounds, which produced more valid and accurate results. The division between Spaniards (40.9%) and foreigners (59.1%) was well distributed. The majority of the foreign respondents (30.9%) had lived on the island for more than ten years, followed by residents of between three and five years (16.1%). The smallest group (12.1%) had been residents for up to 10 years. The predominant age groups were between 26 and 35 years old (35.57%), and between 36 and 45 (25.5%). The other age groups were quite evenly distributed. It is interesting to mention that around 60% of respondents worked either directly or indirectly in tourism, while the rest of the sample, who were either unemployed or worked in an area unrelated to tourism, had previous experience in tourism-related employment (43.2%). This indicates that the tourism industry creates most of the jobs on the island, both for natives and foreigners.

Table 2. Results of the analysis

Rotated Component Matrix									
Item	**Factor loadings**	**Explained Variance %**	**Eigenvalue**	**Cronbach α**	**Mean**		**Std. deviation**		**Description of factor**
Negative activities	.843	44.366	10.648	.90	3.06	2.95	1.22	6.73	*F1: Behavioral impacts on communities*
Negative attitudes	.793				2.93		1.19		
Disturbance caused in neighborhood communities	.745				2.76		1.15		
Damage to the image of the island.	.689				2.74		1.24		
Damage to protected areas	.548				2.81		1.24		
Traffic congestion	.495				**3.81**		1.20		
Disturbance caused by noise	.471				2.57		1.35		
Pollution of sea water	.755	6.634	1.592	.87	3.32	3.25	1.24	4.97	*F2: Physical impacts on nature and landscape*
Consumption of natural resources	.746				3.35		1.19		
Waste problems	.707				**3.69**		1.18		
Tourism architecture and infrastructures	.678				3.07		1.27		
Quality of outdoor recreation	.511				2.80		1.25		
Restrictions to the number of tourists	.842	5.259	1.262	.85	2.82	3.04	1.40	5.01	*F3: Impacts of volume and type of tourism on humans and animals*
Management of tourist volumes and movement	.742				**3.64**		1.22		
Change in type of tourism	.620				3.18		1.26		
Segregation between locals and tourists	.500				2.77		1.25		
Disturbance caused to animal species	.451				2.76		1.22		
Economic contribution to community improvements	.800	4.286	1.029	.73	2.45	3.07	1.31	3.59	*F4: Impacts on employment and the cost of living*
Foreign workforce	.583				2.67		1.21		
Badly paid jobs	.474				3.27		1.22		
Increased cost of living	.424				**3.89**		1.10		
Total variance explained (%)	60.545								

DISCUSSION

Behavioral Impacts on Communities

The dimension with the highest explained variance (44.37%) and Cronbach's coefficient (0.9) was behavioral impacts on communities. It is composed of seven variables, namely: harmful activities, negative attitudes, disturbance of neighborhood communities, damage to the image of the island, damage to prohibited/private and protected areas, noise disturbance and traffic congestion. All of these variables are related to hostile tourist behavior that is reflected through activities such as drunkenness, vandalism, violence, and certain attitudes which are considered negative, such as carelessness, foolishness, or showing off. These factors are often combined with noise from many types of vehicles that serve tourists, loud music played on beaches, in the streets or bars, or other forms of disruptive entertainment. Residents noticed that protected natural areas, especially those of spectacular beauty, such as Mondrago Nature Reserve, Natural Park S'Albufera, natural monument Torrent de Pareis, and other private or prohibited areas, are under threat because of tourists' curiosity and misbehavior. Some of these areas are not easily accessible, and tourists are obliged to rent a car or purchase a group excursion to visit them. Some "wild" beaches of difficult access, such as Caló des Màrmols (Platges de Balears, 2018), require visitors to park their vehicles on the road and walk through private property. This has resulted in problems with parking spaces, overcrowded streets, slower traffic flow, higher rates of traffic congestion and accidents, and last but not least, increased disruption and annoyance for residents.

Residents showed great concern about traffic congestion, which had one of the highest mean values. All the interviewees, and some of the respondents commented that during springtime, in the case of cyclists, and mostly during summertime, in the case of hire cars, cyclists and drivers create traffic congestion around the island and slow down traffic in general. Rossello and Saenz-de-Miera (2011), explained the positive relationship between the increase in the number of tourists and the increase in road accidents on Majorca. More than 70% of trips made by tourists on the roads in the Balearics are motorized, and road density has worsened significantly since 2005 compared to the Spanish mainland (Domblas, 2018). In addition to this, according to the TomTom Traffic Index report (2019) Palma is the second most congested city in Spain, after Barcelona. In 2015, the level of congestion was 27%, which meant that a driver in Palma could be spending an average of 27% of total driving time caught in traffic at any time of day. This percentage was even higher in rush hours, reaching as much as 38% (Europe Press, 2016).

All of these factors directly influence the image of Majorca in the media and in people's minds. Magaluf and El Arenal, popular British and German tourist destinations, are often mentioned in the media in the context of alcohol-related incidents, vandalism and other anti-social behavior by holidaymakers, comparing the ratios of incidents with previous seasons. Although this kind of press probably motivates party tourists to visit, in the long-term, it has tarnished the reputation of these destinations. However, through new investment, new laws and regulations, the local government is slowly trying to change to a different, more qualify type of tourism (Coffey, 2017).

Physical Impact on Nature and Landscape

The second extracted dimension was the physical impact on nature and landscape, with 6.63 percent of explained variance, consisting of the variables: pollution of sea water, waste problems, consumption of natural resources, tourism architecture, and infrastructures, and the quality of outdoor recreation. Much recent research has been focused on the topic of the impacts of tourism on the environment of tourist destinations, as the tourism industry is considered to be one of the major contributors to climate change and global warming (Saenz de Miera & Rossello, 2014). Garau-Vadell et al. (2018) concluded that residents of Majorca perceived the most negative impacts of tourism to be on the environment, above on any other aspect. The results of the present study concur with this, as the mean value of this dimension was higher than the others. This can be explained by the singularity of Majorca, due to its insularity and scarce precipitation in the summer months when the demand for water due to the number of tourists has the greatest impact. Hotels, water parks, swimming pools, and golf courses, are among the primary consumers of water resources in tourism. However, the whole tourism sector in the Balearic Islands accounts for only about 12% of the total annual water consumption, in comparison with the agricultural sector, which consumes 41% (Deyá & Tirado, 2011). Although tourism might not be the primary consumer of water, it is seen as the largest polluter, polluting sea water, producing sewage, garbage, and air pollution (Saenz de Miera & Rossello, 2014). In 2017 a project co-financed by the European Regional Development Fund studied the characterization and seasonal concentration of marine litter on Mediterranean beaches and confirmed the existence of a direct relationship between the accumulation of marine litter and the number of tourists. Among the beaches of Majorca investigated in the project, the popular tourist beach of Torá (Calvià) appeared to be the most affected by the accumulation of marine litter. However, the beach with the highest amount of accumulated marine litter (Es Caragol, Santanyi) was mainly used by locals rather than tourists. The clean coast index for this beach was dirty/very dirty for the low/high season (CCI: 38.4/35.6), while for Torá it was considered dirty in both low and high seasons (CCI: 18.1/17.5). In comparison, the accumulation index for the same beaches was high/extremely high for the low/high season in Torá (4.59/6.10) and high/very high for the low/high season in Es Cargol (5.12/5.99) (Interreg Mediterranean Blue Islands, 2018). Up to 80% of beach garbage was found to be plastic, including cups, bottles, plastic bags, and cotton swabs, which, apart from aesthetic reasons, also poses a threat to animals and humans through the process of decay. For this reason, the Balearic Islands plans to follow the European Commission's strategy to reduce plastic production and promote reusable or easily recyclable products (Galvín, 2018).

Majorca is the Spanish region with the highest average amount of municipal solid waste per capita. Approximately 585.78 kg of solid waste is generated per resident per year; even more in the coastal areas where hotels are located (Arbulu, Lozano, & Rey-Maquieira, 2016). As might be expected, the respondents to this study identified this as a serious issue, as the waste problem variable achieved the highest mean value of all. According to the local environmental association, GOB-Mallorca (2018), with the increased population in summer, more waste is generated, which creates problems in the infrastructure, as the current waste management system is based on incineration, which releases toxins into the atmosphere leading to a problem with air pollution (Galvín, 2018).

Impact of Volume and Type of Tourism on Humans and Animals

As a factor, the impacts of volume and type of tourism on humans and animals explained 5.26% variance. The variables included: restriction on tourist numbers, management of volume and movement of tourists, change in the type of tourism, segregation between locals and tourists, and disturbance to animal species. From this factor, it can be concluded that the number, movement, and diffusion of tourists on the island caused by specific types of tourism such as cruises, cycling, and sea and sun tourism, are related both to segregation between locals and tourists and disturbance to animal species. The management of the volume and movement of tourists was the variable with the highest mean. With regard to the volume of tourists, San Joan airport received 27,970,655 passengers in 2017, a 6.5% increase on the previous year. That is an increase of around 5 million passengers since 2012. Traditionally, the majority of tourists arrive during high season (from April until the end of October), accounting for around 85% of total arrivals in 2018 (Aena, 2018). During the same period, the Balearic ports received 2,130,116 cruise tourists, an increase of 8.56% on 2016, according to data published by the Ministry of Development (Puertos, 2017). Tourist movement is naturally directed towards the same landmarks, attractions, and services, occasionally resulting in overcrowding and reduced tourist satisfaction (Brandajs & Russo, 2019; Seraphin et al., 2018). To exemplify this, reviews on TripAdvisor by tourists who visited Palma Cathedral, the most emblematic symbol of Majorca, (ranked #1 on the website) during high season, reveal that although the majority appreciated the value and beauty of the cathedral, their impression was spoiled by the long queues, the high admission price which did not correspond to the brief guided tour, and overcrowding (TripAdvisor, 2018). On the other hand, other types of tourism, which are not traditionally considered so numerous, such as cycling, sports, or wine tourism, play an essential role in the reduction of seasonality and could be promoted as a substitute for the highly seasonal sun & sea tourism.

Impact on Employment and Cost of Living

This factor consisted of only four variables: economic contribution to community improvements, foreign workforce, poorly paid jobs, and an increase in living expenses. The item with the highest mean of all was the increase in living expenses. It also had the lowest standard deviation, which means that residents were very much coherent in their opinion that living expenses had increased a considerable amount due to tourism.

According to the extensive database compiled from information provided by users and official sources, the cost of living in Palma de Majorca currently has the highest index in Spain, even higher than cities such as Barcelona, Madrid, Zaragoza, Valencia and Sevilla (Numbeo, 2018). Despite having somewhat lower rental prices than Madrid and Barcelona, the cost of groceries and restaurants are the highest on a national level, while at the same time local purchasing power is the lowest (Numbeo, 2018). Unfortunately, the increase in living expenses does not seem to be supported by quality foreign investment, increases in the quality of life, or high purchasing power in the Balearics.

The lower rental prices in Palma de Majorca cannot be considered positive, as this is the tenth consecutive year that the region has remained in the seventh position of the Spanish ranking of GDP per capita. No less concerning is that in 2017, the average income per person in the Balearics (€25,772) was the lowest ever, only 3.1% higher than the Spanish average, while in 2000, it was 2.7% above the national average (INE, 2018). The lowest GDP per capita could be a consequence of the fall in productivity per worker, or, in other words, a loss of value added per worker. According to the Public State Employment Service's Occupational Monitoring Centre (SEPE, 2018), there has been an excessive number of vacancies for low-qualified occupations in food service, hygiene, and cleaning, and in the construction sector, as well as for shop assistants and cashiers. In spite of the excessive number of low qualified occupations, due to the nature of the labor market, there is a high degree of work seasonality and instability. Furthermore, for those job positions involving contact with tourists, knowledge of English, German, Russian, and Scandinavian languages is essential. Therefore, nationals of these countries often have an advantage when seeking employment, especially in the area of support services for the aforementioned markets. In 2017, the foreign workforce represented 19% of the total labor market, an increase of 8% on the previous year (EURES, 2018).

CONCLUSION

This study has attempted to discover the underlying dimensions of tourism-phobia in Majorca, based on the development of a scale which can be applied to other similar destinations saturated by tourism. The study makes several methodological and conceptual contributions to the understanding of those aspects which influence the feeling of tourism-phobia among residents of Majorca. Four underlying dimensions consisting of 21 items measuring residents' tourism-phobia were found: (i) behavioral impacts on communities; (ii) physical impacts on nature and landscape; (iii) the impacts of volume and type of tourism on humans and animals; (iv) and the impacts on employment and the cost of living. The exploratory factor analysis demonstrated an acceptable level of reliability and validity of the scale. Moreover, the results and dimensions are complementary to other similar studies (Lankford & Howard, 1994; Jordan, Vogt, & DeShon, 2015) and recent news reports about Majorca. This proves, therefore, that the main aim and objectives of the study have been achieved.

The most significant variables were an increase in the cost of living, traffic congestion, and waste problems. Even though the majority of respondents work or have worked (in)directly in the tourism industry, their greatest fear is that tourism has led to an increase in the cost of living. With regard to this issue, residents should be given more privileges than tourists when buying or renting flats, and they should be offered discounted prices on entertainment, sport, and leisure. Local authorities should take into consideration the fact that not only tourists, but also seasonal workers who come to Majorca influence the increase in rental prices, especially during the high season, and offer hostels as a possible solution for seasonal workers' accommodation.

The study has also found that traffic congestion influences the feeling of tourism-phobia to a great extent. Therefore, changes and improvements in traffic infrastructure and regulations would be necessary. Furthermore, special regulations, routes, and roads for cycle tourists should be designed to minimize congestion, slow traffic, and accidents.

The third most overwhelming problem for residents of the island concerns waste disposal. Efforts should be made to plan efficient and sustainable waste management, to prevent beaches being polluted, especially with plastic, to maintain the cleanliness of public places, and for the effective separation and recycling of garbage.

However, it is essential to mention that for residents it is not always easy to differentiate tourists from locals, or which sector uses water and electricity the most, for instance, and many general problems can be mistakenly attributed to the tourism industry. Therefore, residents, as well as tourists, should be educated in best practice and advice regarding waste disposal, traffic, prohibited activities, and so on. It is clear that residents are one of the essential stakeholders of any tourist destination, and their opinions and needs must be taken-into account when making plans for tourism development.

Local governments, regional tourism authorities, tourism developers and other stakeholders who have an interest in improving, developing and promoting tourism, while at the same time strengthening the economy, should benefit from this type of research. The appropriate use of the tourism-phobia scale and all future efforts made for residents' welfare should also enhance the quality of tourist services and increase tourist satisfaction. Therefore, research into the level of tourism-phobia in tourism-saturated regions should be a continuous process, which will contribute to improving the competitiveness of the destination.

LIMITATIONS AND FURTHER RESEARCH

Several limitations were identified in the development of the tourism-phobia scale. Firstly, a study of only 149 valid responses is insufficient to provide an entirely accurate overview of the real situation in Majorca. Furthermore, since almost all respondents were contacted through social media platforms, potential respondents who were not on social media were excluded from the study. Therefore, future studies should consider alternative ways of reaching residents (online and offline) in order to achieve a larger sample. Similarly, it would be positively beneficial to extend questionnaire distribution to the winter months, to allow for data to be adequately balanced between high and low season.

Further studies could also include comparative analyses of tourism-phobia scales, and observations from different tourist destinations with other types of tourism, for instance, urban tourism compared to sun and sea destinations.

This study has proven that the greatest fears of residents regarding the impact of tourism are: an increase in the cost of living, and traffic congestion and problems with waste, in line with other institutional reports (Arbulu et al., 2016; Europe Press, 2016; INE, 2018). Bearing in mind that tourism-phobia is perceived not only by residents, but also by tourists, future research could examine this phenomenon from the tourist perspective, and possibly search for discrepancies between the fears of tourists and those of residents.

ACKNOWLEDGMENT

We would like to thank the research group TURNETMK for their advice and economic and logistical support.

REFERENCES

Abril-Sellarés, M., Azpelicueta, M. D. C., & Sánchez-Fernández, M. D. (2015). Turismo sostenible: Lugareños frente a turistas. El caso de la ciudad de Barcelona. *HOLOS*, *3*, 331–337. doi:10.15628/holos.2015.2809

Aena. (2018). *Aeropuerto de Palma de Mallorca*. Retrieved from http://www.aena.es/es/aeropuerto-palma-mallorca/presentacion.html

Arbulu, I., Lozano J. & Rey-Maquieira J. (2016). The challenges of municipal solid waste management systems provided by public-private partnerships in mature tourist destinations: The case of Mallorca. *Waste Management*, (51), 252-258.

Arteaga, J., & Hernández, A. (2017). *Citizen Engagement versus Tourism-phobia: The Role of Citizens in a Country's Brand.* Retrieved from Developing Ideas by Llorente & Cuenca. website: http://www.desarrollando-ideas.com/wp-content/uploads/sites/5/2017/09/170905_DI_Informe_Marca_Pa%C3%ADs_ENG.pdf

Blanco-Romero, A., Blázquez-Salom, M., & Cànoves, G. (2018). Barcelona, Housing Rent Bubble in a Tourist City. Social Responses and Local Policies. *Sustainability*, *10*(6), 2043. doi:10.3390u10062043

Blau, P. M. (1964). *Exchange and Power in Social Life*. Hoboken, NJ: J. Wiley.

Boukas, N., & Ziakas, V. (2016). Tourism policy and residents' well-being in Cyprus: Opportunities and challenges for developing an inside-out destination management approach. *Journal of Destination Marketing & Management*, *5*(1), 44–54. doi:10.1016/j.jdmm.2015.12.004

Boulding, J. (2018). Overtourism a sign of the times, *Canadian Travel Press*, 1. Retrieved from: https://www.travelpress.com/subscription/IDEA/100017_010118_ctp/F912647B177BF06BB74179E561E-A228E/Canadian%20Travel%20Press%20-%20January%204,%202018.pdf

Brandajs, F., & Russo, A. P. (2019). Whose is that square? Cruise tourists' mobilities and negotiation for public space in Barcelona. *Applied Mobilities*, 1–25. doi:10.1080/23800127.2019.1576257

Butler, R. W. (1980). The concept of a tourist area cycle of evolution: Implications for management of resources. *The Canadian Geographer. Geographe Canadien*, *24*(1), 5–12. doi:10.1111/j.1541-0064.1980.tb00970.x

Churchill, G. A. Jr. (1979). A paradigm for developing better measures of marketing constructs. *JMR, Journal of Marketing Research*, *16*(1), 64–73. doi:10.1177/002224377901600110

Coffey, H. (2017). *How Magaluf is shedding its party-town image and appealing to families.* Retrieved from: https://www.independent.co.uk/travel/news-and-advice/magaluf-holidays-clubbing-drinking-family-appeal-upmarket-image-change-mallorca-melia-hotels-calvia-a8080371.html

Coldwell, D. (2017). *First Venice and Barcelona: now anti-tourism marches spread across the Europe.* Retrieved from: https://www.theguardian.com/travel/2017/aug/10/anti-tourism-marches-spread-across-europe-venice-barcelona

De Juan-Vigaray, M. D., Sarabia-Sánchez, F. J., & Garau-Vadell, J. B. (2013). The acculturation of international residential tourists and their shopping behaviours. *Tourism Management*, *36*, 115–118. doi:10.1016/j.tourman.2012.11.017

Deyà Tortella, B., & Tirado, D. (2011). Hotel water consumption at a seasonal mass tourist destination. The case of the island of Mallorca. *Journal of Environmental Management*, *92*(10), 2568–2579. doi:10.1016/j.jenvman.2011.05.024 PMID:21696882

Díaz-Armas, R. J., & Gutiérrez-Taño, D. (2017). Turismofobia. In E. A. González, & F. C. García (Eds.), *La actividad turística española en 2016* (pp. 407–410). Madrid, Spain: Editorial Síntesis.

Domblas, N. (2018). *Las carreteras de Baleares soportan casi tres millones de desplazamientos diarios.* Retrieved from: https://ultimahora.es/noticias/local/2018/04/18/994177/carreteras-baleares-soportan-casi-tres-millones-desplazamientos-diarios.html

Doxey, G. V. (1975, September). A causation theory of visitor-resident irritants: Methodology and research inferences. The Travel Research Association. In *Sixth Annual Conference Proceedings, 'The Impact of Tourism', San Diego, CA* (pp. 8-11).

Duarte, A. D., & Nyanjom, J. (2016). Tourism, quality of life, and residents of a rural town. *Tourism Analysis, 21*(6), 617–629. doi:10.3727/108354216X14713487283165

Egresi, I. (2018). "Tourists Go Home!" – Tourism Overcrowding And "Tourismophobia" In European Cities (Can Tourists and Residents Still Co-Habitate In The City?). En TAW2018 International Scientific Conference (p. 23).

Emerson, R. M. (1976). Social Exchange theory. *Annual Review of Sociology, 2*(2), 335–362. doi:10.1146/annurev.so.02.080176.002003

Europe Press. (2016). *Palma es la segunda Ciudad española con más atascos de tráfico.* Retrieved from: http://www.elmundo.es/baleares/2016/03/22/56f17536e2704e430a8b45b3.html

European Commission. EURES. (2018). *Labour market information – Illes Baleares.* Retrieved from: https://ec.europa.eu/eures/printLMIText.jsp?lmiLang=en&regionId=ES5&catId=457

Galvín, A. (2018). *Baleares marca el camino del adiós a las cápsulas de café, las pajitas y los bastoncillos de usar y tirar.* Retrieved from: https://www.eldiario.es/sociedad/Baleares-capsulas-bastoncillos-mecheros-reciclables_0_730027687.html

Garau-Vadell, J. B., Gutierrez-Taño, D., & Diaz-Armas, R. (2018). Economic crisis and residents' perception of the impacts of tourism in mass tourism destinations. *Journal of Destination Marketing & Management, 7*, 68–75. doi:10.1016/j.jdmm.2016.08.008

GOB-Mallorca. (2018). *Grupo Ornitológico Balear* (GOB-Mallorca). Retrieved June 11, 2019, from https://www.gobmallorca.com/

Goldstein, M. (2017). *Mallorca joins destinations saying #Touristgohome.* Retrieved from: https://www.forbes.com/sites/michaelgoldstein/2017/09/28/mallorca-joins-destinations-saying-touristgohome/#614be1106f6e

Hosteltur. (2018). *España alcanzó los 82 millones de turistas internacionales en 2017 | Economía.* Retrieved June 11, 2019, from Hosteltur: Toda la información de turismo website: https://www.hosteltur.com/126008_espana-alcanzo-82-millones-turistas-internacionales-2017.html

Huete, R., & Mantecón, A. (2018). El auge de la turismofobia ¿hipótesis de investigación o ruido ideológico? *Pasos – Revista de Turismo y Patrimonio Cultural, 16*, 9-19.

Instituto Nacional de Estadística (INE). (2018). PIB per capita - Illes Balers, 2000-2017. Retrieved from: http://www.ine.es/prensa/cre_2017_1.pdf, http://www.ine.es/dyngs/INEbase/es/operacion.htm?c=Estadistica_C&cid=1254736167628&menu=resultados&idp=1254735576581# Interreg

Jordan, E. J., Vogt, C. A., & DeShon, R. P. (2015). A stress and coping framework for understanding resident responses to tourism development. *Tourism Management, 48*, 500–512. doi:10.1016/j.tourman.2015.01.002

King, B., Pizam, A., & Milman, A. (1993). Social impacts of tourism: Host perceptions. *Annals of Tourism Research, 20*(4), 650–665. doi:10.1016/0160-7383(93)90089-L

Lankford, S. V., & Howard, D. R. (1994). Developing a tourism impact attitude scale. *Annals of Tourism Research, 21*(1), 121–139. doi:10.1016/0160-7383(94)90008-6

Lipovčan, L. K., Brajša-Žganec, N., & Poljanec-Borić, S. (2014). What is Good for Tourists Should Be Good for Residents Too: The Relationship Between the Quality of the Touristic Offer and Subjective Well-Being of Residents. *Tourism Analysis, 19*(6), 719–730. doi:10.3727/108354214X14146846679448

Martín, J. M. M., Martínez, J. M. G., & Fernández, J. A. S. (2018). An Analysis of the Factors behind the Citizen's Attitude of Rejection towards Tourism in a Context of Overtourism and Economic Dependence on This Activity. *Sustainability, 10*(8), 2851. doi:10.3390u10082851

Mediterranean BlueIslands. (2018). *Seasonal concentration and characterization of marine litter in selected beaches of 8 Mediterranean islands.* Retrieved from: https://blueislands.interreg-med.eu/fileadmin/user_upload/Sites/Sustainable_Tourism/Projects/BLUEISLANDS/Library/Deliverables/WP3/D3.4.4_Seasonal_concentration_and_characterization_of_ML.pdf

Milano, C., Novelli, M., & Cheer, J. M. (2019). Overtourism and Tourismphobia: A Journey Through Four Decades of Tourism Development, Planning and Local Concerns. *Tourism Planning & Development, 16*(4), 353–357.

Mincotur. (2018). *Previsiones Turísticas 2018 (p. 2).* Retrieved from Mincotur website: https://www.tourspain.es/en-us/Conozcanos/NotasPrensaTemp/2018%2007%2005%20Espa%C3%B1a%20recibir%C3%A1%20este%20verano%2030,4%20millones%20de%20turistas%20internacionales.pdf

Minetad - Ministerio de energía, turismo y agenda digital. (2018.). *España logra un récord de llegadas en 2017 con 82 millones de turistas internacionales.* Retrieved from http://www.minetad.gob.es/es-es/gabineteprensa/notasprensa/2017/documents/180110%20np%20balance%20turismo%202017.pdf

Nozari, E. (2017). *Mallorca locals put pressure on the government to clamp down on tourism.* Retrieved from: https://www.cntraveler.com/story/mallorca-locals-put-pressure-on-government-to-clamp-down-on-tourism?mbid=nl_092817_Intel&CNDID=45233894&spMailingID=12026653&spUserID=MTU4NzkwNjAyOTY0S0&spJobID=1242432308&spReportId=MTI0MjQzMjMwOAS2

Numbeo. (2018) *Cost of living in Spain.* Retrieved from: https://www.numbeo.com/cost-of-living/country_result.jsp?country=Spain

O'Reilly, K. (2007). Emerging tourism futures: residential tourism and its implications. In: C. Geoffroy, & R. Sibley (Eds.), Going Abroad: Travel, Tourism, and Migration. Cross-Cultural Perspectives on Mobility. Newcastle upon Tyne: Cambridge Scholars Publishing, pp. 144-157.

Picornell, M. (2014). Insular identity and urban contexts: Representations of the local in the construction of an image of Palma (Mallorca, Balearic Islands). *Island Studies Journal, 9*(2), 223–238.

Platges de Balears. (2018). *Caló des Màrmols.* Retrieved from: http://www.platgesdebalears.com/caplaibFront/playa_basico.ct.jsp?cPla=65017&cMun=1&cIsl=MA

Puertos. (2017). *Spanish ports exceeded 9.2 million cruise goers in 2017* (p. 2). Retrieved from Spanish Ministery of Public Works and Transport website: http://www.puertos.es/Documents/Notas%20de%20Prensa/NPSeatradeGlobal_02032018[ENG].pdf

Rosselló, J., & Saenz-de-Miera, O. (2011). Road accidents and tourism: The case of the Balearics Islands. *Accident; Analysis, and Prevention*, *43*(3), 675–683. doi:10.1016/j.aap.2010.10.011 PMID:21376854

Royle, S. A. (2009). Tourism changes on a Mediterranean island: Experience from Mallorca. *Island Studies Journal*, *4*(2), 225–240.

Saenz-de-Miera, O., & Rosselló, J. (2014). Modeling tourism impacts on air pollution: The case study of PM10 in Mallorca. *Tourism Management*, *40*(C), 273–281. doi:10.1016/j.tourman.2013.06.012

Sastre, B. (2012). Linguistic integration of the resident German population in Mallorca. *Treballs de Sociolinguística Catalana*, *22*, 249–269.

SEPE. (2018). Informe del Mercado de Trabajo Estatal. Datos 2017. *Catálogo general de publicaciones de la Administración General del Estado* http://publicacionesoficiales.boe.es, 148. Retrieved from http://www.sepe.es/contenidos/observatorio/mercado_trabajo/3017-1.pdf

Seraphin, H., Sheeran, P., & Pilato, M. (2018). Over-tourism and the fall of Venice as a destination. *Journal of Destination Marketing & Management*, *9*, 374–376. doi:10.1016/j.jdmm.2018.01.011

TomTom. (2019). *Traffic congestion ranking | TomTom Traffic Index [Index]*. Retrieved from https://tomtom.com/en_gb/traffic-index/ranking/

TripAdvisor. (2018). *Mallorca Cathedral.* Retrieved from: https://www.tripadvisor.com/Attraction_Review-g187463-d244037-Reviews-or10-Catedral_de_Mallorca-Palma_de_Mallorca_Majorca_Balearic_Islands.html

UNEP & UNWTO. (2005). *Making Tourism More Sustainable - A Guide for Policy Makers*, p. 11-12. Retrieved from: http://www.unep.fr/shared/publications/pdf/DTIx0592xPA-TourismPolicyEN.pdf

Vidal-Folch, I. (2018). *La «fobia» al turismo en Barcelona* | Opinion Home | EL MUNDO. Elmundo. Recuperado de https://www.elmundo.es/opinion/2017/08/08/5988933a268e3e15708b45be.html

Wen, J., Meng, F., Ying, T., Qi, H., & Lockyer, T. (2018). Drug tourism motivation of Chinese outbound tourists: Scale development and validation. *Tourism Management*, *64*, 233–244. doi:10.1016/j.tourman.2017.08.001

ADDITIONAL READING

Almeida García, F., Balbuena Vázquez, A., & Cortés Macías, R. (2015). Resident's attitudes towards the impacts of tourism. *Tourism Management Perspectives*, *13*, 33–40. doi:10.1016/j.tmp.2014.11.002

Christou, P., & Sharpley, R. (2019). Philoxenia offered to tourists? A rural tourism perspective. *Tourism Management*, *72*, 39–51. doi:10.1016/j.tourman.2018.11.007

Garau-Vadell, J. B., Gutiérrez-Taño, D., & Díaz-Armas, R. (2019). Residents' Support for P2P Accommodation in Mass Tourism Destinations. *Journal of Travel Research*, *58*(4), 549–565. doi:10.1177/0047287518767067

Gursoy, D., Ouyang, Z., Nunkoo, R., & Wei, W. (2019). Residents' impact perceptions of and attitudes towards tourism development: A meta-analysis. *Journal of Hospitality Marketing & Management*, *28*(3), 306–333. doi:10.1080/19368623.2018.1516589

Gutiérrez-Taño, D., Garau-Vadell, J. B., & Díaz-Armas, R. J. (2019). The Influence of Knowledge on Residents' Perceptions of the Impacts of Overtourism in P2P Accommodation Rental. *Sustainability*, *11*(4), 1043. doi:10.3390u11041043

Mathew, P. V., & Sreejesh, S. (2017). Impact of responsible tourism on destination sustainability and quality of life of community in tourism destinations. *Journal of Hospitality and Tourism Management*, *31*, 83–89. doi:10.1016/j.jhtm.2016.10.001

Sharpley, R. (2014). Host perceptions of tourism: A review of the research. *Tourism Management*, *42*, 37–49. doi:10.1016/j.tourman.2013.10.007

UNWTO. (2019). 'Overtourism'? – Understanding and Managing Urban Tourism Growth beyond Perceptions, Executive Summary. *Executive Summary*, 12.

Vargas-Sánchez, A., Oom do Valle, P., da Costa Mendes, J., & Silva, J. A. (2015). Residents' attitude and level of destination development: An international comparison. *Tourism Management*, *48*, 199–210. doi:10.1016/j.tourman.2014.11.005

KEY TERMS AND DEFINITIONS

Acculturation: A process of cultural change which occurs when one prevailing culture modifies other cultures.

Anti-tourism: The sentiment and intellectual attitude of residents expressing social anxiety toward tourism.

Irritation index: An index based on local residents' negative attitudes towards the development of tourism.

Overtourism: The subjective belief of residents that there are too many visitors.

Saturated destinations: Overcrowded destinations caused by too many tourists.

Segregation: The separation of two or more groups into different neighborhoods, sorting based on some criteria populations (e.g. nationality, income, race).

Tourism-phobia: The fear, aversion or social rejection that local citizens in a destination feel towards tourists.

Chapter 13
Vacation Rentals, Tourism, and International Migration:
Gentrification in Las Palmas de Gran Canaria (Spain) From a Diachronic Perspective

Josefina Domínguez-Mujica
https://orcid.org/0000-0001-7460-5553
University of Las Palmas de Gran Canaria, Spain

Juan Manuel Parreño-Castellano
https://orcid.org/0000-0002-1082-456X
University of Las Palmas de Gran Canaria, Spain

Claudio Moreno-Medina
University of Las Palmas de Gran Canaria, Spain

ABSTRACT

The growing presence of vacation rentals and international residential migrations are two phenomena determining the recent dynamics and urban structure of most Spanish Mediterranean and island cities. Tourists and migrants tend to be interested in the same urban spaces, and this tends to trigger gentrification, either by changes in the uses of real estate, or driven by the prospects perceived by owners of earning money. This chapter analyses these new mobility flows and urban dynamics in Las Palmas de Gran Canaria, a city with enormous tourism potential located in the outermost regions of European. The chapter analyses the development of tourism in the city and recent transformations in tourism amenities available, including holiday lets. The chapter studies the changes in the resident population paying special attention to foreigners and finally, to reveal the peculiarities and emerging conflicts inherent to tourism gentrification.

DOI: 10.4018/978-1-7998-2224-0.ch013

INTRODUCTION

Between 1996 and 2007, a process of strong economic growth driven by the construction of 6.5 million new homes unfolded in Spain (*Romero, Jiménez & Villoria, 2012*), leading to a high household debt and serious territorial impacts (*Campos Echevarría, 2008; Gaja i Díaz, 2008*). This process was accompanied by a speculative increase in property prices.

The contributing political factors were the de-regulation of mortgages – allowing loans of this kind to be securitised in secondary markets from 1992; the de-regulation of the land market (1998); credit expansion; and immigration policies – more than seven million people settled in Spain between 1994 and 2008. Such factors stimulated the secondary circuit and the accumulation of durable goods for financial purposes, by extending credit to wider social circles.

When Spain was hit by the grave global economic crisis in 2009, funding quickly dried up, real estate projects ground to a halt and the construction industry went into deep recession. The financial and construction crisis spread to the economy as whole and sharp decline in house prices began very swiftly. Consequently, unemployment rates rocketed just when many families began to have problems paying off their mortgages. Against this backdrop, and after a failed policy of stimulating demand with subsidies and social benefits – which led to a substantial increase in the deficit and in Spanish public debt, from 2011 – a programme was rolled out placing strict restrictions on public spending and devaluating wages. At the same time, measures were adopted to promote foreclosures and international investment (*Etxezarreta, Hoekstra, Kol & Cano Fuentes, 2012*), so that financial institutions could make their balance sheets healthier by selling off assets, in many cases to international investors.

From a tourism standpoint, on the other hand, reduced wage costs and other factors pertaining to the European market, enabled the tourist industry to set new records for international tourist arrivals to Spain. Tourist arrivals increased from almost 52m in 2009, to 65m in 2014, and 83m in 2018.

Meanwhile, there was an increase in international investment in some Spanish cities and coastal areas by funds, companies or individuals, as falling prices and evictions created an opportunity to acquire properties at better prices. Such acquisitions were also highly interesting to investors and individuals given the development of holiday lets at the time, especially since 2015. Hence, the cities and tourist areas of the Mediterranean coast and the Canary and Balearic Islands have become some of the most affected by the intensification of tourism, the growth in international investment and new forms of tourism marketing induced by the digital economy, reaching high levels of over-exploitation.

The gentrification of Las Palmas de Gran Canaria took place against this backdrop, especially in district 3 (Isleta-Puerto-Guanarteme), where tourism, holiday rentals and the city's foreign population are concentrated. This chapter focuses on studying this process, based on three main objectives. Firstly, to analyse how tourism has developed in the city of Las Palmas de Gran Canaria and recent transformations in the tourism infrastructure, including holiday lets. The second objective is to study how the resident population of the district has evolved, focusing especially on immigrants. The third and final aim is to characterise the gentrification processes witnessed and the social response that this has generated.

This chapter is divided into the following sections: after the introduction, the first section is devoted to framing the study and the methodology; the second, entitled "tourism supply and demand in Isleta-Puerto-Guanarteme, and the third, "the role of human mobility in the study area", presents changes in tourism and population from a historical perspective. The analysis of the recent gentrification process, its implications for population displacement and the emergence of the first signs of objections are the aims of the fourth section. The text ends with some conclusions that reflect the gentrification model of central neighbourhoods on the fit of the urban-tourism dynamics described.

Before concluding this introduction, some brief notes about Las Palmas de Gran Canaria and the study area are necessary. The city, the most populous in the Canary Islands, had a registered population of 378,517 inhabitants in 2018 (ISTAC). Located in an outermost region of Europe, the economy has clearly specialised in services, many of them relating to tourism and port activities. Tourism became important in the mid-twentieth century, with the appearance of the first coastal tourism accommodation, in the context of the beginning of the mass sun, sea and sand tourism model. Puerto de la Luz, on the other hand, has been the powerhouse of the city economy since construction started 1883. Since then, it has become a hub for connections and provisions for sea traffic between Europe, Africa and America.

These productive activities are located in the southern part of the volcanic peninsula of La Isleta (figure 1), along with most of the port activities, in the neighbourhood of the same name. An isthmus connects this peninsula to the rest of the island, which is fully built up now, but was originally comprised of sand and dunes that remain today in the form of the beaches of Las Canteras and Las Alcaravaneras. The former is especially important because it has a sedimentary reef that shelters it from the waves, to give extraordinary conditions for bathing. The urbanisation of La Isleta and the isthmus of Guanarteme began in the late1880s, at the same time as the port was built, and they are now densely populated by just over 71,000 people.

FRAMING AND METHODOLOGY: INTERSECTION BETWEEN TOURISM AND GENTRIFICATION

Gentrification, and tourism gentrification in particular, is a common phenomenon in many Spanish Mediterranean and island cities, albeit in different stages of evolution. According to *Lees, Shin & López-*

Figure 1. District 3 Isleta-Puerto-Guanarteme (Las Palmas de Gran Canaria). Source: Own elaboration.

Morales (2016), gentrification is a process of social substitution in the urban space through which the population of an area or neighbourhood is displaced and replaced by another with greater purchasing power

The term "gentrification" was initially used in the mid-sixties in relation with the processes of revaluation and social displacement that took place in some neighbourhoods of London (*Glass, 1964*). Ever since the end of the seventies, several theories prevail in the explanation of the process: a) the one that links it with the shift from the industrial society towards a post-industrial society, changes in lifestyle and with the transformation of the structure of households (*Ley, 1986; Zukin, 1989; Hamnett, 1994*); b) the theory that interprets it as a consequence of real estate market activities, as for certain social groups, it is more profitable and beneficial to live in a refurbished house in the centre than in a new one in the suburbs and c) the theory that considers gentrification as the result of the actions of real estate agents in areas where the devaluation of prices generates new business opportunities (*Slater, 2006*).

From an inclusive perspective, gentrification should not be confused, exclusively, with the process of revaluation of an area and its consequent social substitution. *Smith* (1979) establishes a theory of gentrification based on supply and to lesser extent on demand, by which the agents acting in the real estate market gentrify to maximise their profit (rent gap) taking advantage of the market opportunities generated by the life cycle of the properties. Gentrification is conceived as a process in the stages of the life cycle theory of neighbourhoods (growth, abandonment or decline and revitalisation) and is produced by the structural dynamics of Fordist and post-Fordist capitalism. This definition was formulated in the context of the depreciation cycle of central neighbourhoods, but it may be applied to other spaces, as well.

According to this theory, in these neighbourhoods there is an initial stage of new construction and a first cycle of use in which house values rises, unless the other factors, such as changes in the productive structure of the construction and real estate sector, obsolete style or lack of maintenance or rehabilitation impede such a arise. This is followed by a second stage, when properties begin to depreciate, partially and temporarily avoided by investment in refurbishment or by changes in the market structure (housing shortages, changes in the taste of demand...). During this stage, many of the property owners move out and rent them. If the process of falling property values continues, the decline is exacerbated by the sale or rent of properties to increasingly less wealthy social groups, which sometimes leads to the creation of degraded spaces (slums) and to a process of social substitution, including ethnical and nationality changes in the composition of the population, giving rise to widespread abandonment of the properties. This process, known as "filtering", usually precedes the gentrification dynamics, when market conditions become ideal for starting the revaluation process, which requires prices of refurbished or renovated properties to provide a profit margin (rent gap).

This change in market conditions might be favoured by public action, so-called State-led gentrification (*Yetiskul & Demirel, 2018*), or be a consequence of the acquisition strategies of real estate or financial market agents. There is also a form of gentrification that arises from the actions of small owners as a whole, who refurbish their properties to improve their living conditions.

From a broader point of view, the last few years have seen new points of view that relate gentrification to the development of different productive activities, including tourism playing a leading role, since tourism and gentrification intersect in several ways (*Cócola-Gant, 2018*). The gentrifying capacity of tourism has been analysed by different authors (*Gotham, 2005; Lee, 2016*) and, in general terms, has been interpreted as a phenomenon closely related to the "touristification" of the territory. This term identifies the increase of the tourist function within the residential urban space, both under traditional forms of management and through new modalities in the digital economy framework, such as the renting out housing for tourist use.

Tourism and gentrification mutually reinforce each other (*Maitland & Newman, 2008*) in the scenario of advanced capitalist economies and involve the change of the city from a space of social reproduction or support of production to one targeting economic reproduction, transforming the city into merchandise in itself (*Pinkster & Boterman, 2017*).

Taking these general interpretive elements into account, gentrification in some Spanish coastal cities should be approached from a complex point of view. Firstly, in the specific case of new forms of accumulation, the recent transformations of these touristified urban spaces can be examined from the perspective of the succession of different regimes of capital accumulation (*Soja, 1989*), in the context of the structuring thesis (*Lipietz, 1996; Dicken, 1992; Boyer, 2011*), which argues that Fordism has given way to a flexible accumulation regime (post-Fordism or global capitalism) and to the development of the digital accumulation model.

The intervention of investors (*Forrest, Koh & Wissink, 2017; Paris, 2013*) and the development of forms of the collaborative economy, such as "Peer to Peer" (P2P), have encouraged the gentrification process. Furthermore, the crisis has also consolidated new forms of accommodation to guarantee tourist mobility (*Mavrič & Urry, 2009; Hannam, Butler & Paris, 2014; Cohen & Cohen, 2015*) which, at the same time, has caused the gradual expulsion of residents.

Secondly, there is no doubt that the best approach to analyse the severe impact of the real estate crisis in Spain from 2008 has to be Critical Urban Theory, which moots that late capitalism or post-Fordian capitalism facilitates new forms of profits accumulation for specific social groups and economic agents (*Gotham, 2009; Christophers, 2011*). The dispossession or loss of housing is one of its clearest expressions, because of a reduction in the guarantees that provide access to mechanisms of social reproduction (*Harvey, 2003; Hodkinson, 2013*), up to the point that we could talk about gentrification in a scenario of accumulation by dispossession. In the specific case of island geographic contexts with a well-developed tourism footprint, the touristification of the urban space has been interpreted as a specific factor triggering the disposal of homes (*Vives Miró & Rullan, 2017*).

Thirdly, the internationalisation of capital flows and real estate investment is one of the key elements to understanding current urban processes. Spain has implemented policies to attract foreign purchasers since 2013, by enacting the "The Entrepreneur Internationalisation Support Act' (Law 14, 27th September 2013). This Act provided tax breaks and simplified the red tape for international investment companies purchasing properties. Individuals who were not residents in Spain were offered visas and residency permits for spending over €500,000 on a house. This led to some individuals, and especially investments funds (the so-called vulture funds) to start acquiring assets at highly advantageous prices, using vehicles such as Property Investment Companies (FII or SII), Servicers and REITs (in Spain, SOCIMIs) (*Parreño-Castellano et al., 2019*).

Finally, regarding the paradigm of human mobility, it should be remembered that tourism has favoured a growth over time and space (*Hall, 2015*), demonstrating the strength of the fluidity between travelling, leisure and migration (*Cohen, Duncan & Thulemark, 2015*). The mobility associated with the temporary use of second residences, temporary or seasonal lifestyle migrations and the temporary mobility resulting from new forms of work (digital nomads), provide the main lines of the new paradigm of human mobility in these urban areas. However, mobility processes associated with the expulsion or displacement of social groups with low income levels may also be examined from an intra-urban perspective (*Smith, 2002*), based on their exclusion from the housing market. These processes have led to local resistances (*Colomb & Novy, 2016*) especially regarding the proliferation of holiday lets.

The framework described above has driven researchers to consult a wide range of sources on tourism, migration and gentrification. In the first case, the information includes the number of hotel beds from INE (National Statistics Institute, Spain), the Registry of vacation homes that the Gran Canaria Tourist Board is setting up; the figures of cruise tourists compiled by the Port Authority of Las Palmas; data on the supply and tourist use of holiday lets from AirDNA (Airbnb Data and Analytics) and on the number of tourists and tourist spending offered by ISTAC (the Canary Statistics Institute) and the Tourist Movement in Borders (FRONTUR) collected by TURESPAÑA and INE.

Migration and population information comes from the Continuous Register Statistics. For gentrification, we have based our analysis on property price reports drawn up by an appraisal company, Tecnitasa; on the Survey on perception, feeling and attitudes of the population of Las Palmas de Gran Canaria in relation to tourism collected by the City Council; and on the court files of mortgage foreclosures and evictions for non-payment of rent from the Common Service of Notifications and Liens of the Las Palmas de Gran Canaria Judicial District.

All the data was analysed statistically. Some indexes were calculated in accordance with the purposes of the study, such as the foreign index. The figures on mortgage foreclosures and evictions for non-payment of rent and holiday homes was studied from a spatially disaggregated point of view and carried out using a Geographical Information System (GIS). The population data from the Continuous Register Statistics was studied territorially on a census section scale.

TOURISM SUPPLY AND DEMAND IN ISLETA-PUERTO-GUANARTEME

According to Tourist Board of Gran Canaria, Las Palmas de Gran Canaria had a stock of regulated accommodation amounting to 12,502 beds in 2018. This included 57 hotel establishments: urban hotels, pensions, an emblematic hotel and a rural hotel. The self-catering supply included 1,175 establishments, totalling 6,001 beds, slightly less than half of the accommodation available in the borough. This sector included nineteen apartment complexes, eighteen emblematic dwellings, four rural houses and 1,134 holiday lets. Thus, the accommodation supply is highly varied, but with a predominance of hotel, apartment and holiday let beds. These establishments are usually small, with an average of 150 beds; 72 for apartment complexes. Holiday lets, on the other hand, include an average of four beds per establishment. Three quarters of this supply was located within urban district 3 (Isleta-Puerto-Guanarteme), concentrating the sun, sea and sand tourism, as well as a part the port activity-related urban hotels. More specifically, the largest number of beds can be found in the neighbourhood of Santa Catalina-Canteras.

Both business and private individuals showed an interest in this area as a leisure space ever since building started in Isleta-Puerto-Guanarteme. The local and foreign bourgeoisie that began to settle here since the late 19th century because of the port business, built their summer villas on the beachfronts, just when some British operators began to promote it as a destination with an excellent climate and of great therapeutic value and started building the first hotel and spa establishments. After the First World War, the first cruises started arriving and the tourist use increased even further.

It was not until the second half of the 1950s however, that tourist had a clear impact on the structure of this space. It was pioneer in hosting the forms of mass sun, sea and sand tourism. The new hotels used to have between 300 to 600 beds and accommodated tourists from Western Europe attracted by the weather and good conditions for bathing. Much of this growth was based on real estate investment of

Figure 2. Number of hotel beds in Las Palmas de Gran Canaria
Source: CIES, 1974; ISTAC, INE, Tourist Board of Gran Canaria. Own elaboration

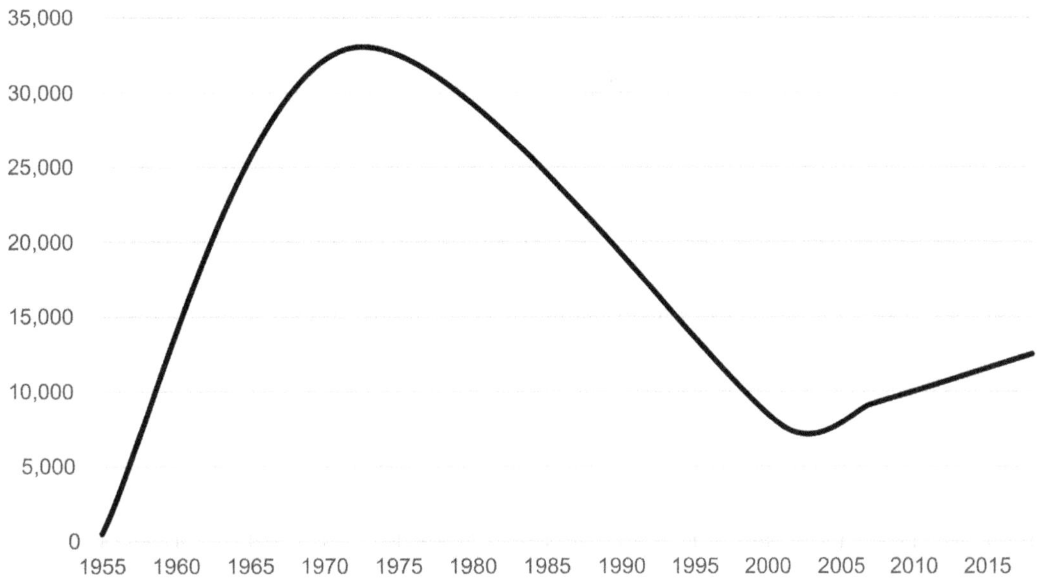

landowners, superimposing the tourism fabric over the residential fabric. However, there were also re-parcelling processes aimed at creating larger establishments (e.g. the "Hotel Don Juan", with 26 floors).

Tourist flows started to decline in the mid-seventies, preferring the new destinations in the south of the island. As a result, some establishments, especially self-catering units and lower-category hotels, began to have profitability problems, causing them to close down or be converted into residential or offices buildings. The progressive decline in tourism continued for over two decades, to the point that, back in 2001, there were only 7,700 registered beds in the city, the lowest figure since the beginning of the Fordist era (figure 2).

There was a certain recovery of tourist accommodation from the end of the nineties, with the renovation of public space in the target area and the strengthening of urban tourism as a market trend. This included 9,139 beds in 2007 (*Domínguez-Mujica, González-Pérez & Parreño-Castellano, 2008*). Since then, against a backdrop of deep economic crisis and subsequent recovery, tourist accommodation supply has grown considerably, especially since 2015. At the same time, there has been a diversification of the kind of lodging available, reaching a figure of 12,502 beds.

Most of this growth is due to the opening of new hotel establishments, which includes new forms of up-market hotels such as boutique hotels, post-hotels, charming or business hotels, and a large number of houses for tourist use, bed and breakfast and guesthouses. As shown in figure 3, most of the tourist accommodation available in 2018 is located, basically, in the Isleta-Puerto-Guanarteme district, where there are 3,263 regulated beds, accounting for 75% of the total holiday lets available in the city.

This is recent supply that has emerged thanks to P2P marketing and other collaborative marketing forms. Based on data on marketing through AirB&B and Homeaway, growth in the number of tourist rentals between the second quarter of 2016 and the first quarter of 2019 is 262.3% (AirDNA). This provides a first impression of the importance of the sudden development of tourist housing in the city

Figure 3. Vacation homes in District 3 Isleta-Puerto-Guanarteme
Source: Tourist Board of Gran Canaria, 2019. Own elaboration

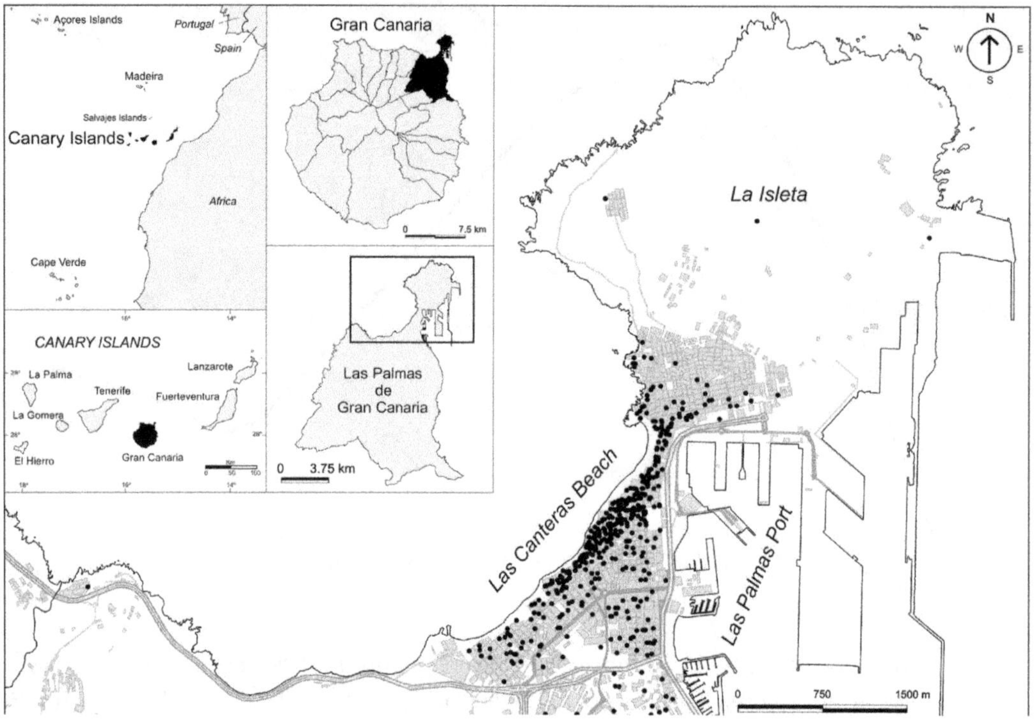

(figure 4). The regulated accommodation offer remained stable until 2017, but the provisional data for 2018 and the openings made and foreseen in 2019 seem to indicate greater growth in supply.

Since 2015, the growth of the tourist accommodation supply has led to an increase in tourist flows and overnight stays. If we focus our interest only on tourists who stayed in hotels and apartments, in 2018, the city received more than 425,000 tourists. This number represents a 52.9% increase against 2009. Overnight stays grown in parallel since 2015, reaching almost one and a half million in 2018, stimulated by a slight increase in the average stay (4.06 days). The most remarkable aspect about the recent growth of tourism is that it occurs while the average daily spend increases, €66.85 in 2017, 13.9% more than in 2014.

There are no reliable data for the flow generated by holiday housing. The average occupancy for the city in 2018, estimated by AirDNA, was around 50%, while some months it would reach around 70%, which leads us to assume that this is a consolidated tourist flow and that it triggers a significant increase in the tourist pressure, if we bear in mind that the average occupancy of regulated accommodation in the city in 2017 was 58.2%. Taking into account the offer and the occupancy rate, we can estimate that approximately 200,000 tourists stayed in holiday lets in 2018, which has meant a substantial growth in tourist pressure.

The number of visitors to the city has also increased because of the arrival of cruise ships. In 2018, the Port of Las Palmas de Gran Canaria received 673,917 visitors, a growth of 175.6% against 2009, according to Las Palmas Port Authority data (figure 5). This figure will continue to grow in the coming years as the Las Palmas Port Authority's plans to build a new cruiser terminal at the Santa Catalina

Figure 4. Holiday Rental Growth in Las Palmas de Gran Canaria
Source: AirDNA. Own elaboration

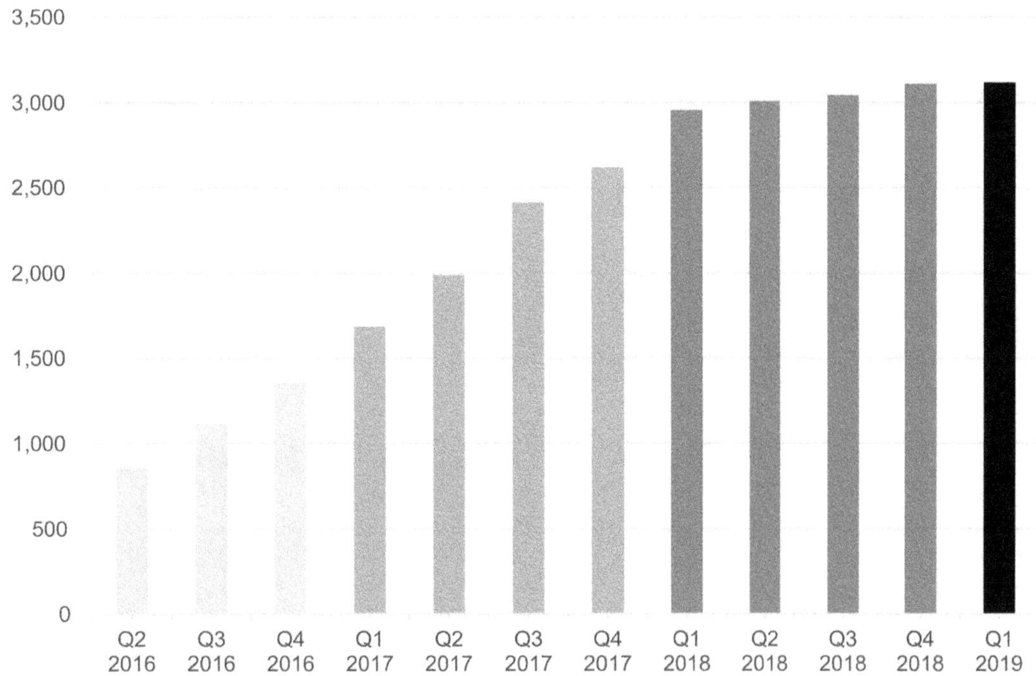

Dock move forwards, since the Puerto de la Luz is becoming one of the most important cruiser bases in the Mid-Atlantic.

Finally, we must not forget that, due to Las Canteras beach, district 3 also receives a significant number of visitors from the tourist resorts of the south of the island. In 2018, Gran Canaria received more than 4.5 million tourists, 3.95 million of whom were foreigners (Frontur). These figures have increased since 2014. Some of the almost four million tourists who stay in the south of Gran Canaria usually spend at least one day visiting Las Palmas de Gran Canaria, which along with the use of the beach and its surroundings by the resident population, generates an over-representational presence of tourist use in this urban space. These facts bring intense pressure to bear on a small space (less than 2 square kilometres), where growing tourism competes with a traditional residential use and the presence of business offices linked to port activities.

THE ROLE OF THE HUMAN MOBILITY IN THE STUDY AREA

The growth of the permanent recorded population of the Isleta-Puerto-Guanarteme district is closely linked to its urbanisation process, which can only be understood in the context of international migratory flows. Although the arrival of foreigners in this area dates back over a century, it has become more noticeable since the 1960s, and above all since the second half of the nineties.

The neighbourhood of Puerto-Santa Catalina undergone an accelerated process of urbanisation at the late 19th century. The labour and business requirements that the port started to create led to the arrival of a large number of migrants from rural areas from all over the province. But at the same time, there

Figure 5. Number of cruise tourists in the port of Las Palmas de Gran Canaria
Source: Port Authority of Las Palmas. Own elaboration

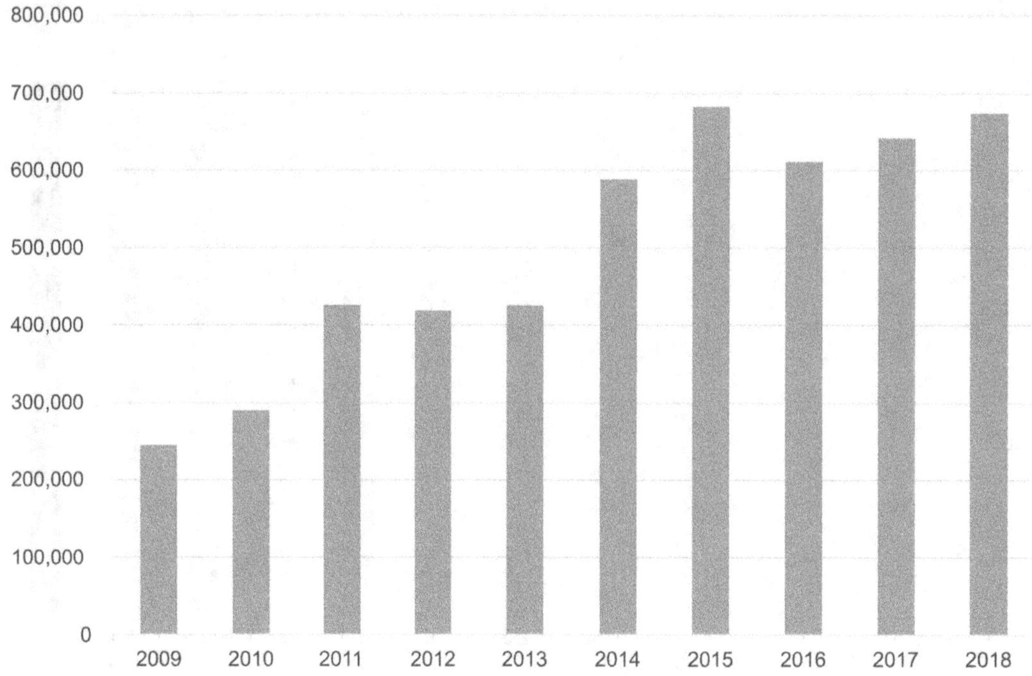

was an already-evident presence of a foreign population from Europe, especially British and German, who came in search of work, business and leisure and settled in the best areas of the isthmus or nearby. Simultaneously, the needs for port workers and operators led to the development of a new urbanisation process in La Isleta and Guanarteme, designed to receive low-income populations. The possibilities of profit driven by the port development and commercial activities also encouraged the formation of foreign communities.

Overall, it can be fairly said that in the mid-fifties, the area of analysis was the most cosmopolitan district of the city, with a population of about 45,000 people, who lived in an urban space of single-family, low-rise buildings, along with some larger buildings – port warehouses and factories – leaving some surviving empty lots. This is when the largest process of urban renewal began, mainly affecting the central area of Puerto-Santa Catalina. The increase in the building density favoured a process of speculative renovation. Single-family houses were often demolished and apartment blocks for residential or mixed use were built on the same plot. The growth of the residential fabric drove the population density of the area, so that more than 60,000 inhabitants lived in the district at the beginning of the eighties, as the same time as there was an increase in foreigners, for business, leisure or lifestyle reasons.

This urban dynamic has changed since the mid-nineties. After fifteen years of demographic and urban stability, a new cycle of real estate and demographic growth began in the area, linked to the economic measures that favoured a strong economic expansion. Economic deregulation and internationalisation measures taken by the Government since 1996, the stimulation of domestic demand through tax reductions, as well as the development of a credit model that facilitated access to investment and consumer funding had an enormous impact on the expansion of the spending capacity of domestic demand and on attracting external demand. On top of this, the 1998 Land Act - clearly aimed at deregulating the market - was enacted, triggering strong economic growth, on the back of the construction industry to a large extent.

This led to the first signs of economic and real estate reactivation in the study area. After the final approval of the revision of the Urban Development Plan of the city in 2000, urban renewal accelerated, and maintained throughout the first decade of the current century. This was accompanied by the construction process and a significant increase in house prices, such that the renovation and refurbishment of residential buildings was aimed at a demand with greater purchasing power.

This growth dynamic was based on the geographical and social mobility of the resident population – thanks to the development of the devolved public administration and the opening of new businesses and commercial spaces – but also due to the arrival of foreign labour. This increased the ratio of foreigners in this area from 11% to 21%, between 2001 and 2010, with the highest figures recorded in 2010, as shown in figure 6, such that, while the population of the district in 2001 61,000, in 2013 it reached 73,281. The correlation between the population and the ratio of foreigners makes it clear that we are witnessing a period of demographic increase based on the incorporation of a population coming from abroad.

The foreigners registering for residence at that time were of diverse origin. As in past decades, the flow of Europeans, mostly from Western Europe, increased in absolute terms and in proportion to the total number of foreigners, in such a way that, while in 2001 they represented 19% of registered foreigners, in 2008, the number increased to 29% (figure 7). Many of them had the economic capacity to buy or rent the medium and high segment housing in the area and, therefore, they have contributed to part of the urban renewal of the district.

Figure 6. Population and Foreigner Ratios in Las Palmas de Gran Canaria
Source: National Statistics Institute. Continuous Register Statistics. Own elaboration

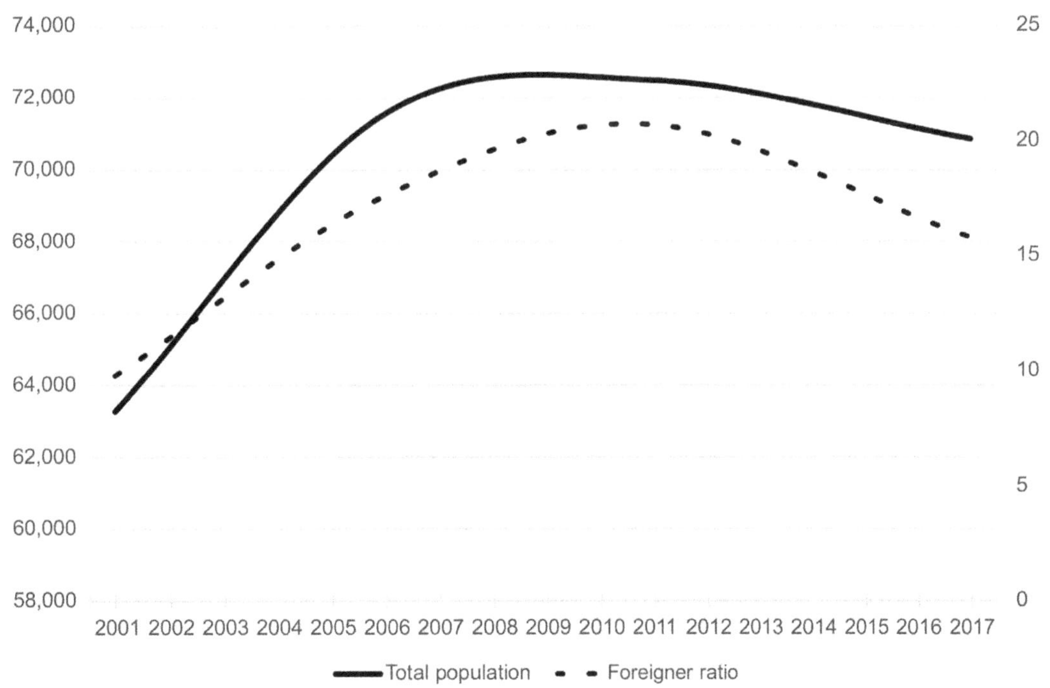

Along with the arrival of Europeans, there was an even higher number of labour migrants from Latin America and Africa (*Domínguez-Mujica, González-Pérez & Parreño-Castellano, 2011*). Although these migrants accounted for 62% of registered foreigners in 2001, by 2008 this figure had fallen to 58% because of the naturalisation processes of immigrants and the greater growth of those of European origin. These labour migrants settled in lower-quality housing and, especially, in buildings of old, un-refurbished tourist apartments, transformed into residential accommodations. Shortly after, in the final years of the economic boom, the increase in house prices and profitability margins in real estate investment drove the refurbishment of some of these properties, reorienting them towards wealthier market segments.

The economic crisis that began in 2009 brought an abrupt halt to real estate renovation and refurbishment. The new situation had a special impact on the purchasing capacity of the lower and middle class and many of them, especially non-EU migrants, were unable to pay mortgages and rents. Since many

Figure 7. Foreign Population by group of nationality in Las Palmas de Gran Canaria
Source: National Statistics Institute. Continuous Register Statistics. Own elaboration

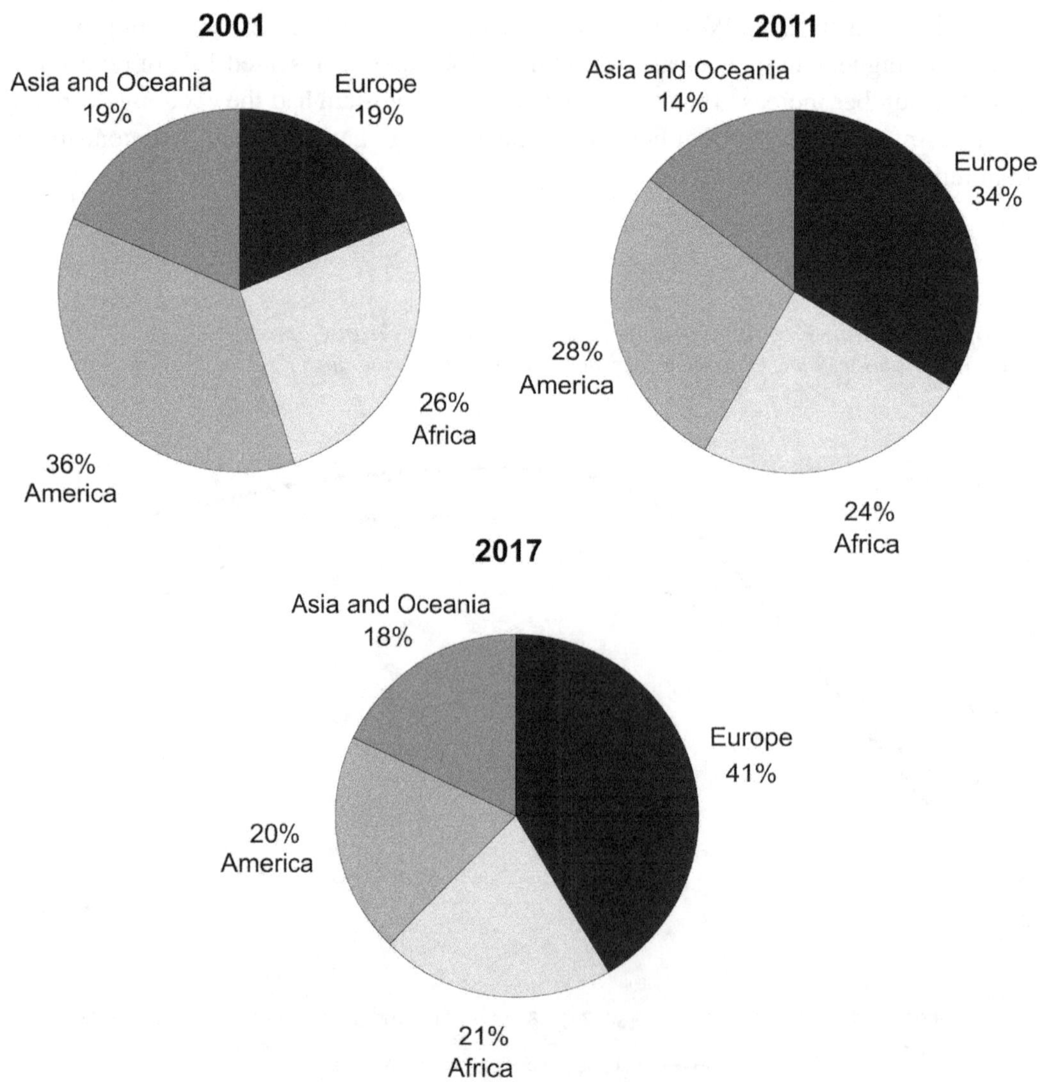

of them lost their houses, the non-EU foreign population either decreased, because they moved to other neighbourhoods where housing was cheaper, or they emigrated again or returned to their home countries.

As shown in figure 7, the decline of this group was constant since 2008, falling by almost a third, such that in 2017 the total population of Africans and Latin Americans represented only 39% of registered foreigners. In contrast, the number of Spanish and other European citizens increased. The steep rise of the EU group stands out between 2008 and 2014, the hardest years of the economic slump of the last decade. The lack of local demand from Spanish and non-EU foreigners was the main factor that attracted the new European population to the district.

Since 2014, the demographic process has been different. With the beginning of economic recovery, there was a sharp decline in the EU population (figure 8). The main causes were the improvement in the economy and the consequent increase in local demand, the growth of the holiday rentals and the arrival of new foreigners. Nonetheless, since 2016 there has once again been an upward trend in the registered community population, especially among nationalities, such as Italians and Germans, because the area has become a consolidated destination for international lifestyle migrants, including the traditional pensioners, and new ones, such as the digital nomads.

Thus, we have witnessed new seasonal migration processes of remote workers, typical of the development of productive structures that characterise digital capitalism. Such flows are related to a working model in which some employees of the telecommunications industry can un-couple their place of residence from their workplace. They can also be freelance workers who choose their place of residence without the physical limitation of the labour market. In other words, there are migratory flows formed by people who want to work from the places they love instead of living in the places where they work.

In the case in question, these flows are recent and started to become evident after the publication of an article in the New York Times back in 2015 entitled "Co-Working on Vacation: A Desk in Paradise", illustrated with the image of some surfers walking along Las Canteras beach. The article dealt with the city as a workplace for a growing number of digital nomads, a phenomenon that is confirmed by the fact that the first annual edition of "Nomad City" was held in 2016. This is an event that brings together both private sector players and the Canary Island administration with a view to enhancing these flows.

Figure 8. Growth rates of Foreign Population by group of nationality (left) and main European nationalities (right) in Las Palmas de Gran Canaria
Source: National Statistics Institute. Continuous Register Statistics. Own elaboration

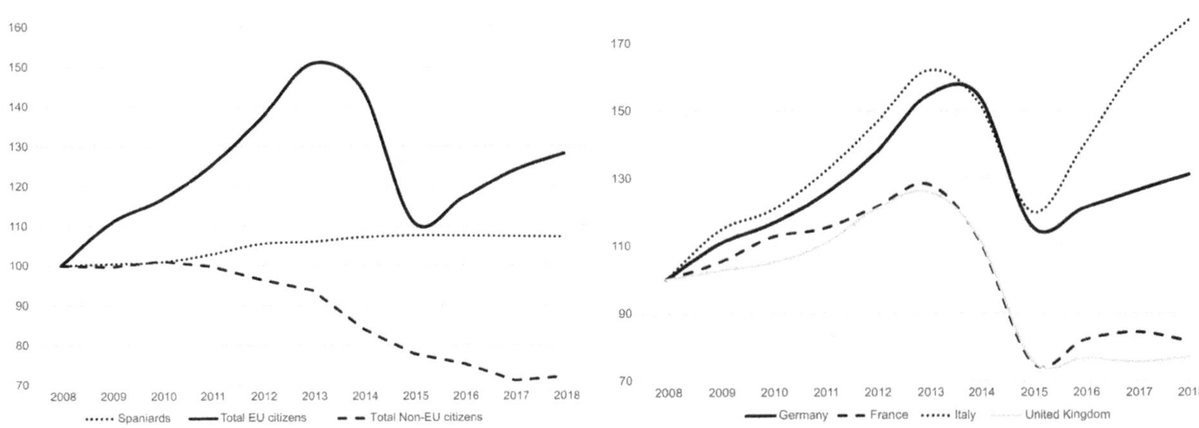

These nomadic migrants are young people with an active lifestyle, who have been living in Las Palmas de Gran Canaria on a temporary basis, combining leisure and work (*Mohn, 2015*). They are usually Western Europeans, although the number of Poles, Czechs and even Russians and Americans has been growing constantly. They have intermediate or high qualifications and are businessmen or mid-level employees. They usually come in winter, with an average stay of about six months. This area attracts them because of its climate, the beach, the cosmopolitan and culturally diverse urban environment, the affordable cost of living, the good food and, in many cases, the possibility of engaging in outdoors activities throughout the year, mostly surfing. At this point, we should point out that Las Canteras is a magnificent city beach with one sector where the conditions are perfect for surfers. Remote working and surfing make an attractive combination for some of these digital workers, leading some surf schools to offer specific services for them.

In addition, the city provides these visitors with the necessary logistical conditions: safety, good international air connections, European-level health services, good connectivity, a growing number of co-working centres, restoration centres adapted to digital work and an increasing number of spaces for co-living. More specifically, Las Palmas de Gran Canaria has about twenty collaborative working centres, some of which are within large international networks. In some cases, the centres offer possibilities for co-living or accommodation and a personalised plan of activities. Furthermore, themed working spaces have been created, where companies form a cluster and entrepreneurs converge. Apart from offering a co-working environment, these centres are also involved in the assessment and orientation of innovative projects. The most prominent case is Marine Park, focused on sea-related business initiatives.

In summary, the urban dynamics of district 3 Isleta-Puerto-Guanarteme have always been determined by international migration, for both business and leisure reasons. The building renovation in the area during the years of the Spanish property bubble cannot be explained without considering international demand and the arrival of investment from abroad. In the same fashion, the housing market was maintained during the crisis because of the growth in the number of Europeans looking for business opportunities. Eventually, with the incipient economic recovery since 2015, the residential dynamic has been determined by the consolidation of the area as a space for European migrants and remote digital workers, having expelled migrant workers with less purchasing power to a certain extent.

RESIDENTIAL DENSIFICATION AND TOURIST GENTRIFICATION

Since Fordist tourism was developed in the area of Las Canteras, there have been four major processes of social change. The first one took place from the sixties to the mid-seventies, coinciding with the period of the "Spanish economic miracle", during Franco's dictatorship. The second emerged from the production crisis of the seventies, with slight nuances, until the mid-nineties, once they had recovered from economic crisis of the time. The third period lasted until the end of the first decade of the current century, with a remarkable real estate boom. Finally, since approximately 2009, we have witnessed a fourth process, characterised by the financial and real estate crisis and by an incipient post-crisis. The latter two stages are the ones that allow us to acknowledge processes of gentrification.

The loss of the district's economic dynamism in the nineties and its consequent property degradation encouraged institutions to roll out intervention plans, aimed at recovering public space. These plans were aimed at creating major infrastructure, the pedestrianisation of streets and the extension of Las Canteras beach promenade. Although these plans began in the second half of the eighties, they showed more significant results in the nineties, with the arrival of European structural funds. In 1997, The "Alfredo Kraus" Opera House was opened in Guanarteme, while the seaside promenade of Las Canteras was extended and refurbished. In 1999, the Elder Science and Technology Museum opened in Santa Catalina Park, along with "Las Arenas" shopping centre in Guanarteme in 1993 and "El Muelle" shopping centre in 2002, the latter close to the Santa Catalina dock.

The re-zoning of the public space and the creation of large public and private infrastructures laid the foundations for the beginning of a real estate revaluation process in the second half of the nineties, which, once again, triggered renovation of buildings aimed at creating new housing for well-off local and foreign demand. This process was accentuated by the revision of the Urban Development Plan of 2000. Unlike the process of the sixties and seventies, the revaluation was more widespread, affecting other areas of La Isleta, especially the Guanarteme neighbourhood.

Part of this social change occurred through the forced displacement of residents because they were unable to pay higher rents, while witnessing the territorial generalisation of an exclusionary dislocation (*Marcuse, 1985*) or, in other words, the impossibility for a growing part of the local population to afford a house in the area. This gentrification was generally boosted by solvent local and foreign demand, in a context of economic growth, but, from a local perspective, the gentrification process was driven by the previous refurbishment of public space.

The beginning of the economic and real estate crisis in Spain in 2009 should have brought about a reduction in housing costs, such that the previous gentrification process would be mitigated in the study area. However, this is one of the places in Spain with the smallest fall in house prices during the economic crisis. According to the independent appraisal company Tecnitasa, between 2008 and 2017, prices in the district fell by 8.8% on average, compared to 35% as the national average, making it the fifth lowest area in Spain for falling house prices. Furthermore, the areas with higher prices, such as Paseo de Las Canteras, held up even better; in 2017 the average value of a square metre was €4,560. Higher prices were only recorded in some Spanish cities (Madrid, Barcelona, San Sebastian, Marbella, Malaga and Santander), while the same was true of rents. In 2017, an average lease was €18.50 euros/square metre on the foreshore, which made Las Palmas de Gran Canaria the fifth most expensive city in Spain.

Firm prices formed the foundation for the consequent elitist drive witnessed in the area. This is a complex process that took place in a context of internationalisation of the housing market and the development of the digital economy. At this stage, the predominance of direct displacement (*DeVerteuil, 2011*) is the most significant aspect. Many residents, unable to pay off their mortgages, in view of the worsening economic conditions, end up losing their houses through judicial processes and out-of-court settlements. Many other impoverished tenants affected by the crisis cannot pay their rents either, and they are evicted relatively quickly under to the provisions of the Urban Leasing Act, approved in 2013. These buildings, in many cases occupied by labour migrants, frequently end up under the control of financial institutions and large international investment funds, which eventually sell or rent the properties at higher prices, given the local and international demand.

There was total of 4,138 foreclosure cases executed between 2009 and 2017 in the Las Palmas de Gran Canaria judicial district (*Parreño et al., 2018*). Figure 9 shows legal proceedings over time, distinguishing between foreclosures and evictions for unpaid rent. The graph reflects the last three economic periods in Spain: real estate crisis (2008-10), debt crisis (2011-13) and post-crisis (since 2014). The crisis meant an increase in these judicial proceedings and they have flattened off since 2015.

Almost three quarters of the legal proceedings executed are concentrated in the area of study. Figure 10 shows the legal proceedings according to the postal address of the affected properties.

Figure 9. Number of Foreclosures and evictions for unpaid rent in District 3 Isleta-Puerto-Guanarteme
Source: Common Service of Notifications and Liens of the Judicial District. Own elaboration

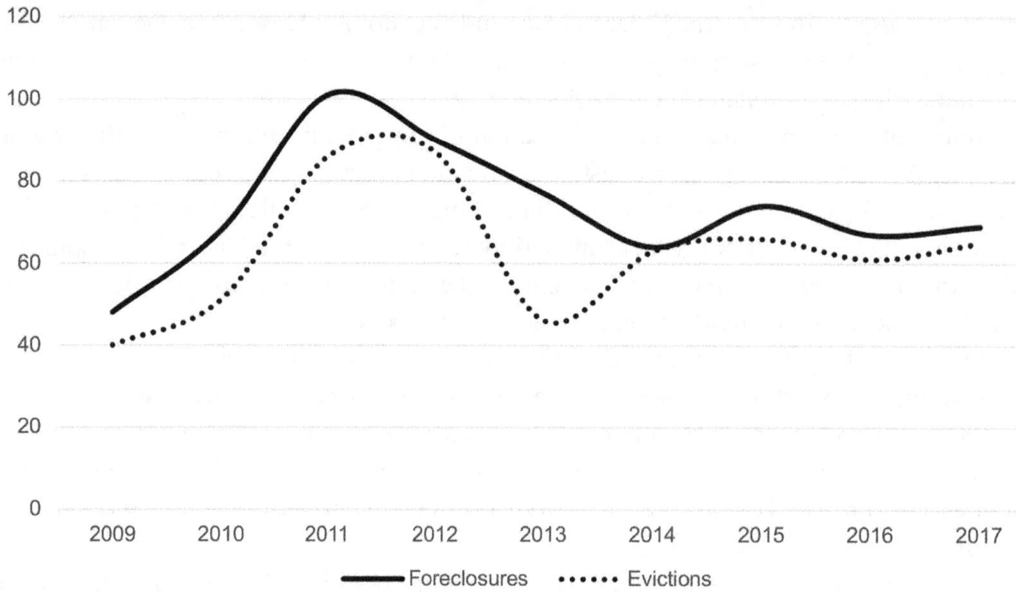

Figure 10. Location of Foreclosures and evictions for non-paid rents in District 3 Isleta-Puerto-Guanarteme
Source: Common Service of Notifications and Liens of the Judicial District. Own elaboration

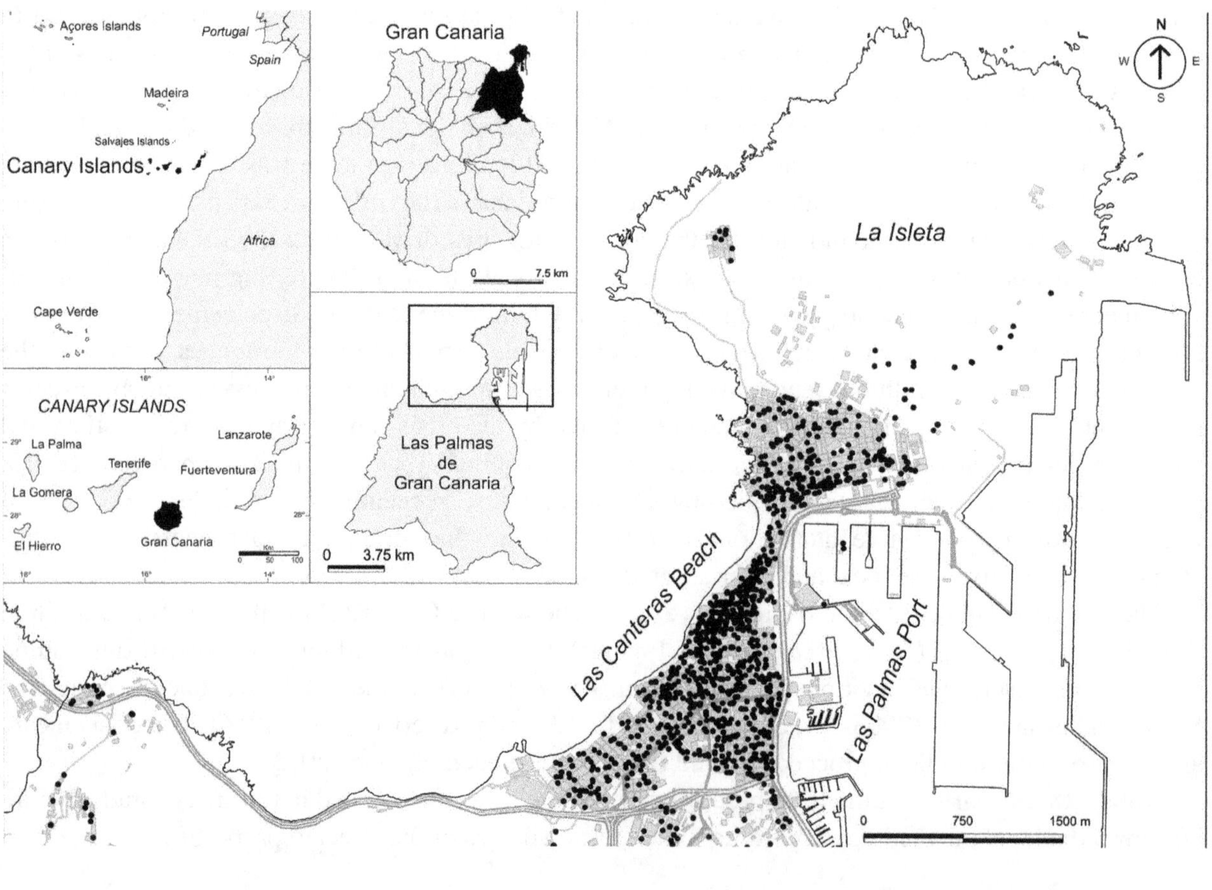

On top of gentrification for real estate reasons, we should add the recent development of tourism. We have already given figures on the rapid development of tourist housing in the district, the increase in the number of tourists staying in hotels and the number of visitors and cruise passengers. These figures corroborate new process of touristification that, apart from the aforementioned displacement processes (direct and exclusionary), have brought further "displacement pressure" (*Marcuse, 1985*), or displacements due to the change of the urban environment, since this adapts increasingly to the needs of the tourists and less and less to the needs of the residents. The appearance of businesses aimed at tourists to the detriment of those aimed at residents, such as gourmet stores, express supermarkets, souvenir shops, catering businesses, etc. is something that characterises the new urban landscape.

While it is true that the development of tourist housing is not a direct factor of eviction in general terms, there is a territorial relationship between this form of tourism and the loss and use of property. A simple visual comparison of the distribution of dispossessions and evictions with the distribution of holiday housings allows us to appreciate that the latter grows were eviction cases are concentrated. Likewise, the spatial correlation of the latter with the presence of a foreign population from Western Europe is also obvious. There is a spatial coincidence between both the growth of the ratio of foreigners from Western Europe and judicial procedures and, therefore, it is logical to think that judicial procedures are a tool for gentrification.

SOCIAL RESPONSE: THE SIMULTANEOUS DEFENSE OF TOURISM AND REJECTION OF HOLIDAY RENTALS

The residential elitism of the area, and especially the current tourist gentrification, coinciding with the period of recent economic recovery, is affecting the housing market, to such an extent that there are the first signs of social oppositions. The rise in real estate values is having an impact on the increase of rents and the price of owned properties, which makes housing prohibitive for segments of the population with lower income. This fact, favoured by the proliferation of holiday housing, is fuelling a certain discomfort that is made visible by protests promoted by platforms of activists such as "*Barrios para vivir*", "*Sindicato de inquilinas*" or "*Ben Magec, ecologistas en acción*" (figure 11).

Because of the protests against the incipient processes of touristification and gentrification in this area, the Tourism Department of the Las Palmas de Gran Canaria City Council[1] took greater heed of the opinion of citizens before drafting their Tourism Management Plan[2] for the city. Thus, it commissioned a survey about the perception, feelings and attitudes of the urban population to tourism, the first one ever to be conducted in Las Palmas de Gran Canaria. The survey included a sample of 1,200 residents from the five urban districts and was conducted in February 2019.

According to the results of the survey, the majority of the population considers that tourism is important for the economy of Las Palmas de Gran Canaria, as stated by, basically, 9 out of 10 respondents. Likewise, 66.4% of these respondents consider that the number of tourists visiting the city is appropriate and only 14.6% consider it excessive.

Furthermore, respondents point out the positive impact that tourism has on creating jobs, wealth and economic growth, improving communication infrastructures and urban furnishing, cultural and leisure activities, public services, public safety and on the quality of life in the municipality in general. On the other hand, a minority of just 9.6% declare that they avoid visiting tourist areas. In addition, 74.9%

believe that tourism has not contributed to the loss of the city's identity and a very low percentage of respondents believe that there is rejection of tourists.

On the contrary, the respondents highlight the damage that tourism causes to the environment (40.9%), the cost of living (56.3%) and, especially, to the availability of affordable housing (68.1%). Regarding the latter, 83.3% consider that tourist housing has increased the price of rents, 78.4% state that it gets harder to find housing and almost half of the respondents believe that tourism sweeps away some jobs. For all these reasons, 61.6% claim that holiday lets have to be better regulated and 15.8% even declare that such activities should be forbidden in neighbouring buildings.

Therefore, this first survey carried out in the city focuses the problem of tourism intensifying gentrification that seems to boost the holiday housing, to the extent that a proportion of the local residents cannot afford a house, due to the increase in rents. In general, no negative opinions about tourism have been recorded. On the contrary, respondents consider that the number of tourists is adequate. The new forms of marketing seem to cause the rejection of the local residents.

CONCLUSION: DYNAMICS AND RESISTANCE IN AN URBAN-TOURIST AREA

District 3 of Las Palmas de Gran Canaria, Isleta-Puerto-Guanarteme, is an urban area of the mid-Atlantic where tourism and residential specialisation coexisted for more than a century. This makes it a privileged observatory for diachronically analysing the urban-tourist dynamics linked to the change in the different accumulation regimes and to the role played by internal and international mobility of the population.

For this reason, in this chapter we have analysed the process of tourism development and the recent transformations in the infrastructure of tourism supply in these neighbourhoods, including holiday rentals; the evolution of the resident population of the district with special attention to people coming from

Figure 11. Protest against holiday homes in Paseo de las Canteras. Source: Canarias7

Western Europe and other countries due to different processes of immigration, as well as the reorganisation of the area as a tourist destination, in an attempt to identify the conflicts and characteristics inherent to the recent process of tourism gentrification.

As we have pointed out, since the end of the fifties, a Fordist-oriented tourist plan has been launched in district 3, while the residential fabric that was fostered by real estate revaluation was becoming denser. Local and foreign investments radically transformed the study area, while tourists and European immigrants had been arriving for leisure or business reasons.

Furthermore, the tourism crisis led to the residentialisation of a large part of the accommodation supply because it has become obsolete since the late seventies, the typical process of mature tourist destinations at a time of restructuring towards post-Fordism. This residentialisation resulted in a new segmentation of real estate products, by offering non-refurbished units to the real estate rental market at low prices, and others owned or rented, after their restoration or complete transformation, to segments of demand with greater purchasing power.

We have also stated that, in the context of the Spanish real estate boom initiated at the end of the nineties, a urban densification and revaluation commenced. Traditional lifestyle migrants from Northern and Western Europe, as well as other local residents of medium and high purchasing power, buy new or refurbished. In a similar fashion, properties not included in this process enter the leasing market, aimed at foreigners, mainly from Latin America, Asia and Africa, with low purchasing power, who arrived in the early years of economic development. Eventually, we have highlighted that today we are witnessing a revaluation and reconsideration of touristification, which began in the last years of the real estate crisis.

Therefore, the area studied follows the model of the depreciation cycle of central neighbourhoods formulated by Smith (1979) and tries to explain urban gentrification from a historical perspective. It is worth recognising a first stage of real estate use of strong revaluation, which, in this case, would coincide with the period between the fifties and seventies. A second period of devaluation, until roughly the end of the 20th century that affects some properties and areas as late as the first decade of this century, led to the increase in rent for more vulnerable social groups, often foreigners ("filtering"). A third stage, in the last years of crisis and post-crisis, in which processes of gentrification or social substitution of these groups by those with greater purchasing power take place. In this last stage, direct displacements are related to the increase in evictions due to unpaid rents or mortgages. In other words, following this model, the process of gentrification is activated by business expectations, not only by local property owners, but also by financial agents and international investors who take ownership of these properties.

It is worth asking why a rent gap has appeared in recent decades. The first answer is obvious: the possibility for financial entities and investors to obtain real estate at low prices, given that Spanish regulations substantially limit the rights of debtors. Thus, "rent gap" is actually a clear expression of accumulation by dispossession, an unfair system of capital loss for the most vulnerable citizens and enrichment for financial entities and investment funds. On top of this, consideration must be given to the impact of an effective demand on the profit margins of the study area. In this case, the existence of international demand to purchase or rent houses is a key factor in the value production circuit.

Moreover, from the supply point of view, institutional action since the nineties must be take into account, focused on recovering public space and re-zoning the district with the construction of large infrastructures. The public initiative enhanced, intentionally or not, an urban scenario for gentrification, so this can be considered the result of the transfer of value from the public to the private sector.

Finally, we cannot omit a fourth essential factor, the tourism potential of the area. The new forms of tourism in the market based on developing housing for tourist use, the digital economy and new forms of marketing, along with new city-tourism housing products are all adapted to the peculiarities of the area analysed. The profitability of these new forms of tourism puts upward pressure on rents and properties, given the greater margin of productivity within which they operate, and favours the expulsion of residents, while not allowing others to have sufficient purchasing power to rent or buy a house in the area. In other words, we are witnessing an accelerated process of tourist gentrification.

The changes referred to are giving rise to social opposition, although no generalised feelings of tourism-phobia have been observed. Island society has probably embraced the idea that the growth of tourism is a guarantee of progress and wellbeing for the Canary Islands, as its origin in the 1960s has modified the very pillars of an impoverished economy that caused periodic crises and led to emigration. The transformation of the recent history of Canary Islands triggered by the impact of tourism on the economy and on a society that had earlier assimilated new uses and customs, is deeply-rooted in the collective mentality. Urban-tourism products, even poorly regulated ones like holiday rentals, are met with suspicion and rejection in certain social fragments that recognise the effect of these on the price of real estate and on the processes of elitism of the urban-tourist heart of the city. We believe that such opposition occurs because larger social groups feel displaced, since they are unable to find affordable rents, while, at the same time, they realise that the supply of traditional rental housing is progressively diminishing.

In summary, Isleta-Puerto-Guanarteme is an interesting example of the life cycle of neighbourhoods and current gentrification dynamics, with the peculiarity that past and present dynamics are closely linked to the development of tourism and the importance of international real estate demand.

ACKNOWLEDGMENT

This research was supported by the State Research Agency (AEI) and the European Regional Development Fund (ERDF) with R&D projects "Crisis and vulnerability in Spanish island cities: transformations in social reproduction spaces" (CSO2015-68738-P) and "Housing and international mobility in cities of the Canary Islands. The emergence of new forms of urban inequality" (RTI2018-093296-B-C21) (Ministry of Science, Innovation and Universities).

REFERENCES

Boyer, R. (2011). Is a finance-led growth regime a viable alternative to Fordism? A preliminary analysis. *Economy and Society*, *29*(1), 111–145. doi:10.1080/030851400360587

Campos Echevarría, J. L. (2008). *La burbuja inmobiliaria española*. Madrid, Spain: Marcial Pons.

Canarias7. (2019, Feb. 2). *Guanarteme, contra la desregulación del alquiler vacacional*. Retrieved from https://www.canarias7.es/siete-islas/gran-canaria/las-palmas-de-gran-canaria/guanarteme-contra-la-desregulacion-del-alquiler-vacacional-GY6504932

Christophers, B. (2011). Revisiting the Urbanization of Capital. *Annals of the Association of American Geographers*, *101*(6), 1347–1364. doi:10.1080/00045608.2011.583569

CIES. (1974). Introducción a un estudio socioeconómico del turismo en la provincia de Las Palmas. Las Palmas de Gran Canaria: CIES (Centro de Investigación Económico y Social).

Cócola-Gant, A. (2018). Tourism gentrification. In L. Lees, & M. Phillips (Eds.), *Handbook of Gentrification Studies* (pp. 281–293). Cheltenham, UK: Edward Elgar. doi:10.4337/9781785361746.00028

Cohen, E., & Cohen, S. A. (2015). A mobilities approach to tourism from emerging world regions. *Current Issues in Tourism, 18*(1), 11–43. doi:10.1080/13683500.2014.898617

Cohen, S. A., Duncan, T., & Thulemark, M. (2015). Lifestyle mobilities: The crossroads of travel, leisure and migration. *Mobilities, 10*(1), 155–172. doi:10.1080/17450101.2013.826481

Colomb, C., & Novy, J. (2016). Urban tourism and its discontents: an introduction. In C. Colomb, & J. Novy (Eds.), *Protest and Resistance in the Tourist City* (pp. 1–30). London, UK: Routledge. doi:10.4324/9781315719306

DeVerteuil, G. (2011). Evidence of Gentrification-induced Displacement among Social Services in London and Los Angeles. *Urban Studies (Edinburgh, Scotland), 48*(8), 1563–1580. doi:10.1177/0042098010379277

Dicken, P. (1992). *Global shift: The internationalization of economic activity*. London, UK: Chapman Publishing.

Domínguez-Mujica, J., González-Pérez, J. M., & Parreño-Castellano, J. M. (2008). Transformaciones recientes en barrios turísticos maduros. Los casos de Palma de Mallorca y Las Palmas de Gran Canaria, España. *Scripta Nova. XII, 270 (93).*

Domínguez-Mujica, J., González-Pérez, J. M., & Parreño-Castellano, J. M. (2011). Tourism and human mobility in Spanish Archipelagos. *Annals of Tourism Research, 38*(2), 586–606. doi:10.1016/j.annals.2010.11.016

Etxezarreta Etxarri, A., Hoekstra, J., Dol, K., & Cano Fuentes, G. (2012). De la burbuja inmobiliaria a las ejecuciones hipotecarias. *Ciudad y Territorio, 174*, 597–613.

Forrest, R., Koh, S. Y., & Wissink, B. (2017). Hyper-divided cities and the 'immoral' super-rich: Five parting questions. In *Cities and the Super-Rich* (pp. 273–287). New York: Palgrave Macmillan. doi:10.1057/978-1-137-54834-4_14

Gaja i Díaz, F. (2008). El "tsunami urbanizador" en el litoral mediterráneo. El ciclo de hiperproducción inmobiliaria 1996-2006. *Scripta Nova, XII, 270 (66).*

Glass, R. (1964). *London: aspects of Change*. London, UK: Centre for Urban Studies & MacGibbon and Kee.

Gotham, K. F. (2005). Tourism Gentrification: The case of New Orleans' Vieux Carre (French Quarter). *Urban Studies (Edinburgh, Scotland), 42*(7), 1099–1121. doi:10.1080/00420980500120881

Gotham, K. F. (2009). Creating liquidity out of spatial fixity: The secondary circuit of capital and the subprime mortgage crisis. *International Journal of Urban and Regional Research, 33*(2), 355–371. doi:10.1111/j.1468-2427.2009.00874.x

Hall, C. M. (2015). On the mobility of tourism mobilities. *Current Issues in Tourism, 18*(1), 7–10. doi:10.1080/13683500.2014.971719

Hamnett, C. (1994). Socio-economic change in London: Professionalisation not polarization. *Built Environment, 31*(3), 192–203.

Hannam, K., Butler, G., & Paris, C. M. (2014). Developments and key issues in tourism mobilities. *Annals of Tourism Research, 44*, 171–185. doi:10.1016/j.annals.2013.09.010

Harvey, D. (2003). *The 'New' Imperialism: Accumulation by dispossession*. Oxford, UK: Oxford University Press.

Hodkinson, S. (2013). The new urban enclosures. *City, 16*(5), 500–518. doi:10.1080/13604813.2012.709403

Lee, D. (2016). How Airbnb short-term rentals exacerbate Los Angeles's affordable housing crisis: Analysis and policy recommendations. *Harvard Law & Policy Review, 10*, 229–253.

Lees, L., Shin, H. B., & López-Morales, E. (2016). *Planetary gentrification*. Cambridge, MA: John Wiley & Sons.

Ley, D. (1986). Alternative Explanations for Inner-City Gentrification: A Canadian Assessment. *Annals of the Association of American Geographers, 74*(4), 521–535. doi:10.1111/j.1467-8306.1986.tb00134.x

Lipietz, A. (1996). Warp, woof, and regulation: A tool for social science. In G. Benko, & U. Strohmayer (Eds.), *Space and Social Theory* (pp. 250–283). Cambridge, MA: Blackwell.

Maitland, R., & Newman, P. (2008). Visitor-host relationships: conviviality between visitors and host communities. In B. Hayllar, T. Griffin, & D. Edwards (Eds.), *City Spaces–Tourist Places: Urban Tourism Precincts* (pp. 223–242). New York: Elsevier. doi:10.1016/B978-0-7506-8195-7.00011-1

Marcuse, P. (1985). Gentrification, abandonment, and displacement: Connections, causes, and policy response in New York City. *Journal of Urban and Contemporary Law, 28*, 195–240.

Mavrič, M., & Urry, J. (2009). Tourism studies and the new mobilities paradigm. In T. Jamal, & M. Robinson (Eds.), *The Sage Handbook of Tourist Studies* (pp. 645–657). Los Angeles, CA: Sage. doi:10.4135/9780857021076.n36

Mohn, T. (2015). Co-Working on Vacation: A Desk in Paradise. The New York Times (January 19). Retrieved from https://www.nytimes.com/2015/01/20/business/co-working-on-vacation-a-desk-in-paradise.html?action=click&contentCollection=Business%20Day&pgtype=imageslideshow&module=RelatedArticleList%C2%AEion=CaptionArea&version=SlideCard-4

Paris, C. (2013). The homes of the super-rich: Multiple residences, hyper-mobility and decoupling of prime residential housing in global cities. In I. Hay, & M. Flinders (Eds.), *Geographies of the super-rich* (pp. 94–109). Cheltenham, UK: Edward Elgar. doi:10.4337/9780857935694.00013

Parreño-Castellano, J. M., Domínguez-Mujica, J., Armengol-Martín, M. T., Boldú-Hernández, J., & Pérez-García, T. (2019). Real estate dispossession and evictions in Spain: A theoretical geographical approach. *Boletín de la Asociación de Geógrafos Españoles, 80*(2602), 1–25.

Parreño-Castellano, J. M., Domínguez-Mujica, J., Armengol-Martín, M. T., Pérez-García, T., & Boldú-Hernández, J. (2018). Foreclosures and Evictions in Las Palmas de Gran Canaria during the Economic Crisis and Post-Crisis Period in Spain. *Urban Science, 2*(4), 109, 1-15.

Pinkster, F. M., & Boterman, W. R. (2017). When the spell is broken: Gentrification, urban tourism and privileged discontent in the Amsterdam canal district. *Cultural Geographies*, *24*(3), 457–472. doi:10.1177/1474474017706176 PMID:29278248

Romero, J., Jiménez, F., & Villoria, M. (2012). (Un) sustainable territories: Causes of the speculative bubble in Spain (1996–2010) and its territorial, environmental, and socio-political consequences. *Environment and Planning. C, Government & Policy*, *30*(3), 467–486. doi:10.1068/c11193r

Slater, T. (2006). The eviction of critical perspectives from gentrification research. *International Journal of Urban and Regional Research*, *32*(1), 192–194.

Smith, N. (1979). Toward a theory of gentrification. A back to the city movement by capital, not people. *Journal of the American Planning Association*, *45*(4), 538–548. doi:10.1080/01944367908977002

Smith, N. (2002). New globalism, new urbanism: Gentrification as global urban strategy. *Antipode*, *34*(3), 427–450. doi:10.1111/1467-8330.00249

Soja, E. W. (1989). *Postmodern geographies: The reassertion of space in critical social theory*. London, UK: Verso.

Vives-Miró, S., & Rullan, O. (2017). ¿Desposesión de vivienda por turistización? Revalorización y desplazamientos en el Centro Histórico de Palma (Mallorca). *Revista de Geografía Norte Grande*, *67*(67), 53–71. doi:10.4067/S0718-34022017000200004

Yetiskul, E., & Demirel, S. (2018). Assembling gentrification in Istanbul: The Cihangir neighbourhood of Beyoğlu. *Urban Studies (Edinburgh, Scotland)*, *55*(15), 3336–3352. doi:10.1177/0042098017746623

Zukin, S. (1989). *Loft Living. Culture and Capital in Urban Change*. New Brunswick: Rutgers University Press.

KEY TERMS AND DEFINITIONS

Digital Capitalism: A phase of capitalism developed over the internet, creating a wide network economy supporting corporate business processes.

Digital Nomads: people who work remotely, location independent and using technology to accomplish their jobs.

Holiday Rental: Renting out a furnished apartment, house or similar, to tourists on a temporary basis.

International Residential Migrations: Forms of international mobilities for residential purposes involving a variety of different processes, from investor migration programmes, to amenity and retirement migrations.

Lifestyle Migrants: Migrants who move for quality of life reasons and who are, on average, affluent, older and retired or semi-retired.

Post-Fordism: A system of economic production and consumption found in most industrialised countries since the late 20th century, which moved away the industrial bases of huge factories towards specialised markets based on small flexible manufacturing units.

Tourism Gentrification: A process of changing the personality of a traditional neighbourhood through the influx of tourists and tourism-related businesses.

Touristification: The process of making an urban, rural or natural space suitable for tourists.

ENDNOTES

1 Concejalía de Turismo del Ayuntamiento de Las Palmas de Gran Canaria

2 Plan Director de Turismo de la ciudad

Chapter 14
Alternative Tourism Strategies to Enhance Local Sustainable Tourism Development:
The ALTER-ECO Project in Gandia (Spain)

Rafael Temes-Cordovez
https://orcid.org/0000-0002-5604-4892
Polythecnic University of Valencia, Spain

Begoña Serrano-Lanzarote
Polythecnic University of Valencia, Spain

Juan José Tuset Davó
Polythecnic University of Valencia, Spain

Ruth De León-Rodríguez
European University of Valencia, Spain

ABSTRACT

This chapter establishes the relationship between the ALTER-ECO project and the problem of overtourism. Background reviews the phenomenon of excess tourists based on recent research. The measures implemented in Gandia (Spain) as part of the ALTER-ECO project are described. Monitoring influx of tourists, installing Free Wi-Fi at Gandia Beach, and creating a mobile app are three measures that have contributed to easing the effects of overtourism. The debate proposes monitoring, public Wi-Fi, and gamification as three strategies for dispersing tourist load. Conclusions include measures and strategies that coincide with UNWTO 2018 recommendations.

DOI: 10.4018/978-1-7998-2224-0.ch014

INTRODUCTION

In the last two decades Europe has undergone an intense urban planning process, especially intense in the Mediterranean regions and the Spanish coasts. The typical Mediterranean city known for its compactness, density and continuity shows a clear deconfiguration which is especially intense in coastal areas. Nowadays, these compact city centres are accompanied by large, dispersed and low-density peri-urban areas, causing a "dissolution" (Colaninno & Roca, 2012) of the city which has become a recurring theme in urban planning analyses and reflections that seek to quantify today's development phenomena (Gaja, 2012). The Comunitat Valenciana (Region of Valencia) can be used as a paradigm for this change in *urban form* process boosted by economic development based on clearly unsustainable construction. Documents such as the European Parliament "Auken Report, Recommendation 2002/413/EC on Integrated Coastal Zone Management in Europe or "*A toda costa*" (2018) by Greenpeace, coincide in showing their concern for the impact of extensive urban development in Spain, with special emphasis on the situation in the Comunitat Valenciana. This over-exploitation and over-occupancy did not come out of nowhere, but was the result of three expansive economic cycles associated with three property bubbles (Miralles, 2014): the first cycle from 1959 to 1972 created the first tourist developments on the coast; the second from 1985 to 1990 generated significant urban development on the Mediterranean coast; and the third from 1997 to 2006 caused a new avalanche of investments in the property construction sector. Unlimited growth in the shape of a "virtuous" speculative property cycle came to an abrupt halt in 2007 when the real estate bubble burst. Simultaneous to this rapid urban development, growth in tourism encouraged by lower transport costs, easier travel and a growing middle class in advanced and emerging economies, have made cities increasingly more popular destinations for business and leisure tourism. City centres have become especially attractive and scenic (Milano, 2019). In this sense, the so-called urbanisation of tourism (Mullins, 1991) is described as the moment cities are built around the offering and consumption of a wide variety of goods and services for fun, pleasure, relaxation and recreation and not exclusively around basic needs such as housing, healthcare and education. According to the World Travel & Tourism Council, city tourism represents 45% of global international tourism. Furthermore, 54% of the world's population lives in urban centres and cities; this number is expected to rise to 68% in 2050. Therefore, observing the effects tourism may have for urban life is extremely important (WTTC, 2018). Not in vain, almost 50 years ago city planner Peter Hall said, regarding the growing importance of tourism for economies and urban planning in European cities, that "the era of mass tourism is the most important factor in the last 30 years of this century for change in major cities in Europe and also many smaller, historical cities" (Hall, 1970). This increase in city tourist attractions, which in the case of Spain has been especially relevant in cities such as Barcelona, Madrid and Valencia (Urbantur, 2016), has been accompanied by an increased perception of the negative external factors of tourism which translate into a feeling of overcrowding, noise and annoyance with protests in some cities and terms such as *overtourism* and *tourismphobia* becoming widespread as synonymous with mass growth, saturation and tourist pressure (UNWTO, 2018, 2019).

As shown in the report "Facing the challenges of overtourism through open/big data. Integrated Territorial Project "HERIT-DATA" (TCV, 2019), authors such as Postma A. and Schomuecker D. (2017) note that the impact of tourism has evolved since World War II, varying in its characteristics and perception over time. They affirm that during the first phase from 1960-1970, the emphasis of tourist impact in studies focused on a positive approach to economic impact. Tourism was thus seen as a means for strengthening economies. However, from 1970-1980, the focus gradually switched towards social aspects, with a more negative approach in some cases. Attention was paid to cultural and environmental impacts, influenced by urban contexts that were still marked by excessive industrialisation, sustainability and the need for quality of life of citizens. From 1980-1990, interest generated by tourist impact in studies focused on a holistic integration between the economic perspective, social vision and environmental balance. This trend remains in force with the turn of the new century. However, the concerns of the main tourism leaders and major operators include capturing and selectively diversifying tourist flows, leveraging the daily presence and evolution of new technologies today, accessing and handling data and, in short, controlling information. Deery et al. (2012), note that tourism impact studies have become a well-developed, almost mature field. However, Williams (2009) states that there is still a lack of understanding between tourism and destination communities due to the number of inconclusive or conflicting empirical studies which still have a controversial underlying conceptual base.

In last 2018, prestigious scientific journal "Tourism Geographies. An International Journal of Tourism Space, Place and Environment" published a special edition dedicated to urban tourism planning in European cities, with special interest in analysing tourist pressure, planning and heritage. In the introductory report of this special edition, Shoval (2018) reviews initial research on urban tourism, the distribution of tourism in cities, the attraction of historical city centres and heritage, and the adverse effects of overtourism in urban environments. According to recent studies (Milano, 2018; Simancas & Peñarrubia, 2019; Perkumienė et al, 2019), although it may seem like a recent phenomenon, since the decade of 1970-1980 various social scientists have been developing theories such as the Irritation Index by George Doxey (1975), Tourism Life Cycle by Richard Butler (1980) and Tourism Carrying Capacity by Ainsley O'Reilly (1986), which warned of deterioration, antagonism and limits to constant tourist growth. *Overtourism*, since it was first defined (Ali, 2016), describes a situation linked to the above which has grown and spread. According to the definition adopted in the report "Overtourism? - Understanding and Managing Urban Tourism Growth beyond Perceptions" and endorsed by various tourism research centres and universities, Overtourism can be defined as "the impact of tourism on a destination, or parts thereof, that excessively influences perceived quality of life of citizens and/or quality of visitor experiences in a negative way" (p. 4). Meanwhile, Goodwin (2016) refers to overtourism as the opposite of "Responsible Tourism", which attempts to use tourism to create better places to live and visit. Tourism should not be seen as a burden on the host society. Managing tourism to benefit visitors and residents has always been a key issue for the sector. Using the UNWTO doctrinal definition of carrying capacity (UNWTO,1983) and, above all, the widely accepted definition offered by Mathieson and Wall (1982), this can be understood as "the maximum number of people that may visit a tourist destination at the same time, without causing destruction of the physical, economic, socio-cultural environment and an unacceptable decrease in the quality of visitors' satisfaction". Then and now, carrying capacity is a key concept for understanding sustainable tourism that can only be developed and managed by taking both visitors and local communities into account.

However, a hasty definition of overtourism can lead us to trivialise the problem. We cannot make the mistake of simplifying the concept of overtourism by considering it as a quantitative problem based simply on a specific number of excess visitors. This problem is more complex and must be addressed from a broader perspective, discarding stereotypes as a result of a hurried diagnosis. Table 1 summarises this idea, highlighting four reflections to better understand the phenomenon (UNWTO, 2018, 2019).

Another aspect that makes it difficult to analyse overtourism lies in a lack of a series of commonly accepted indicators to analyse the various different profiles. The study "Overtourism: impact and possible policy responses", a research document commissioned by the European Parliament Committee on

Table 1. Myths associated with overtourism (Prepared based on UNWTO 2018)

Type of problem	Main idea	Contextualisation of the problem
QUANTITATIVE PROBLEM	Mass tourism is not only related to the number of visitors, but also with the capacity to manage them	Some cities can manage a large number of visitors, while other have problems absorbing a much smaller number of visitors.
PROBLEM WITH TYPE OF USE	Mass tourism is not exclusively a problem of tourism.	Mass tourism can be perceived when the resources and infrastructures of a city, or part of the city, suffer excess pressure. However, not only tourists or other visitors use these services and infrastructures. Residents also compete to use the space and services with temporary visitors to the city.
PROBLEM OF SPACE	Mass tourism is generally very localised rather than a problem throughout the city	Tourist pressure is linked to the most popular areas of a city or its main attractions. But, it is possible to find areas with little or no visitors in even the most popular cities.
TECHNOLOGY PROBLEM	Technology or smart solutions are important but they cannot resolve the problem of mass tourism	Smart technologies are deemed to be the most effective to address managing congestion. Nevertheless, and despite the infinite potential benefits of these tools, they are not enough on their own.

Transport and Tourism (TRAN) (Peeters et al., 2018), analyses 290 regions, including 53 with at least one destination facing overtourism. The diagnosis in this paper highlights the Comunitat Valenciana, Andalusia and the Canary Islands as the three regions at the greatest risk of overtourism in Spain. The specific case of the Comunitat Valenciana is addressed with two pilot experiences in Valencia and Gandia as part of the ALTER-ECO project.

THE ALTER-ECO PROJECT

The main objective of the "Alternative tourist strategies to enhance the local sustainable development of tourism by promoting Mediterranean Identity" (ALTER-ECO) Project, approved by the "Interreg MED[1]" Programme, was to improve the balance between attracting tourism as a source of economic growth, and conserving the traditional model of the Mediterranean city as an example of sustainability. Improving the balance of tourist attractions directly refers to overtourism, attempting to seek strategies and measures that ensure an adequate visitor/resident ratio. Preserving the Mediterranean identity also affects this point indirectly. The classic Mediterranean city model essentially combines these two complementary concepts, making it more accessible as a human habitat, while fostering lower consumption of natural resources: sufficient levels of urban compactness and high complexity in the combination of uses and functions. Both refer to spaces rich in density and activities that require a suitable balance to avoid distortion.

It is important to emphasise the fragility of the Mediterranean city model and the crisis it is undergoing today. The text "*Problemas urbanos y paradigmas disciplinares en los territorios de la urbanística actual*", Font (2005) suggests six paradigms for updating the urban project based on personal research and on authors such as Giuseppe Dematteis, Francesco Indovina, Nuno Portas, Roberto Camagni and Gabriel Dupuy, among others. These paradigms are: generalisation of mobility and location of new activities; considering a polynuclear central structure; combination of uses compared to functional specialisation; urban improvement and refurbishment compared to new land use; free spaces changing from urban reserves to spaces for balance; and finally, the preference for a suitably compact city. Two of the six measures for updating the disciplinary discourse (suitably compact models and a combination of uses compared to specialisation) directly refer to the essence of the Mediterranean identity this project aims to reinforce. Meanwhile, statements are often found in some forums correlating compactness and complexity as synonyms for saturation or congestion. Compactness and complexity are not determined by the number of individuals or density of homes or visitors, rather they are based on diversity and the proximity and price of transport between built-up areas and people.

In this context, the ALTER-ECO project aims to test innovative measures in the six pilot areas[2] (Figure 1) to protect or promote the Mediterranean identity of cities, while seeking a balance between visitors and residents. The specific objectives established were:

1. Reduction and improved management of the impact of tourist activities on the environment, and in areas exceeding carrying capacity.
2. Improved social sustainability of Mediterranean tourist destinations by reinforcing and promoting the Mediterranean identity.
3. Improving cooperation, joint and integrated planning among public and private agents, and management of conflicts of interest, in order to create new business opportunities.

Figure 1. ALTER-ECO project pilot areas (Authors, 2019)

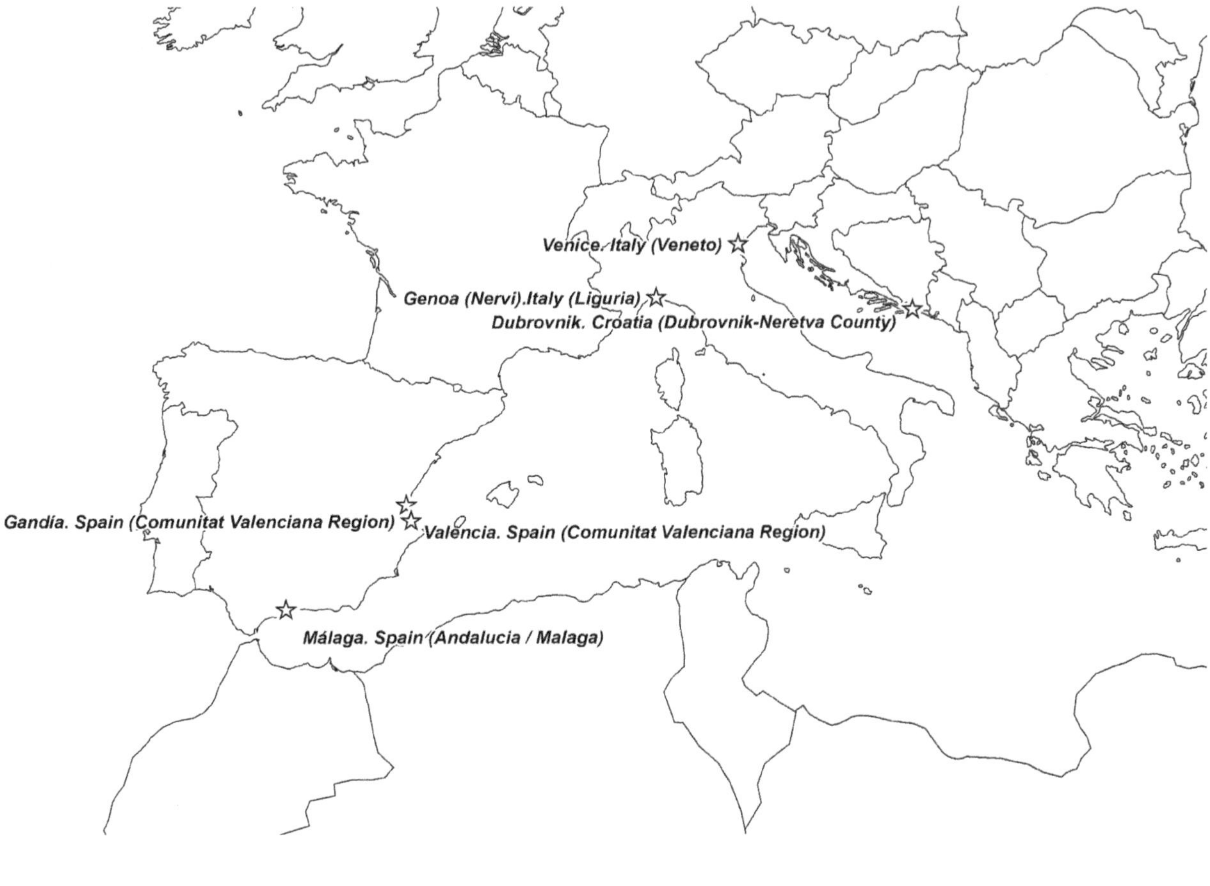

Characteristics of Other ALTER-ECO Project Pilots

It is interesting to at least have a brief history of the strategies implemented in the other ALTER-ECO project pilots so as to better contextualise the initiatives of the case study. Tables 2 and 3 offer a brief summary of the main characteristics of each project; the threats and weaknesses initially detected; and the measures developed to diversity the tourist offering and combat mass tourism.

Having presented the basic characteristics of the ALTER-ECO Project, we believe that the experience in Gandia (Spain) is an interesting case study that applies different measures and strategies to combat overtourism. The main objective of this chapter is therefore to describe, assess and debate the measures and strategies carried out in the case of the city of Gandia (Spain) to reduce mass tourism based on the ALTER-ECO Project experience. This chapter will end by verifying to what extent the strategies adopted in the case of Gandia to address overtourism coincide with or are different from the 11 strategies and measures to address visitor growth in cities (UNWTO, 2018, 2019).

Overtourism in Gandia (Case Study)

Gandia is a city in the Comunitat Valenciana, to the southeast of the province of Valencia and capital of the Safor region. The municipal district lies at the centre of the Gulf of Valencia and is 60.8 km^2. A significant aspect of the morphology of Gandia are its two urban areas: the Grau port area and the beach, and the city centre. They lie 3.2 km apart but tourist activity is mainly on the coast. The city centre, meanwhile, has most of the shopping areas, public offices, symbolic monuments and, therefore, is the venue for most cultural events.

Gandia has a registered population of 74,814, but the floating population is estimated at around 100,000-120-000, making it the 7th most populated city in the Comunitat Valenciana (due to its significant tourist influence in summer).

As for tourism, the predominant model in Gandia is "sun and beach" tourism. Tourist activity is concentrated mostly in the months of July and August, with lower occupancy levels in June and September.

The main weaknesses of the tourist model in Gandia are:

- Excessive concentration of tourists and tourist activity during the summer period.
- The effect of the progressive expulsion of local population as a result of increased housing prices (holiday homes) and the inconveniences of overcrowding due to tourism.
- A lack of connections between tourist areas and other parts of the city and the old town, which prevents dissipating tourist pressure and load especially in the beach area.
- One highlight of the Gandia tourist model is the high percentage of accommodation in apartments (the highest in the Comunitat Valenciana). This unique feature is accompanied by low accommodation quality and maintenance, which attracts low economic profile tourism.
- Deteriorated image of Gandia related to the television programme "Gandia Shore".
- Lack of real data on the influx of tourists and their behaviour, which hinders the design of effective strategies.

The main strengths include:

Table 2. Main characteristic the pilot site and weakness and threats emerged (Project ALTER-ECO. Testing Report, 2019)

Pilot Area	Main characteristic	Weakness and threats emerged in the preliminary study
Valencia and Gandia	Urban destination consolidated for leisure break trips with WHS sites (Valencia) and a classic sun and beach destination (Gandia).	• Lack of environmentally sustainable indicators tracking • Better positioning of rival Spanish destinations (Valencia) • Digital presence in Valencia is not homogeneous • Non-regulated lodgement offer • Excessive budgetary reliance of Valencia's City council • Seasonality, overcrowding and pressure in Gandia
Malaga	Urban tourist destination pivot of a strong sun and sea region (Costa del Sol).	• Slowness and apathy from the local administration (public spaces quality, new regulations, ...) • The need to expand and connect the tourist areas with other zones of the historic centre as well as with the city • House pricing in the historic center and population ageing • Not enough shops, businesses and services on offer during holidays that coincide with cruise arrivals • Overcrowding and pressure • Difficulty to abandon the over-dependence on the use of the car and to fully support alternative mobility schemes
Venice	Affected by overcrowding together with its Lagoon, is listed as a cultural WHS. On-going modifications of city structure	• Shortage of funding for restoration and urban maintenance • Reduction of the resident population in the historic city and in the other lagoon historical settlements. • Concentration of tourist flows • Inadequate awareness of the universal values of the Site • Irreversible change of the morphological characteristics of the lagoon eco-system, of its hydraulic dynamics, and tide levels • Transformed usage of the building heritage • Unlimited and uncontrolled tourist development causing damage to heritage properties due to excessive use
Dubrovnik	Historic WHS with old walled city suffering concentration of demand and offer. Increasing international demand. Modifications of morphological structure	• Inadequate differentiation of products and service standards. • Inefficient destination management model- lack of coordination in tourism sector • Lack of management plan with UNESCO WHS • Lack of clear mobility strategy in relation to tourism management • Mass tourism frustration and resistance by local society • Fragmented tourist offer and low cooperation level between stakeholders • Abandoning of destination management concept
Rhodes	After Crete, the island of Rhodes is the most visited destination in Greece. Long average of stays and high seasonality. Overdependance on T.O. activity. Medieval city is WHS	• Over-dependences (few markets, few operators, few months); • Increased competition; • The migratory crisis in the Aegean Sea has also affected Rhodes.
Nervi (Genoa)	The pilot of Genoa will be implemented in sea resort of Nervi (high quality destination)	• The centre and some areas of the city have become attraction points but there are some areas of the city undiscovered by tourism. • Need to improve the tourism in the city by working on city marketing, on social networks and on worldwide promotion.

- Comfortable climate for outdoor physical activity throughout the year due to suitable temperatures and relative humidity.
- Gandia beach has excellent physical and functional conditions. Good water quality for swimming.
- Natural resources of high ecological value, accessible and well-preserved (l'Ahuir beach, Mondúver mountains, Gandia reservoir, Parpalló-Borrell cave, etc.)
- Good air connections; excellent train (high speed) and road connections.

Table 3. Feasible tool(s) pilots may adopt (Project ALTER-ECO. Testing Report, 2019)

Pilot Area	Feasible tool(s) pilots may adopt
Valencia and Gandia	• ICT tools - counting people through Wi-Fi Access Points to detect mobile phones • ICT tools - update of a digital platform supporting • Method - incentives and promotion of alternative routes and point of interest • Deepening - tourism hospitality offer structure (quality labels development)
Malaga	• ICT tools - Mobile phone-led crowd management technique tests (including mobile enabled sensors). • Method - Decision Support System (DSS) • Deepening - B2B platform related to cruise schedule and other tourism events (from adapting opening times to better tourism management flows)
Venice	• Method - Decision Support System (DSS) • Method - develop ETIS indicators • ICT tools - Mobile phone-led crowd management technique tests (including mobile-enabled sensors).
Dubrovnik	• ICT tools - counters through cameras (including Wi-Fi equipment) and cloud system to recollect, analyze and share results in real time. • ICT tools - system dedicated to mobility of tourists (connections more efficient and more sustainable) • Method - Decision Support System (DSS) • Deepening - daily cruise tourism relevance vs. general mass tourism
Rhodes	• ICT tools - system dedicated to mobility of tourists (connections more efficient and more sustainable) • Method - develop ETIS indicators • Deepening - daily cruise tourism relevance vs.sun sea tourism
Nervi (Genoa)	• Tourist Flow Model (TFM) for understanding how to improve the local public transport offer. • ICT tools - hi-quality system to integrate local offer (from 6 public museum to 4 historical parks) • ICT tools - Augmented reality for natural offer (cultural heritage and botanical collections) • Method - develop ETIS indicators

Reports

The ALTER-ECO project has launched innovative measures in the city of Gandia. Before their implementation, and in order to gain sound knowledge of the relationship between tourist activity -the city's main economic source- and conserving the classic Mediterranean city model as an example of economic, environmental and social sustainability, two technical reports were prepared.

The Report on the Analysis of the city of Gandia was prepared by the Universitat Politècnica de València as it is associated with the city in matters related to tourism. It consisted in analysing the current situation of tourism in the city of Gandia from three areas:

- Compilation of tourist indicators.
- Study of accommodation on offer, focusing particularly on privately-owned apartments offered as tourist accommodation.
- Analysis of companies working locally that could be interested in managing privately-owned apartments.

The report findings revealed the benefits of having a flexible system of sustainable tourism indicators that can adapt to market fluctuations; the annual update of indicator system data to assess evolution and trends in the destination; and, finally, to motivate the administration to implement and develop a system of indicators that local companies see as an opportunity to improve the quality of their services.

The second technical report contained advice on developing the Mediterranean identity in Gandia. It was drafted by researchers from the Universitat de València, who focused on two areas:

- Determining the concept of transferable Mediterranean identity and diagnosis of Gandia.
- Determining territorial resources in Gandia that create a Mediterranean identity.

The report established the follow key factors that must be included in a transferable Mediterranean identity: geographic aspects/climate; Mediterranean diet; cultural interaction; source of openness/mobility; poetic sense of life, architecture and landscape; hospitality; family and friends as key groups and a way of life. The report also offered specific recommendations for managing Gandia as part of the Mediterranean Identity.

The ALTER-ECO project has attempted to develop a series of measures which lead to creating a better connection between the beach area (where tourism is currently concentrated) and other areas of the city, such as the old town or some highly valuable natural landscapes found in the district. Measures included monitoring influx of people at Gandia beach (phase one and phase two), Free Wi-Fi at Gandia Beach, and the Gandia Tour & Play mobile application. Figure 2 describes the time and duration of each measure implemented.

Monitoring

The first measure aimed to act on seasonality in the beach area during the summer months. It consisted in monitoring the area indicated in order to obtain data on volume of visitors to the beach area and see which measures could be proposed to end seasonality in tourism, proposing Gandia City Centre as a destination to complement sun and beach tourism.

This problem of seasonality, which even leads some businesses to close throughout the rest of the year, is not inherent to Gandia (Aguiló & Saster, 1984; Ramón & Abellán, 1995), but has also been detected in other cities around the Mediterranean Basin. Studying this phenomenon is especially interesting as solutions may be replicated.

Figure 2. Timeline of the measures (Authors, 2019)

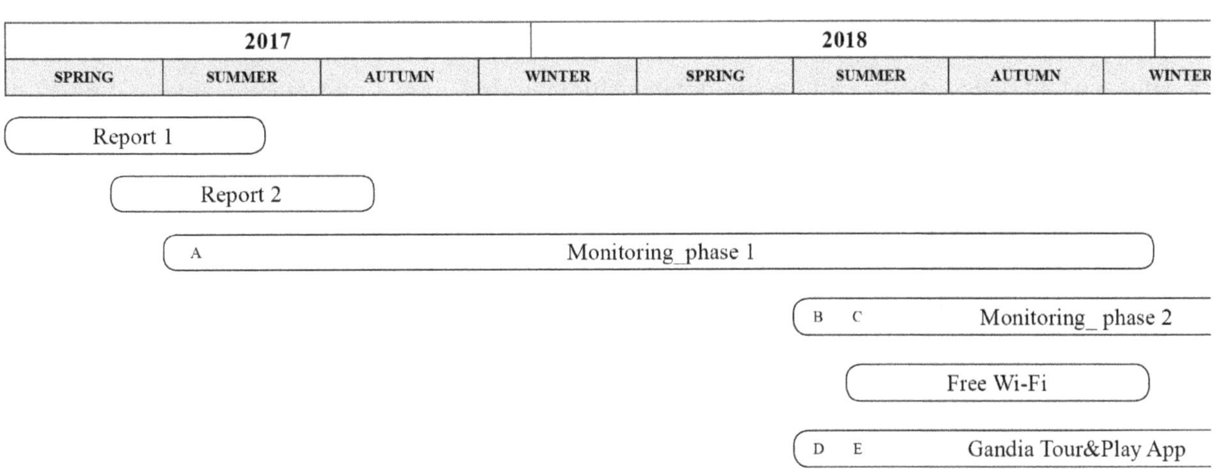

Monitoring was conducted in two phases; the first 1-year phase only concerned Passeig Maritim de Neptú in Gandia, along a 2 km stretch of the beachfront promenade. Five devices were placed for monitoring based on the availability of lampposts nearby and in an attempt to spread the devices evenly so as to obtain a significant sample of the number of people along the promenade. These data were obtained using the visitors' mobile devices and Smartphones.

The second monitoring phase consisted in installing six more data collection points in Gandia City Centre, and connecting equipment at both locations to obtain data on volume of visitors and verify whether visitors move between the two different areas of study.

The first monitoring phase identified and recorded user and tourist mobile devices. The data obtained were the number of visitors detected during different time slots and distribution of number of visitors according to time (e.g., < 5min, 5-15 min, etc.). On the other hand, data obtained during the second phase were more specific as not only were the data of greater quality (more reliable and stable) but no data were lost when there was a disconnection, unlike the first phase. The equipment could store the information collected and transfer it to a server (http://rapido3.uv.es:8020) when the connection was restored.

Data obtained were divided into 5-minute intervals and monthly reports showed:

- Occupancy by location: These graphs represent daily occupancy for each location by week. On them we can observe morning and afternoon peaks, as well as the different behaviour on weekdays and at the weekend.
- Total daily occupancy per location: A matrix shows total occupancy per location and day.
- Stay graphs: These graphs show stay times in absolute figures. Three stay ranges were established: People who stay under 15 minutes, people who stay between 15 minutes and 4 hours and, finally, those who stay more than 4 hours.
- Daily origin - destination matrices: A table is shown with the origin-destination matrix between all study locations. Each box on the table shows two figures: the figure in percentage of the matrix and the absolute figure. This absolute figure is the number of devices detected with the origin and destination established by the matrix pair.
- Loyalty table: These tables show the general average stay of visitors, in days. In other words, how many days a unique visitor has passed by one of the sensors. Most visitors usually stay between one and three days.

Using new technologies in monitoring equipment at Gandia beach and city centre enabled us to discover volume of visitors and whether or not there is mobility between both areas. One of the main conclusions obtained was the care required when studying diversification strategies due to the risk of transferring the problem to another area. The case of the Florence (Generalitat Valenciana. Turisme, 2019), with the Data Science for Social Good Europe 2017, is a good example of how the use of new technologies and monitoring saturated areas can allow us to diversify offerings by informing visitors of the busiest areas on their mobiles, proposing alternative routes or spaces.

Free Wi-Fi at Gandia Beach

The second measure was a result of the first. Internet connectivity and hotspots were installed at Gandia beach based on the existing sensor and influx meter infrastructure installed by the ALTER-ECO project.

This consisted of including a public Free Wi-Fi network at five points along the promenade so as to offer free Internet access to beach visitors and residents so they could obtain tourist information.

The purpose of this action was to transfer sun and beach tourists to the city centre, informing them of a different tourist offering on their mobile devices and smartphones, as well as promoting a visit to the city of Gandia outside the summer months. One of the goals of the Free Wi-Fi network was to offer neighbours and visitors to the beach direct access to municipal tourist information despite the limited Internet service capacity. Specifically, information was offered on tourist attractions in Gandia city centre and transport options to the area. The action also aimed to obtain big data on tourists at the beach to complement them with details of their origin and information on mobility and influx.

This second measure provided Internet connectivity for Gandia beach and promenade by installing a dedicated radio link with unrestricted data consumption in accordance with Spanish legislation on bandwidth limitations established by the CNMC (National Commission on Markets and Competition) (Figure 3). The Wi-Fi network installed at the five points had two purposes: a) to control visitor's connection to the Wi-Fi system, managing the band width and connection times of each user, and b) to have a captive portal with an access screen featuring a homepage image to advertise the cultural offer related to the cultural heritage of the Borja, then leading to the Gandia tourism website.

Figure 3. Gandia Beach Free Wi-Fi Network (ALTER-ECO, 2019)

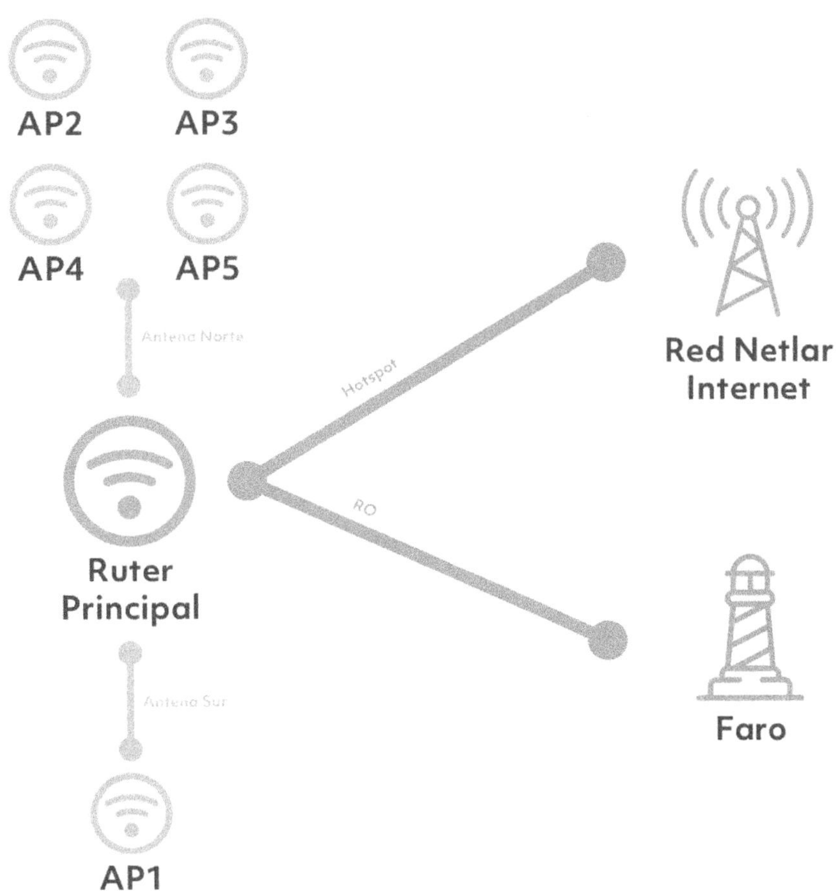

Gandia Council Tourism Department took part in this action from the outset in light of the possibility of improving tourist offerings, and accepted it as a response to a real demand for extending services at the beach to make it more attractive and competitive. With this measure, Gandia attempted to overcome the excess residents and visitors at the beach during the summer. In its attempt to reposition the tourist destination, Gandia has progressed in its modernisation and tourist innovation. As a tourist destination it not only aspires to be a welcoming destination, but also aims to make a visitor's stay easier and more comfortable with access to new technologies. Adapting Gandia to the smart city model will help it stand out from other nearby tourist destinations.

Measure 2 provided data offering daily information on:

- Identification of Free Wi-Fi users' countries of origin (Figure 4)
- Number of network accesses (Figure 5)

Figure 4. Users' countries of origin (August 2018) (Authors, 2019)

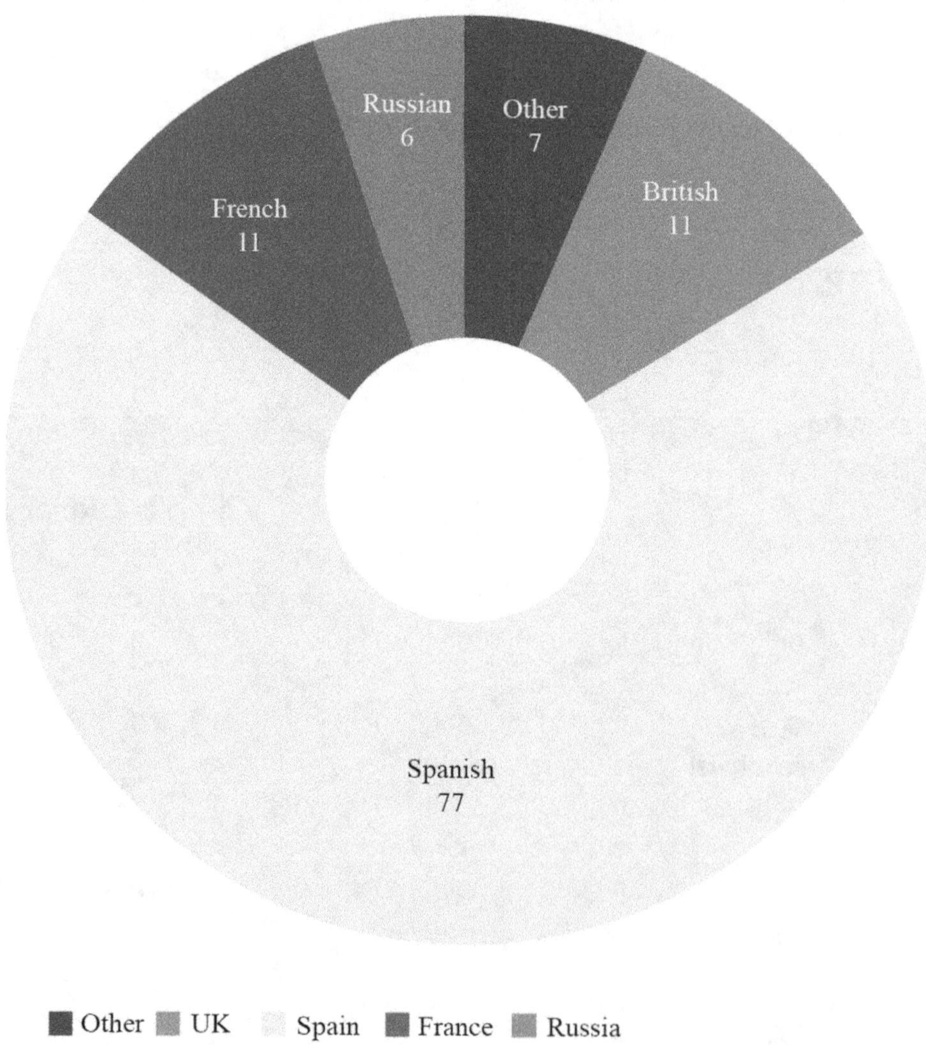

Figure 5. Number of network accesses (August 2019) (ALTER-ECO, 2019)

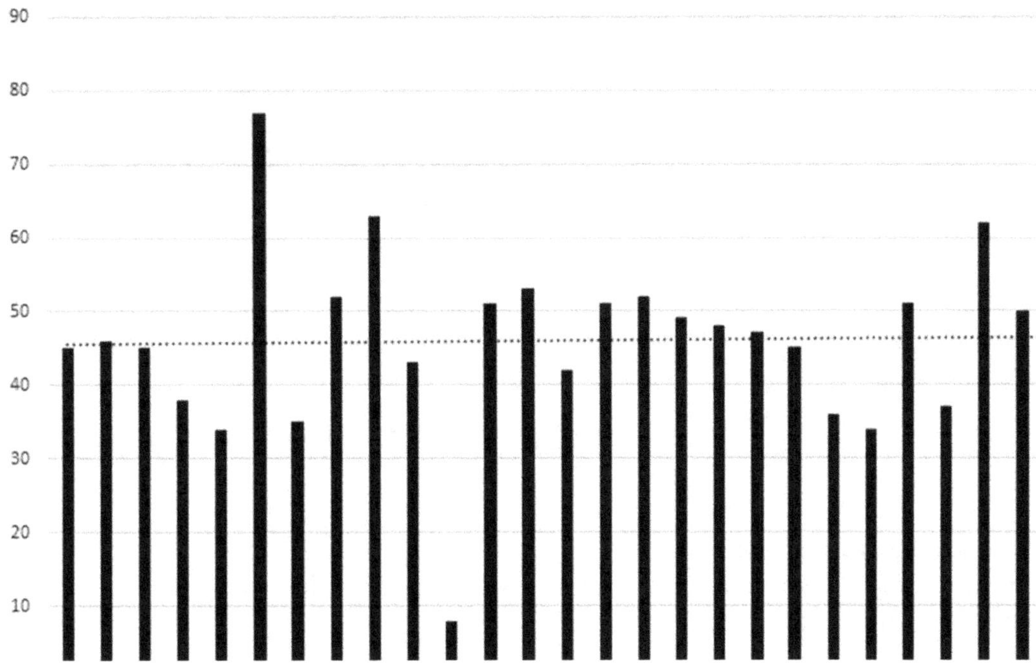

Data collected during the time the measure was active have helped make effective decisions on tourist strategy planning in Gandia to alleviate the effects of overtourism in the beach area. And also to improve the visiting experience for tourists and residents at Gandia beach.

Gandia Tour&Play App

The Gandia Tour&Play app was designed to take tourists visiting Gandia from higher density areas during the summer period (the beach) to areas with more natural and cultural attractions and less density (l'Ahuir beach and the old town). This app is based on the ALTER-ECO app, designed by start-up Play&Go Experience for the city of Valencia, also as part of the ALTER-ECO project. But for Gandia the app was adapted to the specific demands and needs of the Gandia Council Tourism Department. A promotional campaign was conducted on Google Ads during the app launch and dissemination period; posters, leaflets and banners were designed (Figure 6).

The Gandia Tour&Play app uses augmented reality and gamification techniques to guide visitors around three areas of Gandia: the beach, the old town and the natural environment of l'Ahuir beach, in order to make tourist activity more dynamic in these areas. The app dynamics are based on geolocating tourist resources, heritage and infrastructures in the three selected areas in the city, using game dynamics to encourage visitors to discover them. Tourists visit the locations using 3D maps they can use as a guide and to locate missions spread around the places to visit, completing these missions to win prizes. The gaming experience offers users:

- A complete digital guide to the most important points of interest on the map.
- Information on the main events with a geolocated agenda.

Figure 6. Gandia Tour&Play app banner (ALTER-ECO, 2019)

- Interaction with the location.
- They can discover the town in a different way with sustainable tourism.
- Visit gamification.
- Complete missions with different objectives, and win prizes.

The results obtained by implementing this measure offer different types of information:

- Data on space. Geolocated information shared by visitors on social networks and the areas visited can be represented on maps.
- App downloads details. The online store provides information on the number of users using the app as a geolocated tourist information tool.

Debate and Recommendations

The measures implemented in Gandia have shown that predominant tourism model causes excessive concentration of tourists and tourist activity during the summer. Its main weakness is the lack of connections between tourist areas and other parts of the city and the old town, this prevents tourist pressure and load being dissipated, especially in the beach area.

The ALTER-ECO project has provided a boost for Gandia to correct its tourism model. Based on the concepts of monitoring, free Wi-Fi and gamification we propose a debate on the actions undertaken and establish recommendations for improvement.

Monitoring

The purpose of monitoring spaces is to obtain data on volume of visitors in certain areas or spaces of a city. In the specific case of Gandia, the aim of the second phase proposed was to take a step further in the search for strategies to redistribute tourist flow to other areas that are less crowded than the beach, leveraging the rich heritage in Gandia city centre. However, one aspect not considered and which is essential for achieving a more comprehensive and targeted action is to also make residents and locals aware of these measures. They must be informed of the steps and measures the destination is taking to manage tourist flows. This is an important detail to promote participation and social cohesion, while also agreeing capacity for psychosocial support (TCV, 2019). Meanwhile, Peeters et al. (2018), in their work for the European Parliament Committee on Transport and Tourism (TRAN), highlight monitoring tourism as one measure to alleviate the impact of tourism. However, they specify that "Such methods should include not only volume and tourism density/intensity measures, but also measures related to data collection on the number of tourists and day-visitors, Airbnb and other new forms of accommodation and transport mode shares" (p. 112). Nevertheless, the monitoring action carried out in Gandia did not take these measures into account, which is one factor to be considered in future actions.

Free Wi-Fi

To understand the impact of information and communication technologies (ICT) on the relationship between businesses and tourists in Gandia as a tourist destination, we must consider concepts such as "smart tourism" and "smart destination", (Ivars Baidal et al., 2018a). Implementing Free Wi-Fi in Gandia has contributed to transforming the behaviour of "smart tourists". We have observed what Femenia-Serra (2019) calls critical managerial implications for both destination management organisations and the companies that make up the smart tourism ecosystem (STE).

In light of the unquestionable collaboration of public and private companies in an STE, one of the fundamental requirements is to provide top-quality free public Wi-Fi so that residents, visitors, companies and administrations can communicate (SEGITTUR, 2015). The success stories of Palma and Benidorm (Spain) have made this type of communication possible with mobile devices permanently connected to the Internet (Ivars Baidal et al., 2018b).

Gandia's evolution towards a smart city model must seek a large number of services in the urban environment. Most of these services are based on omnipresent, continuous and real time access to the information of users on the move, as well as a continuous exchange of data between millions of individual devices in the urban area. Rivano (2018) indicates the need to have different wireless access networks to empower future smart cities. These networks must be complementary and interoperable. This issue could not be specified in the case of free Wi-Fi in Gandia.

Tourists want Wi-Fi connection from the moment they arrive at their destination, although they often complain about how slow and unreliable it is. Research into the hotel sector (Mellinas, 2017) indicates that speed seems to be the most important aspect determining how satisfied travellers are with Wi-Fi. Although the relationship between connection speed variables and user satisfaction is very weak. The lack of success of Free Wi-Fi at Gandia beach suggests we should seriously consider questions that are complementary to proper management, supervision and maintenance of technology equipment.

Gamification

One side of the ALTER-ECO Project in Gandia worth highlighting is that it fosters *m-tourism*. Tourists use mobile applications to get information on their travels before, during and after their trip. In this sense, Saura et al., (2017) recommend applying a marketing approach to designing and developing this type of application. As tourists-consumers are better informed, they oblige companies to offer products and services that adapt to their needs, tastes and preferences. Soares et al. (2019) indicate that real time

IoT content notification systems have security limitations and require a greater band width to support multiple communications. This could not be implemented in Gandia due to the short time the mobile app has been operating.

The evolution of Information and Communication Technologies has led to the development of mobile devices and applications to better manage, promote and market tourism. Gamification applied to tourism has contributed to this (Xu et al., 2017). Weber (2014) classified different types of gamification practices. As there were augmented reality apps and bonuses were given for taking part, the case of Gandia is included in the category of gaming-based apps focusing on knowledge of local heritage and city spaces with a gamified tourist guide.

Gamification in tourism is an opportunity to design memorable experiences for visitors which can influence their mood or cause an intrinsic motivation with the visit (Bulencea & Egger, 2015). It also contributes to a more gratifying interaction and greater tourist satisfaction, as well as increasing brand knowledge and loyalty to the destination (Xu et al., 2017). Gandia has two tourist mobile applications: Gandia Tour&Play and Territori Borgia. They have been used as a starting point for expanding tourism industry support functions with significant potential. Based on this possibility, Gandia must equip administrations and local businesses to create policies based on advanced knowledge and services for tourists in order to correct imbalances caused by overtourism. Lamsfus et al. (2015) suggests it is the destination's "intelligence". This is not only related to intensely deploying information and communication technology infrastructures, but associates it to using technology to seek a deeper understanding of the characteristics and significance of human mobility.

FUTURE RESEARCH DIRECTIONS

Launching the three strategies: monitoring, Free Wi-Fi and gamification, in Gandia has offered a series of lessons learned so as to amend or reconsider some future lines of work, which will be applicable not only to Gandia, but also to other cities with similar problems.

- Applying new technologies such as monitoring sensors, Wi-Fi networks or mobile apps has involved technology companies, start-ups or businesses external to the traditional tourist sector. Medium-sized cities often have insufficient and/or inadequate companies to carry out this type of specialised tasks. A regional server or directory would be advisable for registering companies with guaranteed quality certificates so that local administrations could contact them for this type of work.
- International, national and regional legislative restrictions determine that contracts must normally be awarded to the best economic bid, with experience and/or technical quality as secondary criteria. Therefore, inexperienced companies often win the contract, entailing inefficient management and low maintenance. Again, this problem could be solved with a platform registering companies that comply with minimum requirements of experience, equipment and guarantees.
- Involving residents and the local trade and business sector is one key point for the success of the strategies carried out, avoiding the problem of *tourismphobia*. Residents must be invited to take part in the process, applying policies that motivate them with direct benefits and tax advantages. Traditional or local trade must also be involved in the policies implemented so they can generate new business opportunities related to visitors.

CONCLUSION

Developing and applying the ALTER-ECO project in the pilot city of Gandia has determined to a varied and complementary type of results of the real scope the measures implemented had while they lasted. Below we describe the results obtained as proposals which, although they confirm to a great extent compliance with the objectives proposed in the ALTER-ECO Project, must also be seen as proposals for action against overtourism, which affects the city of Gandia and numerous similar tourist destinations along the coast of the Comunitat Valenciana.

- Use of new communication technologies. Implementing the measures in Gandia enabled us to test new communication technologies such as: measurement sensors, public Wi-Fi network and mobile apps in the scope of tourist activities. This is part of a regional government drive to promote innovation in tourist destinations by converting them into "Smart Tourist Destinations". And also the purpose of the ALTER-ECO project which, in its pilot in the Comunitat Valenciana and more specifically the city of Gandia, has equipped two urban areas with this type of technology (old town and beach). The first step in achieving the goal of becoming a Smart Destination.
- Data collection. Information on the number, mobility, frequency and influx of tourists in the defined areas generates datasets that make up a tourist information system. Processing this information properly offers regional and local officials in charge of organising and planning tourism greater control over problems related to excess tourism. Measuring by monitoring influx, controlling access to public Wi-Fi, or the use of tourist apps and their synchronisation with social networks, tells us, although somewhat inaccurately, about the real behaviour, patterns and conduct of tourists. Big data offers an overview of the spatial organisation of tourists, almost in real time, to detect areas of greater tourist intensity, reduce loads and promote the end of seasonality. Data from Gandia reveal that the Marquesa area is a potential attraction which, if developed, will help redistribute tourist load and end the seasonality of the destination.
- Social education. The measures implemented are based on the use of mobile devices to amplify the range of communication channels between visitors, residents, tourist companies and the city through its public institutions. This progressively consolidated Gandia as a "Smart tourist destination" while offering a direct communication channel between the destination and tourists. Offering instant access to the cultural agenda, tourist activities and services advertised, and simultaneously the chance to leave comments and opinions in real time, empowers the tourist consumer, putting them on a level with residents and neighbours. The measures implemented have made it possible to influence the conducts and behaviours, and alter the habits, of residents and visitors by improving leisure and recreation offers of cultural, landscape and environmental interest in other tourist areas. This is not only a measure to relieve some areas in favour of others. It also creates new alternative routes and, together with the geolocated cultural agenda, offers tourists and residents a new way of discovering the city.
- Business opportunities. The ALTER-ECO project had access to know-how from tourist companies and start-ups for redesigning alternative routes or adapting tourist apps to specific cases. This has promoted creativity in the sector and introduced design and tourist sector management companies, as well as companies specialising in reading and interpreting gross tourist data to convert them into relevant information or recommendations for public administrations. And also collaboration with public universities on initial issues of consultancy and drafting reports, and

subsequent research from studying and monitoring data obtained during the project. This type of public-private collaboration can not only catalyse and boost institutional collaboration, but they open up new business opportunities to promote the industry associated with the destination.

- New Governance. ALTER-ECO project measures have shown that public and local administrations are committed to collaborating to set up, manage and maintain technical equipment and prepare content. One example of this is monitoring influx and developing mobile applications. The direct involvement of regional and local government in designing and developing measures was key to their success. In Gandia, the council was directly involved in promoting and launching the measures, offering its coverage and perspective from the outset; this factor helped improve the usefulness of the results.

Finally, we wanted to conduct a double comparison, corroborating the measures and strategies carried out in the case study of Gandia with the other pilot areas of the ALTER-ECO Project and, in turn, with the proposed strategies and measures developed by UNWTO (2018) and which in a short time have become a reference for addressing problems of overtourism. In Table 4, of the 11 original strategies we have selected those found in each of the pilot areas. Of the 68 original measures, Table 5 highlights those applied in the pilot areas. From this we can draw the following conclusions:

We can identify 8 strategies and a total of 19 measures developed in the pilot areas. Of these, 7 strategies and 16 measures were developed in Gandia.

The strategies most developed were Strategy 2. Promote time-based dispersal of visitors; Strategy 3. Stimulate new visitor itineraries and attractions; Strategy 3. Stimulate new visitor itineraries and attractions y Strategy 11. Set monitoring and response measures. These strategies aim to reduce seasonality and reduce city hotspots by dispersing visits through the use of new technologies.

The least-used strategies were Strategy 5. Enhance visitors' segmentation; Strategy 8. Improve city infrastructure and facilities and the Strategy 9. Communicate with and engage local stakeholders. The least-used strategies can be summarised as joint actions to change traffic according to tourist influx, and organising communication and participation platforms with communities and agents.

Overall we can say that the ALTER-ECO Project is well-aligned with the strategies and measures to address visitors' growth in cities developed by UNWTO (2018), and that Gandia and Valencia in particular are the pilot areas that used a greater diversity of strategies.

Table 4. Compliance in pilot areas with strategies and measures to address visitors' growth in cities. (Prepared based on UNWTO 2018)

Pilot Area	St.1	St.2	St.3	St.4	St.5	St.6	St.7	St.8	St.9	St.10	St.11
Gandia	X	x	X			X	x			x	X
Valencia a	X	X	X			x	X			X	X
Malaga	X	X	X								X
Venice				X							X
Dubrovnik	X	X	X	X							X
Rhodes	x	X	x								
Nervi (Genoa)											X

Table 5. Strategies and measures to address visitors' growth in cities. Adaptation to the pilot areas of the ALTER-ECO Project. (Prepared based on UNWTO 2018)

Strategies	Measures
Strategy 1 Promote the dispersal of visitors within the city and beyond	→ Develop and promote visitor attractions and facilities in less visited parts of the city and in its surroundings → Create joint identity of city and its surroundings
Strategy 2 Promote time-based dispersal of visitors	→ Promote experiences during off- peak months → Stimulate events in off-peak months → Use new technologies (apps and others) to stimulate dynamic time-based dispersal
Strategy 3 Stimulate new visitor itineraries and attractions	→ Promote new itineraries at the city entry points and through the visitor's journey, including at tourist information centres → Produce city guides and books highlighting hidden treasures → Stimulate development of guided tours through less-visited parts of the city → Develop virtual reality applications to famous sites and attractions to complement onsite visits
Strategy 4 Review and adapt regulation	→ Define the carrying capacity of the city and of critical areas and attractions etc. → Ensure visitors use parking facilities at the edge of city
Strategy 6 Ensure local communities benefit from tourism	→ Engage local communities in the development of new tourism products → Conduct an analysis of supply-demand potential of the local communities and promote their integration in the tourism value chain → Improve quality of infrastructure and services considering residents and visitor
Strategy 7 Create city experiences that benefit both residents and visitors	→ Develop tourism experiences and products that promote the engagement of residents and visitors
Strategy 10 Communicate with and engage visitors	→ Educate visitors on local values, traditions and regulations → Provide adequate information about traffic restrictions, parking facilities, fees, shuttle bus services, etc.
Strategy 11 Set monitoring and response measures	→ Monitor key indicators such as seasonal fluctuations in demand, arrivals and expenditures, patterns of visitation to attractions, visitor segments, etc. → Advance the use of big data and new technologies to monitor and evaluate tourism performance and impact

The United Nations New Urban Agenda and the 17 Sustainable Development Goals, more specifically goal no. 11 "Make cities and human settlements inclusive, safe, resilient and sustainable", must be a priority for both cities and tourist destinations in the coming years. Projects such as ALTER-ECO contribute to achieving these goals, stressing the need for balance between a destination's identity and its capacity to welcome and distribute visitors with a diversified offer.

ACKNOWLEDGMENT

The project ALTER-ECO has been led by the Ministry of Housing, Public Works and Territorial Structuring, through the Valencian Institute of Building (IVE). This research has been supported by the IVE.

REFERENCES

Aguiló, E., & Sastre, A. (1984). La medición de la estacionalidad del turismo: El caso de Baleares. *Estudios Turísticos*, *81*, 79–88.

Ali, R. (2016). *Exploring the Coming Perils of Over tourism.* Skift. Retrieved July, 7, 2019 from https://skift.com/2016/08/23/exploring-the-coming-perils-of-overtourism/

Auken, M. (2009). *Informe sobre el impacto de la urbanización extensiva en España en los derechos individuales de los ciudadanos europeos, el medio ambiente y la aplicación del Derecho comunitario.* Parlamento Europeo, Comisión de Peticiones, 20. Estrasburgo, Francia. Retrieved from http://www.europarl.europa.eu/sides/getDoc.do?pubRef=-//EP//TEXT+REPORT+A6-2009-0082+0+DOC+XML+V0//ES

Bulencea, P., & Egger, R. (2015). *Gamification in tourism: Designing memorable experiences.* Norderstedt: Bod. Books on demand.

Butler, R. W. (1980). The concept of a tourist area cycle of evolution: Implications for management of resources. *Canadian Geographer, 24*(1), 5–12. doi:10.1111/j.1541-0064.1980.tb00970.x

CEC (Commission of the European Communities). (2002). *Recommendation of the European Parliament and Council of 30 May 2002 concerning the implementation of integrated coastal zone management in Europe 2002/413/EC.*

Colaninno, N., & Roca Cladera, J. N. (2012). Un desarrollo metropolitano policéntrico: dispersión y compactación en los procesos de crecimiento urbano, a escala territorial. In F. Gaja-Díaz (Ed.), Deconstruction. La desconfiguración del litoral mediterráneo español. Colección Manual de referencia (pp. 14-75). Valencia, Spain: Editorial UPV.

Deery, M., Jago, L., & Fredline, L. (2012). Rethinking social impacts of tourism research: A new research agenda. *Tourism Management, 33*(1), 64–73. doi:10.1016/j.tourman.2011.01.026

Doxey, G. (1975). A Causation Theory of Visitor–Resident Irritants: Methodology and Research Inferences. In *Proceedings of the Impact of Tourism Sixth Annual Conference* (pp.195-198). San Diego, CA: The Travel Research Association.

España, G. (2018). *A toda costa.* Retrieved from https://es.greenpeace.org/es/sala-de-prensa/informes/a-toda-costa/

Femenia-Serra, F., Neuhofer, B., & Ivars-Baidal, J. (2019). Towards a conceptualisation of smart tourists and their role within the smart destination scenario. *Service Industries Journal, 39*(2), 109–133. doi:10.1080/02642069.2018.1508458

Font, A. (2005). Problemas urbanos y paradigmas disciplinares en los territorios de la urbanística actual. In A. Font, M. Corominas, & J. Sabaté (Eds.), *Los territorios del urbanista, 10 años del Master en Proyectación Urbanística*. Barcelona, Spain: Ediciones UPC.

Gaja-Díaz, F. (Ed.). (2012). *Deconstruction. La desconfiguración del litoral mediterráneo español. Colección Manual de referencia.* Valencia, Spain: Editorial UPV.

Goodwin, H. (2016). *Responsible tourism: Using tourism for sustainable development.* London, UK: Goodfellow Publishers. doi:10.23912/978-1-910158-84-5-3101

Hall, P. (1970). Horizon of hotels. *New Society, 15*(389), 389–445.

Ivars-Baidal, J., Femenia-Serra, F., Giner Sánchez, D., & Celdrán-Bernabeu, M. (2018a). *Las OGDs ante el ecosistema turístico inteligente: oportunidades y riesgos*. Paper presented at the 2nd UNWTO World Conference on Smart Destinations. Oviedo (Spain), June 25-27, 2018.

Ivars-Baidal, J., Navarro, E., Celdrán, M., Perles, J., & Pérez-Medina, M. (2018b). El enfoque integral de los destinos turísticos inteligentes (DTI) en áreas litorales: alcance, progresos y limitaciones. In *Actas IV Congreso Ciudades Inteligentes* (pp. 471-476) Madrid, Spain: Tecma.

Lamsfus, C., Martín, D., Alzua-Sorzabal, A., & Torres-Manzanera, E. (2015). Smart Tourism Destinations: An Extended Conception of Smart Cities Focusing on Human Mobility. In I. Tussyadiah, & A. Inversini (Eds.), *Information and Communication Technologies in Tourism. Proceedings of the International Conference in Lugano, Switzerland*, February 3-6, 2015 (pp. 363-375). Springer International Publishing. 10.1007/978-3-319-14343-9_27

Mathieson, A., & Wall, G. (1982). *Tourism: Economic, Physical and Social Impacts*. London, UK: Longman.

Mellinas, J. P. (2017). La relativa importancia de la velocidad en las conexiones WI-FI. In J. F. Vera-Rebollo, J. A. Ivars-Baidal, & M. A. Celdrán Bernabeu (Eds.), Actas del Seminario Internacional Destinos Turísticos Inteligentes: nuevos horizontes en la investigación y gestión del turismo. (pp. 38-51) Sant Vicent del Raspeig: Publicacions de la Universitat d'Alacant. doi:10.14198/Destinos-Turisticos-Inteligentes.2017.02

Milano, C. (2018). Overtourism, malestar social y turismofobia. Un debate controvertido. *Revista Pasos*, *16*(3), 551–564. doi:10.25145/j.pasos.2018.16.041

Milano, C. (2019). *Ciudades y turismo urbano: ¿una relación indestructible?* Barcelona, Spain: Ostelea School of Tourism & Hospitality.

Miralles, J. L. (2014). *El darrer cicle immobiliari al País Valencià o El progrés de la misèria*. Valencia, Spain: Fundació Nexe.

Mullins, P. (1991). Tourism urbanization. *International Journal of Urban and Regional Research*, *15*(3), 326–342. doi:10.1111/j.1468-2427.1991.tb00642.x

O'Reilly, A. M. (1986). Tourism carrying capacity: Concept and issues. *Tourism Management*, *7*(4), 254–258. doi:10.1016/0261-5177(86)90035-X

Peeters, P., Gössling, S., Klijs, J., Milano, C., Novelli, M., Dijkmans, C., ... Postma, A. (2018). *Research for TRAN Committee - Overtourism: impact and possible policy responses*. Brussels, Belgium: European Parliament, Policy Department for Structural and Cohesion Policies.

Perkumienė, D., & Pranskuunienė, R. (2019). Overtourism: Between the Right to Travel and Residents' Rights. *Sustainability*, *11*(7), 21–38. doi:10.3390u11072138

Postma, A., & Schmuecker, D. (2017). Understanding and overcoming negative impacts of tourism in city destinations: Conceptual model and strategic framework. *Journal of Tourism Futures*, *3*(2), 144–156. doi:10.1108/JTF-04-2017-0022

Ramón, A., & Abellán, M. J. (1995). Estacionalidad de la demanda turística en España. *Papers de Turisme*, *17*, 45–73.

Rivano, H., Augé-Blum, I., Bechkit, W., Boussetta, K., Fiore, M., Stanica, R., & Valois, F. (2018). Wireless Access Networks for Smart Cities. In I. Management Association (Ed.), Smart Technologies: Breakthroughs in Research and Practice (pp. 476-507). Hershey, PA: IGI Global. doi:10.4018/978-1-5225-2589-9.ch022

Saura, J. R., Palos-Sánchez, P. R., & Reyes-Menéndez, A. (2017). Marketing a través de aplicaciones móviles de turismo (m-tourism): Un estudio exploratorio. *International Journal of World of Tourism*, *4*(8), 45–56. doi:10.12795/IJWT.2017.i08.04

Segittur. (2015). *Informe destinos turísticos inteligentes: construyendo el futuro*. Retrieved from: https://www.segittur.es

Shoval, N. (2018). Urban planning and tourism in European cities. *Tourism Geographies*, *20*(3), 371–376. doi:10.1080/14616688.2018.1457078

Simancas, M., & Peñarrubia, M. P. (2019). Analysis of the Accommodation Density in Coastal Tourism Areas of Insular Destinations from the Perspective of Overtourism. *Sustainability, MDPI. Open Access Journal*, *11*(11), 1–19.

Soares, N., Vairiinhos, M., Beça, P., & Ribeiro, T. F. (2019). Real-Time Notifications: Marketing Applied to the Internet of Things. In D. Simões, B. Barbosa, & S. Filipe (Eds.), *Smart Marketing with the Internet of Things* (pp. 114–122). Hershey, PA: IGI Global; doi:10.4018/978-1-5225-5763-0.ch006

TCV. (2019). *Afrontando los retos del overtourism a través de open/big data. Proyecto Territorial Integrado "HERIT-DATA"*. Retrieved from http://www.turisme.gva.es/turisme/es/files/pdf/HD_Informe_ejecutivo_BENCHMARKING.pdf

Tourism, R. (n.d.). *OverTourism*. Retrieved from http://responsibletourismpartnership.org/over tourism.

Urban, T. U. R. (2016). *Monitor de competitividad turística de los destinos urbanos españoles*. Retrieved from https://www.exceltur.org/urbantur/#

Valenciana, G. T. (2019). Afrontando los retos del overtourism en destinos turísticos patrimoniales y naturales a través de open/big data. Valencia: Generalitat Valenciana.

Weber, J. (2014). Augmented Reality Gaming: A new Paradigm for Tourist Experience. In *Information and communication technologies in tourism, eProceedings of the ENTER 2014 PhD Workshop* (pp. 57-67), Dublin, Ireland: Springer.

Williams, S. W. (2009). *Tourism Geography: a new synthesis*. New York: Routledge.

World Tourism Organization. (1983). *Risks of saturation of tourist carrying capacity overload in holiday destinations. Report of the secretary-general on the execution of the general programme of work for the period 1982-1983*. New Delhi, India: UNWTO.

World Tourism Organization. Centre of Expertise Leisure, Tourism & Hospitality, NHTV Breda University of Applied Sciences & NHL Stenden University of Applied Sciences (Eds.). (2019). 'Overtourism'? – Understanding and Managing Urban Tourism Growth beyond Perceptions, Volume 2: Case Studies. Madrid, Spain: UNWTO. doi:10.18111/9789284420629

World Tourism Organization (UNWTO); Centre of Expertise Leisure, Tourism & Hospitality; NHTV Breda University of Applied Sciences; and NHL Stenden University of Applied Sciences (2018). *'Overtourism'? – Understanding and Managing Urban Tourism Growth beyond Perceptions, Executive Summary*. Madrid, Spain: UNWTO, doi:10.18111/9789284420070

WTTC. (2018). *City Travel & Tourism Impact 2018*. Retrieved from https://wttc.org/-/media/files/reports/economic-impact-research/cities-2018/city-travel--tourism-impact-2018final.pdf

Xu, F., Buhalis, D., & Weber, J. (2017). Serious games and the gamification of tourism. *Tourism Management*, *60*, 244–256. doi:10.1016/j.tourman.2016.11.020

KEY TERMS AND DEFINITIONS

Carrying Capacity: The maximum number of people that can visit a tourist destination at the same time without causing destruction of the physical, economic, socio-cultural environment and an unacceptable decrease in the quality of visitors' satisfaction.

Dataset: A collection of data. A group of related sets of information that is composed of separate elements but can be manipulated as a unit by a computer. When this collection is extremely large is called big data.

Gamification: Understood as the use of games in non-leisure environments and settings in order to motivate, and encourage effort, loyalty and other positive values. Using technology increases possibilities and participants. Applied to tourism, it can promote tourism marketing and change player-tourist behaviour.

Internet of things (IoT): Refers to a digital interconnection of daily objects with the Internet.

Monitoring: Using devices to count the quantity or volume of visitors to a specific area.

m-tourism: The use of mobile applications for tourism.

Overtourism: The impact of tourism on a destination, or parts thereof, that excessively influences perceived quality of life of citizens and/or quality of visitors experiences in a negative way.

ENDNOTES

1 Interreg MED is a European transnational cooperation programme for the Mediterranean adopted by the European Commission and co-funded by the European Regional Development Fund. It covers 57 regions in 13 European countries (including 3 pre-EU accession countries) from Portugal to Cyprus. For more information: https://interreg-med.eu/

2 Pilots include four cities: Málaga, Genoa, Venice and Dubrovnik, and two regions: the Comunitat Valenciana (Region of Valencia) and the South Aegean Region

Section 4
Case Studies:
Historical Centers

Chapter 15
Lisbon Historic Quarters: Identity vs. Overtourism

Cláudia Helena Henriques
University of Algarve, Portugal

ABSTRACT

This chapter contributes to the emergent debate about sustainable tourism versus overtourism, in the context of urban development. The study underlines, on one hand, the growing importance of tourism in the historic quarters of Lisbon, and on the other hand the rise of overtourism and its effects on residents' quality of life. Consequently, there is the analysis and debate regarding the policies responses of the Lisbon Council in the framework of an integrated urban rehabilitation in Lisbon's historic quarters and the Lisbon Council Strategy.

INTRODUCTION

Travel & Tourism (T&T) is one of the world's largest and fastest-growing economic sectors (WTTC, 2019). In 2018, it represented 3.2% of World GDP (direct contribution) and is forecast to rise by 3.6% in 2019. Additionally, it is expected to continuing to grow by 2029 (WTTC, 2019).

The tourism importance, it's exponential growth and recent overcrowded territories, enhance several impacts both in local 'community and business', and tourists. Accordingly, "overtourism" – as a phenomenon created by "the perfect storm of budget airlines, short-term rental websites, and social media" (Loudis, 2018) led to tourist demand concentration associated with crowded/overcrowded territories. Once the problem, "has largely been ignored" (Neuts & NijKamp, 2012; in Jacobsen, Iversen & Hem, 2019: 53) an emphasis in tourism planning and management is urgently required.

In this context, this paper aims at investigating the growing importance of overtourism in Lisbon, the capital city of Portugal, and its central historic quarters, namely Alfama, Mouraria e Bairro Alto. Regarding the methodological framework, firstly it begins discussing concepts such as "crowded/overcrowed" as well as "overtourism". Consequently, it examines the tourism policy/strategy in European

DOI: 10.4018/978-1-7998-2224-0.ch015

cities, namely related to the rise of "tourismophobia" or "tourismophilia". Additionally, it also highlights the main challenges of tourism planning and management in order to overcome the negative impacts of overtourism.

Regarding the methodological framework, it is based on a case study analysis. Firstly, it puts in evidence Lisbon tourism market characteristics. Secondly, it discusses local policy concerns regarding tourism and historic quarters "overtourism" impacts. In accordance, documentary research on Lisbon's tourism policy is discussed fundamentally based on Municipality documents. Finally, the main conclusions are presented.

LITERATURE REVIEW

Overtourism and the Increased Tourismophobia

Growth in international tourist arrivals (5% in 2018) and receipts continues to outpace the world economy and both emerging and advanced economies are benefiting from rising tourism income (UNWTO, 2019: 2). Europe accounts for half of the world's international arrivals (51%), 5% of growth, and almost 40% of international tourism receipts (UNWTO, 2019: 6)[1]. "Southern and Mediterranean Europe led results [+8% of tourist arrivals], with most destinations enjoying double-digit growth. Among the larger destinations, Italy, Greece, Portugal and Croatia saw robust performance" (UNWTO, 2019: 10).

Cities, and namely the "300 of the largest city travel destinations accounts for over half a billion trips per year, which amount to 45% of global international travel" (Oxford Economics's Global City Travel; in WTTC, 2018: 18). Simultaneously, "travel to city destinations has grown more rapidly than total international travel demand over the past decade and account for a rising share of overall travel demand (…)" (WTTC, 2018: 18).

Consequently, many destinations are presenting high levels of tourism demand concentration and intensity, usually associated to concepts such as "crowded", "overcrowded", "overtourism" (any destination suffering the strain of tourism) (Richardson, 2017), "hotspot destinations" ("attract many 'site seers' associated with popularity and overall destination appraisal), "tourist bubble" (Cohen, 1972), "(excessive) commercialization" (Ponting & McDonald, 2013), among others. The term is also related to "carrying capacity" (Borg, Costa & Gotti, 1996) that is the maximum limit to tourism development. "Overtourism" has become a phenomenon created by "the perfect storm of budget airlines, short-term rental websites, and social media" (Loudis, 2018).

Crowding is associated with several meanings compatible with its own complexity, subjectivity and dynamic. On one hand, it could be understood as "indication of tourism destination popularity and fame" (Petr, 2009), on another hand, perceived as a "source of negative traveller reactions (Dowling, 2006; Jacobsen, 2000; Krebs, Petr & Surbled, 2007; in Jacobsen, Iversen & Hem, 2019: 53) and resident community reactions (Seraphin, Sheeran & Pilato, 2018: 374).

As several studies highlight, "crowding as both perceived social density and as physical density in relation to desired or accepted levels" (in Jacobsen, Iversen & Hem, 2019: 5) and consequently is associated with "approach reaction" or "avoidance reaction". "Overcrowding" could also be associated with "success", although the term is imprecise (McKinsey & Company and WTTC, 2017: 8). The referred authors' assumption is that: - overcrowding is not exactly alike in any two destinations; - overcrowding is currently more a matter of sensibility than data; - overcrowding is easier to prevent than to recover from.

"The concerns about overtourism expressed in both academic and popular literature include: too many tourists in one place, rowdy and other inappropriate behaviour by tourists, antagonism between residents and tourists, crowding, strains on infrastructure, loss of authenticity, loss of amenity and reduction in the quality of life of residents and reduced enjoyment of experiences by tourists" (Dodds & Butler, 2019: i).

Dodds and Butler (2019: i) also point out that the characteristics and externalities of overtourism "are not all new". They underline that although "through the past four decades, the potential and actual deteriorating relationship between hosts and guests has been the subject of much study and discussion (Doxey, 1975; Butler, 1980; Krippendorf, 1987; Boissevan, 1996)", and that "issues related to tourism at individual destinations and resorts were discussed earlier still (Ogilvie, 1993; Christaller, 1963; Young, 1973)" (Dodds & Butler, 2019: i), demand and supply perspectives should be more deeply considered.

Following the tourism demand perspective, it's important to state that "tourists resent too many fellow travellers arriving in the same places" (Jacobsen, 2000). Bresson and Logossah (2011) point out that "Some people in socially congested environments want to stand out from the crowd". "Some tourists [also] believe that the possibility of enjoying a unique experience is inversely proportional to the number of visitors" (Jacobsen, 2000). However, as Jacobsen, Iversen and Hem (2019: 53-54) underline, "impacts of crowding on appraisals and behaviour are complex, with studies showing mixed results". Crowding evaluation may vary according to different types of visitor types, contexts, times, and territories. It also can reduce people's satisfaction, attitude, and loyalty regarding destinations.

Regarding the local community and local business, Seraphin, Sheeran and Pilato (2018: 374) highlight that "tourists are impacting negatively on the quality of life of locals", "the environmental sustainability of destinations is being jeopardised" and also that "positive contributions of tourists (day trippers) to local business being limited" (Buckley, 2017; Leadbeater, 2017; Paris, 2017; in Seraphin, Sheeran & Pilato, 2018: 374).

In a context of a changing paradigm, "locals are now more interested in their quality of life than income generated by the tourism industry" (Seraphin, Sheeran & Pilato, 2018: 374). The "excessive" number or intensity of tourists preclude a symbiotic and integrated relationship between residents and tourists, generating conflicts, once the value and uses of the territory are not the same to the different stakeholders (Casanovas, 2017: 136).

Further, changes associated with the "new economy" and "digital economy", result in new tourism trends, such as new forms of tourist accommodation and low-cost travels (Casanovas, 2017: 136). In its turn, and regarding new types of tourist accommodation, destinations are growingly offering accommodation on private floors at lower prices (usually through digital platforms) consequently enhancing the rising housing prices (sale and lease) often associated with real estate speculation, gentrification, overcrowded tourism destinations, namely in city centres. For instance, Lestegás et al. (2019: 1), put in evidence "the relationship between the spatial concentration of short-term rentals in Lisbon's historic centre and the phenomena of uneven development and tourism gentrification", contributing the study of the new processes of neoliberal urbanization in the crisis-ridden countries of Southern Europe. The assumption is that "the great share of whole-home rentals and the expansion of the short-term rental market over the housing stock are symptoms of the commodification of housing in the neoliberal city".

In consequence, it turns out the growing number of voices, frequently associated with anti-tourism movements or "resistant identities" (Routledge, 2001), associated with concepts, such as "tourismophobia" (aversion to tourism and tourists) (Casanovas, 2017). The "Irritation Index" (in the 1990s) (created in order to determine residents' feelings about unwanted visitors), and *The Landscape Eaters article*

(Krippendorf, 1975; in Casanovas, 2017) (about the ecological consequences of excessive tourism in the Alps) are relevant milestones that precede the current reality).

Casanovas (2017: 138-142), in an attempt to determine what are the most common features associated with tourismophobia-generating sites, put forth what is called the "tourismophobic" ranking. The main features are: a) tourism growth in a short time; b) concentration of tourist icons/tourist attractions; c) limited public space to be shared; d) gentrification; e) mobility saturation; f) primacy of promotion over management; g) tolerant administration; h) problematic tourist labour market. Following McKinsey & Company and World Travel & Tourism Council (2017: 21) "leaders can begin with nine metrics to understand potential risks: overall context (metric: importance of tourism; arrival growth); alienated local residents; degraded tourist experiences; overloaded infrastructure; damage to nature; and threats to culture and heritage".

Along with "tourismophobia" emerges the concept of "tourismophilia" (Vallés, 2017) ("philia" rather than "phobia", focusing on valuing responsible tourism and responsible tourist behaviour). "Tourismophilia" refers to the appreciation of projects in favour of educating and creating responsible cities.

Accordingly, urban tourism planning and management are determinant to tourism sustainable development. Richardson (2017: 46) points out, that there is a need for a holistic approach, once tourism should be considered in the framework of the interconnectivity between economic, social and environmental dimensions.

Overtourism and The Challenge of Sustainability: Tourism Planning and Management

The European Commission (EC, 2019: 66), in the Document "Towards a Sustainable Europe by 2030", highlights its determination to achieve the 2030 SDGs (Sustainable Development Goals) (see Agenda 2030), which are "deeply rooted in the European project and its treaties" (in EU, 2019: 4). According to the EC (2019: 7), monitoring over the past five years has shown progress in virtually all SDGs.

In the context of "overtourism", it should be highlighted the need for responsibility in ensuring effective destination management that minimises the negative impacts of the tourism sector. As Zurab Pololikashvili (Secretary-General World Tourism Organization (in UNWTO, 2019: 2) states: "We need to grow more in value rather than just in volume".

Two management approaches stand up, namely the Ambidextrous Management ("exploitation and exploration") approach and Trexit (a term to signify exit - suggested by the anti-tourism movements) approach.

The first one, as Seraphin, Sheeran and Pilato (2018: 374) underline, "has been hailed to be innovative, sustainable and at the basis of success, implies putting together two opposites: first, exploitation (focusing on existing customers and/or markets). The focus is on the present (exploiting and improving the existing products, services, competencies and operating strategies)". Regarding "exploration (creative and innovative aspects), [it has a focus] on the future (experimenting with new products, services, competencies and operating strategies)".

The second, Trexit, consists of "forbidding the access to destinations during a certain period of time" (Ibidem, 2018: 374).

Stokes, Moore, Smith, Larson and Brindley (2016; in Seraphin, Sheeran & Pilato, 2018: 374) "add to this by arguing that organisations in advanced-market economies tend towards exploratory management and policies while emerging market economies are associated with exploitation".

In order to achieve sustainable development of tourism, Mckinsey & Company and WTTC (2017: 8) also suggest a set of good management practices, namely: 1. Build a comprehensive fact base and update it regularly; 2. Conduct rigorous, long-term planning to encourage sustainable growth; 3. Involve all sections of society - commercial, public, and social; 4. Find new sources of funding.

In this framework, practical actions are put in evidence, such as: imbalances of visitors during certain seasons, days of the week, and times of day, as well as during headline events: 1. Smooth visitors over time; 2. Spread visitors across sites; 3. Adjust pricing to balance supply and demand; 4. Regulate accommodation supply; 5. Limit access and activities (McKinsey & Company and WTTC, 2017: 99).

Seraphin, Sheeran and Pilato (2018: 375) present a sample of recommendations to overcome overtourism (in Europe). Among them: "repression plus education of tourists" (Sicsic, 2017), "teaching visitors how to appreciate" the city; "provide tourists an authentic experience (eat and drink with locals (Gordon, 2017), "increase taxes" (Bressanello, 2017), "respect campaign" (about telling tourists how to behave) (Buckley, 2017).

Many of these measures are being implemented in several European cities. Some of them are controversial. For instance:

- Restraint of the tourist volume/limit entrance [limit the rental through platforms such as Airbnb (e.g. London (limit of 90 days/year, for each house) and Berlin]; limit the number of tourist entrance in destinations (e.g. Santorini only permits eight thousand visitants/day, mainly due to cruises)
- Limit garbage (e.g. Milan plastic bottles were banned in some of the city's most iconic areas)
- Application of fee payment/tourism tax on individuals (or accommodation taxes) (e.g. Barcelona (has imposed taxes on those who stay more than 12 hours in the city); Austria; Lisbon; Rome (tourists pay between three and seven euros a night to stay in the city); Croatia (Croatia' "*sojourn tax*" apply higher fees on tourists who visit during peak season)
- Impossibility to sell houses to foreign people (e.g. New Zeland) in order to control housing rental problems and prices speculation
- Prohibition of new construction of accommodations units (e.g. Barcelona)/limit or impossibility to obtain building licenses or to operate more accommodation units or Fast Food restaurants (e.g. Veneza)
- Regulation of the housing market (e.g. Barcelona; Palma de Mallorca; Amsterdam, Venice; Lisbon).

RESEARCH METHODOLOGY

The present study's general objective was to determine the emergence of overtourism in Lisbon, namely in its central historic quarters and the sequent "political" and policy answer of the Municipal Council (*Câmara Municipal de Lisboa*) in the context of sustainable tourism development.

The specific objectives were to investigate:

- The importance of tourism in Lisbon

- The main factors responsible for the emergence of Lisbon's overtourism or high tourism intensity, such as the geographic distribution of LA (Local Accommodation/*Alojamento Local - AL*) apartments in Lisbon
- The recent policies or measures within the framework of Lisbon's economic and social strategy as well as an integrated urban rehabilitation process.

Given these objectives, the conceptual framework was based on a theoretical discussion regarding overtourism. On one hand, there is research regarding/concerning overtourism and a tourismophobia relationship. On the other hand, a review and analysis of overtourism urban management policies.

Secondly, the paper considered the case study approach associated with Lisbon's overtourism and related planning and management policies.

Thirdly, there is a discussion about a set or composite of indicators that contribute to assessing the growing tourism dynamic and policy. Two main domains were considered – the quantitative and qualitative.

The quantitative analysis relied on data from three main sources, namely indicators provided by IMPACTUR (2019), National Institute of Statistics and Tourism Observatory Surveys (*Turismo de Lisboa/Observatório de Turismo de Lisboa* [TL OTL] – Visit Lisbon), Inside Airbnb on the number and characteristics of Airbnb listings (Inside Airbnb 2018), and official data available from the National Registry of Local Accommodation (*Registo Nacional de Alojamento Local*) (RNAL, 2019).

The qualitative content analysis (Bardin, 2015) was based on planning documents (obtained from the central government, Lisbon city council, among others). Tourism planning or policy documents were also reviewed at the local, regional and national level.

CASE STUDY: TOURISM PLANNING AND MANAGEMENT CHALLENGES IN THE LISBON HISTORIC QUARTERS

Lisbon Tourism and Its Historic Quarters: Brief Characterization

Lisbon city is the capital and largest city of Portugal in the south of Europe. A brief characterisation of the T&T sector in Lisbon (Table 1) shows its relevant tourism importance in terms of GDP, GDP growth, employment and employment growth.

The "Annual report of 2018 – Lisbon" (Lisbon region annual analysis), shows that regarding "Overnight Stays and guests in lodgings" indicators, Lisbon region records 14,283,582 (2017), 14,285,454 (2018) overnight stays, 6,176,156 (2017) and 6,236,963 (2018) "guest". The respectively overnight stays and guest homologous variation rate 2018/17 is 0.0% (overnight stays) and 1.0% (guests). The homologous variation rate 2018/16 is respectively 9,5% and 11.2% (IMPACTUR, 2019). Lisbon region growth rate between 2018/17 (%) and 2018/2016 (%) is generally above the country average in terms of overnight stays and guests (for instance, 4.5% and 11.0% Portugal homologous variation rate 2018/16) (IMPACTUR, 2019). Lisbon's fast growth (ou growth increase) became with the HUb of Airport Portela (transatlantic connections) and low-cost airlines operations in 2012, that following CML (2018) "put Lisbon on the Map".

The 'Global Destination Cities Index – 2019' (Mastercard, 2019) regarding "Consistent and Steady Growth" underlines that "over the past decade, the one constant has been continual change. Each year, more people are travelling internationally and spending more in the cities. Between all of the destinations

Table 1. Characterization T&T sector in Lisbon

Indicators	Lisbon
• Population	540,000 (Census 2011) (Municipality) 2,800,000 (Metropolitan Area)
• Area (km^2) (Municipality – *concelho*)	100,05
• T&T direct GDP (2017)	Rank: 45° % of city: 5.7% % of country: 31.8%
• Inbound spending US$bn (real 2017 prices)	7.6
• T&T direct GDP – average 10 year growth (2007-2017)%	3.4%
• T&T direct GDP (2027)	% of city: 6.1% % of country: 28.3%
• T&T Direct Employment (2017)	% of city: 7.4% % of country: 27.2%

Source: Own elaboration based on WTTC (2018: 18) and Euromonitor International (2018)

within the Index, arrivals have grown on average 6.5 per cent year-over-year since 2009, with expenditure growing on average 7.4 percent". According to Mastercard Index (Mastercard, 2018) Lisbon occupies the 2nd place in the ranking of Europe's top fastest-growing cities in leisure tourism (first place: Bucarest and Santa Cruz de la Palma).

Lisbon's city main motivation segment is "city and short breaks". Lisbon Tourism Survey (TL, 2016) considers "city and short breaks" the most important foreign tourism motivational segment with a weight of approximately 70% (TL, 2016: 2) and grew to 90% in 2018 (TL OTL, 2018: 4).

By "purpose of trip" the "visit to monuments and museums" stands out, namely associated with knowledge of Portuguese culture, gastronomy and wine, local customs, among others. The main purpose of visit – visit monuments and museums – is based on the visit to the following "attractions, monuments and museums" (TL, 2016: 3): Torre de Belém (85,2%), Mosteiro dos Jerónimos (82.4%), Padrão Descobrimentos (76.2%), CC Belém (55.1%), Museu Chiado (44.7%), Museu Gulbenkian (44.0%), Sé de Lisboa (41.7%), among others.

Additionally, Lisbon's characteristics, following the European Commission (European Union Science Hub Home, 2018; in JRC, 2018), are associated to Lisbon's rich cultural heritage that includes two UNESCO World Heritage Sites (namely the Belém Tower and the Jerónimos Monastery). Other areas of central and historic Lisbon are being targeted as valuable with the approval of two applications by the Lisbon City Council to the indicative list of Portugal as world heritage: "Historic Lisbon, Global City" and "Pombalina Lisbon" (CML, 2016). The Lisbon classification as the European Capital of Culture, and relevant international events (such as "Expo 98", "Euro 2004") that have taken place in the city, have been boosting the local cultural infrastructure and activities.

In recent years, the local authorities have taken new action to strengthen the economic potential of arts, culture and creativity. Co-working spaces, fab labs and start-up incubators are now available in the rehabilitated urban areas. The historic quarters and their rehabilitation process is also highlighted, as well as important cultural and creative events (Lisbon Architecture Triennale; Lisbon Fashion Week, Lisbon & Estoril Film Festival; European Creative Hubs Forum; Festivals, Festivals for Europe - labelled festivals), among others.

Lisbon has also received a significant number of awards. For instance, the Portuguese capital has already won the Best European Destination (EBD, 2019) award (in the first edition of the contest in 2010 and 2015) and the World Leading City Destination 2018 (World Travel Awards Winner), Europe´s Leading Tourist Board 2019 (TL OTL, 2019), etc.

The 'Satisfaction and Image Survey - 2018' (TL OTL, 2018: 22) shows the image of 'Lisbon-Lisbon – Lisbon region´ as a place of 'interactive people' (90.8% agreement), 'ancient city with history' (90.5%), 'human city (91,0%), 'nice, peaceful city' (90.9%), 'creative and trendy city' (92.0%), 'unique city' (91.7%) 'city with diverse offer', among others.

Overtourism and Local Accommodation: How to Solve the Problem?

Local accommodation (LA) [*Alojamento local*/AL] is a growing phenomenon that is associated with "overtourism" in Lisbon's historic quarters, as price, localisations and commodities attract tourists (Publituris Hotelaria, 2018).

As CML (2018: 4) highlights, "The localisation of tourist establishments in the municipality of Lisbon began by privileging the main avenues of the city and surrounding areas. In the middle of the last century, *Baixa*, *Av. da Liberdade*, *Av. Almirante Reis* and *Rua Castilho* were the places of installation of most establishments. After 1998, with EXPO 98, new hotels were set up around *Oriente* Station in *Parque das Nações*. From 2012, the first hostels appeared and then the LA establishments in apartments. The locations became predominantly the historical neighbourhoods of Castle Hill (*Colina do Castelo*), *Chiado*, *Principe Real*, *Bairro Alto*, *Bica*, *S. Paulo* and *Madragoa* and, only recently, a dispersion through the city with a slight concentration in *S. Vicente*, on the Hill. Santana and the parish of *Ajuda,* near the Monumental Zone of *Belém* (CML, 2018: 4).

Lisbon presents a high level of LA, with "Accommodation capacity in LA (guests)" of 107, 927 and "N.º guest (users) of LA per 1000 km^2" of 1 078,68 (Table 2). However, the historic quarters of Lisbon, namely those belonging to the Parishes ("freguesias") of Santa Maria Maior (includes Mouraria, Alfama, Castelo, Baixa and Chiado) and Misericórdia (Bairro Alto and Príncipe Real, among others), reveal the highest tourist intensity and tourist foreign demand in terms of rental or acquisition of housing (Table 3, figure 1) (CML, 2018).

A relevant concentration of Airbnb listings stands up in a small portion of the Lisbon council. As Inside Airbnb (2018) and Instituto Nacional de Estatística (INE, 2018) point out, there is a concentration of 21.1% (of a total of 14,722) in only 3.0% of the municipal area. *Santa Maria Maior* alone contains almost as many listings as the aggregate of the 17 civil parishes with less than 500 each, which provide 22.2% of the total but represent 81.6% of the municipal area. The Lisbon's historic core aggregates the five civil parishes with more listings each - *Santa Maria Maior*, *Misericórdia*, *Arroios*, *Santo António*, and *São Vicente* - concentrate 67.8% of the total number of accordance listings in just 10.8% of the municipal area (Inside Airbnb, 2018; Instituto Nacional de Estatística, 2018; RNAL, 2019, 2018).

The Jornal de Negócios (JN, 2018) states that about 41% of the Lisbon historic centre (namely the parish of *Santa Maria Maior*) is oriented to tourists. Each km^2 of this parish has approximately 3.000 LA with capacity to host more than 15,000 tourists. In summer months, when occupancy rates are higher, it is expected that for every 100 residents the parish attracts 184 to 197 accommodation guests daily. Additionaly, Anthena Insight (2016), "Airbnb Market Report - Lisbon August 2016", states that "looking at the top ten performing listings from the period of June 2015 to July 2016, the most profitable Airbnb

Table 2. Local Accommodation (LA) Establishments (ET)| Registration 2019 [Estabelecimentos de Alojamento Local (AL) | Registo 2019]

Lisbon (Council)	**Total of LA: 18336**
Accommodation capacity in Tourist Establishments (beds / users) | *[Capacidade de alojamento em Empreendimentos Turísticos (camas/utentes)]*	45 871
Accommodation capacity in LA (guests) *[Capacidade de alojamento em Alojamento Local (utentes)]*	107 927
Total accommodation capacity for tourists in ET and LA (beds / users) *[Capacidade total de alojamento a turistas em ET e AL (camas/utentes)]*	153798
Lisbon	
Nº beds/guests(users) in ET per km^2	458,46
Nº guest (users) of LA per 1000 km^2	1078,68
Nº beds/guests (users) in ET and LA per km^2	1537,14
Lisbon Metropolitan Area	
Relative concentration of accommodation offer in Tourist Establishements (%)	22,74%
Relative concentration of accommodation offer in Local Accommodation (%)	30,99%
Relative concentration of accommodation offer to tourists (%)	27,28%

Source: RNAL (2019)

were in the historic centre of Lisbon. The actual report (Insideairbnb, 2019) points a similar situation with Lisbon with 22,242 listings (100%).

The situation of the high density of the local accommodation industry enhances demographic pressure, as well as territorial pressure. This pressure is compatible with negative impacts that, on one hand, tourists do not benefit from a good experience and on the other hand, local community' quality of life is threatened. Additionally, it could generate perverse effects such as gentrification[2] and "a conflict between housing as home and as real estate" (Madden & Marcuse, 2016: 4; in Lestegás et al., 2019: 2). As Lestegás et al. (2019: 2) point out, due to the commodification of housing and the proliferation of

Figure 1. Parishes of Lisbon and the tourism pressure
Source: JN (2018)

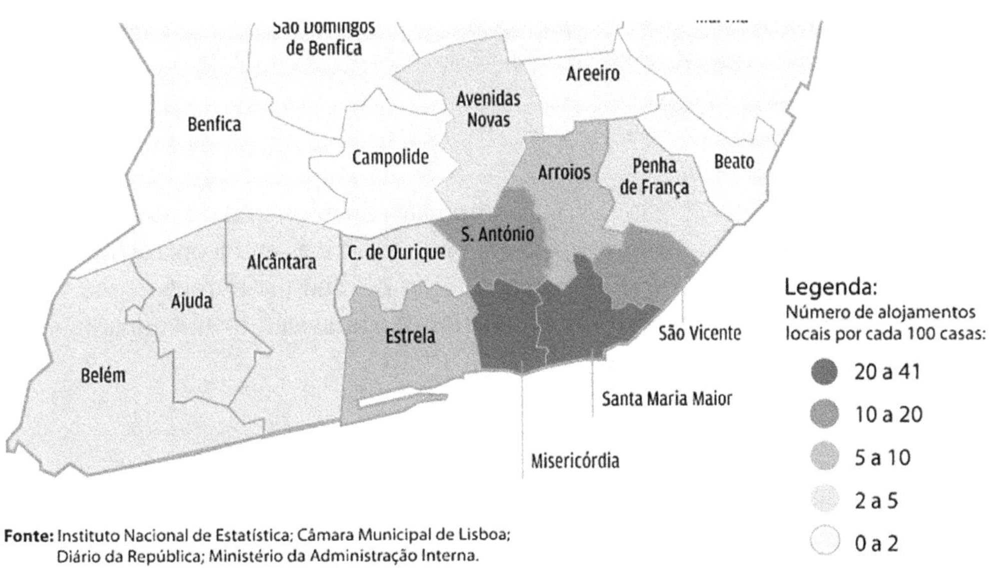

Table 3. Parishes of Lisbon and tourism pressure

Parishes	Nº BG	Nº GLA	Nº ULA	RCATE(%)	RCALA(%)	RCTA(%)
Ajuda	0	59,61	59,61	0	0,86	0,61
Alcantara	64,98	119,85	184,82	1,98	1,55	1,68
Alvalade	46,8	44,76	91,57	3,25	1,32	1,89
Areeiro	66,91	75,16	142,07	2,94	1,4	1,86
Arroios	97,72	468,08	565,79	6,74	13,73	11,64
Av. Novas	458,17	229,04	687,21	21,6	4,59	9,66
Beato	0	34,39	34,39	0	0,41	0,28
Belém	20,39	105,88	126,27	0,73	1,62	1,36
Benfica	111,83	8,44	120,26	9,02	0,29	2,89
Campo de Ourique	3,75	104,16	107,91	0,18	2,13	1,55
Campolide	202,72	55,24	257,96	6,83	0,79	2,59
Carnide	0	9,31	9,31	0	0,17	0,12
Estrela	14,1	294,56	308,58	0,61	5,49	4,04
Limiar	0	13,35	13,35	0	0,56	0,4
Marvila	0	9,87	9,87	0	0,35	0,24
Misericódia	54,89	1468,18	1523,08	1,56	17,74	12,92
Olivais	19,15	22,64	41,79	1,41	0,71	0,92
Parque das Nações	119,57	101,4	220,98	5,48	1,98	3,02
Penha de França	2,79	112,53	115,31	0,17	2,92	2,1
Santa Clara	0	1,82	1,82	0	0,04	0,03
Santa Maria Maior	383,48	1920,14	2303,62	10,72	22,81	19,21
Santo António	905,88	908,8	1813,96	23,37	9,96	13,96
São Domingos Benfica	44,43	25,97	70,39	3,2	0,79	1,51
São Vicente	6,13	548,21	554, 34	0,2	7,79	5

Legend: Nº beds/guests (users) ET per 1000 inhabitants (Nº BG); Nº guest (users) in LA per 1000 inhabitants (Nº GLA); Nº of beds / guests (users) in ET and LA per 1000 inhabitants (Nº ULA); Relative concentration of accommodation offer in Tourist Establishements (%) (RCA); Relative concentration of accommodation offer in Local Accommodation (%) (RCALA); Relative concentration of tourist accommodation offer (%) (RCTA)

Source: Own elboration based on RNAL (2019)

vacation, we are assisting a "global rent gap [that] stimulates the proliferation of vacation rentals at the expense of the supply of residential housing, fueling property prices and jeopardizing housing affordability". "With Portugal being a peripheral member of the EU and the Eurozone, the vulnerability of local households to the impacts of tourism gentrification is aggravated by the remarkable income gap with their counterparts of the core" (Lestegás et al., 2019: 2).

Policy Framework

"Portugal 2027" and "Tourism Strategy 2027 - Portugal" (Estratégia Turismo 2027 – Portugal) show the concern with tourism sustainability in Portugal.

The Tourism Strategy 2027 sets priorities and strategic axis with their respective lines of action. The priorities are as follows: Valorize the territory; Boost the economy; Enhance knowledge; Generate networks; Position Portugal (LET, 2016). In this strategic approach, Lisbon is seen as "a multicultural destination with a strong international vocation" (LET, 2016). Lisbon is recognised as having a set of assets that attract tourists, namely, "active differentiators" (climate and light; history and culture; sea, nature and biodiversity; and water), "active qualifiers" (gastronomy and wines, artistic and cultural events, sports and business), "emerging assets" (well-being, living in Portugal) and "unique transversal asset" (people) (LET, 2016).

In this framework, the historic quarters tourism policy incidence is determinant. It is important to note that the municipality of Lisbon, until the 1990s, lacked medium and long-term planning, which was largely responsible for the degradation of many historical buildings in various HQs. Fortunately, this trend was reversed after the 1990s based on the Strategic Plan of Lisbon (1992), the Municipal Director Plan (1994) and Urbanisation Plans (UPs) (1996), whose guidelines were followed up by the Lisbon City Council's activity plans (Henriques, 2003).

In 1994, through the Urban Rehabilitation Process (URP), five Integrated Projects (IPs) did much to ensure the URP´s continued integration. The IPs sought to create local cultural facilities as an important investment that could contribute to the HQs' tourism development (EGEAC, 2018).

In its "Strategic Letter 2010–2024", the Lisbon City Council states that culture (and cultural tourism) is associated with three main neighbourhoods, which include "the downtown area of *Chiado* (which extends into all the historic areas, with a specific emphasis on Castle, *Alfama* and *Bairro Alto*)". The other two areas are the monumental area of *Belém-Ajuda* and the urban renewal area of the *Parque das Nações* (CML, 2009: 83).

The Urban Rehabilitation Strategy of Lisbon 2011–2024 (*Estratégia de Reabilitação Urbana de Lisboa 2011–2024*) (CML, 2011) also recognises that an integrated intervention strategy is essential for the continued competitiveness of Lisbon and its neighbourhoods.

In the Lisbon City Council Programme 2013–2017 *(Programa de Governo da Cidade de Lisboa* 2013–2017), the Lisbon City Council sets out initiatives along five major axes, namely, a closely integrated city that is enterprising, inclusive, sustainable and global. The programme's strategic objective is to create a city with better lifestyle opportunities (CML, 2013: 82). Within the framework of a "global" Lisbon, the goal is to develop a "city of culture and creativity; a city of intercultural dialogue, tourism and the sea; and a metropolitan city (CML, 2013: 86).

The "Strategic Plan for Tourism in the Lisbon Region 2015–2019" (Roland Berger/ERTRL/TL, 2014; TP, 2015) seeks to place Lisbon in the "city/short breaks" market and highlights the city's culture as an across-the-board "qualifier" of Lisbon's tourism offer. With clear impacts in the management of historic quarters, the recent documents titled respectively "Strategic Vision for the Urban Planning of Tourism in Lisbon - 2018 CML - City Planning" ("Visão Estratégica para o Ordenamento Urbanístico do Turismo em Lisboa – 2018) (CML, 2018), Law (Lei) n.º 62/2018 (22 August - changes the regime of LA units exploration authorization) and the "Regulamento Municipal do Alojamento Local" (CML, 2019) putting in evidence the municipality vision to promote a better interconnection between Lisbon quarters urbanization and tourism[3]. However, as the Associação do Turismo de Lisboa (TL OTL, 2019) points out, Lisbon's residents' majority (86%) perceives tourism as a positive activity to the city and consequently highlights the importance of managing the territory use in order to promote its economic, social and environmental values.

Lisbon's "Strategic Vision" highlights its concern to a) defend the stock of permanent housing; b) limit the settlement of new tourist establishments in several historic quarters (where their presence has an excessive weight in relation to the total available houses); c) limit the gentrification with intervention in habitational market; (d) improve the maintenance of public space and urban services in places where tourist pressure is highest, in particular urban cleaning and waste collection; e) increase the supply of public transport in general; f) improve the attractiveness of new tourist attractions in the city outside the historic centre; g) encourage the dispersion of tourism throughout the metropolitan territory with visits to the beaches of Caparica and the Cascais Line, the Tagus Estuary Reserve and the natural parks of Sintra, Cascais, Arrábida and Sado Estuary; h) streamline the Tagus Estuary as a large natural space (CML, 2018: 7).

From October 2018 till present, some measures have been implemented. Following CML (2019) it stands up:

- Legal changes in "LA" regime. Municipality and assembly of condominiums can intervene in the authorisation of the LA activity, namely through the mechanism named "containment areas" ("áreas de contenção" - in Lisbon they are in Alfama, Mouraria, Castelo, Bairro Alto and Madragoa). The objective is to preserve the social reality of neighbourhoods and places. "Containment Areas", in which the opening of local accommodation needs the permission of the City Council and assembly of condominiums[4]
- Payment of fees by tourists. Lisbon charges a fee of 1 euro per night up to a maximum of 7 euros (Municipality tourist tax was first applied in 2016). The rate is applied per night and per guest over the age of 13
- Housing policy measures. To increase the supply of affordable housing in the historic centre, the municipality of Lisbon has launched an Affordable Income Program [AIP – *Programa de Renda Acessível* (PRA)] designed to create 7,000 housing units distributed in 15 intervention areas
- Limits to tourist buses (except scenic buses) circulation in historic quarters (city centre, namely in the area of Sé and Castelo, "Cais do Sodré, Rua do Alecrim, Príncipe Real, Largo do Rato) as well as tuk-tuks (prohibited to circulate in the vicinity of the Castle, with hours limited from 9h to 21h)
- Homogeneous Tourist Areas/zones ["homogeneous tourist zones" (Zonas Turísticas Homogéneas (ZTH)] is associated with a georeferencing of tourist enterprises and LA establishments and the delimitation of neighbourhoods enabling the possibility that new local accommodation registers should be suspended.

This framework complemented with the market autoregulation process is enhancing a slow down in the LA demand licences (more than 60% in the 2019 first trimester) in the majority of the Lisbon parishes (ALEP, 2019).

CONCLUSION

Many cities are facing a growing number of voices against unsustainable tourism development, frequently associated with "overtourism". In accordance, the raise of anti-tourism across European cities associated with "tourismophobia" is enhancing the need for adequate planning and management, once this situation can cause much damage and disruption.

Regarding the Lisbon study case, the city recognises the importance of tourism to its sustainable development. However, the exponential growth of tourism in the city centre, and namely in some historic quarters (such as Alfama, Mouraria, Castelo, Bairro Alto), associated with new models of hospitality, such as LA - boosted by the rise of peer-to-peer digital platforms, such as Airbnb - is creating some anti-tourism behaviours in residents.

Lisbon's historic quarters offer the most important attractions of the city, and consequently, the emergent tourism pressure could erode the quality of the tourist attractions and offerings and the quality of life of residents. Simultaneously, harmful negative impacts, such as gentrification expansion and the short-term rental market transformation that pushes the city at the expense of residential housing, should be prevented.

Consequently, in order to foster Lisbon's tourism sustainable development, Lisbon Municipal Council envisages several challenges, such as the excessive concentration of tourism in the central historic quarters associated not only with the growth of European overtourism, but also with the emergent concern regarding the rise of anti-tourism residents movements.

Overtourism is in inadequacy with an authentic tourist experience, and it does not promote the sustainable development of values and goals. Accordingly, once urban planning and management is determinant to the sustainable tourism development in Lisbon, it is important, on one hand, to promote and preserve the identity of the city and, on the other hand, the quality of life of its residents.

Lisbon Tourism Strategy and Urban Rehabilitation in Historic quarters based on documents, such as "Strategic Vision for the Urban Planning of Tourism in Lisbon - 2018 CML - City Planning" and Law (Lei) n.º 62/2018 - 22 August, are impacting in concentration levels of LA (*Alojamento local*). Simultaneously, measures, such as the payment of fees by tourists, housing regulation, limitation of transports circulation, among others, are trying to regulate tourism carrying capacities in order to develop sustainable urban tourism.

ACKNOWLEDGMENT

This research was funded by national funds through the FCT- Foundation for Science and Technology under the project UID/SOC/04020/2013

REFERENCES

ALEP [Associação de Alojamento Local em Portugal]. (2019). *Alojamento local em Portugal*. Retrieved from http:www.alep.pt

Anthena Insight. (2016). *Airbnb Market Report – Lisbon August 2016*. Retrieved from http://insideairbnb.com/lisbon/?neighbourhood=&filterEntireHomes=false&filterHighlyAvailable=false&filterRecentReviews=false&filterMultiListings=false#

Bardin, L. (2015). *Análise de Conteúdo. Editora: Edições 70*. Coimbra: Extra Coleção.

Borg, J. V. D., Costa, P., & Gotti, G. (1996). Tourism in European heritage cities. *Annals of Tourism Research*, *23*(2), 302–321.

Casanovas, O. (2017). Turismo y convivência ciudadana. In L. *Coma, & J. Santacana, (Coords.), Ciudad educadora y turismo responsable* (pp. 135–145). Ediciones Trea.

CML. (2009). *Estratégias para a Cultura em Lisboa, CML – Pelouro da Cultura – Site da CML: Estudos de Caracterização.* Retrieved from http://www.cm-lisboa.pt/fileadmin/VIVER/Urbanismo/urbanismo/Documentos/estrategias_para_a_cultura.pdf

CML. (2011). *Estratégia de Reabilitação Urbana de Lisboa – 2011/2024.* Site da CML. Retrieved from http://www.cm-lisboa.pt/fileadmin/VIVER/Urbanismo/urbanismo/Reabilitacao/estrat.pdf

CML. (2013). *Programa de Governo da Cidade de Lisboa 2013/2017.* Retrieved from http://issuu.com/camara_municipal_lisboa/docs/programa_gov_cidade_de_lisboa_2013

CML. (2016). *Historic Lisbon, Global City and Pombalina Lisbon – Heritage of Humanity Proposal.* Retrieved from http://www.cm-lisboa.pt/viver/urbanismo/planeamento-urbano/planodiretor-municipal/pdm-em-vigor

CML. (2018). *Visão Estratégica para o Ordenamento Urbanístico do Turismo em Lisboa – 2018* [Strategic Vision for the Urban Planning of Tourism in Lisbon - 2018 CML - City Planning]. Retrieved from http://www.cm-lisboa.pt/fileadmin/VIVER/Urbanismo/urbanismo/alojamentolocal/4_Vis%C3%A3o_estrat%C3%A9gica_Zonas_Tur%C3%ADsticas_Homog%C3%A9neas_Cidade_de_Lisboa_2018-10-15.pdf

CML [Câmara Municipal de Lisboa]. (2019). *Regulamento Municipal do Alojamento Local.* Retrieved from http://www.cm-lisboa.pt/viver/urbanismo/alojamento-local

Dodds, R., & Butler, R. (2019). *Overtourism: Issues, realities and solutions. The Gruyter Studies in Tourism.* Walter de Gruyter GmbH Editors, Berlin/Boston. Retrieved from https://books.google.pt/books?id=qoKVDwAAQBAJ&pg=PT11&hl=pt-PT&source=gbs_toc_r&cad=3#v=onepage&q&f=false

EBD [European Best Destinations]. (2019). *European Best Destinations Awards.* Retrieved from https://www.europeanbestdestinations.com/destinations/search/search-results/#e=external&q=lisbon%20awards

EC. (2019). *Reflection paper:* Towards a Sustainable Europe by 2030, Highlights its determination to achieve the 2030 SDGs (Agenda 2030), COM(2019)22.

EGEAC. (2018). *Relatório e Contas – 2018.* Retrieved from http://www.egeac.pt

Euromonitor International. (2018). Top 100 City Destinations 2018. Retrieved from https://go.euromonitor.com/white-paper-travel-2018-100-cities.html?utm_source=blog&utm_medium=blog&utm_campaign=CT_WP_18_12_04_100%20Cities&utm_content=organic

Henriques, C. (2003). *Turismo, Cidade e Cultura – Planeamento e Gestão Sustentável.* Edições Silabo.

IMPACTUR [Indicators for the Monitoring and Forecast of Tourist Activity]. (2019). *Informação estatística turística de Lisboa.* Retrieved from http://www.ciitt.ualg.pt/impactur/dt_anuais.asp?nut=17

INE [Instituto Nacional de Estatística]. (2018). Estatísticas de Portugal. Retrieved from https://www.ine.pt/xportal/xmain?xpgid=ine_main&xpid=INE&xlang=pt

Insideairbnb (2019, 2018). *Inside Airbnb. Adding Data to the Debate*. Retrieved from http://insideairbnb.com/lisbon/

Jacobsen, J. (2000). *Tourism in Peripheral Areas: Case Studies*. Frances Brown, Derek D. Hall, Derek R. Hall.

Jacobsen, J., Iversen, N., & Hem, L. (2019). Hotspot crowding and over-tourism: Antecedents of destination attractiveness. *Annals of Tourism Research*, *76*, 53–66. doi:10.1016/j.annals.2019.02.011

JN [Jornal de Negócios]. (2018). *As freguesias de Lisboa mais pressionadas pelo alojamento local*. Retrieved from https://www.jornaldenegocios.pt/economia/rendas/detalhe/mapa-quais-sao-as-freguesias-de-lisboa-mais-pressionadas-pelo-alojamento-local. https://www.publiturishotelaria.pt/2018/12/20/alojamento-local-teve-um-impacto-economico-de-412-milhoes-de-euros-em-2018/

JRC. (2018). *The Cultural and Creative Cities Monitor*. JRC/European Commission.

Law (Lei) n.º 62/2018 - 22 august (changes the regime of Alojamento Local units exploration authorization).

(2019). LET [Laboratórios Estratégicos para o Turismo] (2016). *Estratégia Turismo 2027* – Portugal (Portugal–Laboratório Estratégico Lisboa. Lestegás, I., Seixas, J. & González, R. (2019). *Commodifying Lisbon: A Study on the Spatial Concentration of Short-Term Rentals. Soc. Sci.*, *8*, 33. Retrieved from https://www.researchgate.net/publication/330640248_Commodifying_Lisbon_A_Study_on_the_Spatial_Concentration_of_Short-Term_Rentals

Loudis, J. (2018). *'Overtourism' Is Driving Europeans Crazy*. Retrieved from https://www.thenation.com/article/overtourism-is-driving-europeans-crazy/

Mastercard. (2018). *The 'Global Destination Cities Index – 2018'* (Mastercard's 2018 Global Destination Cities Index). Retrieved from https://newsroom.mastercard.com/press-releases/big-cities-big-business-bangkok-london-and-paris-lead-the-way-in-mastercards-2018-global-destination-cities-index/

Mastercard. (2019). *The Global Destination Cities Index 2019*. Retrieved from https://newsroom.mastercard.com/wp-content/uploads/2019/09/GDCI-Global-Report-FINAL-1.pdf

McKinsey & Company and World Travel & Tourism Council. (2017). *Coping with success Managing Overcrowding in Tourism Destinations*. McKinsey & Company and World Travel & Tourism Council. Retrieved from https://www.wttc.org/-/media/files/reports/policy-research/coping-with-success---managing-overcrowding-in-tourism-destinations-2017.pdf

Publituris Hotelaria. (2018). Alojamento Local teve um impacto económico de 412 milhões de euros em 2018. *Publituris Hotelaria*, 20-12-2018

Richardson, D. (2017). Suffering the strain of tourism. Retrieved from: TTG@wtm

RNAL [Registo Nacional de Turismo]. (2019, 2018). *Alojamento Local - Registo Nacional de Turismo*. Retrieved from https://tdpindicadores.maps.arcgis.com/apps/MapSeries/index.html?appid=d2e682ede e1248a79d15b7b43a68ca2d

Roland Berger/ERTRL/TL. (2014). *The Strategic Plan for Tourism in the Lisbon Region 2015–2019* [Plano Estratégico para o Turismo na Região de Lisboa 2015-2019 - Posicionar a Região de Lisboa num novo patamar de excelência turística]. Retrieved from http://ertrl/TL%20

Routledge, P. (2001). 'Selling the rain' resisting the sale: Resistant identities and the conflict over tourism in Goa. *Social & Cultural Geography*, *2*(2), 221–240. doi:10.1080/14649360120047823

Seraphin, H., Sheeran, P., & Pilato, M. (2018). Over-tourism and the fall of Venice as a Destination. *Journal of Destination Marketing & Management*, *9*, 374–376. doi:10.1016/j.jdmm.2018.01.011

TL. (2015). *Satisfaction and Image Survey - 2015 – Lisbon Region*. Observatório de Turismo de Lisboa. Retrieved from http://www.visitlisboa.com/sites/default/files/2016-10/SATISFACTION%20AND%20IMAGE%20SURVEY%202015.pdf

TL. (2015 a). *Visitor Activities and Information Survey – Lisbon Region - 2015*. Observatório de Turismo de Lisboa. Retrieved from http://www.visitlisboa.com/sites/default/files/2016-10/VISITOR%20ACTIVITIES%20AND%20INFORMATION%20SURVEY%202015.pdf

TL OTL. (2018). *Inquérito Motivacional – Cidade de Lisboa*. Retrieved from https://www.visitlisboa.com

TL OTL. (2018). *The Satisfaction and Image Survey - 2018*. Retrieved from https://www.visitlisboa.com

TL OTL [Turismo de Lisboa /Observatório de Turismo de Lisboa]. (2018). *Inquérito Motivacional 2018 – Cidade de Lisboa.*

TL OTL [Turismo de Lisboa /Observatório de Turismo de Lisboa]. (2019). *Inquérito Motivacional 2018 – Cidade de Lisboa*. Retrieved from https://www.visitlisboa.com

TL [Turismo de Lisboa]. (2016). *Site do Turismo de Lisboa. Lisbon Tourism Survey*. Retrieved from http://www.visitlisboa.com/see-and-do/heritage-and-culture

TP/Governo de Portugal. (2015). *Turismo 2020: Plano de Ação para o Desenvolvimento do Turismo em Portugal*. Retrieved from http://www.turismodeportugal.pt/Portugu%C3%AAs/turismodeportugal/apresentacoes/Documents/Turismo-2020-Sintra.pdf

UNWTO. (2019). *International Tourism Highlights 2019*. Edition UNWTO (2019) International Tourism Highlights 2019 Edition.

Vallés, M. (2017). No es turismofobia, es turismofilia, *Diario de Mallorca*, 2 de agosto del 2017.

WTTC. (2018). *City Travel & Tourism Impact 2018*. Retrieved from http:www.wttc.org.com

WTTC [World Travel & Tourism Council]. (2019). *Travel & Tourism Economic Impact - World*. Retrieved from https://www.wttc.org/economic-impact/country-analysis/country-data/

KEY TERMS AND DEFINITIONS

Overtourism: Associated to concepts such as "crowded" or "overcrowded" destinations.

Sustainability: Associated to principles such as conservation, preservation, equity, long term, partnerships (see Agenda 2030).

Tourismophobia: Aversion to tourism and tourists.

ENDNOTES

1 The top 10 destinations (countries) receive 40% of worldwide arrivals and the top 10 tourism earners account for almost 50% of total tourism receipts (UNWTO, 2019: 9).

2 ALEP (2019) points out that there are no figures regarding this phenomenon of 'emptying out' residents of certain neighborhoods caused by 'turistification'. There were many vacant, abandoned, or empty buildings in the historic quarters, some of which came to be requalified with the new wave of tourism. In the parish of Santa Maria Maior there were 3,498 vacant houses and 1,290 which were second homes.

3 See Law n.º 62/2018, 22 Agoust, amending the system of authorization to operate local accommodation establishments. Amends the New Urban Leasing Regime 2012 (promoted by Law No. 31/2012, of 14/08).

4 See "Instalação de novos estabelecimentos de Alojamento Local 1. Áreas de Contenção e suspensão de novos registos" (regime jurídico de exploração dos estabelecimentos de alojamento local, conforme alterado pela Lei n.º 62/2018, de 22 de agosto ("RJEAL") (CML, 2019).

Chapter 16
In Search of Overtourism Indicators in Urban Centres

Manuel De la Calle-Vaquero
Complutense University of Madrid, Spain

María García-Hernández
https://orcid.org/0000-0002-8469-6150
Complutense University of Madrid, Spain

Sofía Mendoza de Miguel
Complutense University of Madrid, Spain

Elena Ferreiro-Calzada
Complutense University of Madrid, Spain

ABSTRACT

Urban tourism is in constant growth. The increase in the number of tourists has a special impact on historic centres. Some problems related to overcrowding arise in these spaces, which represent important challenges for urban management. This chapter reflects on the need to define overtourism indicators that allow dimensioning the phenomenon and its impacts. But it also involves a deep reflection on the limits of application of these indicators. These limits derive from the absence of reference values and the operational difficulties to obtain data. First of all, the state of the art regarding the indicators is made. Secondly, based on a review of the existing bibliography, the next section raises some indicators of activity and tourism specialization. The focus is on European cities and the application of these indicators is shown in the historic centre of Madrid. Another section also looks at the perception of the phenomenon by different local stakeholders due to the absence of commonly accepted overtourism values, referring to these perceptions as valuation criteria.

DOI: 10.4018/978-1-7998-2224-0.ch016

INTRODUCTION

Tourism is in constant growth. Many cities, and especially those European towns with rich historical-cultural heritage and good airport connections, are experiencing very high rates of tourism growth. Growth is not only affecting large capitals and traditional historic cities, but is extending to cities that until a few years ago were not part of the international map of urban tourism destinations. In addition, the increase in the number of tourists has a special impact on historic centres, places of high heritage and/or symbolic value that make up the majority of tourist activity. Some problems related to overcrowding arise in these spaces, which represent important challenges for urban management.

For some time now, attention has been drawn to the problems of tourist pressure on the resources and most popular urban destinations, in general spaces linked to historical heritage (Borg, 1998; García Hernández, 2003; Russo, 2002). However, in recent times these problems have worsened and we talk about overtourism. A new scenario appears in which the tourist pressure acquires a new dimension. Firstly, because of its intensity, associated with the accelerated growth of tourist flows. Secondly, due to its effect on urban spaces that until now have not been affected by the arrival of massive flows of visitors. Thirdly, the proliferation of housing for tourist use in city centres, a factor that threatens the very residential condition of these spaces. And finally due to its notoriety from, among other things, the popularization of the term overtourism in the media and the incorporation of the problem in the agenda of social movements and political parties. It was in 2017 and 2018 that the first documents promoted by international institutions (WTTC & McKinsey&Company, 2017; ECM, 2018; Peetrs et al., 2018; WTO, 2018; WTO, 2019) were disseminated and academic papers began to be published (Koens et al, 2018; Milano, 2018; Milano et al, 2018). Broadly speaking however, the lack of conceptual precision of the term overtourism, which often appears as a counterpart to the term 'Tourism-phobia' to describe residents' discomfort or indignation in response to the excessive growth of tourism (Huete & Mantecón, 2018; Milano, 2018; Colomb & Novy, 2016), is striking.

The World Tourism Organization defines overtourism as "the impact of tourism on a destination, or parts thereof, that excessively influences the perceived quality of life of citizens and/or the quality of visitor experiences in a negative way" (WTO, 2018). In this sense, some authors point out the similarities with research topics and approaches already raised years ago in the academic field. Diana Dredge (2017) has come to metaphorically qualify the phenomenon as "*old wine in new bottles*" and Koens et al (2018) link it to the tradition of tourism impact studies and previous conceptual developments related to tourist carrying capacity or limits of acceptable change. In fact, the terms overcrowding and overtourism are closely related to the concept of carrying capacity, since this implies defining thresholds or maximum limits for tourist use. In this sense, one would think that the phenomena of overcrowding and/or overtourism appear normally when those limits are exceeded (Peetrs et al, 2018). But how do you measure this phenomenon of tourist congestion? How much is too much? At what point does a destination go from having many visitors to defining states/situations as one of overcrowding or overtourism? The *a priori* deployment of indicators could help to detect such overtourism situations. From a theoretical point of view, the use of indicators would make it possible to know the situation and dynamics of tourist activity with respect to the optimal or maximum carrying capacity limits or thresholds pre-defined for each destination; giving continuity to a line of work with a long trajectory such as the analysis and proposal of sustainable tourism indicators.

As well as many existing works on tourism indicators, there are also a number of studies dedicated to identifying, typifying and/or analysing the effects of tourism on reception spaces (Mathieson & Wall,

1982). These effects may be positive or negative and have different dimensions; there are environmental, social, economic and functional impacts, etc. The extent and magnitude of these impacts varies over time and is determined by variables linked to the characteristics of both the tourist affluence and the reception spaces.

In the case of urban tourism, there has traditionally been a very positive view of the effects of tourism. Its contribution to the local economy and employment has been highlighted. It has also been used as a vector for the physical regeneration of degraded spaces. Since the 1990s, however, there have been some initial reflections on the negative effects associated with the high influx of tourists in some urban destinations, such as historic cities (Glasson et al, 1995). In the current context, the negative effects identified a long time ago are intensifying and extending to other types of urban destinations. Among other aspects, references to processes such as the following are common:

- Changes in commerce and hospitality, with a general reorientation towards the foreign public;
- Decline in residential function, due to the expansion of hotels and the transformation of housing into housing for tourist use;
- Road mobility congestion;
- Congestion of public space;
- Trivialisation and standardisation of the urban landscape, with a growth of the tourist iconography;
- Problems of coexistence between residents and visitors, leading to a deterioration in the quality of life of local communities;
- Local irritation towards tourism resulting in tourism activity rejection movements (tourism-phobia).

The interpretation of these phenomena and their relationship to overtourism is complex. On some occasions, we talk about tourist gentrification, which implies a displacement of the popular classes of the neighbourhoods of the historical centres towards peripheral neighbourhoods due to an excess of tourist pressure (Arias-Sans, 2018; Cocola, 2018; Gotham, 2005; Gravari-Barbas & Guinand, 2017; Hiernaux & González, 2014). In others, tourism is referred to as a general process of functional change that implies a reorientation towards leisure and tourism activities (Freytag & Bauder, 2018; Romero, 2018; Sequera & Nofre, 2018). In any case, the existence of overtourism situations is recognized, that is to say, an excess of tourism that implies an increasing deterioration of the urban quality of life and even implies a possible loss of competitiveness of certain cities as tourist destinations. Addressing the study of these processes raises the need to define tourism overload indicators that allow dimensioning the phenomenon and its implications in urban spaces. But it also involves a deep reflection on the limits of application of these indicators. These limits derive from the absence of reference values and the operational difficulties to obtain data.

To meet these objectives, this chapter is organized into five sections. The first serves as an introduction to the topic. The second presents the state of art regarding the indicators, with special reference to indicators of tourism sustainability and overtourism. Based on a review of the existing bibliography, the third section raises a series of indicators of activity and tourism specialisation. The focus is on European cities and the application of these indicators is shown in the historic centre of Madrid. The fourth section looks at the perception of the phenomenon by different local stakeholders due to the absence of commonly accepted overtourism values, referring to these perceptions as valuation criteria. Finally, a final section of conclusions is drawn up.

INDICATORS OF SUSTAINABILITY AND TOURISM. THE STATE OF THE ART

There is a wide range of literature on indicators, applied or applicable to both tourism development and urban management. In its most basic definition, an indicator is an element that offers qualitative or quantitative information about a phenomenon and can be made up of one or more data referring to numbers, perceptions, facts, opinions or measures. In theory, the indicators make it possible to specify the magnitude, intensity, evolution and prognosis of a phenomenon or process, as well as to carry out their evaluation. They must comply with the fundamental premises of generating information, being methodologically consistent, scientifically valid, and easy to apply and communicate (López Palomeque et al, 2018).

Depending on the information they provide (level of elaboration and combination of data) there are two basic types of indicators: simple and complex or synthetic indicators (Sánchez Rivero & Pulido Fernández, 2008). The former provides information with a low level of treatment from secondary sources (published statistics) or from their own compilation (surveys e.g.). Whereas synthetic or index indicators arise from combining several simple indicators by means of a weighting system that places the components in a hierarchy. Some authors also differentiate the so-called "indicator systems" as a compilation of simple indicators whose result is interpreted jointly (López Palomeque et al, 2018; Torres-Delgado & López Palomeque, 2014; Torres-Delgado & Saarinen, 2013).

There are different classifications of indicators according to the purpose of what they measure or indicate. There are indicators of compliance, effect, impact, pressure, etc. In this sense, the typology proposed by the Organisation for Economic Co-operation and Development (OECD) in 1993, in its system of environmental indicators according to a Pressure-State-Response model, had a great projection. It was disseminated internationally and was re-adapted and reused in terms of sustainable tourism development. The model includes three types of indicators: a) Pressure indicators (reflecting direct and indirect pressures on the environment); b) State indicators (which describe the environmental conditions at a given time, that is, the quantity and quality of natural resources); and c) Response indicators (which correspond to the degree to which society responds to environmental changes and integrates the policies of the different territorial and sector administrations, as well as the actions of companies and social agents in the face of certain problems).

However, for these pressure, state and response indicators to be able for example to indicate levels of sustainability of tourism activity, their application requires the establishment of reference values. These values would describe the sustainability threshold for the different variables that make up the indicator system by comparing the state of the conditions of a given tourist destination at a given time (environmental, economic and social conditions resulting from anthropogenic pressures and natural processes) with the desirable state. However, it is almost impossible to find documents and practical cases in which the reference values that make up this desirable state are delimited, as this is a complex task, subject to value judgements and difficult to objectify for some dimensions of the development of tourism activity (perception of social impacts for example) or even overtourism as a global phenomenon.

Indicators of Sustainable Tourism Development

Indicators are attributed a key role in tourism planning and management processes. The World Tourism Organization itself has been working since the early 1990s on the development and application of indicators for sustainable tourism development at the destination level (WTO, 1995; WTO, 2004). It defines an indicator as "that quantitative and synthetic instrument that facilitates the analysis and evaluation of information so that, together with any other type of instrument, the decision-makers reduce the probability of inadvertently making unfortunate decisions" (WTO, 1995). It also raises the need to use different types (WTO, 2004; 14):

- Early warning indicators (e.g. decrease in the number of tourists intending to return);
- Indicators of pressure on the system (e.g. water scarcity or evidence of crime);
- Measures of the situation of the sector (e.g. occupancy rate, satisfaction of tourists);
- Measures of the impact of tourism development on biophysical and socio-economic environments (e.g. deforestation rates, changes in consumption patterns and income levels of local communities);
- Measures of management activities (e.g. cost of cleaning up coastal pollution);
- Measures of the effect, the results or performance of the management or response indicators (for example, change in pollution levels, greater number of returning tourists, etc.).

Although the development of indicators included in its documents is very complex, by way of summary it proposes 29 indicators referring to 12 basic issues (WTO, 2014; 244): satisfaction of residents with tourism, effects of tourism on the community, satisfaction of tourists, seasonality of the activity, economic benefits, energy management, management and quality of drinking water, and wastewater treatment, solid waste management, control of development and control of the intensity of tourist use.

With the motivation to improve tourism statistics in the area of sustainability, in 2015 UNWTO commenced, jointly with the United Nations Statistics Division, the initiative Statistical Framework for Measuring the Sustainability of Tourism (SF-MST). In its initial design, it was conceived as an organizing structure for integrating statistics on the economic, environmental and social dimensions of sustainable tourism, framed in Sustainable Development Goals (SDGs) y the 2030 Development Agenda. This is a long-term work, with the contribution of a large number of entities and experts. The system, which is still being developed, will allow a homogeneous recording of data on the economic, environmental and social dimensions, and applicable at different scales. (global, supra-national, national, regional, municipal or city-region, and local). The final objective is to establish indicators that can measure the sustainability of tourism and the contribution of tourism to the fulfilment of the SDGs.

The European Commission has also launched a *European System of Tourism Indicators (ETIS)* (European Commission, 2013 and 2016). This is a voluntary management tool designed to work with a common approach comparable at European level. Monitoring results are based on self-evaluation, observations, data collection and analysis by the destinations themselves. However, it does not set minimum values to be attained nor does it provide any type of certification. The system of indicators is intended only as an information tool that would allow destinations to monitor sustainability and, consequently, manage tourism activity more effectively. To this end, the European Commission's documents propose the use of 27 core indicators covering four areas: a) destination management (indicators related to public policy and visitor satisfaction), b) economic value (indicators of tourist flow, performance of tourism enterprises, employment and the functioning of the tourism supply chain), d) social and cultural impact (health and safety, gender equality, inclusion and accessibility, protection of cultural heritage and identity); and d) environmental impact (use of transport, climate change, solid waste management, wastewater treatment, water management, energy consumption and protection of diversity and landscape).

In addition to institutional initiatives, scientific publications have long reflected on the type and scope of sustainable tourism development indicators and the need to propose synthetic or aggregated indicators with empirical applications in different types of tourist spaces. Vera et al. (2001) elaborates a proposal for sustainable tourism development indicators adapting the PSR models of the OECD (Pressures, States, Responses) and the DPSIR of the Environment Agency of the European Union (Driving forces, Pressures, States, Impacts, Responses). They work with a structure that proposes four types of indicators: indicators related to the territorial or urban model, pressure indicators, state (quality) indicators and indicators of political and social response. López Palomeque et al. (2018) adapt the model of the European ETIS system and apply it to the municipalities of the province of Barcelona (Spain) proposing a synthetic index called the *Tourism Sustainability Index (ISOST)*, based on the aggregation of simple indicators, comparable at inter-municipal level and using diachronic logic (historical evolution). Navarro Jurado et al. (2012) develop for the Costa del Sol (Spain) a methodology to evaluate the growth limits of tourist destinations through a mathematical formulation (multi-criteria analysis) that takes as reference synthetic indicators applied to two scenarios: one of weak sustainability and the other of strong sustainability. And Schianetz & Kavanagh (2008) also develop a systemic indicator system and test it using as a case study a holiday eco-village project near Lamington National Park in Queensland, Australia. The list of bibliographical references on the subject is extensive, both in relation to different types of destinations, different forms of tourism and to the methodologies used to define the indicators (Miller, 2001; Choi & Sirakaya, 2006; Bossel, 1999, among others). However, the underlying question continues to address the gap between theoretical demands for knowledge and practical achievements (Céron & Dubois, 2003).

Overtourism Indicators

Although in recent years there has been a remarkable production of documents relating to overtourism, the absence of specific references to the measurement of the phenomenon and, therefore, to the subject of indicators is striking. In addition, the greatest efforts in this area correspond to the work of different types of public and private entities. In some cases, these entities use the most renowned research centres and authors in the study of this subject. In others, these are documents produced by some of Europe's leading tourism consultancies. In general, the use of long-established indicators is used, although efforts to apply them to urban spaces are particularly noteworthy. These are the spaces in which the problems of excessive tourist pressure and social response to tourism have been most recently recognised.

One of the first documents is *Coping with success: Managing overcrowding in tourism destinations*, conducted by the World Travel & Tourism Council in 2017. It seeks to understand the nature of the problem of tourism congestion and to identify specific solutions for its management. For this reason, it develops a series of indicators by which to understand the situation of the destinations. Specifically, it develops nine indicators: two related to the importance of tourism for the destination (average contribution of direct tourism to GDP and employment; and average annual growth rate of international and national arrivals) and seven related to the main challenges caused by congestion (tourist density -number of visitors per square km-, tourist intensity -number of visitors per resident-, negative TripAdvisor reviews, seasonality of arrivals, concentration of attractions, air pollution and predominance of historical sites). It applies these indicators to 68 tourist cities and organizes the results by quintiles for each indicator. Each group of indicators is associated with one of the manifestations of overtourism (threats to heritage, degraded tourism experience, etc.), although this association has many limitations. When a city's results are in the top quintile, it is considered to be at risk of overcrowding in that manifestation.

More recently, the European Parliament's Committee on Transport and Tourism has launched a report entitled *Overtourism: impact and possible policy responses* (Peeters et al, 2018). This is work carried out at European level using two areas of analysis: 1. Regions at NUTS level 2, where work is mainly done with available statistics; and 2. Specific destinations using statistics and surveys of members of the European Travel Commission. The report analyses management approaches in different locations and assesses policy responses. It also examines a series of quantitative indicators to establish whether they provide thresholds for overtourism. Some of the indicators considered were already mentioned in the previous report, such as tourist density, tourist intensity, seasonality or the contribution of tourism to GDP. However, it also raises other indicators such as the intensity of air transport, position of destinations with respect to airports, the growth of annual air transport, the intensity of cruises, the positioning of the destination with respect to a cruise port and the positioning of the destination with respect to World Heritage. The main contribution is the use of indicators aimed at measuring the effects of certain collaborative economy activities in the destination (Airbnb's share of the combined bed capacity of Airbnb and booking.com; Airbnb average shortest distance to booking.com; Airbnb's growth rate; and Google searches on 'Name of a city+Airbnb' averaged per month). The conclusion is very relevant: it is considered that a common set of indicators cannot be defined for the diagnosis of overtourism due to the complexity of the phenomenon.

In addition, European Cities Marketing, an association of European Destination Marketing / Management Organizations (DMOs), and the destination marketing agency Toposhophy produced the *Managing Tourism Growth in Europe* report in 2018. *The ECM Toolbox* (ECM, 2018). As in previous documents, a list of indicators for the measurement and monitoring of overtourism is proposed. However, it does not delve into the conceptual description of these indicators. In addition to the indicators mentioned above, others are proposed such as local population increase or decrease in specific areas; number of hotel and short term private accommodation rentals beds; property prices per square metre; local resident sentiment; visitor satisfaction analysis; etc. Although this is a battery of interesting indicators, its application is limited by the lack of data. It is also questionable whether these indicators are strictly linked to tourism (e.g. the price of real estate).

Also in 2018 the Austrian Hotel Association and the consultancy Roland Berger published the report *Protecting your city from overtourism. European city tourism study 2018.* They place 52 European cities in relation to two indicators: income per available room (RevPAR) in hotels and tourist intensity (in this report they refer to this indicator as tourist density). The two indicators give rise to a matrix, where cities are grouped into six clusters: 1. "Peak performance": cities with high RevPAR and tourist intensity; 2. "Under pressure", cities that have a high and increasing tourist intensity but a reduced RevPAR; 3. "Sustainable quality", cities with a comparatively low tourist intensity but with a high RevPAR; 4. "Unused potential", cities where both parameters are below European average levels; 5. "Shining stars" includes cities with sustainable levels of tourism, right in the middle of the matrix; and 6. "Mass Trap", those cities with high tourist intensity and low RevPAR.

Although the quality of the contributions is variable, the profusion of documents in such a short time is particularly striking. It is also necessary to consider the work of the WTO mentioned above (WTO, 2018 and 2019) and some contributions of a more academic nature that include more or less partial references to the subject of indicators (Muler et al, 2018; Weber et al, 2017).

INDICATORS OF TOURIST ACTIVITY AND SPECIALISATION

A basic conclusion can be drawn from the review of the scientific and institutional literature on indicators of tourism overload: there is a great deal of theoretical reflection that contrasts with its problems of practical application. The greatest difficulties derive from the absence of data that can be fed into the indicators, data that allow situations to be compared over time and in different places. And if the data exist, there are no commonly accepted values that mark the moments of tourist overload.

The theoretical models for the development of indicators for application in urban spaces state that these indicators should be associated with the different factors or dimensions of the central concept to be evaluated, in this case, overtourism. Based on studies of tourist carrying capacity, the literature on the subject proposes a broad set of common management dimensions. On the other hand, most of the work carried out to date uses very general indicators that refer to the initial idea of a tourist impact whose scope and meaning is determined by two vectors: firstly the volume and characteristics of the influx of visitors; and secondly the conditions specific to the destinations, in this case the cities and more specifically the urban centres.

As a work proposal, "indicators of tourist activity" are those that focus on the first impact vector and basically include magnitudes related to the influx of visitors. In addition, the "specialisation indicators" aim to establish the scope and meaning of tourism at the local level. The simplest are basic indices that handle a single magnitude relative to the influx of visitors. Complex indices are more sophisticated, as they relate two different magnitudes: one related to visitors and the other related to the city. As noted above, there are no commonly accepted values about the moment when overtourism situations occur. Therefore, the values resulting from these indicators have only a comparative value. In one place, the dynamics of the values are compared over a year and at a year-on-year level, establishing trends. Values are also compared between different spaces, at the level of cities or in relation to different areas of the city. As has been pointed out repeatedly, tourist activity tends to concentrate on certain sectors of urban space.

Indicators of Tourist Activity

These are the simplest and least expressive indicators, as they only reflect magnitudes related to the influx of visitors. Although in some cities estimates have been made of this influx, in most cases the number of travellers staying in tourist establishments and overnight stays is used. Comparisons between countries are difficult because each national statistical system uses classifications of different facilities and collects information from one or another type of facility. The territorial unit referring to the destination also varies, even more so when it comes to urban centres.

The tourist activity figures provided by TourMIS, drawn up on the basis of contributions from local DMOs, reflect these difficulties well. Within what the system considers to be the best available option, the categories used vary significantly: NGS = Bednights in hotels and similar establishments in greater city area; NAS = Bednights in all forms of paid accommodation in greater city area; NG = Bednights in hotels and similar establishments in city area only; NA = Bednights in all forms of paid accommodation in city area only.

On other occasions, more partial activity indicators are used, reflecting a single facet of visitor influx. For business tourism, data from the International Congress and Convention Association (ICCA) on the number of meetings are used. For cruises, the Cruise Lines International Association (CLIA) provides data on the number of vessels and passengers per port. In general, this information comes from official sources in each country. The International Air Transport Association (IATA) provides information on air traffic (flights, passengers, connections, etc.), although data for each city and airport are published by national statistical authorities.

The ability of these indicators to show overload situations is limited. However, their interest increases if the data are analysed from two perspectives. Firstly, the growth dynamics shown by the variable used: number of tourists, cruisers, dwellings for tourist use and others. According to TourMIS data, in 2018 the number of overnight stays in the 15 most visited cities in Europe amounted to 373.6 million. With respect to 2011, the year Paris is incorporated into the system, more than 73 million overnight stays have been earned, 25.42%. Moreover, this figure has been surpassed in a large number of cities: Amsterdam (71.05%), Berlin (47%), Budapest (57.15%), Hamburg (52.46%), Istanbul (73.95%) or Lisbon (67.62%), among others.

The second perspective corresponds to the temporal distribution of the variable. Seasonal concentration implies an unequal distribution of pressure throughout the year, with moments that concentrate tourist activity and potentially overtourism situations. As before, it can be applied to the number of tourists, overnight stays, cruise passengers, etc. In general, the difference between the months with the highest and lowest values of each variable is taken into account. The distribution of overnight stays collected by TourMIS points to a high seasonality in the large capitals of Central Europe, while in cities such as Madrid the tourist inflow is quite stable throughout the year. At a general level, the increase in tourist influx and seasonality favour situations of overtourism. When both processes come together the risk of these problems increases. Therefore, it is especially necessary to analyse whether the current growth of urban tourism maintains or changes the patterns of temporal distribution of the influx in each city.

Until recently, most indicators were based on data from official statistics. Today, big data are also captured and processed using data mining techniques and complex statistical operations. In many cases, these data are geolocated, which allows an approximation with a high level of detail to the real areas of tourist concentration, the places where overtourism situations appear. It is very common to capture data from different social networks, especially those that allow sharing comments and photos. Based on previous work (Kádár & Gede, 2013), Kádár (2014) uses the photographs available on Flickr.com to record the distribution of tourist activity in three Central-Eastern European capital cities: Vienna, Prague and Budapest. From a broader perspective, Salas Olmedo et al (2018) compare the digital fingerprint left by tourists in Panoramio (as an approximation to the spaces of the visit), Foursquare (spaces of consumption) and Twitter (interactions with enough association to the spaces of the accommodation). In addition to identifying the main tourist areas in the city of Madrid, they establish areas of tourist specialisation such as destinations, shopping and hospitality consumption or accommodation.

Data mining is also used to quantify and characterize the supply of tourist housing. The influx of visitors to a city has always been greater than tourists staying in commercial tourist establishments, mainly hotels. This situation has been aggravated by the recent expansion of the houses for tourist use, with very large amounts both of an illegal or non-legal nature. This expansion implies a very notable change in urban tourism, in the growing process of “Airbnbization” (Richards, 2017). It puts direct pressure on the residential market, turning housing into a tourist commodity to the detriment of the local population. Given that the large shared hosting platforms do not provide information on supply and demand at the level of each city, different groups are using web scraping techniques, with downloads of large volumes of data repeated from time to time, to which data mining is applied to obtain structured information with the data that support the public denunciation of the evolution of the phenomenon. For example, Inside Airbnb extracts information from Airbnb advertisements, which allows the offer of dwellings for tourism purposes to be quantified and analysed on the basis of variables such as room type, activity or availability. On its website, it provides an approximate location of the dwellings and a grouping of data at the level of areas of the city that uses the administrative boundaries in force in each country. The work that uses this and similar sources is substantial (among others, see Freytag & Bauder, 2018; Gutierrez et al, 2017), in many cases from very critical perspectives of the phenomenon (among others: Cócola, 2016; Lee, 2016).

The availability of georeferenced data for different dates allows a wide range of analysis from different methods of cartographic representation. These methods may have different levels of sophistication in data processing. At the most basic level, the location of the elements to be represented on a point map is simply shown. These elements can also be added according to a superimposed grid and represented in three dimensions so that the height provides information about the number of elements in each cell, as can be seen in Figure 1. Other methods involve the use of spatial analysis tools such as kernel density, which allow the visualization of heat maps. This technique has been used, for example, to represent the expansion of tourist housing in the historic centre of Palma de Mallorca (Yrigoy, 2017), and is shown in Figure 2 for the city of Madrid.

These images have been generated following the methodology used by Kadar and Gede (2013) for the analysis of the distrubtion patterns of photos uploaded to Flickr by tourists and resident. In this case the georeferenced photos taken between 2015 and 2018 in the historic centre of Madrid have been downloade generating a layer of points. Tourist have been classified as those users who have taken photos

Figure 1. 3D model of Flickr photos taken by tourists and residents in Madrid

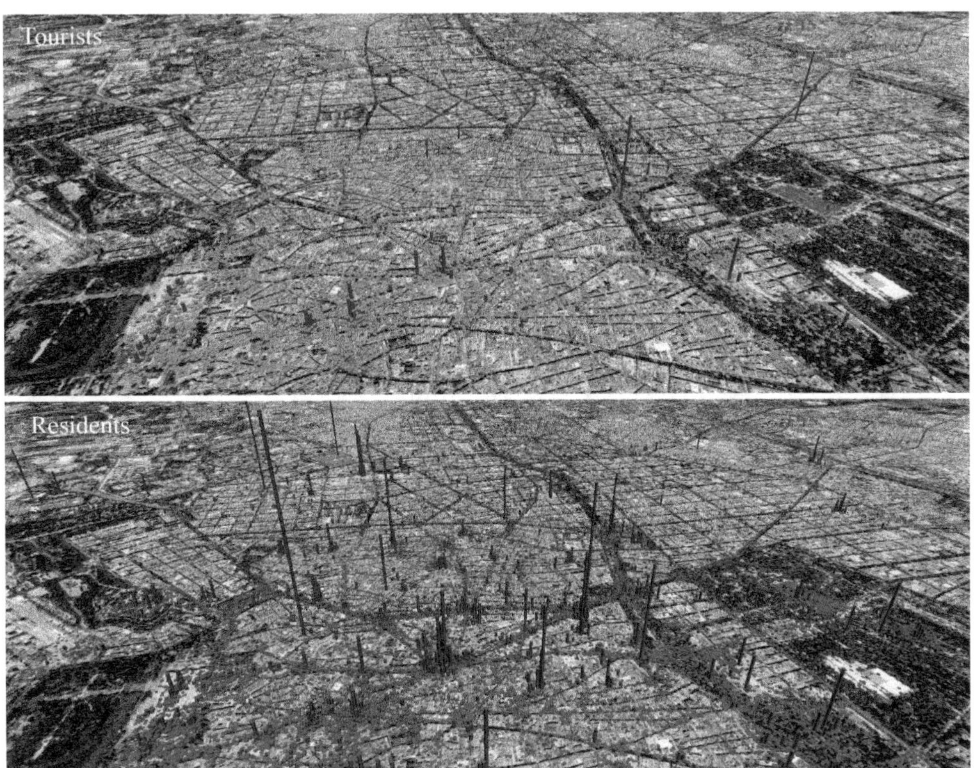

for a maximum of three consecutive days. The point layer has been regrouped according to a 100 m2 hexagonal tessellation that has been afterwards extruded in ArcGlobe according to the number of photos. The comparison of the two images shows how the tourist use of the city is much more concentrated around a few landmarks while resients make more extensive use and show interest in local elements that do not appear in the tourist pattern.

Tourism Specialisation Indicators

This type of indicator establishes a relationship between a variable related to tourism and another that has to do with an aspect of the local reality. In their simplest formulation they correspond to simple indices that work with a single magnitude. For example, the weight of tourism in the GDP and/or employment of the city, usually expressed as a percentage. Although estimates of this weight are common in national

Figure 2. Evolution of AirBnB listings density in Madrid

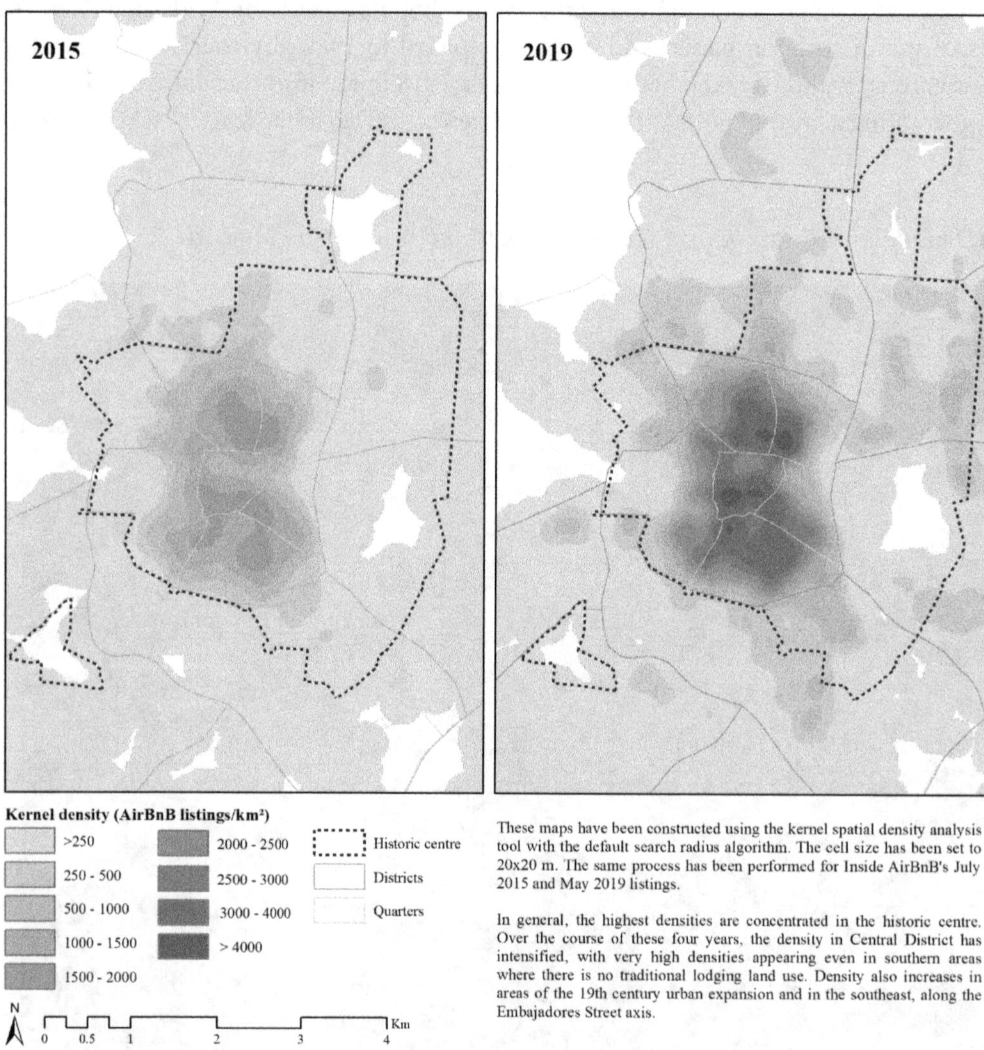

These maps have been constructed using the kernel spatial density analysis tool with the default search radius algorithm. The cell size has been set to 20x20 m. The same process has been performed for Inside AirBnB's July 2015 and May 2019 listings.

In general, the highest densities are concentrated in the historic centre. Over the course of these four years, the density in Central District has intensified, with very high densities appearing even in southern areas where there is no traditional lodging land use. Density also increases in areas of the 19th century urban expansion and in the southeast, along the Embajadores Street axis.

and regional studies, they are less frequent at the urban level as there are usually no statistics with this degree of territorial disaggregation. Given the growing importance of tourist accommodation and its association with some discomfort in urban centres, indicators are also beginning to be used which measure the proportion of accommodation capacity in tourist accommodation over total accommodation capacity, or the number of tourist accommodation over the total number of dwellings or the number of homes.

Complex indices are more sophisticated, as they relate two different magnitudes. As a result, the difficulties in obtaining data are even greater. Indicators of tourist intensity are usually based on the resident population. Variables related to tourism include the number of travellers, overnight stays or, failing this, accommodation capacity. At the European level, TourMIS relates the number of overnight stays and inhabitants. As indicated above, the territorial units of statistical reference and the types of accommodation considered vary according to destination. In addition, the population calculation year also differs. Taking these caveats into account and taking as a reference the overnight stays recorded in 2018, the European cities with the highest tourist intensity were Venice (31.4 overnight stays per inhabitant), Lisbon (25.6), Salzburg (24.4), Amsterdam (19.7) and Bruges (18.5). For Rome, the figure is 9.9, above Berlin (9.1), Vienna (8.7), London (6.6), Budapest (5.9) and Paris (4.1). In Spain, Barcelona and Madrid register a similar number of overnight stays: 19.2 million and 19.8 million, respectively. On the other hand, the tourist intensity of Barcelona amounts to 11.9 while in Madrid it is only 6.2 overnight stays per inhabitant, as it is a municipality with a much larger resident population.

Approximations at more detailed scales tend to use the number of accommodation places, as data on tourists and overnight stays are generated at a municipal level. The level of disaggregation of data on resident population varies from country to country. For example, in Spain there is free access to the number of inhabitants at the level of districts, neighbourhoods and census sections. There is also this information on the number of dwellings and households, so it is common to establish indicators that attempt to measure the pressure of the tourist dwelling on the housing stock. Figures 3 and 4 show the application of tourist intensity indicators by census section in Madrid.

Within the complex indices, indicators of tourist density are also widely used. These indicators refer to the surface area. At the city level, values are managed on the number of tourists, overnight stays or lodging locations per unit area (hectare, km2). Since visitors tend to concentrate on the central spaces of each city, the values obtained barely reflect tourist pressure. Alternatively, WTTC & McKinsey&Company (2017) work uses a density indicator calculated as 2016 arrivals divided by the number of square kilometres in the area encompassing TripAdvisor's top 20 attractions for the destination.

If information on the exact location of the accommodation floor is available, it is possible to carry out analyses at a higher level of detail. Administrative divisions and conventional statistics are sometimes used. In their work on Madrid, Sorando & Ardura (2018) use an indicator of the number of Airbnb listings per hectare at the census section level. A simple comparison between the 2015 and 2017 results reflects an increase in density in the central areas and an expansion of this accommodation formula in neighbouring neighbourhoods that until now did not have a commercial tourist accommodation offer. Since administrative boundaries sometimes present problems of adjustment to the distribution patterns of tourist activity over the city, one can resort to the use of geometric figures that result from superimposing a mesh on the urban space, as shown in Figure 5.

The use of tourist intensity and density indicators based on the number of tourists or overnight stays implies some distortion. Unlike population figures, which reflect the number of residents at any given time, tourist influx figures are the sum of tourists or overnight stays recorded over a period of time, usually a year. In order to minimise this distortion, Weber et al (2017) introduce a time component in

Figure 3. Madrid. Tourist intensity: total places of accommodation/1000 inhabitants

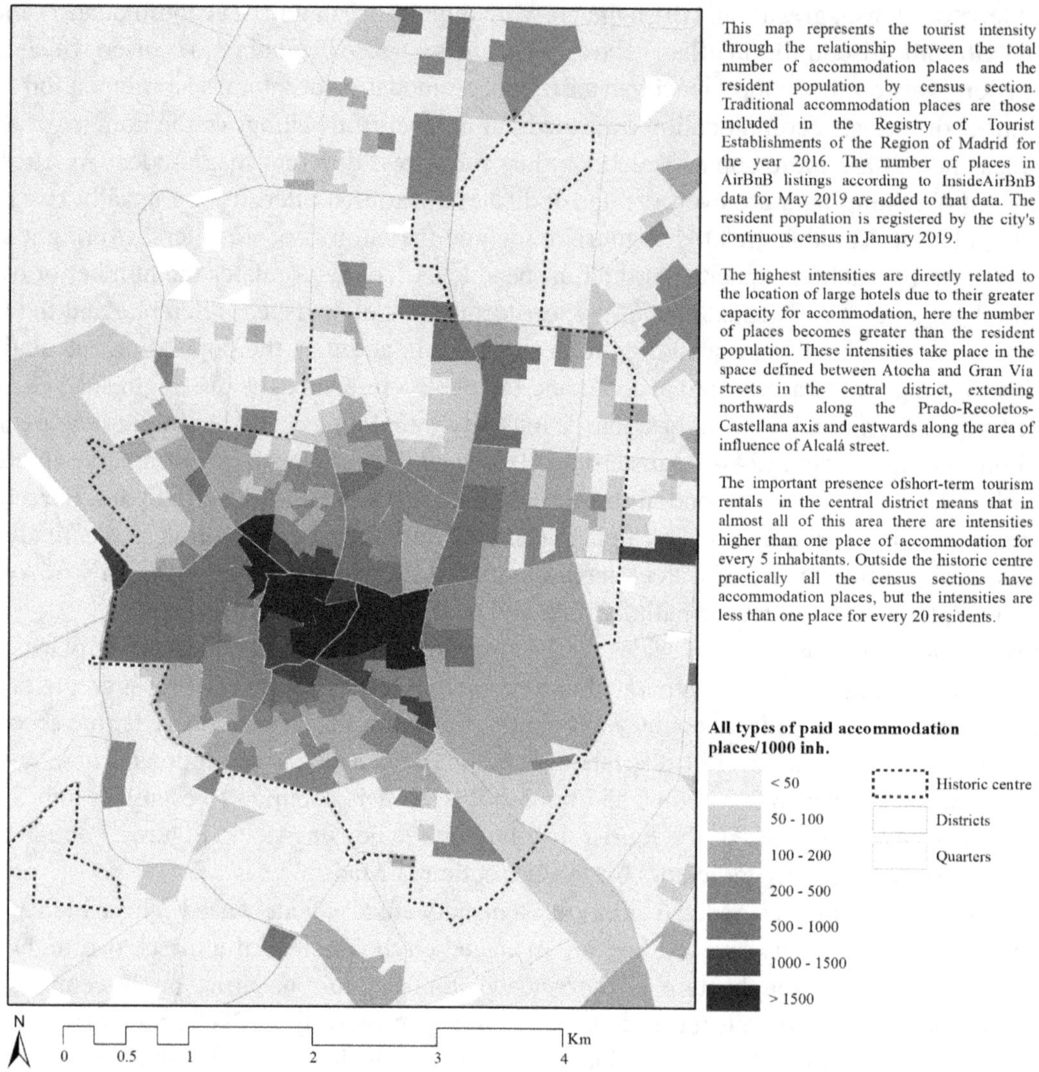

two of their pressure indicators: 1) Tourist Penetration Rate (TPR) - the number of tourists in the city per day and per 100 inhabitants on average; and 2) Tourist Density Rate (TDR) - the number of tourists in the city per day and per 1 km^2.

STAKEHOLDERS AND PERCEPTION OF OVERTOURISM SITUATIONS

The indicators set out in the previous sections allow situations to be compared at a temporal level as well as between places. However, at the present time, we are not able to set limits to determine thresholds for overtourism situations. These thresholds have a strong value component: it implies being able to answer the question of when many tourists are too many. And there is no single answer to this question. Contemporary urban societies are intrinsically heterogeneous, made up of different stakeholders who

Figure 4. Madrid. Evolution of tourist intensity: AirBnB listings/households

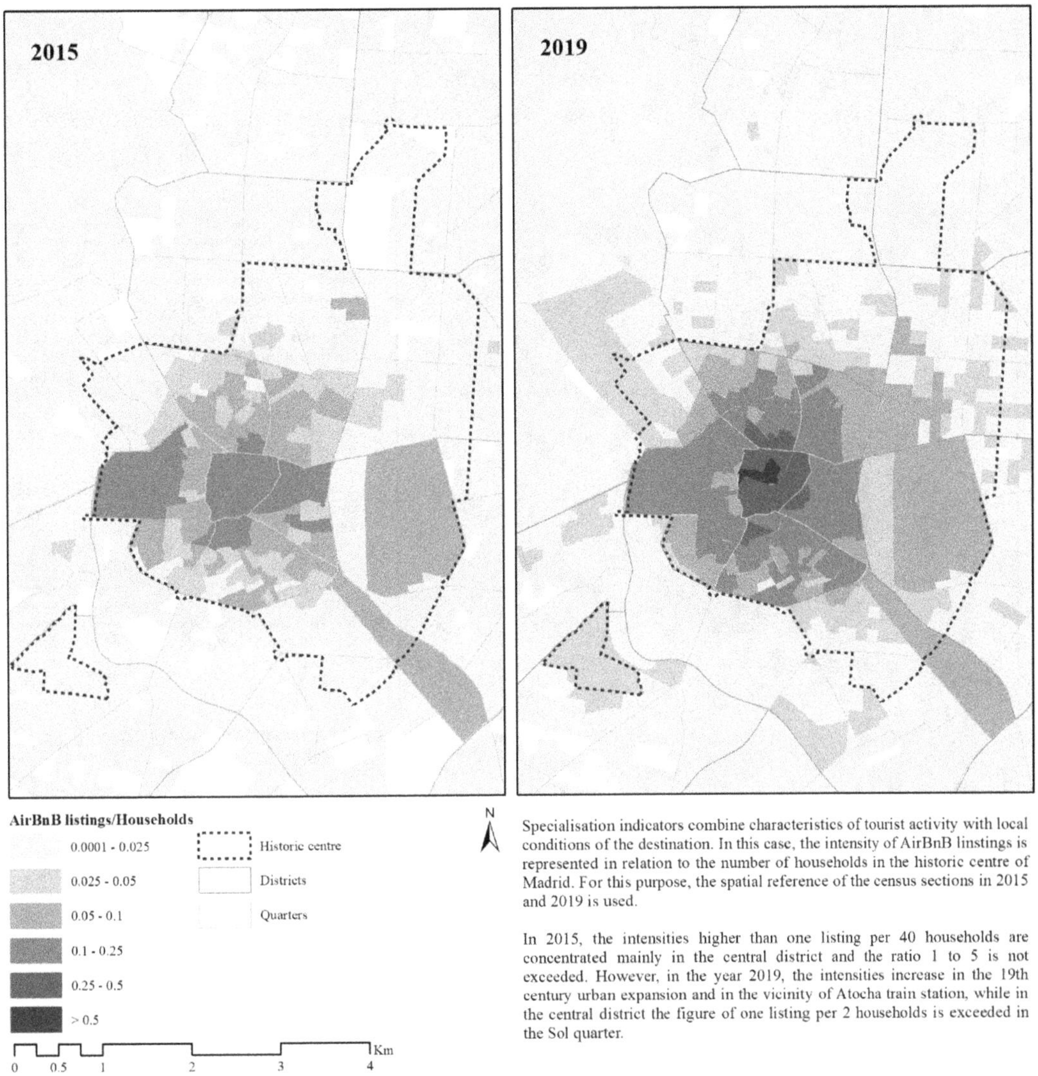

Specialisation indicators combine characteristics of tourist activity with local conditions of the destination. In this case, the intensity of AirBnB linstings is represented in relation to the number of households in the historic centre of Madrid. For this purpose, the spatial reference of the census sections in 2015 and 2019 is used.

In 2015, the intensities higher than one listing per 40 households are concentrated mainly in the central district and the ratio 1 to 5 is not exceeded. However, in the year 2019, the intensities increase in the 19th century urban expansion and in the vicinity of Atocha train station, while in the central district the figure of one listing per 2 households is exceeded in the Sol quarter.

will respond differently according to their interests (Becken & Simmons, 2019). In the recent WTO (2018) work on overtourism, Koens identifies five stakeholders whose interests and positions need to be considered: 1. Residents and other local users of the city (temporary residents and commuters); 2. Domestic and international visitors; 3. Tourist industry players; 4. Tourism policy-makers and DMOs; and 5. Wider city context stakeholders (policy departments outside of tourism, non-tourism businesses, representative organizations/unions of residents, representative organizations of natural environment/cultural heritage protection, etc.).

Local government entities play a key role. If those responsible for these entities identify situations of overtourism, it is more likely that an explicit policy will be adopted to combat the negative effects of tourism growth. In the case of Spain, the authors of this text are currently working on an on-going investigation into the perception of the technical levels of the town councils of the main Spanish tourist cities of the problems generated by tourism in urban centres. In the first phase, an online questionnaire is

Figure 5. Madrid. Density of accommodation places from hexagonal grid

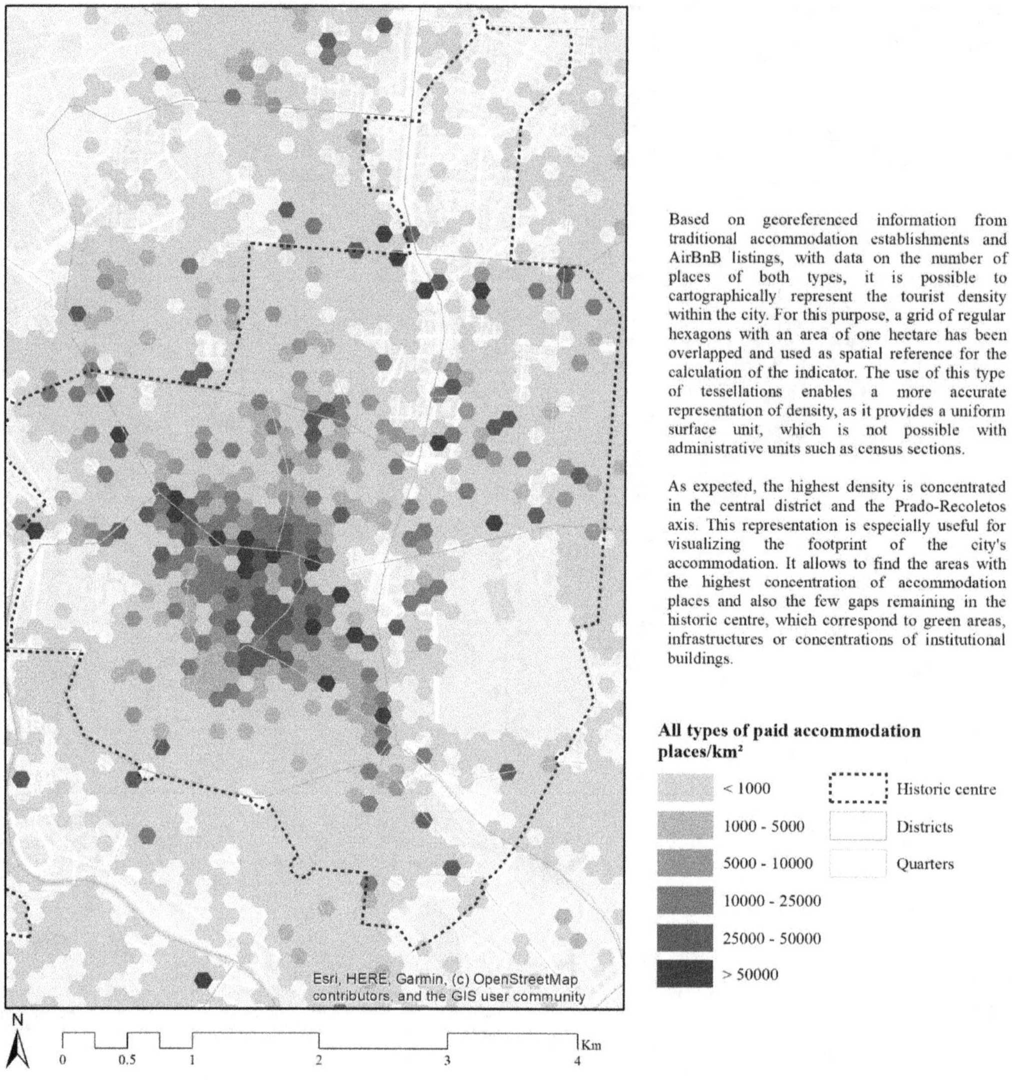

used to gather as many responses as possible. Among other questions, it is asked whether they have the perception that there is excessive tourist pressure in the city or, as the case may be, in the urban centre or in one of the main tourist spaces. Additionally, some of the problems that usually affect the centres of the Spanish tourist cities are pointed out and the respondent is asked to indicate the degree of responsibility that they place on tourism regarding the development of this problem. Preliminary results indicate that 70% of those interviewed indicated the existence of excessive tourist pressure in their city. At a second level, the spatial areas affected by this pressure are researched. Of the total responses obtained, 10% considered that the pressure was limited to some very emblematic buildings, 34% to certain streets and squares in the centre, 21% to the entire urban centre and 5% to the city as a whole. Once this first phase of the research is completed, a series of face-to-face interviews will be conducted to contextualize the results obtained in the general survey.

Business associations have also begun to incorporate the topic of overtourism into their institutional discourses. There is notable concern about the expansion of a negative view of tourism that is spreading in many mass media. This view contrasts with the traditional image of urban tourism, which was generally considered to be a very beneficial activity for cities. The institutional positions of the tourist associations allude to the need to sustainably manage the tourism success story, a management in which their participation is necessary. Problems associated with overtourism are recognized, but these problems do not derive from the simple growth of urban tourism but from its deficient management. Their argument thus appears to be aligned with the official approaches of UNWTO and many DMOs. Specifically, the consolidated tourism sector blames overtourism on the recent expansion of holiday homes driven by platform companies such as Airbnb. Thus, in a recent report of the Asociación Empresarial Hotelera de Madrid (2018), two major challenges of tourism in the city are pointed out: 1. The disproportionate growth in the supply of unregulated accommodation; and 2. Overcrowding of destination and problems of coexistence. And they consider both problems to be fully related.

Business organizations accuse social movements of encouraging local rejection of tourism, a phenomenon known as "tourism-phobia. As part of a general approach to the defence of the "right to the city", these movements have recently incorporated a very negative view of tourism as a manifestation of the commercialisation of the city favoured by the public authorities. Gentrification and touristification appear as parallel phenomena. The reaction to these phenomena is widespread in many European cities, as reflected in the compendium of Colomb & Novy (2016). Within critical studies, it is common to map urban conflicts, often in a participatory manner. This technique can be used to develop an indicator of tourism-related conflicts, indicating their weight in relation to the total and their specific nature. In this sense, Romero's work (2018) shows the recent growth of conflicts linked to the implementation of sharing economies, both in terms of accommodation (Airbnb) and transport (Uber, Cabify). Also in Madrid most of the protests are directed against the expansion of housing for tourist use (Gil & Sequera, 2018), with those central residential neighbourhoods which, until recently, have been off the beaten track, being especially virulent.

Social movements tend to present themselves as the voice of the citizenry, but their representativeness is very limited. In general, the attitude of the resident population towards tourism development is considered to be one of the critical dimensions of a territory's social carrying capacity. This attitude, which will never be homogeneous, is determined by a multiplicity of aspects (Almeida et al, 2015). Among other things, it is common to cite the level of economic dependence on tourism, the sense of attachment to the place and the degree of direct contact with tourists and travellers. The majority use surveys, in which the interviewee is asked to take a position on certain issues.

In general, this form of work is maintained in the most recent research on the local perception of overtourism. In their work for WTO (2018), Postma & Papp analyse residents' perceptions of tourism growth in eight major European cities. The results indicate that the local population recognizes positive and negative effects both at the personal level and in relation to the community as a whole. The majority of respondents consider that there is still room for further tourism growth, although there is a significant minority, for whom tourism growth is an issue. However, this group is not unified on where the emphasis of tourism growth or lack thereof should lie. Perhaps this minority nurtures local anti-tourism movements and encourages a certain media environment of tourism-phobia. Perceptions also vary widely between cities. In Barcelona, municipal surveys have long placed tourism as one of the main problems of the city in the opinion of its inhabitants.

The deterioration in tourist satisfaction can also be considered an indicator of overtourism. The growth of the tourist influx also has disadvantages for the following: increased waiting time to access museums and monuments, difficulties in finding space in bars and restaurants, increases in the prices of goods and services, crowded streets and squares, etc. and even the feeling of loss of authenticity of the places visited, a particularly sensitive aspect in urban heritage sites. The current wave of tourism deteriorates the objective conditions in which the visit takes place. However, it is not clear whether this affects the global competitiveness of the destination. The studies carried out in Florence by Popp (2012) reflect the complexity of the subject. Using a qualitative methodology, it captures the moments and places where tourists perceive congestion. In these occasions, the congestion by an excessive tourist density does deteriorate the experience of the place. But it also identifies situations of positive agglomeration: visitors understand that a certain bustle is part of a living city and the very presence of other tourists creates a sense of security for the less experienced traveller. In any case, it remains to be determined how the perception of agglomeration affects the selection of the destination, the subsequent travel recommendations or the possibility of repeating the visit.

Most of the work on stakeholders' perception of tourism is based on surveys. In addition, semi-structured interviews and analysis of documents generated by the more formal organisation groups are carried out. These techniques are more expensive. In addition, the studies carried out are very focused on detecting problems in each city, so the results are difficult to compare. As a recent alternative, it is possible to use the capture of massive data from social networks and other platforms that collect comments from different stakeholders. Semantic analysis of messages posted on Twitter, for example, can help identify certain states of opinion about tourism in a city. These opinions reflect the perceptions and positions of organized groups such as social movements, business organizations and political parties, but also of citizens and Twitter visitors. The geolocation of the messages itself makes it possible to identify the places and moments where dissatisfaction with tourism is highest. In the case of TripAdvisor, it only collects ratings and comments from tourists, although the analysis of the most negative ratings can help identify a number of negative aspects commonly associated with overtourism situations.

CONCLUSION

For some time now, problems related to tourist overload have been identified in different types of places, in many cases spaces with important heritage values. There is already a remarkable academic tradition of carrying capacity studies and an important doctrinal corpus from international bodies such as ICOMOS, UNESCO's World Heritage Centre, UNEP and UN-Habitat. In 2005 the WTO published a manual entitled *Tourism Congestion Management at Natural and Cultural Sites*, with a strong orientation towards planning and management. Within this framework, a large number of tourism sustainability indicators were designed, to be applied to different types of resources and destinations.

Tourist pressure is currently increasing, especially in city centres. This increase stems from two factors. Firstly, the growth in the influx of visitors, especially in cities with very good air accessibility and cruise ports. And secondly, the expansion of holiday homes in the residential neighbourhoods of urban centres, a circumstance which threatens the residential status of these places. The preservation of this condition has been a central objective of the urban management of many European cities over the last forty years. As a reaction to the increase in tourism pressure, social movements have incorporated the denunciation of the most negative effects of tourism on the city into their discourse.

In this context, the need arises again to establish indicators that allow measuring the magnitudes and dynamics of the phenomenon. A considerable effort has been made by different public and private entities to identify these indicators in order to achieve a more sustainable management of tourism growth. The list of proposed indicators is very extensive, although it is difficult to attach them to specific dimensions of a site's carrying capacity and, in most cases, there are no homogeneous spatial and temporal data with which to feed these indicators. As has been pointed out repeatedly, it contrasts the efforts of theoretical reflection with the possibilities of practical application.

Based on the review of documents and the search for data, it is possible to identify a series of indicators of common application to urban destinations. These indicators are grouped into the following types:

- Indicators of tourist magnitude, which only indicate the tourist importance of a city in relation to others. The number of visitors, tourists staying in hotel establishments, cruise passengers, airline passengers or, otherwise, accommodation location is used.
- Indicators of tourism dynamics, which reflect the dynamics of tourism over a series of time. The capacity of a place to assimilate tourism is much greater when it comes to gradual growth. Whereas currently it is a matter of accelerated growth, especially regarding the expansion of housing for tourist use.
- Indicators of seasonality in tourism, alluding to the temporary distribution of the influx in a year. If this affluence is very concentrated, the tourist pressure in these periods is much greater. Among other aspects, it is a key issue for the design of urban infrastructures and services.
- Indicators of tourist intensity, which relate figures of tourist magnitude and values with magnitudes of local order, such as the number of residents, dwellings or homes. They reflect the importance of tourism in that city and, internally, in its different areas and neighbourhoods.
- Tourism density indicators, produced similarly to tourist intensity indicators but based on surface magnitudes.

With respect to the traditional line of study of indicators, the most recent reflections on overtourism in urban destinations incorporate two novelties. In the first place, the importance given to the monitoring of housing for tourist use. This is a totally disruptive factor at the level of urban tourism, which greatly favours the processes of residential displacement associated with tourist gentrification. These processes are affecting, above all, spaces that are off the beaten track, neighbourhoods in the process of touristification due to the expansion of this accommodation formula. Secondly, the use of web scraping techniques and the application of data mining to obtain supplementary data to feed possible overtourism indicators is also new. These techniques are applied, for example, to understand the magnitude and dynamics of the supply of housing for tourist use marketed through platforms such as Airbnb. The fact that the data are geolocalized also allows for a much more detailed spatial analysis, identifying the places where tourist pressure is concentrated at a much lower cost than conventional systems used in the past.

However, all the recent reflection on overtourism and indicators leads us to the same problem as with previous work on carrying capacity in urban destinations. In essence, the problem lies in the inability to establish values based on which many visitors can be considered too many. In other words, values that establish the point of equilibrium from which the disadvantages of tourism outweigh its benefits. Responding to these questions implies investigating the local perception of tourism since similar figures on magnitude, dynamics, seasonality, intensity and density of tourism can give totally disparate local responses. This is done through surveys, interviews and, more recently, the capture of opinions in social networks with the techniques of web scraping and data mining mentioned above. In any case, contemporary urban societies are far from homogeneous, so each stakeholder will perceive the advantages and disadvantages of tourism according to their principles and interests.

In addition, the spatial concentration of tourist activity in urban centres implies an unequal distribution of benefits and burdens. The growth of tourism benefits the urban community in terms of GDP, employment and overall prosperity. On the other hand, the most negative effects are concentrated in urban centres, in such a way that their residents accumulate the problems of tourist pressure when they only receive a reduced part of the potential benefits that tourism generates for the city. Solving this gap is one of the challenges of urban management in the coming years, and will only be achieved through an agreement between the different stakeholders based on the model of urban centre, tourist destination and city that is desired in the medium and long term. The definition of this model must serve as a reference scenario to determine the sustainability or unsustainability of current tourism growth, therefore establishing the carrying capacity thresholds that give a sense of value to all indicators designed or yet to be designed. An exciting challenge that must give its support to a new research agenda.

REFERENCES

Almeida García, F., Balbuena Vázquez, A., & Cortés Macías, R. (2015). Resident's attitudes towards the impacts of tourism. *Tourism Management Perspectives*, *13*, 33–40. doi:10.1016/j.tmp.2014.11.002

Arias-Sans, A. (2018). Turismo y Gentrificación. Apuntes desde Barcelona. *Papers. Regió Metropolitana de Barcelona. Territori, estratègies, planejament*, 60, 130–139.

ASOCIACIÓN EMPRESARIAL HOTELERA DE MADRID. (2018). *Análisis estratégico de los riesgos y escenarios de futuro del sector turístico y de hospedaje de la Comunidad de Madrid. Propuestas de actuación 2018.* Available at https://aehm.es/wp-content/uploads/2018/04/Propuestas_AEHM_2018vf.pdf

Austrian Hotelier Association & Roland Berger. (2018). *Protecting your city from overtourism. European city tourism study 2018.* Munich, Germany: Roland Berger GmbH.

Becken, S., & Simmons, D. G. (2019). Stakeholders management: different interest and different action. In R. Dodds, & R. W. Butler (Eds.), *Overtourism. Issues, realities, and solutions* (pp. 234–249). Berlin, Germany: De Gruyter; doi:10.1515/9783110607369-016

Bossel, H. (1999). Indicators for sustainable development: theory, method applications. A report to the Balaton group. Winnipeg, Canada: International Institute for Sustainable Development (IISD).

Céron, J. P., & Dubois, G. (2003). Tourism and sustainable development indicators: The gap between theoretical demands and practical achievements. *Current Issues in Tourism*, *6*(1), 54–75. doi:10.1080/13683500308667944

Choi, H. C., & Sirakaya, E. (2006). Sustainability indicators for managing community tourism. *Tourism Management*, *27*(6), 1274–1289. doi:10.1016/j.tourman.2005.05.018

Cocola Gant, A. (2016). Holiday Rentals: The New Gentrification Battlefront. *Sociological Research Online*, *21*(3), 10. doi:10.5153ro.4071

Cocola-Gant, A. (2018). Tourism gentrification. In A. L. Lees, & M. Phillips (Eds.), Handbook of Gentrification Studies (pp. 281-293). Cheltenham, UK: Edward Elgar. doi:10.4337/9781785361746.00028

Colomb, C., & Novy, J. (Eds.). (2016). *Protest and resistance in the tourist city*. New York, NY: Routledge. doi:10.4324/9781315719306

del Romero Renau, L. (2018). Touristification, Sharing Economies and the New Geography of Urban Conflicts. *Urban Science*, *2*(4), 104. doi:10.3390/urbansci2040104

Dredge, D. (2017). "Overtourism" Old wine in new bottles? Retrieved from https://www.linkedin.com/pulse/overtourism-old-wine-new-bottles-dianne-dredge

ECM-European Cities Marketing. (2018). *Managing Tourism Growth In Europe The ECM Toolbox*, 30.

EUROPEAN COMMISSION. (2013). *The European tourism indicator system. Toolkit for sustainable destinations*. Luxembourg: European Union.

EUROPEAN COMMISSION. (2016). *The European Tourism Indicator System. ETIS toolkit for sustainable destination management*. Luxembourg: Publications Office of the European Union; doi:10.2873/983087

Freytag, T., & Bauder, M. (2018). Bottom-up touristification and urban transformations in Paris. *Tourism Geographies*, *20*(3), 443–460. doi:10.1080/14616688.2018.1454504

García Hernández, M. (2003). *Turismo y conjuntos monumentales. Capacidad de acogida turística y gestión de flujos de visitantes*. Valencia, Spain: Tirant lo Blanch.

Gil, J., & Sequera, J. (2018). Expansión de la ciudad turística y nuevas resistencias. El caso de Airbnb en Madrid. *EMPIRIA. Revista de Metodología de Ciencias Sociales*, 41, 15-32.

Glasson, J., Godfrey, K., Goodey, B., & Absalom, H. (1995). *Carrying Capacity and Management Responses in Europe's Historic Towns and Cities*. Aldershot: Avebury.

Gotham, K. F. (2005). Tourism gentrification: The case of new Orleans' vieux carre (French Quarter). *Urban Studies (Edinburgh, Scotland)*, *42*(7), 1099–1121. doi:10.1080/00420980500120881

Gravari-Barbas, M., & Guinand, S. (Eds.). (2017). *Tourism and gentrification in contemporary metropolises: international perspectives*. New York, NY: Routledge. doi:10.4324/9781315629759

Gutiérrez, J., García Palomares, J. C., Romanillos, G., & Salas Olmedo, M. H. (2017). The eruption of Airbnb in tourist cities: Comparing spatial patterns of hotels and peer-to-peer accommodation in Barcelona. *Tourism Management*, *62*, 278–291. doi:10.1016/j.tourman.2017.05.003

Hiernaux, D., & González, C. I. (2014). Turismo y gentrificación: Pistas teóricas para una articulación. *Revista de Geografía Norte Grande*, *8*(58), 55–70. doi:10.4067/S0718-34022014000200004

Huete, R., & Mantecón, A. (2018). El auge de la turismofobia ¿hipótesis de investigación o ruido ideológico? *Pasos (El Sauzal)*, *16*(1), 9–19. doi:10.25145/j.pasos.2018.16.001

Kádár, B. (2014). Measuring tourist activities in cities using geotagged photography. *Tourism Geographies*, *16*(1), 88–104. doi:10.1080/14616688.2013.868029

Kádár, B., & Gede, M. (2013). Where do tourists go? Visualizing and analyzing the spatial distribution of geotagged photography. *Cartographica*, *48*(2), 78–88. doi:10.3138/carto.48.2.1839

Koens, K., Postma, A., & Papp, B. (2018). Is overtourism overused? Understanding the impact of tourism in a city context. *Sustainability (Switzerland)*, *10*(12), 1–15. doi:10.3390u10124384

Lee, D. (2016). How Airbnb short-term rentals exacerbate Los Angeles's affordable housing crisis: Analysis and policy recommendations. *Harv. L. & Pol'y Rev.*, 10, 229. Retrieved from https://harvardlpr.com/wp-content/uploads/sites/20/2016/02/10.1_10_Lee.pdf

López Palomeque, F., Torres-Delgado, A., Font Urgell, X., & Serrano Miracle, D. (2018). Gestión sostenible de destinos turísticos: La implementación de un sistema de indicadores de turismo en los destinos de la provincia de Barcelona. *Boletín de la Asociación de Geógrafos Españoles*, *77*(77), 428–461. doi:10.21138/bage.2547

Mathieson, A., & Wall, G. (1982). *Tourism economic, physical and social impacts*. Harlow: Longman.

Milano, C. (2018). Overtourism, malestar social y turismofobia. Un debate controvertido. *PASOS Revista de Turismo y Patrimonio Cultural*, *16*(3), 551–564. doi:10.25145/j.pasos.2018.16.041

Milano, C., Novelli, M., & Cheer, J. M. (2018). *Overtourism and Tourismphobia: A journey through four decades of tourism development, planning and local concerns*. Tourism Planning and Development Journal; doi:10.1080/21568316.2019.1599604

Miller, G. (2001). The development of indicators for sustainable tourism: Results of a Delphi survey of tourism researchers. *Tourism Management*, *22*(4), 351–362. doi:10.1016/S0261-5177(00)00067-4

Muler Gonzalez, V., Coromina, L., & Galí, N. (2018). Overtourism: Residents' perceptions of tourism impact as an indicator of resident social carrying capacity - case study of a Spanish heritage town. *Tourism Review*, *73*(3), 277–296. doi:10.1108/TR-08-2017-0138

Navarro Jurado, E., Tejada Tejada, M., Almeida García, F., Cabello González, J., Cortés Macías, R., Delgado Peña, J., ... Solís Becerra, F. (2012). Carrying capacity assessment for tourist destinations. Method for the creation of synthetic indicators applied in a coastal area. *Tourism Management*, *33*(6), 1337–1346. doi:10.1016/j.tourman.2011.12.017

OECD-ORGANISATION FOR ECONOMIC CO-OPERATION AND DEVELOMPENT. (1993). OECD core set of indicators for environmental performance reviews. *Environment Monographs*, 38. Paris, France: Organisation for Economic Co-operation and Develompent.

Peetrs, P., Gossling, S., Klijs, J., Milano, C., Novelli, M., Dijkmans, C., & Postma, A. (2018). *Research for TRAN Committee - Overtourism: impact and possible policy responses*. Brussels, Belgium: European Parliament, Policy Department for Structural and Cohesion Policies.

Popp, M. (2012). Positive and negative urban tourist crowding: Florence, Italy. *Tourism Geographies*, *14*(1), 50–72. doi:10.1080/14616688.2011.597421

Richards, G. (2017). The new geographies of tourism: Space, place and locality. In D. Dredge, & S. Gyimóthy (Eds.), *Tourism and Collaborative Consumption: Perspectives, Politics, Policies and Prospects* (pp. 169–184). Springer. doi:10.1007/978-3-319-51799-5_10

Russo, A. P. (2002). The "vicious circle" of tourism development in heritage cities. *Annals of Tourism Research*, *29*(1), 165–182. doi:10.1016/S0160-7383(01)00029-9

Salas-Olmedo, H., Moya-Gómez, B., García-Palomares, J. C., & Gutiérrez, J. (2018). Tourists' digital footprint in cities: Comparing big data sources. *Tourism Management*, *66*, 13–25. doi:10.1016/j.tourman.2017.11.001

Sánchez Rivero, M., & Pulido Fernández, I. (2008). *Medida de la sostenibilidad turística. Propuesta de un índice sintético*. Madrid, Spain: Editorial Universitaria Ramón Areces.

Schianetz, K., & Kavanagh, L. (2008). Sustainability indicators for tourism destinations: A complex adaptive systems approach using systemic indicator systems. *Journal of Sustainable Tourism*, *16*(6), 601–628. doi:10.1080/09669580802159651

Sequera, J., & Nofre, J. (2018). Debates Shaken, not stirred. New debates on touristification and the limits of gentrification. *CITY*, *22*(5-6), 843–855. doi:10.1080/13604813.2018.1548819

Sorando, D., & Ardura, A. (2018). Procesos y dinámicas de gentrificación en las ciudades españolas. *Papers. Regió Metropolitana de Barcelona. Territori, estratègies, planejament*, 60, 35–47. Retrieved from https://www.raco.cat/index.php/PapersIERMB/article/view/339239/430199

Torres-Delgado, A., & López Palomeque, F. (2014). Measuring sustainable tourism at the municipal level. *Annals of Tourism Research*, *49*, 122–137. doi:10.1016/j.annals.2014.09.003

Torres-Delgado, A., & Saarinen, J. (2013). Using indicators to assess sustainable tourism development: A review. *Tourism Geographies*, *16*(1), 31–47. doi:10.1080/14616688.2013.867530

TourMIS. (n.d.). Retrieved from https://www.tourmis.info/index_e.html

UNWTO Committee on Tourism Statistics and TSA. (2018). *Statistical Framework for Measuring Sustainable Tourism: Consultation draft*. UNWTO.

van der Borg, J. (1998). Tourism management in Venice, or how to deal with success. In D. Tyler, Y. Guerrier, & M. Robertson (Eds.), *Managing tourism in cities. Policy, Process, and Practice*. West Sussex: John Wiley & Sons.

Vera, J. F. (2001). Planificación y gestión del desarrollo turístico sostenible: Propuestas para la creación de un sistema de indicadores. Alicante: Instituto Universitario de Alicante.

Weber, F., Stettler, J., Priskin, J., Rosenberg-Taufer, B., Ponnapureddy, S., Fux, S., . . . Barth, M. (2017). Tourism destinations under pressure. Challenges and innovative solutions. Lucerne, Switzerland.

WTO-ORGANIZACIÓN MUNDIAL DEL TURISMO. (1995). *Lo que todo gestor turístico debe saber. Guía práctica para el desarrollo y uso de indicadores de turismo sostenible*. Madrid, Spain: UNWTO.

WTO-WORLD TOURISM ORGANIZATION. (2004). *Indicators of Sustainable Development for Tourism Destinations. A Guidebook*. Madrid, Spain: UNWTO.

WTO-WORLD TOURISM ORGANIZATION. (2005). *Tourism Congestion Management at Natural and Cultural Sites*. Madrid, Spain: UNWTO.

WTO-WORLD TOURISM ORGANIZATION. (2018). *'Overtourism'? – Understanding and Managing Urban Tourism Growth beyond Perceptions*. Madrid, Spain: UNWTO; doi:10.18111/9789284419999

WTO-WORLD TOURISM ORGANIZATION. (2019). 'Overtourism'? – Understanding and Managing Urban Tourism Growth beyond Perceptions.: Vol. 2. *Case Studies*. Madrid, Spain: UNWTO; doi:10.18111/9789284420629

WTTC & McKinsey & Company. (2017). *Coping with Success: Managing Overcrowding in Tourism Destinations*. London, UK: World Travel & Tourism Council.

Yrigoy, I. (2018). Rent gap reloaded: Airbnb and the shift from residential to tourist rental housing in the Palma Old Quarter in Mallorca, Spain. *Urban Studies (Edinburgh, Scotland)*. doi:10.1177/0042098018803261

Chapter 17
Perceptions and Opinions of the Host Community Regarding Overtourism in the Tourist–Historic City:
A Case Study in Toledo (Spain)

Luis Alfonso Escudero Gómez
https://orcid.org/0000-0002-7954-4064
University of Castilla-La Mancha, Spain

ABSTRACT

In the last few years, the number of visitors in historic cities has grown, resulting in a situation labeled 'overtourism'. In these tourist-historic cities, tourism is one of the main local economic foundations. This chapter asks whether the social carrying capacity of the host community has been exceeded, through a case study in Toledo, Spain, a UNESCO World Heritage Site since 1986. The methodology employed is a survey to residents in the city. The main findings show that the residents do not perceive the overtourism situation in the city as something serious. However, some impacts such as the historic centre becoming a museum for tourists, or the traffic congestion derived from overcrowding in both the traffic and pedestrian flow are clearly perceived by the host community. It is the residents in the historic centre those who manifest a more negative opinion of touristic development. This chapter may be of interest to academics, decision makers, and those responsible for tourism in historic cities.

DOI: 10.4018/978-1-7998-2224-0.ch017

INTRODUCTION

I, as a resident in Toledo, cannot see the benefit that tourism brings me. Ordinary Toledoans do not see the benefits of tourism (male, has lived in the historic centre since 1975). (Blanco & Arroyo, 2018, p. 41)

Tourism affects the lives of community residents by influencing their quality of life (Andereck & Nyaupane, 2011). Tourism disrupts the life of residents at destination (Dioko, 2017) and entails social costs to destinations (Pizam, 1978). Today there are more tourist than ever before. In recent years there has been an explosive development of urban tourism that could qualify as a tourist boom (Namberger, Jackisch, Schmude, & Karl, 2019). As a result, the number of tourists per inhabitant increases in many city destinations. As the number of visitors to certain cities has multiplied, so have the pressures of tourism on these cities. There is talk of 'overtourism' —and other similar terms such as 'overcrowding' (McKinsey & Company, & World Travel & Tourism Council, 2017)— and the situation is such that it has provoked protests against tourism in several cities. An increase in the number of visitors and constant growth are identified as distinctive elements of overtourism (Milano, 2018). Nonetheless, it is worth noting that many problems attributed to overtourism are not as much about the number of tourists as about visitor behaviour (Smith, Sziva, & Olt, 2019) and management (UNWTO, 2018).

The phenomenon of negative impacts of mass tourism in a destination was detected in the past (Boissevaint, 1996), but today it is of much relevance. 'Overtourism' is a term recently used to contextualize the potential hazard of the negative impacts of excessive tourism to many popular tourist destinations worldwide (Cheung & Li, 2019). Thus, the term 'overtourism' has rapidly gained traction across multiple sectors, including academia, policy formulation, social movements and the media (Milano, 2018; Dodds & Butler, 2019b). Destinations of all kinds face, and almost certainly will continue to experience overtourism (Dodds & Butler, 2019a). Historic centers, being destinations for heritage or cultural tourism, have also experienced a clear growth in their tourist flows. This is the case in small and medium-sized European cities that retain an important patrimonial legacy. These urban centers have been classified as tourist-historic cities (Ashworth & Tunbridge, 1990) and are currently first line tourist destinations. A cultural commodification has happened, by which heritage has become merchandise for mass consumption, for tourist consumption (Timothy, 2011). For this reason, the urban landscapes of the historic cities have been through a fast process of touristification (García, de la Calle, & Yubero, 2017) since they are destinations for mass tourism (González, 2019). At the same time, it is obvious that in these urban communities of small and medium-sized cities there is an awareness of the importance of tourism for the local economy. They are cities that live off tourism. This dichotomy raises an important question that is also the primary objective of the research and the first research question (RQ1) which is given an answer to in this case study about Toledo, Spain; how do the host societies of the tourist-historic cities value the current touristification process?

Destinations undergo an overtourism situation when the hosts feel that there are too many visitors and that the quality of life in the area has deteriorated unacceptably (Goodwin, 2017). Therefore, it is very significative to understand the residents' attitudes towards tourism, especially the host community's perception when it is under much pressure from tourism and tourists (Katarzyna, Six, & Vanneste, 2017). Deery, Jago, & Fredline (2012) point out that "understanding the social impacts of tourism on communities is extremely important for government at all levels so that action can be taken to reduce the likelihood of a community backlash against tourists and tourism development" (p. 64).

In the academic world, the study of the host community's attitudes towards tourists has been established since the late 1970's and in the 1980's with papers such as Pizam's (1978) and Brougham & Butler's (1981), and has developed mainly from the 1990's on, in a number of publications and based on research such as Perdue, Long, & Allen's (1990). Currently, the academic literature about the residents' attitudes, perceptions and opinions on tourism development is very broad. Most of these works are based on the perceptions of the impacts of tourism by the host community (Gerritsma & Vork, 2017). In addition, nowadays, specifically, articles are emerging aimed to understanding how this overtourism phenomenon affects the residents, in similar research to this chapter's. Thus, Barcelona is an interesting case for analysing rejection towards tourism in a context of overtourism (Martín, Guaita, & Salinas, 2018). Also in Catalonia, but at a completely different territorial level, Muler, Coromina, & Galí (2018) analyse the residents' perceptions of overtourism impacts in a small heritage town, Besalu, with under 2,500 inhabitants, in the province of Gerona. Namberger et al. (2019) have studied the relationship between overtourism and local level disturbance in the German urban tourist destination of Munich. Smith et al. (2019) carry out a survey among the residents in the Hungarian capital, Budapest, in order to analyse how the negative impacts of growing overtourism are perceived, and Kuščer & Mihalič (2019) have published an article about the residents' attitudes towards overtourism in the Slovenian capital, Ljubljana.

When it comes to current growth of overtourism, additional research is needed in terms of the relationships between the residents' satisfaction and the perceived tourism impacts (Kuščer & Mihalič, 2019). In the destinations where there is heritage tourism, it is especially important to get to know the residents' attitudes towards tourism development (Chen & Chen, 2011). Dines (2016) points out the need for a more in-depth analysis of how the illogical consequences of tourism in historic urban areas are perceived. At the same time, Su & Wall (2013) remark that it is necessary to carry out more case studies in order to better understand the relationship between tourism and heritage. At present, case studies are needed to assess the perceptions and opinions of host communities regarding overtourism in tourist-historic cities.

The author suggests an empirical analysis based on the example of the city of Toledo in Spain. Toledo, whose historic center and cultural landscapes have been a UNESCO World Heritage Site since 1986, is one of the main destinations for cultural tourism in Spain (Escudero, 2018). It has experienced a considerable increase in the number of visitors, especially after 2014, due to the success of the 400th anniversary celebration of El Greco's death and the tourist boom that Spain is currently experiencing. Toledo is an ideal case for the proposed investigation. The present study focuses on the city as a whole with its possible distinct overtourism phenomena.

Dodds & Butler (2019a) state that "overtourism is more of a relative problem than an absolute one, in that much depends on characteristics of the affected destinations, their residents and the relative scale of tourism in those communities" (p. 262). Externalities through tourism are complex and exert negative as well as positive impacts on residents' welfare (Marsiglio, 2018). The chapter's objective is to analyse the perceptions and opinions of the host communities in tourist-historic cities, as well as to compare whether or not such attitudes are dependent on demographic and sociodemographic variables The hypothesis of this chapter is that the perceptions and opinions of residents in medium-sized cities where tourism is a fundamental part of the local economy, do not reflect the degree of tourismphobia portrayed by the media (Huete & Mantecón, 2018) and academic works (Goodwin, 2017). Conversely, they place emphasis on tourismphobia in certain well-established sun and beach destinations, such as Majorca (González, 2019), and large cities (Colomb & Novy, 2016), such as Barcelona (Goodwin, 2019), Amsterdam (Gerritsma & Vork, 2017), Budapest (Smith et al., 2019), Paris (Gravari-Barbas & Jacquot, 2016) and Prague (Pixová & Sládek, 2017). That is to say, the hypothesis is set out as follows:

Hypothesis: In small and medium-sized European tourist-historic cities, save for significant exceptions, overtourism does not contribute to general situations of tourismphobia owing to the important economic profit of tourism for the local community.

According to this hypothesis, the Social Exchange Theory (SET) –a theory commonly used as framework (Andereck & Vogt, 2000) when carrying out studies on the residents' attitudes— comes into effect in the tourist-historic cities. According to the SET, the residents in a tourist destination will support tourism development as long as they perceive that the benefits outweigh the costs (Rasoolimanesh, Jaafar, Kock, & Ramayah, 2015). In the tourist-historic cities, tourism contributes significantly to socioeconomic development because of the income and jobs it creates. Following this hypothesis, the SET is a valid explanation of why this process of conflict and rejection to tourism does not yet happen in small and medium-sized historic cities (Muler et al., 2018), except that high levels of overtourism have already been surpassed, which happens in well- known cases such as Venice's (Séraphin, Sheeran, & Pilato, 2018; Nolan & Séraphin, 2019) or Dubrovnik's (Panayiotopoulos & Pisano, 2019). From this hypothesis, then, two questions are answered. The first research question (RQ1) has been previously formulated, and it is the one that can give an answer to the hypothesis, focusing on what the inhabitants in Toledo perceive and think of the negative impacts of overtourism in their city. The second research question (RQ2) questions whether the residents' perceptions and opinions of the negative impacts of tourism differ based on several demographic and socioeconomic variables, such as sex, gender or personal dependency on tourism, among others.

The chapter is structured as follows. After this introduction and a section dedicated to background, which delves deeper into overtourism in the tourist-historic cities, there is a research design where the use of Toledo as an important case study is contextualized and justified, and the methodology is explained. Next, the main results of the research and a conclusion are presented.

Although the theoretical framework will be a fundamental pillar of the work, the results are mainly empirical and will be of interest to academics, decision-makers and those responsible for tourism in historic cities.

BACKGROUND

The notion of 'tourist-historic city' makes reference to an area of older cities where the urban structure, architecture and artifacts are used to create a place-based heritage touristic product (Asworth & Tunbridge, 2000). The tourist-historic cities are becoming touristic monoculture enclaves by a functional touristification process. Thus, inside the tourist-historic cities, private houses and small shops have been replaced by restaurants, hotels and souvenir shops (Munsters, 2005). The increase in tourist activity affects the host society's quality of life in tourist-historic cities in aspects such as mobility, loss of identity, excessive tourist marketing, increased prices, the expansion of tourist housing and its consequences, etc. This causes magnification in the perception and unrest towards an excessive touristic specialization (Milano, 2018).

Above everything, growing touristic pressure (García et al., 2017) is becoming more and more incompatible with sustainable development, a thesis supported by Montanari and Staniscia (2010) after carrying out a case study in Rome. According to Jansen-Verbeke y Lievois (1999) the touristification process in the historic cities is irreversible and its negative impacts do not have enough answers yet despite having introduced the principles of sustainability theoretically and in good practices.

At present, locations crowded with tourists are common in compact old towns of European cities (Namberger et al., 2019). Even though the impact of tourism on heritage cities is inherently place-specific (Simpson, 1999), the overcrowding in the tourist-historic cities gives place to common processes of social and space transformation. Changes in the morphology of the locations, in their physical structures, in the functional patterns and the use of public spaces contribute to this transformation process (Jansen-Verbeke & Lievois, 1999). Besides, many aspects of daily life in these spaces are affected by tourist activity. Rončák (2019) explains how the small town centre in Prague is affected by constant tourist flow until it becomes what the author calls a commercialised "tourism ghetto". In the same destination, Pixová & Sládek (2017) describe how this has created discontent among the residents.

A fundamental theoretical concept developed in academic literature to understand excessive tourism is 'tourist carrying capacity' (O'Reilly, 1986; McCool & Lime, 2001). UNWTO (2018) defines this as the maximum number of people that may visit a tourist destination at the same time, without causing destruction of the physical, economic and sociocultural environment and an unacceptable decrease in the quality of visitors' satisfaction. In fact, according to UNWTO (2018) 'overtourism' is a new buzzword for existing concepts of tourism congestion management and tourism carrying capacity. In any case, carrying capacity is a concept and an analysis tool to capture overtourism (Namberger et al., 2019). It is a key challenge for tourism developers and managers (UNWTO, 2018). According to Séraphin et al. (2018), this concept should be associated with the term overtourism. Every tourist destination and attraction could support a limited carrying capacity. This turns the question of where overtourism starts into a difficult one to answer (Kuščer & Mihalič, 2019) and demands case studies to find out the answer, such as the one carried out by Feliziani and Miarelli (2012) that aims to find out how many visitors Rome should receive.

The concept of tourist carrying capacity has evolved and broadened (Watson & Kopachvesky, 1996). It can be expressed in environmental, economic and social-cultural terms, as well as with regard to technological limits and health and safety issues (Dodds & Butler, 2019b). Álvarez-Sousa (2018) points out that in cultural cities it is especially relevant that the society's carrying capacity or social carrying capacity derives from how the locals and the visitors perceive the effects of tourism in different assets of their social space. A continued situation of overtourism would give place to overloading that social carrying capacity. Precisely, analysing the residents' perceptions of the touristic impacts, as will be done in this chapter, is an important indicator of the resident social carrying capacity (Muler et al., 2018).

When tourism growth and its derived negative impacts overloads the social carrying capacity, there is a situation of overtourism which affects the local community. In that moment, overtourism would cause resistance movements to arise, which vary from positions from which the current tourism model is questioned, to positions which are completely and truly anti-touristic (Colomb & Novy, 2016; Papathanassis, 2017) or tourismphobic.

At present, tourismphobia is a key topic in public discussion (Huete & Mantecón, 2018). Milano (2018) also points out that there is political exploitation and a media bubble on tourismphobia. In 1987, Liu, Sheldon, and Var already stated that as tourism started to become first line in public discussion and in the media, the general knowledge of its negative impacts increased. Today, many anti-tourism movements or 'resistant identities' are voicing their concern regarding the continuously growing number of tourists (Séraphin et al., 2018). This way, there are more situations of rejection of tourism than there are in environments that are traditionally dependent on tourism (Martín et al. 2018). In a number of destinations, different types of actions are the result of social protests against tourism development; graffiti, rallies, production of audiovisual and printed media and even attacks on the different infrastructures and touristic equipment. There is a mobilised or organized movement of residents in the destination, upset by tourism development (Kuščer & Mihalič, 2019). The residents have protested visibly and vociferously (Dodds & Butler, 2019a) against the consequences of mass tourism and the mobilisation of the tourism industry, under the motto 'Tourists go home' (Hughes, 2018). In this chapter, we would like to look into what happens when these manifestations are not evident yet. That is to say, find out which are the perceptions and opinions of the resident host community on overtourism in a tourist-historic city with no tourismphobic social movements such as Toledo (Spain).

RESEARCH DESIGN

Study Area: Toledo, A Tourist-Historic City, Main Heritage Tourist Destination in Spain

As a case example to study the perceptions and opinions of the host community regarding overtourism in the tourist-historic city, the city of Toledo, Spain, was chosen for several reasons. Toledo is one of the main heritage cultural tourist destinations in Spain (Escudero, 2018). The population of Toledo is 84,282 in 2018. Its historic centre, of medieval Arab origins, encloses a labyrinth-like morphology of steep, narrow streets which contain a great number of historic architectural landmarks (convents, palaces, churches, etc.). Among these, there are some remarkable ones such as the Cathedral, the Alcazar, the Synagogue of El Transito and the Monastery of San Juan de los Reyes. Besides, its unique location in a crag surrounded by river Tajo offers some extraordinary cultural landscapes. The historic centre and its cultural landscapes were declared UNESCO World Heritage Site in 1986. It is this heritage that has made Toledo into a touristic destination, with its realities and its problems —previously addressed by the author (Escudero, 2018)—, and which have attracted a growing number of visitors to its historic centre to the point of overtourism, or at least to a point of high risk of overtourism. The widespread increase in the number of tourists in the streets and landmarks is evidence to this phenomenon (see Figure 1). As in other destinations, it is the high tourism growth that justifies an overtourism risk analysis (Kuščer & Mihalič, 2019).

Understanding tourism in a destination requires accurate data. Destination Marketing Organizations (DMOs) must work with these data in order to understand tourism and to design suitable policies (Dodds & Butler, 2019a). In the case of Toledo, it is the local government that is in charge of managing, planning and promoting the tourist destination. However, there are no available data about questions as basic as the total number of visitors. There is no tracking of key indicators such as number of arrivals, number of tourists and day-trippers, or the seasonal fluctuations in the demand. Granted, no advance has been achieved in the use of big data and new technologies to monitor and evaluate tourism performance and impact. The source usually cited is the Hotel Occupancy Survey of the National Institute of Statistics, which amalgamates a sample of people who stay in regulated hotel accommodation. According to this source, Toledo received 600,545 travelers and 924,246 overnight stays in 2018 and, in comparison, in 2000 it received 398,856 travelers and 568,308 overnight stays (National Statistics Institute, 2019). However, no data has been compiled of tourists who stay in apartments such as those offered by unregulated marketplaces like Airbnb.

Figure 1. Pictures of tourism overcrowding in Toledo's historic centre.
Source: Photographs taken by the author on 9/28/2017, top left; 4/4/2017, right; and 4/24/2017, bottom left.

On the other hand, the results shown by the Hotel Occupancy Survey are always incomplete as, fundamentally, they do not take into account one-day visitors. Day-trippers are not included in the statistics, but they occupy and roam around the same space, visit the same places and behave similarly to tourists. Cheung & Li (2019) show that more day trips causes a relationship of permanent negative cointegration with the residents' sentiment. In fact, Nolan and Séraphin (2019) show that in the issue of overtourism in Venice a key issue is the high number of day visitors. In Toledo, the number of day-trippers is also very high, mainly due to how close Madrid, the capital of the country, is. Residents and tourists alike travel to Toledo from Madrid daily, and that brings about one of the main problems of the city as tourist destination (Escudero, 2018).

When the number of tourists surpasses the number of residents, these can feel as though their city is being taken and changed in a way they do not particularly like (Nolan & Séraphin, 2019). If we calculate the tourism intensity —the visitor to resident ratio (Muler et al., 2019)— the value obtained for 2018 is 7.12 tourists per resident, yearly, according to the figures shown by the imperfect Hotel Occupancy Survey. If we calculate based on the conservative estimate of 3 million visitors per year, the obtained value is 35.59 visitors per resident in a year.

METHODOLOGY

The methodology utilized to carry out the research includes, first of all, a documentary observation by means of a bibliographic review of the subject. Subsequently, the technique that will be applied is the analysis of a survey following the model of previous investigations (Snaith & Haley, 1999; Namberger et al., 2019). Surveys are usual research tools in tourism studies of proven suitability (Saveriades, 2000). The survey was administered between November and December of 2018 to 442 residents of Toledo chosen throughout the city, both in the historic centre and in the residential areas. With this distribution it is possible to compare hypothetical differences in the perceptions and opinions of the inhabitants of the area of greatest tourist pressure, the historic center, and the rest of the people of Toledo. According to Di Grino (1986), the required number of answers to achieve a representative sample in a population of 25,000 or more people, is 348. This figure allows trust of 95% inside a margin of error of ± 2.5%.

The survey was applied randomly and in a complementary manner, following the snowball effect method where respondents took us to relatives and close friends. Questions in the survey were asked to evaluate the opinions of the resident community through the use of the Likert scale from 0 to 5, with 1 meaning 'I totally disagree' and 5 meaning 'I totally agree'. The questions were chosen after reading recent literature on overtourism and building a total number of items not too high and not too low, 13 (for instance, Muler et al. (2018) use 12), which was considered to be manageable and significant. Thus, the questionnaire includes references to problems of touristification, overcrowding, mobility, increased prices, increased taxes, and environmental and patrimonial degradation. The reliability of the factors has been checked using Cronbach's alpha, which showed that all factors had sufficient reliability since the obtained value was 0.847, clearly above the required minimum value of 0.7 (Hair, Black, Babin, & Anderson, 2018). The data were analysed using descriptive methods, which helps describe and summarise data in a meaningful way.

The questionnaire also collected basic demographic, social and economic characteristics of the respondents which will make it possible to qualify the results and observe whether there are significant differences depending on gender, age, the length of residence in the city, the place of residence, the ownership of housing, whether respondents or their relatives work in the tourism sector, and monthly income. These independent variables have been chosen based on previous studies on the host community's opinions of touristic development in heritage cities (Snaith & Haley, 1999; Muler et al., 2018). One of the goals of the studies on the residents' attitudes is, in general, to explore the relationship between the independent variables and their perceptions and opinions (Muler et al., 2018). Thus, one of the most utilized parametric tests in the research about the residents' attitudes towards tourism has been carried out in this research by applying an analysis of variance (ANOVA) (Sharpley, 2014). Data normality was checked by the Kolmogorov-Smirnov test. Levene's test of homogeneity of variance was applied, which confirmed that it is possible to accept the equality of variances and, therefore, that the ANOVA may be used (O'Brien, 1979).

RESULTS

Residents' expectations regarding how they think they have benefited from and been impacted by tourism, determine whether it is seen as something positive or negative (Dodds & Butler, 2019a). Based on previous studies which assess the residents' attitudes towards overtourism through their perspectives of the impacts of tourism (Kuščer & Mihalič, 2019), a similar case study is performed in the tourist-historic city of Toledo, Spain, aimed to understanding the perceptions and opinions of the host community regarding overtourism.

The Residents' Perceptions and Opinions Regarding Overtourism in Toledo

This study thus focuses on the residents' perceptions and opinions of different impacts derived from touristic development in Toledo in order to check whether the increase in tourism in the city has surpassed or not the resident social carrying capacity. The aim is to understand to what extent the host community is of the opinion that the tourist-historic city of Toledo is in an overtourism situation. With this in mind, a questionnaire was developed for them to assess 13 hypothetical impacts of overtourism in the city through a five-point Likert scale (see Table 1). The final mean is 3.59, with a standard deviation of 0.48 among the means of the 13 question items. Katarzyna et al. (2017) show that the acceptable change limits have not been overcome in the Belgian tourist-historic city of Bruges. The same result was obtained by Namberger et al. (2019) and by Munich and Muler et al. (2018) for Besalu. The obtained results in this case study agree with these precedents and we can conclude that Toledo has not yet surpassed its resident social carrying capacity towards the negativity of the impacts of overtourism in the city. Nevertheless, this is a high enough score that it has to be taken into account and there should be concern about the future attitudes of the host community towards tourism in Toledo.

In some Spanish cities some local movements against tourism have arisen, located especially in the historic centre, like in San Sebastian's case (de la Calle, García, & Yubero, 2017). In Toledo there has not been a social protest beyond some exceptional actions and comments on social media. For example, there is no graffiti against tourism or tourists. However, just like in Kuščer & Mihalič's research (2019), the survey has confirmed some negative variables that influence irritation of local residents with tourism impacts and present a potential overtourism risk.

When analysing the question items individually, we observe that in some of the negative impacts of overtourism the residents present a higher degree of agreement, while others are perceived as lower. The most remarkable impact is the fact that Toledo is a museum for tourists. This item gets the highest score at 4.27 and also the highest percentage of people who give a 5 (57.2%). Overtourism brings about the transformation of identity features in urban spaces (Huete & Mantecón, 2018). In historic centres, there is an excess musealisation (Fessler & Berenstein, 2006). The Toledo host community clearly perceives that the historic centre in Toledo is going through a process of musealisation.

Secondly, there is high agreement inside the host community, a 4.0 in the Likert scale, on the fact that tourist rental accommodation has raised the cost of renting in Toledo. This question can be linked to that of Tourism unfairly increases the cost of real estate in Toledo, which also gets a high score, above mean, 3.78. In previous works, a remarkable negative impact was the increase in house prices (Huete & Mantecón, 2018; UNWTO, 2018; Smith et al., 2019). In fact, Goodwin (2009) points out that accommodation is one of the major areas of conflict. In this process, something fundamental is the growing integration of houses in an informal economy based on regulated tourist accommodation offer (Huete & Mantecón, 2018). For this reason, we opted for analysing the residents' perception and acceptance of providers such as Airbnb, like Namberger et al. (2019) did. We could confirm that the residents relate the excessive concentration of tourist apartments in Toledo and the increase of real estate price, repeating the findings by Lestegás, Lois, & Seixas (2018) and by Namberger et al. (2019). The residents appoint Airbnb as the one responsible for the rise in the cost of renting in the city. However, other more common problems derived from the proliferation of tourist rental accommodation, such as those related to noise and coexistence, and that have caused strong social protests in cities like Barcelona (Hughes, 2018), receive a lower score in our survey, 3.35. We also need to point out that the Airbnb phenomenon is still a novelty in Toledo and there is a high percentage of people (over 17%) who give no answer to the two questions about this topic in the survey. We assume that it happens so because they know nothing about it or because they feel unable to give an opinion.

Other problems also do receive a higher score than mean. It is the case of the questions related to mobility. They are the opinion that Tourism in Toledo creates problems of pedestrian congestion (at a 3.97 score), Tourism increments the traffic flow in Toledo (3.91), and Tourism makes it more difficult

Table 1. Findings for residents' opinions of overtourism impacts

Question Items	Likert scale (%)						Mean	Standard deviation
	1	2	3	4	5	No answer		
Q1. Tourism increments the traffic flow in Toledo	4.5	10.6	17.0	24.2	42.8	0.9	3.91	1.20
Q2. Tourism makes it more difficult to park in Toledo	4.3	12.0	21.5	15.6	45.9	0.7	3.87	1.24
Q3. Tourism in Toledo creates problems of pedestrian congestion	3.2	8.6	19.2	24.9	43.0	1.1	3.97	1.13
Q4. Toledo is a museum for tourists	2.5	5.2	12.4	20.6	57.2	2.0	4.27	1.04
Q5. There are too many day-trippers in Toledo	7.5	10.9	28.3	27.4	24.7	1.4	3.52	1.20
Q6. There are too many tourist-oriented shops in Toledo	5.2	10.2	25.8	20.8	36.7	1.4	3.75	1.21
Q7. Tourism produces more waste in Toledo	7.5	11.1	25.6	28.1	25.3	2.5	3.54	1.21
Q8. Tourism development raises local taxes	6.6	8.4	24.2	20.4	12.9	27.6	3.34	1.17
Q9. Tourism unfairly increases the cost of real estate in Toledo	4.8	6.6	22.4	23.5	30.8	12.0	3.78	1.16
Q10. Tourist rental accommodation has raised the cost of renting in Toledo	2.7	6.8	14.5	21.5	3.64	18.1	4.00	1.12
Q11. Tourist rental accommodation cause noise and coexistence nuisance in Toledo	7.5	13.3	24.0	18.3	19.4	17.6	3.35	1.25
Q12. Tourism affects negatively Toledo's cultural heritage	27.6	24.2	24.0	12.4	7.5	4.3	2.46	1.25
Q13. Tourism reduces the use of Toledo by the residents	19.5	16.7	26.7	15.8	18.8	2.5	2.98	1.38
Mean for the scale							**3.59**	**0.48**

to park in Toledo (3.87). The high concentration of visitors affects mobility and the use of public spaces (García et al., 2017). When visitors take over the main streets, they create congestion and a feeling of overcrowding. In the study in Besalu there is high agreement in that tourists get in the way of residents in town (Muler et al, 2018). The residents in Toledo perceive the negative impact of overtourism in these cases.

The question (Q6) identifying that there are too many tourist-oriented shops in Toledo also receives a score above the mean at 3.75. Indeed, Escudero (2018) indicates that there is an excess of tourist orientated businesses in Toledo's historic centre. Here we come across another problem related to overtourism and touristification raised by Huete & Mantecón (2018): the excessive amount of tourist-oriented businesses.

The rest of the questions get a score under the mean. Such is the case of the statement that There are too many day-trippers in Toledo, which gets 3.52. This proves that the resident social carrying capacity has not been surpassed yet in Toledo. In this sense, the low score of the question that Tourism reduces the use of Toledo by the residents, is also significant, at 2.98. It is yet another evidence that the social impacts of overtourism in this case are not perceived as extreme and the host community, even though they are of the opinion that there is a musealisation of the historic centre, considers that this does not prevent the city from being enjoyed by the residents, too.

The opinion that Tourism produces more waste in Toledo gets an excessively high score, 3.54. The tourist mass causes congestion and pollution (dirt, odor and noise nuisance) in the little streets in the historic centres (Munsters, 2005). In previous research the opinions of the host community showed this (Sommer & Helbrecht, 2017; Muler et al., 2018; Smith et al., 2019), but in the case of Toledo this impact is considered to be lower than others derived from overtourism.

The perception that reaches a lowest score in the survey is the one about how Tourism affects Toledo's cultural heritage negatively, at 2.46. There is broad disagreement on this question. In our case, this study coincides with that by Muler et al. (2018). The host community in Toledo thinks that tourism does not affect heritage conservation negatively. The same finding was obtained by UNWTO (2018) in eight European cities, reaching the conclusion that residents think that tourism protects the historical parts of the city.

The Residents' Opinions on Overtourism According to Their Demographic and Socioeconomic Characteristics

The residents' perceptions are far from uniform. The opinions on the impacts of tourism change from segment to segment of the population since they are influenced by many variables (Nunkoo & Ramkissoon, 2012). We will analyse here the relationship between demographic and socioeconomic independent variables of the host community and their opinions on the overtourism phenomenon in the city, like in previous studies (Muler et al., 2018). For this, we calculated the mean for each question item and for the mean of the scale in each group of every independent variable and, fundamentally, an ANOVA test was performed in order to find out whether there were significant differences between the groups in relation to opinions of the impacts of overtourism in Toledo (see Tables 2 and 3)

The results obtained show great independence between the perceptions and opinions of the residents in Toledo of overtourism, and the demographic and socioeconomic variables. Of all the analysed variables, only one shows significant differences according to the analysis of variance (ANOVA). It is the Place of residence. In this variable it is important to differentiate between the historic centre and other residential areas, like Namberger et al. (2019) point out for Munich. The residents in the historic centre of Toledo show a more negative evaluation of the negative impacts of overtourism in the city (3.68) than those who live in other areas of the city (3.54). Tourist activity concentrates in the historic centre of the city and the residents who live there and come into daily contact with tourists are the ones most affected by the negative impacts of tourism development and are also the ones who have a more negative opinion of this. In particular, according to the ANOVA test, the differences between the residents in the historic centre and the residential areas show up in the opinion that Tourism causes problems of pedestrian congestion, in the opinion that There are too many day-trippers, that There are too many tourist-oriented shops and that Tourist rental accommodation causes noise nuisance and coexistence problems. Since they are the most directly affected by these problems, their opinions are more negative than those of the residents in other areas of the city.

The other analysed variables do not show significant differences, as opposed to what happens in other studies when it comes to questions such as gender (Regoeczi, 2008) or occupation in the tourism industry (Muler et al., 2018). However, we can also point out that the opinions on overtourism are more negative in older people —ages 18-40 score at 3.54, ages 41-60 at 3.65 and ages over 60 at 3.70—. Toledo is the opposite case to Munich where the older the respondents, the less they perceived the negative impacts of overtourism (Namberger et al., 2019). There are also differences between the homeowners, 3.57, and those who are not owners, 3.65. However, both variables do not show a significant value in the ANOVA test.

Table 2. Findings for residents' opinions of overtourism impacts according to gender, age, and length of residence.

Question items	Independent variables										
	Gender			Age				Length of residence			
	Men	Women	ANOVA	18-40 years	41-60 years	+60 years	ANOVA	1-5 years	6-15 years	+15 years	ANOVA
Q1. Tourism increments the traffic flow in Toledo	3.95	3.88	F value: 0.391 p: 0.532	3.82	4.00	4.09	F value: 1.571 p: 0.209	3.91	3.76	3.93	F value: 0.424 p: 0.655
Q2. Tourism makes it more difficult to park in Toledo	3.88	3.87	F value: 0.002 p: 0.966	3.78	3.96	3.97	F value: 1.143 p: 0.320	3.94	3.76	3.88	F value: 0.288 p: 0.750
Q3. Tourism in Toledo creates problems of pedestrian congestion	4.02	3.93	F value: 0.565 p: 0.453	3.97	3.93	4.23	F value: 1.020 p: 0.362	3.92	4.00	3.97	F value: 0.062 p: 0.940
Q4. Toledo is a museum for tourists	4.26	4.29	F value: 0.80 p: 0.778	4.22	4.34	4.39	F value: 0.847 p: 0.429	4.23	4.29	4.28	F value: 0.058 p: 0.944
Q5. There are too many day-trippers in Toledo	3.61	3.44	F value: 1.977 p: 0.160	3.47	3.55	3.66	F value: 0.521 p: 0.594	3.69	3.41	3.50	F value: 0.756 p: 0.470
Q6. There are too many tourist-oriented shops in Toledo	3.79	3.71	F value: 0.502 p: 0.479	3.67	3.83	3.86	F value: 0.991 p: 0.372	3.74	3.69	3.76	F value: 0.075 p: 0.928
Q7. Tourism produces more waste in Toledo	3.51	3.56	F value: 0.173 p: 0.678	3.45	3.67	3.46	F value: 1.809 p: 0.165	3.63	3.26	3.57	F value: 1.599 p: 0.203
Q8. Tourism development raises local taxes	3.24	3.42	F value: 1.927 p: 0.166	3.31	3.30	3.76	F value: 1.753 p: 0.175	3.20	3.00	3.42	F value: 2.509 p: 0.083
Q9. Tourism unfairly increases the cost of real estate in Toledo	3.81	3.76	F value: 0.147 p: 0.702	3.71	3.85	3.88	F value: 0.697 p: 0.498	3.96	3.76	3.76	F value: 0.576 p: 0.563
Q10. Tourist rental accommodation has raised the cost of renting in Toledo	3.96	4.04	F value: 0.492 p: 0.483	4.02	3.99	4.03	F value: 0.026 p: 0.974	3.98	4.11	3.99	F value: 0.246 p: 0.782
Q11. Tourist rental accommodation cause noise and coexistence nuisance in Toledo	3.39	3.31	F value: 0.330 p: 0.566	3.28	3.44	3.34	F value: 0.658 p: 0.519	3.49	3.20	3.35	F value: 0.639 p: 0.528

continues on following page

Table 2. Continued

Question items	Independent variables										
	Gender			Age				Length of residence			
	Men	Women	ANOVA	18-40 years	41-60 years	+60 years	ANOVA	1-5 years	6-15 years	+15 years	ANOVA
Q12. Tourism affects negatively Toledo's cultural heritage	2.48	2.44	F value: 0.118 p: 0.731	2.40	2.51	2.56	F value: 0.543 p: 0.582	2.35	2.42	2.48	F value: 0.245 p: 0.783
Q13. Tourism reduces the use of Toledo by the residents	2.97	2.98	F value: 0.002 p: 0.968	2.93	3.05	2.82	F value: 0.580 p: 0.560	2.94	2.84	3.00	F value: 0.315 p: 0.730
Mean of the scale	**3.60**	**3.58**	**F value: 0.024 p: 0.878**	**3.54**	**3.65**	**3.70**	**F value: 1.273 p: 0.281**	**3.61**	**3.50**	**3.60**	**F value: 0.382 p: 0.683**

Table 3. Findings for residents' opinions of overtourism impacts according to place of residence, home ownership, employed in the tourism industry, and income.

Question items	Independent variables												
	Place of residence			Home ownership			Employed in the tourism industry			Monthly income (€)			
	Historic centre	Residential areas	ANOVA	Yes	No	ANOVA	Yes	No	ANO-VA	0-1000	1000-2500	+ 2500	ANO-VA
Q1. Tourism increments the traffic flow in Toledo	3.97	3.87	F value: 0.685 p: 0.408	3.89	3.94	F value: 0.119 p: 0.731	3.80	3.95	F value: 1.335 p: 0.249	3.95	3.91	4.27	F value: 1.343 p: 0.262
Q2. Tourism makes it more difficult to park in Toledo	3.99	3.80	F value: 2.280 p: 0.132	3.84	3.94	F value: 0.535 p: 0.465	3.82	3.89	F value: 0.264 p: 0.608	3.89	3.89	4.09	F value: 0.404 p: 0.668
Q3. Tourism in Toledo creates problems of pedestrian congestion	4.14	3.86	F value: 6.497 p: 0.011*	3.96	4.00	F value: 0.150 p: 0.699	4.00	3.96	F value: 0.107 p: 0.743	3.96	4.05	3.73	F value: 1.290 p: 0.276
Q4. Toledo is a museum for tourists	4.21	4.31	F value: 0.971 p: 0.325	4.25	4.32	F value: 0.346 p: 0.557	4.23	4.29	F value: 0.283 p: 0.595	4.30	4.31	4.22	F value: 0.123 p: 0.885
Q5. There are too many day-trippers in Toledo	3.68	3.42	F value: 4.935 p: 0.027*	3.47	3.61	F value: 1.379 p: 0.241	3.50	3.52	F value: 0.045 p: 0.833	3.55	3.57	3.36	F value: 0.441 p: 0.644
Q6. There are too many tourist-oriented shops in Toledo	3.92	3.64	F value: 5.464 p: 0.020*	3.67	3.90	F value: 3.285 p: 0.071	3.68	3.77	F value: 0.428 p: 0.514	3.70	3.78	3.73	F value: 0.199 p: 0.820
Q7. Tourism produces more waste in Toledo	3.63	3.48	F value: 1.477 p: 0.255	3.50	3.63	F value: 1.204 p: 0.273	3.66	3.50	F value: 1.499 p: 0.222	3.56	3.56	3.70	F value: 0.203 p: 0.817

continues on following page

Table 3. Continued

Question items	Independent variables												
	Place of residence			Home ownership			Employed in the tourism industry			Monthly income (€)			
	Historic centre	Residential areas	ANOVA	Yes	No	ANOVA	Yes	No	ANO-VA	0-1000	1000-2500	+ 2500	ANO-VA
Q8. Tourism development raises local taxes	3.35	3.33	F value: 0.023 p: 0.880	3.41	3.21	F value: 2.071 p: 0.151	3.38	3.32	F value: 0.160 p: 0.689	3.36	3.26	3.46	F value: 0.491 p: 0.613
Q9. Tourism unfairly increases the cost of real estate in Toledo	3.90	3.71	F value: 2.570 p: 0.110	3.80	3.75	F value: 0.217 p: 0.642	3.85	3.76	F value: 0.384 p: 0.536	3.84	3.76	3.93	F value: 0.403 p: 0.669
Q10. Tourist rental accommodation has raised the cost of renting in Toledo	4.13	3.92	F value: 3.199 p: 0.075	4.03	3.94	F value: 0.531 p: 0.466	3.94	4.03	F value: 0.435 p: 0.510	4.14	3.95	4.04	F value: 0.956 p: 0.386
Q11. Tourist rental accommodation cause noise and coexistence nuisance in Toledo	3.51	3.24	F value: 4.143 p: 0.043*	3.28	3.50	F value: 2.409 p: 0.122	3.47	3.30	F value: 1.302 p: 0.255	3.35	3.29	3.53	F value: 0.478 p: 0.620
Q12. Tourism affects negatively Toledo's cultural heritage	2.42	2.48	F value: 0.277 p: 0.599	2.41	2.55	F value: 1.097 p: 0.296	2.39	2.48	F value: 0.373 p: 0.542	2.41	2.57	2.27	F value: 1.152 p: 0.317
Q13. Tourism reduces the use of Toledo by the residents	2.99	2.97	F value: 0.017 p: 0.896	2.91	3.12	F value: 2.289 p: 0.131	2.80	3.04	F value: 2.430 p: 0.120	2.98	3.05	3.03	F value: 0.123 p: 0.884
Mean of the scale	**3.68**	**3.54**	**F value: 5.652 p: 0.018***	**3.57**	**3.65**	**F value: 1.988 p: 0.158**	**3.58**	**3.60**	**F value: 0.004 p: 0.949**	**3.61**	**3.61**	**3.64**	**F value: 0.004 p: 0.996**

* p< 0.05. Significant difference.

FUTURE RESEARCH

The research demonstrates the limitation of conducting a survey of the community of residents. The results cannot be considered definitive since they are a sample. Equally, overtourism is a rapidly expanding process within Toledo, and so the survey's findings are provisional. For this reason, future research ought to continue to investigate the phenomenon of overtourism in this and other tourist-historic cities. Moreover, there needs to be focus on the fundamental problems of overtourism such as the problems of urban mobility and transport or the 'museumification' of historic centres.

CONCLUSION

The research done in the case study of Toledo, Spain, confirms the hypothesis. In spite of the growth of tourism in the destination to the point of overtourism, the residents and their perceptions and opinions give a final score of 3.59 in the five-point Likert scale on the negative impacts of tourism. Despite the visitor overcrowding and the evidences of overtourism in the city, the host society does not fully agree in their opinions and perceptions in that this is a serious problem. We can conclude that they abide by the Social Exchange Theory and that the social carrying capacity is not overloaded.

Nevertheless, the survey proves that there is risk that this situation will actually happen in the future. There is clear evidence of irritation among the residents when analysing the RQ1 about how the residents value the touristification process in Toledo and what their perceptions and opinions are regarding overtourism. This way, the residents consider Toledo to be a museum for tourists. At the same time, they are concerned about the problem of traffic and mobility, on foot and by car, due to the congestion created by excess of visitors. The residents are also clearly concerned about the raise in the cost of houses for sale and for rent, related to tourist rental accommodation. At the same time, the opinion that there are too many tourist-oriented businesses is remarkable. Thus, Toledo is not living a tourismphobia situation but it is very obvious that some overcrowding is perceived.

RQ2 questioned whether the residents' perceptions and opinions on the negative impacts of tourism are different depending on several demographic and socioeconomic variables. We found out that the residents in the historic centre, the area that concentrates most of the tourist activity, have a more negative opinion of the impacts of overtourism than the residents in other residential areas in the city. For the rest of the analysed variables, despite the answers being heterogeneous, there are no significant differences.

The trends of worldwide tourism, such as the increase of mobility, more affordable transport and new accommodation options, will continue to stimulate a higher demand for urban tourism (UNWTO, 2018). For this reason, there is a high chance that in the future, overtourism will continue to exist and to grow. This brings some risk that Toledo might end up surpassing the social carrying capacity and reaching tourism and tourist-rejection situations which are already obvious in some European urban centres.

The higher the tourism pressure, the greater the effort has to be in order to keep a normal civic life. The cultural and urban essence of the tourist-historic cities must be preserved, to keep them as cities with tourism, not touristic cities. It is not the aim of this chapter to point out which are the measures which must be taken. To find a current sample of recommendations to overcome overtourism in Europe, the monograph by the World Travel & Tourism Council (McKinsey & Company, & World Travel & Tourism Council, 2017), the UNWTO report (2018), the European Parliament TRAN Committee report (2018) or Séraphin et al. (2018)'s work can be consulted. In any case, in tourist-historic cities such as Toledo, resilience measures must be necessary in order to achieve sustainable tourism development (Cheung & Li, 2019) and for that, politicians, the tourism industry and the local community must work together (Cheer & Lew, 2018; Nolan & Séraphin, 2019).

Residents play a vital role in the sustainable tourism development of a destination. For this reason, this study and other similar can be helpful for the political and the tourism managers in the tourist-historic cities. The methodological approach of this investigation can be applied to similar tourist destinations. The results obtained offer a perspective of the perceptions and opinions of the local community regarding overtourism. This allows for more sustainable planning and management of tourist activity in the destination. Taking up the challenges that tourism faces in the tourist-historic cities calls for an intense cooperation between multiple stakeholders for the long run, a complicated but necessary and ideal goal for sustainable tourism (Waligo, Clarke, & Hawkins, 2013). The implementation of sustainability demands every relevant stakeholders' participation in the destination, their cooperation and consensus, a strong political leadership, a proper governing and the support of residents and visitors.

ACKNOWLEDGMENT

The researcher wants to thank his students in the "Tourism and Heritage" course of the 4th year of the Degree in Humanities and Heritage from the School of Humanities of Toledo, their collaboration and participation to carry out the survey under his management.

This research received no specific grant from any funding agency in the public, commercial, or not-for-profit sectors.

REFERENCES

Álvarez-Sousa, A. (2018). The problems of tourist sustainability in cultural cities: Socio-political perceptions and interests management. *Sustainability*, *10*(2), 503. doi:10.3390u10020503

Andereck, K. L., & Nyaupane, G. P. (2011). Exploring the nature of tourism and quality of life perceptions among residents. *Journal of Travel Research*, *50*(3), 248–260. doi:10.1177/0047287510362918

Andereck, K. L., & Vogt, C. A. (2000). The relationship between residents' attitudes toward tourism and tourism development options. *Journal of Travel Research*, *39*(1), 27–36. doi:10.1177/004728750003900104

Ashworth, G. J., & Tunbridge, J. E. (1990). *The tourist-historic city*. London, UK: Belhaven Press.

Ashworth, G. J., & Tunbridge, J. E. (2000). *The tourist-historic city. Retrospect and the prospect of managing the heritage city*. Oxford, UK: Pergamon.

Blanco Izquierdo, R., & Arroyo Romero-Salazar, L. (2018). Geografía humana del Casco Histórico de Toledo. Estilos de vida y su transformación. In J. Manzano (Ed.), *Casco Histórico de Toledo. Construir la convivencia* (pp. 7–9). Toledo: Toletum Revolutum.

Boissevaint, J. (Ed.). (1996). *Coping with tourists: European reactions to mass tourism* (Vol. 1). Providence: Berghahn Books.

Brougham, J. E., & Butler, R. W. (1981). A segmentation analysis of resident attitudes to the social. impact of tourism. *Annals of Tourism Research*, *7*(4), 569–590. doi:10.1016/0160-7383(81)90042-6

Cheer, J. M., & Lew, A. A. (2018). Understanding tourism resilience: Adapting to social, political, and economic change. In J. M. Cheer, & A. A. Lew (Eds.), Routledge advances in tourism: Vol. 42. Tourism, resilience, and sustainability: Adapting to social, political and economic change (pp. 3–17). London, UK: Routledge.

Chen, Ch.-F., & Chen, P.-Ch. (2011). Resident attitudes toward heritage tourism development. *Tourism Geographies*, *12*(4), 525–545. doi:10.1080/14616688.2010.516398

Cheung, K. S., & Li, L.-H. (2019). Understanding visitor–resident relations in overtourism: Developing resilience for sustainable tourism. *Journal of Sustainable Tourism*, 1–20. doi:10.1080/09669582.2019.1606815

Colomb, C., & Novy, J. (Eds.). (2016). *Protest and resistance in the tourist city*. London, UK: Routledge. doi:10.4324/9781315719306

De la Calle Vaquero, M., García Hernández, M., & Yubero Bernabé, C. (2017). Las dinámicas del sector del alojamiento en la ciudad de Donostia-San Sebastián: implicaciones turísticas y urbanas. In *Actas XXV Congreso de la AGE. 50 años de congresos de Geografía. Naturaleza, territorio y ciudad en un mundo global* (pp. 1553-1562). Madrid, Spain: Asociación de Geógrafos Españoles.

Deery, M., Jago, L., & Fredline, L. (2012). Rethinking social impacts of tourism research: A new research agenda. *Tourism Management*, *33*(1), 64–73. doi:10.1016/j.tourman.2011.01.026

Di Grino, B. N. (1986). Community surveys: Purchasing confidence and accuracy. *Journal of Park and Recreation Administration*, *4*(1), 61–69.

Dines, N. (2016). An irreconcilable first-place: The precarious life of tourism and heritage in a southern European historic-centre. *International Journal of Heritage Studies*, *24*(2), 142–153. doi:10.1080/13527258.2016.1195428

Dioko, L. A. N. (2017). The problem of rapid tourism growth: An overview of the strategic question. *Worldwide Hospitality and Tourism Themes*, *9*(3), 252–259. doi:10.1108/WHATT-02-2017-0005

Dodds, R., & Butler, R. W. (2019a). Conclusion. In R. Dodds, & R. W. Butler (Eds.), *Overtourism. Issues, realities, and solutions* (pp. 262–276). Berlin, Germany: De Gruyter. doi:10.1515/9783110607369-018

Dodds, R., & Butler, R. W. (2019b). Introduction. In R. Dodds, & R. W. Butler (Eds.), *Overtourism. Issues, realities, and solutions* (pp. 1–5). Berlin, Germany: De Gruyter. doi:10.1515/9783110607369-001

Escudero Gómez, L. A. (2018). Realities and problems of a major cultural tourist destination in Spain, Toledo. *Pasos. Revista de Turismo y Patrimonio Cultural*, *16*(3), 617–636. doi:10.25145/j.pasos.2018.16.045

European Parliament TRAN Committee. (2018). *Overtourism: Impact and Possible Policy Responses*. Brussels, Belgium: European Parliament.

Feliziani, V., & Miarelli, M. (2012). How many visitors should there be in the city? The case of Rome. *Review of European Studies*, *4*(2), 179. doi:10.5539/res.v4n2p179

Fessler Vaz, L., & Berenstein Jacques, P. (2006). Contemporary urban spectacularisation. In J. Monclús, & M. Guàrdia (Eds.), Culture, Urbanism, and Planning. Routledge. AZW ebook.

García Hernández, M. C., de la Calle Vaquero, M., & Yubero Bernabé, C. (2017). Cultural heritage and urban tourism: Historic city centres under pressure. *Sustainability*, *9*(8), 1346. doi:10.3390u9081346

Gerritsma, R., & Vork, J. (2017). Amsterdam residents and their attitude towards tourists and tourism. *Coactivity: Philosophy. Communication*, *25*(1), 85–98. doi:10.3846/cpc.2017.274

González Pérez, J. M. (2019). The dispute over tourist cities. Tourism gentrification in the historic centre of Palma (Majorca, Spain). *Tourism Geographies*, 1–21. doi:10.1080/14616688.2019.1586986

Goodwin, H. (2017). The challenge of overtourism. *Responsible Tourism Partnership Working Paper 4*. Retrieved from https://haroldgoodwin.info/pubs/RTP'WP4Overtourism01'2017.pdf.

Goodwin, H. (2019). Barcelona – crowding out the locals: a model for tourism management? In R. Dodds, & R. W. Butler (Eds.), *Overtourism. Issues, realities, and solutions* (pp. 125–138). Berlin, Germany: De Gruyter. doi:10.1515/9783110607369-009

Gravari-Barbas, M., & Jacquot, S. (2016). No conñict? Discourses and management of tourism-related tensions in Paris. In C. Colomb, & J. Novy (Eds.), *Protest and Resistance in the Tourist City* (pp. 45–65). London, UK: Routledge.

Hair, J. F., Black, W. C., Babin, B. J., & Anderson, R. E. (2018). *Multivariate data analysis* (8th ed.). London, UK: Cengage Learning EMEA.

Huete Nieves, R., & Mantecón Terán, A. (2018). The rise of tourismphobia. Research hypothesis or ideological noise? *Pasos. Revista de Turismo y Patrimonio Cultural, 16*(1), 9–19. doi:10.25145/j.pasos.2018.16.001

Hughes, N. (2018). 'Tourists go home': Anti-tourism industry protest in Barcelona. *Social Movement Studies, 17*(4), 471–477. doi:10.1080/14742837.2018.1468244

Ixová, M., & Sládek, J. (2017). Touristification and awakening civil society in post-socialist Prague. In C. Colomb & J. Novy (Eds.), *Protest and resistance in the tourist city* (pp. 73–89). London, UK: Routledge.

Jansen-Verbeke, M., & Lievois, E. (1999). Analysing heritage resources for urban tourism in European cities. In D. G. Pearce, & R. W. Butler (Eds.), *Contemporary Issues in Tourism Development* (pp. 81–107). New York: Routledge. doi:10.4324/9780203380307_chapter_6

Katarzyna, J., Six, S., & Vanneste, D. (2017). Building tourism-resilient communities by incorporating residents' perceptions? A photo-elicitation study of tourism development in Bruges. *Journal of Tourism Future, 3*(2), 127–143. doi:10.1108/JTF-04-2017-0011

Kuščer, K., & Mihalič, T. (2019). Residents' attitudes towards overtourism from the perspective of tourism impacts and cooperation—The case of Ljubljana. *Sustainability, 11*(6), 1823. doi:10.3390u11061823

Lestegás, I., & Lois González, R. C. (2018). The global rent gap of Lisbon's historic centre. *International Journal of Sustainable Development and Planning, 13*(4), 683–694. doi:10.2495/SDP-V13-N4-683-694

Liu, J. C., Sheldon, P. J., & Var, T. (1987). Resident perception of the environmental impacts of tourism. *Annals of Tourism Research, 14*(1), 17–37. doi:10.1016/0160-7383(87)90045-4

Marsiglio, S. (2018). On the implications of tourism specialization and structural change in tourism destinations. *Tourism Economics, 24*(8), 945–962. doi:10.1177/1354816618784788

Martín Martín, J. M., Guaita Martínez, J. M., & Salinas Fernández, J. A. (2018). An analysis of the factors behind the citizen's attitude of rejection towards tourism in a context of overtourism and economic dependence on this activity. *Sustainability, 10*(8), 2851. doi:10.3390u10082851

McCool, S. F., & Lime, D. W. (2001). Tourism carrying capacity: Tempting fantasy or useful reality. *Journal of Sustainable Tourism, 9*(5), 372–388. doi:10.1080/09669580108667409

McKinsey & Company, & World Travel & Tourism Council. (2017). *Coping with success. Managing overcrowding in tourism destinations*. Retrieved from https://www.mckinsey.com/industries/travel-transport-and-logistics/our-insights/coping-with-success-managing-overcrowding-in-tourism-destinations

Milano, C. (2018). Overtourism. Social Unrest and Tourismphobia. A controversial debate. *Pasos. Revista de Turismo y Patrimonio Cultural*, *16*(3), 551–564. doi:10.25145/j.pasos.2018.16.041

Montanari, A., & Staniscia, B. (2010). Rome: A difficult path between tourist pressure and sustainable development. *Rivista di Scienze del Turismo*, *1*, 301–316.

Muler González, V., Coromina, L., & Galí, N. (2018). Overtourism: Residents' perceptions of tourism impact as an indicator of resident social carrying capacity - case study of a Spanish heritage town. *Tourism Review*, *73*(3), 277–296. doi:10.1108/TR-08-2017-0138

Munsters, W. (2005). Cultural Tourism in Belgium. In G. Richards (Ed.), *Cultural Tourism in Europe* (pp. 80–92). Association for Tourism and Leisure Education; Retrieved from http://www.tram-research.com/cultural_tourism_in_europe.PDF

Namberger, P., Jackisch, S., Schmude, J., & Karl, M. (2019). Overcrowding, overtourism, and local level disturbance: How much can Munich handle? *Tourism Planning & Development*, *16*(4), 452–472. doi:10.1080/21568316.2019.1595706

National Statistics Institute. (2019). *Hotel Occupancy Survey*. Retrieved from http://www.ine.es/dynt3/inebase/es/index.htm?padre=238&dh=1

Nolan, E., & Séraphin, H. (2019). Venice: capacity and tourism. In R. Dodds, & R. W. Butler (Eds.), *Overtourism. Issues, realities, and solutions* (pp. 139–151). Berlin, Germany: De Gruyter. doi:10.1515/9783110607369-010

Nunkoo, R., & Ramkissoon, H. (2012). Power, trust, social exchange and community support. *Annals of Tourism Research*, *39*(2), 997–1023. doi:10.1016/j.annals.2011.11.017

O'Brien, R. G. (1979). A general ANOVA method for robust tests of additive models for variances. *Journal of the American Statistical Association*, *74*(368), 877–880. doi:10.1080/01621459.1979.10481047

O'Reilly, A. M. (1986). Tourism carrying capacity: Concept and issues. *Tourism Management*, *7*(4), 254–258. doi:10.1016/0261-5177(86)90035-X

Panayiotopoulos, A., & Pisano, C. (2019). Overtourism dystopias and socialist utopias: Towards an urban armature for Dubrovnik. *Tourism Planning & Development*, *16*(4), 393–410. doi:10.1080/21568316.2019.1569123

Papathanassis, A. (2017). Over-tourism and anti-tourist sentiment: An exploratory analysis and discussion. *Ovidius University Annals. Economic Sciences Series*, *17*(2), 288–293.

Perdue, R. R., Long, P. T., & Allen, L. (1990). Resident support for tourism development. *Annals of Tourism Research*, *17*(4), 586–599. doi:10.1016/0160-7383(90)90029-Q

Pizam, A. (1978). Tourism's impacts: The social costs to the destination community as perceived by Its residents. *Journal of Travel Research*, *16*(4), 8–12. doi:10.1177/004728757801600402

Rasoolimanesh, S. M., Jaafar, M., Kock, N. F., & Ramayah, T. (2015). A revised framework of social exchange theory to investigate the factors influencing residents' perceptions. *Tourism Management Perspectives*, *16*, 335–345. doi:10.1016/j.tmp.2015.10.001

Regoeczi, W. C. (2008). Crowding in context: An examination of the differential responses of men and women to high-density living environments. *Journal of Health and Social Behavior*, *49*(3), 254–268. doi:10.1177/002214650804900302 PMID:18771062

Rončák, M. (2019). Prague and the impact of low-cost airlines. In R. Dodds, & R. W. Butler (Eds.), *Overtourism. Issues, realities, and solutions* (pp. 152–1638). Berlin, Germany: De Gruyter. doi:10.1515/9783110607369-011

Saveriades, A. (2000). Establishing the social tourism carrying capacity for the tourist resorts of the east coast of the Republic of Cyprus. *Tourism Management*, *21*(2), 147–156. doi:10.1016/S0261-5177(99)00044-8

Séraphin, H., Sheeran, P., & Pilato, M. (2018). Over-tourism and the fall of Venice as a destination. *Journal of Destination Marketing & Management*, *9*, 374–376. doi:10.1016/j.jdmm.2018.01.011

Sharpley, R. (2014). Host perceptions of tourism: A review of research. *Tourism Management*, *42*, 37–49. doi:10.1016/j.tourman.2013.10.007

Simpson, F. (1999). Tourism impact in the historic center of Prague: Resident and visitor perceptions of the historic built environment. *The Geographical Journal*, *165*(2), 173–183. doi:10.2307/3060415

Smith, M. K., Sziva, I. P., & Olt, G. (2019). Overtourism and resident resistance in Budapest. *Tourism Planning & Development*, *16*(4), 376–392. doi:10.1080/21568316.2019.1595705

Snaith, T., & Haley, A. (1999). Residents' opinions of tourism development in the historic city of York, England. *Tourism Management*, *20*(5), 595–603. doi:10.1016/S0261-5177(99)00030-8

Sommer, C., & Helbrecht, I. (2017). Seeing like a tourist city: How administrative constructions of conflictive urban tourism shape its future. *Journal of Tourism Futures*, *3*(2), 157–170. doi:10.1108/JTF-07-2017-0037

Su, M. M., & Wall, G. (2013). Community involvement at Great Wall World Heritage sites. Beijing, China. *Current Issues in Tourism*, *18*(2), 137–157. doi:10.1080/13683500.2013.823917

Timothy, D. J. (2011). *Cultural heritage and tourism: an introduction*. Bristol: Channel View Publications.

Waligo, V. M., Clarke, J., & Hawkins, R. (2013). Implementing sustainable tourism: A multi-stakeholder involvement management framework. *Tourism Management*, *36*, 342–353. doi:10.1016/j.tourman.2012.10.008

Watson, G. L., & Kopachevsky, J. P. (1996). Tourist carrying capacity: A critical look at the discursive dimension. *Progress in Tourism and Hospitality Research*, *2*(2), 169–179. doi:10.1002/(SICI)1099-1603(199606)2:2<169::AID-PTH35>3.0.CO;2-R

World Tourism Organization (UNWTO). (2018). *'Overtourism'? Understanding and managing urban tourism growth beyond perceptions*. Madrid, Spain: UNWTO.

ADDITIONAL READING

Bouchon, F., & Rauscher, M. (2019). Cities and tourism, a love and hate story; towards a conceptual framework for urban overtourism management. *International Journal of Tourism Cities*. doi:10.1108/IJTC-06-2019-0080

Capocchi, A., Vallone, C., Pierotti, M., Amaduzzi, A., & Capocchi. (2019). Overtourism: A literature review to assess implications and future perspectives. *Sustainability*, *11*(12), 3303. doi:10.3390u11123303

Hargrove, C. M. (2017). *Cultural heritage tourism: five steps for success and sustainability*. Lanham: Rowman & Littlefield.

Marsiglio, S. (2017). On the carrying capacity and the optimal number of visitors in tourism destinations. *Tourism Economics*, *23*(3), 632–646. doi:10.5367/te.2015.0535

Milano, C., Cheer, J. M., & Novelli, M. (2019). *Overtourism. Excesses, discontents and measures in travel and tourism*. Wallingford: CABI. doi:10.1079/9781786399823.0000

Pechlaner, H., Innerhofer, E., & Erschbamer, G. (Eds.). (2020). *Overtourism. Tourism management and solutions*. New York: Routledge.

Perkumienė, D., & Pranskūnienė, R. (2019). Overtourism: Between the right to travel and residents' rights. *Sustainability*, *11*(7), 2138. doi:10.3390u11072138

Tournois, L., & Djeric, G. (2019). Evaluating urban residents' attitudes towards tourism development in Belgrade (Serbia). *Current Issues in Tourism*, *14*(14), 1670–1678. doi:10.1080/13683500.2018.1465030

KEY TERMS AND DEFINITIONS

Cultural Commodification: The production and 'packaging' of heritage for tourist consumption.

Cultural Tourism: All movements of persons to specific cultural attractions, such as heritages sites, artistic and cultural manifestations, arts and drama outside their normal place of residence.

Heritage Tourism: Tourist activity in destinations with preserved buildings, conserved cityscapes and morphological patterns, as well as places associated with historical events and personalities.

Overtourism: The excessive growth of visitors leading to overcrowding in areas where residents suffer the consequences and guests feel that the quality of the experience has deteriorated.

Social Carrying Capacity: The number of tourists above which the social and cultural changes caused by tourism are no longer accepted by the destination's residents.

Tourismphobia: Rejection attitudes towards tourism, with protests and actions that manifest it publicly.

Tourist-historic City: An area of older cities where the urban structure, architecture and artifacts are used to create a place-based heritage touristic product.

Section 5

Case Studies: Natural and Rural Areas

Chapter 18
Perceived Impacts of Running Events in Protected Areas:
The Case of Trans Peneda-Gerês at Peneda Gerês National Park, Portugal

Goretti Silva
Polytechnic Institute of Viana do Castelo, Portugal & UNIAG, Portugal

Alexandra I. Correia
Polytechnic Institute of Viana do Castelo, Portugal & UNIAG, Portugal

Elvira Vieira
Polytechnic Institute of Viana do Castelo, Portugal & UNIAG, Portugal & Research Group of ISAG, European Business School, Portugal

Luís André Soares
Polytechnic Institute of Viana do Castelo, Portugal & UNIAG, Portugal

ABSTRACT

This chapter reflects upon the reality associated with running sports events in protected areas, in particular at the Peneda Gerês National Park (PNPG), with a particular focus on the Trans Peneda-Gerês (TPG), and understands its impacts within a sustainable tourism development perspective. With the diversity of landscapes and natural resources, and its characteristic villages, PNPG, as many other protected areas in Portugal, has a strong potential for Outdoor Sports/Tourism, offering some of Portugal's best running trails. As such, a considerable number of running events are currently being hosted. Based on a research process which includes both questionnaires and interviews, the authors propose to identify the profile and behavior of 2019 TPG edition's participants, and to assess local stakeholders' perception of economic, environmental, and social impacts, and to contribute to the discussion from a sustainable development perspective of the creation and management of tourism-related sports events, namely running, in protected areas.

DOI: 10.4018/978-1-7998-2224-0.ch018

INTRODUCTION

In an increasingly competitive global marketplace destinations need to find ways to differentiate themselves (Hudson & Ritchie, 2009). As most destinations´ attributes, as scenery, attractions, friendly people, and a unique culture and heritage, may be perceived as common by visitors, these factors may no longer be differentiators and destinations are focusing more on the tourist experience, and creating events, both culture and sport-related, to increase their recognition and appeal (Richards & Wilson, 2004; Wilson 2006).

Sport and tourism are amongst the most sought-after leisure experiences (Ritchie & Adair, 2009). Traveling due to sport-related activities and events has become increasingly popular as has been demonstrated by a growing number of studies (Gammon & Robinson, 2003; Gavrili-Alexandris & Vakalakis, 2016; Koronios & Kriemadis, 2018). Sports events are acknowledged as having an important contribution to positioning (Gibson, 1998) as well as to economic development and regeneration of host cities, regions, and countries (Getz & Fairley, 2004; Lee & Taylor, 2005; Wilson, 2006). The main economic benefit for a region derives, not only from the consumption by visitors during the sport-related event but also from increased tourism flows in the post-event period (Preuss, 2005).

However, and despite the positive impacts, it is important to keep in mind that sports events may not always be perceived as positive, particularly when taking place in natural, sensitive areas, with outstanding environmental features. In such cases, tourism flows and sports activities, as is the case of outdoor running, are also likely, to some extent, to cause negative environmental impacts. Adventure racing events are an example of sports events, which occur across a range of land tenures, including protected areas, and that may be commercially sponsored and involve hundreds of competitors and spectators. Thus, and although adventure race is becoming a well-established recreational use in many countries, the question of whether it is an appropriate activity for a particular protected area, should be considered (Newsome, 2014).

Concerns regarding the continuously increasing number of tourists visiting destinations, whether they are urban, or rural, and the unsustainable mass tourism practices (often referred to as *overtourism*, are growing, not only in society, expressed by the motives of "anti-tourism" movements, but also in academia (Milano et al., 2018; Seraphin, et al., 2018). A high concentration of visitors might put places at risk due to the destruction of natural and cultural resources, compromising the place sustainability (Milano et al., 2018). Furthermore, there are concerns that visitors are impacting negatively on the quality of life of locals, contributing to noise and to trash that is often left behind, for example, while the positive economic contribution to local businesses is limited. In this context, the idea of responsible tourism has gained attention in the last few years, both from scholars and practitioners. It has been highlighted that responsible tourism has the potential to create positive tourism impacts by allowing the achievement of businesses', visitors' and local communities' objectives and expectations, and to, simultaneously, contribute to the protection of resources in tourism destinations (Pan et al., 2018).

With the diversity of landscapes and natural resources and a significant part of the territory classified as protected areas, the North region has a strong potential for Outdoor Sports/Tourism, offering some of Portugal's best running trails. As such, a considerable number of running events, organized by various entities, and with very diverse profiles and scopes, from the local to the international, are currently being hosted. This paper aims to reflect upon the impacts of running sports events, with a particular focus on the Trans Peneda-Gerês (TPG), in the Peneda Gerês National Park (PNPG), within a sustainable tourism development perspective. Based on a mixed-methods research process, which includes both questionnaires and interviews, we propose to identify the profile and behavior of 2019 TPG edition's participants, and to assess local stakeholders' perception of economic, environmental and social impacts, as well as to contribute to the discussion regarding the development and management of tourism-related sports events, namely running, in protected areas.

This study is a step ahead in this kind of investigation since both sides of the market, demand, and supply, are being linked. The paper both describes and, more importantly, explains participants' perceptions and behavior as sport tourists within the context of a running event, and stakeholders' perspectives about its impacts, with an emphasis on overtourism perceptions. Reflections provided at the end of the chapter also seek to contribute to the understanding of the sports events, and running activities in particular, and to the formulation of policy guidelines and management strategies to target and prevent mass tourism and overtourism phenomena.

LITERATURE REVIEW

The continued growth in tourism and deepening diversification is well evidenced in the number of international travel and expenditures. Modern tourism is closely related to a growing number of new destinations, as is the case of natural and rural peripheral areas. Costa & Chalip (2005) argue that more peripheral and isolated rural communities tend to offer greater resistance to tourism changes. They are characterized by aging populations with a high predominance of youth migration, who have greater difficulty accepting differences and the need to join forces to capitalize on the benefits of tourism. These communities are linked to tradition and leave it to others to exploit the opportunities created by these sporting events. The lack of involvement and local enthusiasm in exploring new opportunities in a rural context is also explored by Perrin-Malterre (2018). The seasonality of tourism in this area of greater environmental sensitivity, facilitates overtourism phenomena, especially in regions excessively dependent on one type of activity (Bordeau, Corneloup & Mao, 2002). The diversification of supply makes it possible, on the one hand, to minimize the large discrepancy between high and low seasons of tourist activity (Perrin-Malterre, 2018). In this context, interest in sports and sports events, as a driver for local development has been expanding over the last two decades, both in its competitive and recreational aspects (Schmidt, 2013).

Sport tourism, as a research area, is very recent and the first studies carried out to characterize and measure its impact was developed as part of major sporting events, with large infrastructures and complex organizational and logistic webs, which main promoters were the organizing countries (e.g. Olympic Games, Football World Cups, etc). Mega or large sports events are capable of generating large cash flows, locally, nationally and even internationally, but often are also acknowledged as suffering from several negative side effects (Coleman & Ramchandani, 2010; Kennelly, 2017), But sports and sport-related events taking place in rural, natural areas, are often of a small-scale, like trail running, hiking or other, mainly performed in natural environments. Unlike mega-events, small scale destination events (Getz & McConnell, 2014), do not have the same level of economic impacts but do not have the same risks either. Small scale events are organized and implemented more sustainably, building on existing local resources. The development of a sporting event in peripherical regions, or protected areas, with innate natural attributes, increases the challenge of ensuring its sustainability and ecological/environmental protection, respecting local culture and population's lifestyle. Duglio & Beltramo (2017) point out that minor competitor/spectator events are one of the main ways to exploit the tourism potential of nature reserves, without endangering their environmental balance. This kind of event tends to attract tourists with environmental consciousness, who choose the event they intend to take part in according to the destination's natural characteristics (Davies et al., 2018). Chalip (2004) identifies several short and long term benefits for local communities: the former can be reached by enticing visitor spending, increasing visitor length of stay, retaining event expenditures and expanding business relationships, while the latter can be obtained by showcasing the region through the event and using the event in the local marketing strategy. Likewise, Richelieu (2018) argues that small scale sporting events favor a horizontal social capital and should not be seen as an end in itself, but as a way to achieve socio-economic development and legacy, regarding the destination branding. On the other hand, small scale sports activities, and sport tourism events contribute to the dispersion of the tourist flows by reducing the burden on certain support structures, especially linked to the maintenance of the sites and the preservation of their natural balance. It also ensures conditions of financial sustainability, of accommodation and catering activities, especially at the level of the fixed costs, which otherwise could only be maintained during the high season.

Nowadays the traveling motivations of sports tourists are acknowledged as increasingly varied. Kaplanidou & Vogt (2010) emphasize the importance of identifying drivers that condition people's decisions to participate in a particular sports event or activity, to better identify their impacts on the local community. At this level, based on the motivational literature, it is possible to identify more than one approach to the classification of visitors' motivations. A particularly interesting approach is that of Singh et al. (2016), which segment tourists regarding both ecotourism and sport tourism dimensions, identifying four distinct groups of tourists, depending on their sensitivity to issues related to the sustainability of tourism, development of sporting activities or local certification as an ecological destination. These action-oriented tourists (Singh et al., 2016), while participating in sporting events, especially when held in protected natural areas, are stimulated to have a pro-environmental behavior accordingly with the event's rules. Actions like waste collection, no permanent route markings or cleaning of areas of high concentration of athletes are just some of the measures highlighted by Peric & Salvic (2017) in three different trail running events. Regarding the Trans Peneda-Gerês (TPG) trail event, regulation determines disqualification in case of deterioration or contamination of the environment, holds participants responsible for transporting all-natural and unnatural wrappings and wastes to the nearest depot and does not provide liquid supplies through plastic bottles or cups. In the same way, Bordeau, Corneloup & Mao (2002) see mountain sports, as running, as soft tourism generators regarding their minor social and environmental impact comparing with other kinds of sports.

The growing interest in trail running is mainly due to the saturation of urban environment and road racing, along with an increase in the desire to experience peace and contact with the natural environment, and improved accessibility and equipment, overall (Shmidt, 2013). Although there is no established and widely accepted definition due to the possibility of associating different types of environment, pavement, extension, difficulty and purposes (Chase & Hobbs, 2010) and some lack of regulation (Schmidt, 2013), trail running is defined, by the International Trail Running Association (ITRA) (2019), as a pedestrian competition open to all, in a natural environment, and that is practiced on all naturally irregular terrain, such as mountains, forests, countryside, deserts. The course can extend for a few kilometers or go well beyond 80 kilometers for ultra-trails, with the minimum possible of cemented or asphalted roads (20% maximum). Thus, trails as paths can be considered as a tourist attraction and trail running as a racing model that, whatever its purpose, takes place in a predominantly natural environment and creates a link between the practitioner and Nature (Santiago, 2019). Trail running is a sporting activity carried out by very involved athletes, who take the performance of the activity seriously and thus mobilize friends and family to participate directly and indirectly in their travels. Moreover, due to their level of commitment and involvement, trail runners and other outdoor niche sports tourists, tend to have more environmental awareness and leave a smaller ecological footprint.

Overall there are three main factors that, combined, allow maximization of small-scale events' economic return, both directly and indirectly: participants' willingness to pay, increased length of stay, and direct and indirect effects on local branding. Coleman & Ramchandani (2010) argue that most of the small-scale events are financially self-sufficient because runners are willing to pay to participate. Besides the race registration fee, non-local sports tourists tend to stay longer than the event day, leading them to spend in accommodation and catering, the most significant economic revenue for local communities (Kennelly, 2017). Getz & McConnell (2014) point that in particular running amateur athletes have a more mature and serious approach to the events, reflected in the level of investment they are willing to make to participate (preparation, traveling and accommodation). Although most of the motivations' set is similar for many types of niches, either urban or rural running events, in the last ones participants tend to be more motivated by adventure and values related to local sustainability and social responsibility, especially in events held in places like national parks (Davies et al., 2018).

To increase the sport tourists' length of stay, local stakeholders should work together regarding the convenient event schedule (right after a weekend), organization of collateral public activities or development of articulated actions with the local tourism industry (Peric & Salvic, 2017; Duglio & Beltramo, 2017). The set of actions that can be developed under the pretext of the sporting event (Roche et al., 2013), along with some kind of tourism sustainability certification (Singh et al., 2016), important to potential visitors in nature-based sport tourism (Davies et al., 2018), allows the creation of a local branding or marketing strategy (Roche et al., 2013; Richelieu, 2018).

In spite of the acknowledged benefits, Singh et al. (2016) highlight the danger of cultural massification and loss of identity, leading to changes in value systems, traditional lifestyle, family relationships, and individual behavior or community structure, problems usually connected with overtourism. Despite the social advantages that the exploitation of a rising sport brings, trail running, or some other sport and adventure activities, taking place in Natural environments, the social benefits are often conditioned, on the one hand, by the level of resistance and skepticism of local communities towards outside visitors (Costa & Chalip, 2005) and, on the other hand, by the lack of interaction between local stakeholders to seize a common opportunity of tourism development (Costa & Chalip, 2005; Perrin-Malterre, 2018). According to Perrin-Malterre (2018), local stakeholders often resist coexisting socially and spatially to create favorable conditions to improve tourist identity as a trail running destination. Many local communities are too jealous of their benefit, forgetting that creating conditions for a community benefit can be a path towards increasing their well-being. If local communities are socially unprepared for this, trail running will never be a tourist product capable of increasing the region's attractiveness and encouraging loyalty.

Moreover, many authors identify the need for policies designed to better adapt rural populations to the cultural and social shock and grant local sustainability, either cultural, social, economic or environmental. The environmental factor is a sensitive one for these rural populations, especially regarding outdoor activities that use the natural environment for their development, like trail running. But also, the revitalization of the rural economy (Costa & Chalip, 2005), the regeneration of local structures (Wise & Peric, 2017) and the diversification of activities (Bordeau, Corneloup & Mao, 2002; Perrin-Malterre, 2018). Harnessing always depends on how local populations embrace the concept and integrate it into their lifestyle. That is why the articulation between local stakeholders is so important, especially of public entities, acting in order to condition and limit the negative effects of events, without, however, impairing their potential economic and social return (Kennelly, 2017; Davies et al., 2018; Peric & Salvic, 2017; Singh et al., 2016; Richelieu, 2018; Van Dyck et al., 2017).

METHODOLOGY

Geographical Context

The Parque Nacional da Peneda-Gerês (PNPG), located in the Northern Region of Portugal (NUTS II), is the only existing National Park in Portugal, and encompasses territories of three districts and five municipalities: Melgaço, Arcos de Valdevez, Ponte da Barca (district of Viana do Castelo), Terras de Bouro (district of Braga) and Montalegre (district of Vila Real). The PNPG borders Spain at about 100 km being also an integral part of the Gerês-Xurés Transfrontier Biosphere Reserve, classified as a World Heritage Site by the United Nations Educational Scientific and Cultural Organization (UNESCO).

The PNPG has five park entrance doors that serve as a priority PNPG visitation site, where tourists should drive when they arrive to seek information, get support, request services, and purchase products. The five doors of the PNPG are the 'Portas de Lamas de Mouro', 'Portas do Campo do Gerês', 'Portas de Montalegre', 'Portas do Mezio', and 'Portas do Lindoso e Paradela'.

The best way to know and to contact with the natural and cultural values of the National Park is through the walking paths. There are several signposted walking paths, most of which have printed folding support for their interpretation. In addition to these, it is possible to travel other routes, which are not marked on the ground, and which generally correspond to rural paths or shepherds' trails.

Tourism at PNPG

Nature Tourism and ecotourism are considered some of the most promising and fast-growing markets worldwide, particularly in European protected areas as they offer unique and highly valued natural and cultural resources, which lend themselves to such forms of tourism. In Portugal, the number of domestic visitors to natural and rural spaces for tourism and recreation, and protected areas, in particular, has raised in the last years (Marques et al., 2010). Five distinct segments of visitors were identified by Marques et al. (2010), based on the motivation for their visit and further characterized by the perceived importance of activities, facilities, and services, frequency of visit and socio-demographics. According to the study results (p.1), "three segments are nature-focused and two are focused on activities or events. Only one segment shows specific interests and motivations associated with ecotourism and similar characteristics to international eco-tourists; moreover, the reasons for the visit in one segment are unrelated to any of the motivations of nature-based tourism".

The Peneda Gerês National Park (PNPG), is such an example of a nature tourism destination, having tourism-related activities an increasing weight in its economy. The PNPG territory, with a long history as a national tourist destination, has a consolidated offer which has indirectly contributed to its social and economic dynamics. Walking trails and routes, whether large, or small ones are the most relevant Outdoor Tourism and Nature Tourism activities within PNPG, evidencing increased demand over the last decade. Being the only national park in Portugal, widely acknowledge by its biodiversity, gastronomy, cultural and historical factors (Nogueira & Pinho, 2015), makes it a major tourist attraction. In addition to the wide range of services directly related to the tourism sector (hospitality, catering, quality handicraft, and agri-food outlets and entertainment), natural resources and values, tourists will also find material and intangible heritage, infrastructures, equipment, and services, contributing to make each visit a unique and very positive experience. As a result, the region has witnessed a growing number of small-sized leisure and animation-related companies developing activities targeting national and international markets. The most predominant offered activities are hiking, cycling, activities related to fauna and flora observation, along with water-related. PNPG is a park with settlements and/or very small villages where tradition still lives, and this is a unique feature that has also been used by companies to promote and differentiate their offer. In their study, Nogueira & Pinho (2015) concluded that in PNPG, local stakeholders interact in marketing information communication, which is considered very important in the region.

As mentioned before, tourists visiting the region, apart from practicing outdoor activities, can also experience direct contact with the local community, which in many cases, allows communities to be directly linked to the tourism offer. It is extremely important that these communities benefit from the tourism activity and that the outdoor tourism developed in the park does not conflict with communities' way of life. This concern is a critical issue in PNPG that has been raised over the years. According to Marques et al. (2010, p.12), "the perception of the emergent demand for protected areas as tourist destinations together with the pressure on the increasing use of preserved spaces by nature-based companies have compelled the Portuguese Government to set rules and restrictions for tourism in protected areas whilst also promoting forms of visitor occupation that contribute to local development e.g. offering recreation and leisure activities compatible with the conservation objectives and the resident population's lifestyle". This issue is gaining and increasing importance at a time when a sustainable development approach is considered essential for the development of tourism activity, with economic, social, cultural positive impacts, avoiding negative environmental impact and cultural degradation. Furthermore, there is the perception that tourism in protected areas requires the conservation of natural values. Therefore, and according to the Governmental Agency for the Conservation of Nature and Biodiversity, ICNF (cited Marques et al. 2010), conserving nature and landscapes is important to Portugal's image and to the increased value of the tourism supply itself.

The Trans Peneda-Gerês (Tpg)

In Portugal, there were 624001 federated athletes in 2017, of which 18766 practiced athletics. According to PORDATA, in the last decade, federated athletics practitioners, including trail running practitioners, went from 2.5% of the total federated athletes to 3.0%. Only between 2016 and 2017, the growth rate of these federated practitioners was around 29%, having been the fastest growing sport at the national level.

One of the best known and successful trails running athletes in Portugal is Carlos Sá, who states that "In Portugal, we have been able to race all year round". According to him, over the last years, there has been a democratization of the sport, since only 20% of participants race to win. The rest of the trail runners participate to have fun and observe the landscape. He is the organizer of the Trans Peneda-Gerês (TPG) – Trail World Championships Experience, through Carlos Sá Nature Events, with the support of Montalegre and Terras de Bouro city councils and the seal of the Portugal Trail Running Association (ITRA) and the International Running Association (ITRA). TPG was born for the 2016 World Trail Championship, with a distance of 85 km (besides shorter paths with 16, 35 and 55 km), right in the heart of the National Park. In total, in 2016, participated 1.061 athletes, of which 346 foreigners, from 38 different countries.

At present, the racecourse runs through the territory of four of the five municipalities that make up the Peneda-Gerês National Park (PNPG). Most of these trails were already used by the Peneda-Gerês Trail Adventure (PGTA), a multi-day race in its sixth edition. The number of athletes enrolled in PGTA has grown at a consistent pace since 2016 and confirms the growing popular interest in trail running, particularly concerning female athletes (Table 1).

Driven by success and enthusiasm for PGTA, the launching of TPG in 2016 attracted 1.061 participants (of which 346 were international and 715 were Portuguese). The high number of participants in 2016 comes from the fact that the 85km run as part of the Trail World Championships. That was no longer the case in 2019, lowering the number of registrations, but still with a significant 657 entries (602 were national and 55 were international).

The 2019 TPG offered four distances (11km, 25km, 45km, and 80km), depending on the starting point and path followed, but all finished at Gerês Village; TPG 80 Km starts at Montalegre's center, TPG 45 Km and 25 KM both start at Mizarela's Bridge, but follow different paths and TPG 11 Km starts at Fafião.

Already many Portuguese and international athletes participated in the events organized by Carlos Sá Nature taking place within the National Park, besides the TPG, namely the Peneda Gerês Trail Adventure, already mentioned above, and the Gerês Extreme Marathon. According to Carlos, participants are mostly attracted by the path's natural beauty and traditional villages, offering excellent potential for nature sports fans and a great opportunity to contribute to the local economic development. Carlos Sá intends to take this project further, in order to run, in one step, the entire park, linking the two extremes, Castro Laboreiro (in the municipality of Melgaço) with Montalegre, as it happens in PGTA but in 7 days, similar to the installation of the PNPG Great Pedestrian Route (scheduled for 2019). The total distance of PGTA this year, for the Advanced level, was 212 km.

The Organization has adopted the motto "leave no more than footprints and take no more than photographs" and stimulates environmental and social responsibility awareness in all participants. Preservation of nature and respect for local communities are highly recommended to all participants, and who disrespects is penalized. For instance, someone caught leaving trash behind, should be disqualified (Singh et al., 2016; Peric & Salvic, 2017; Davies et al., 2018).

Data Collection

Given the aim and objectives of this study, empirical data was gathered using a survey through self-administered questionnaires and semi-structured interviews. The questionnaire aimed at identifying the profile and motivations of participants, as well as their perception of the event in its different dimensions. Questionnaires were applied in April 2019, during Trans Peneda-Gerês event. The research

Table 1. Number of participants at Peneda Gerês Trail Adventure

	Participants (total)	Male	Female
PGTA 2016	154	107	47
PGTA 2017	148	97	61
PGTA 2018	178	113	66
PGTA 2019	171	100	71

Source: authors

team approached the participants near the finish line after finishing the race. Given the data collection circumstances, a non-probability convenience method was used for sampling. In the first part of the questionnaire, participants were asked about the event, which included questions about typology of the event they were participating in; the number of previous events, if was not the first time they were participating in the event; about where they had gathered information about the event; the importance of different elements in their decision to participate in the event; the length of stay, the type of accommodation and travel party size; how they evaluated the event according to different features; and also about their spending patterns for lodging, food, transportation, shopping, for example. In the second part of the questionnaire, questions on participants' demographic profile were included, such as age, gender, education background, residence, and income. The questionnaire was administered in the Portuguese and English languages. The results of 152 questionnaires were analyzed using the Statistical Package for the Social Sciences (SPSS) 25.

Also, and to assess local stakeholders' perception of economic, environmental and social impacts, semi-structured interviews were conducted to key stakeholders, in early June 2019. Respondents were selected through a convenience sampling method and they were applied to a local business associated to the event, to the organizing entity of the event, and a regional development association (Adere-PG), aiming at identifying and interpreting the impacts of this type of events, based on different viewpoints. In total, three interviews were conducted, and they took place at Peneda-Gerês. The interviews lasted an average of 45 minutes each and were transcribed and analyzed manually, systematically and objectively. The interview design was aligned to key aspects of qualitative research interviews (Smith, 2017) and the themes addressed were concerned with the main research aim and objectives mentioned previously. One set of questions focused on specific reasons that led to the event organization and /or support. The second group of questions aimed at eliciting perceived impacts both, positive and negative. Additional questions were also asked, about the profile of participants and/or customers.

Data Presentation and Discussion

Of the approximately 600 athletes who participated in the 2019 edition of TPG, 152 answered the questionnaire (a response rate of 25%, as shown in Table 2). Of these, 76% were Portuguese and 24% were international.

Because attracting non-local sports tourists, particularly international ones, is assumed as likely to have a stronger impact on the local economy (Duglio & Beltramo, 2017), throughout the rest of the analysis, results for both groups (Portuguese and International respondents) are presented separately (Table 3). Qui-square test and Mann Whitney test were used to verify any differences between the two groups. When existing, statistically relevant differences will be highlighted and discussed.

Table 2. Nationality of respondents

	n	%
Portuguese	115	75,7
International	37	24,3
Total	**152**	100,0

Source: authors

Table 3. Profile of respondents

		Nationality			
		Portuguese		International	
		n	%	n	%
Gender	Female	35	30,4	16	43,2
	Male	80	69,6	21	56,8
Marital Status	Live alone	51	44,3	13	35,1
	Live with someone	64	55,7	24	64,9
Children	No children	52	45,2	26	70,3
	1 or 2	56	48,7	7	18,9
	3 or more	7	6,1	4	10,8
Education	<= High School	59	51,3	12	32,4
	Higher Education	56	48,7	25	67,6
	X^2 = 3,283; p=,162; sig.= 0,045				
Income of Household	Between 600€ e 1200€	45	40,9	4	10,8
	Between 1201 e 1800€	27	24,5	3	8,1
	Between 1801€ e 2400€	17	15,5	4	10,8
	Between 2401 e 3000€	14	12,7	7	18,9
	3001€ or more	7	6,4	19	51,4
	(x^2=44,033; p=0,547; sig.=,000)				
Age (grouped)	<=45	97	84,3	16	43,2
	46+	18	15,7	21	56,8
	(x^2=22,688; p=0,404; sig.=,000)				

Source: authors

As for their age, international participants seem to be older with 57% above 45 years, against 84% of the Portuguese under this age limit (x2=22,688; p=0,404; sig.=,000), being the average age 47 for the first group and 37 for the second (Mann-Whitney Test U= 3364,000, Z= 5,31, *p*=0,000). In general, there were a few more men than women answering the questionnaire, which is following the proportion of gender of participants taking part in the TPG event, considered as a very difficult trail path, more in accordance with a male physical complexion. Respondents are, in both cases, balanced between living alone and with someone, but most international have no children (70%) as opposed to Portuguese (just 45%).

International participants have a higher level of education, with 68% holding a higher education diploma, against only 49% of the Portuguese. This fact is relevant because the level of education is associated with higher income, *willingness* to *spend more* and *stay longer*. As shown in Table 3, the income of 51,4% of international respondents' households, is above 3.000€ whereas 67% of the Portuguese is up to 1.800€.

These results are in accordance with the literature which stresses that higher education and higher income are associated with higher levels of environmental awareness and more responsible tourists, both at the environmental and social levels when it comes to making choices for leisure activities.

Based on Mann-Whitney tests (Table 4), one can say that there are significant differences between national and international tourists regarding the following motivations to participate in the event: 'For the whole experience', 'Natural context /Landscape', 'Cultural context', and Acknowledgment of Carlos Sá Nature Events company'. Portuguese participants place greater emphasis on the experience as a whole, and the quality of cultural and natural context where the running event takes place, which likely is associated with them being more familiar with Peneda Gerês National Park. Portuguese also acknowledge to a greater extent the reputation of Carlos Sá as the organizer of running events, which would be expected, as they were likely to be more familiar with their work. On the other hand, international participants had greater expectations concerning the location of the event (and the quality of natural resources). These results reveal different motivations to participate in this type of event, which is in accordance with the literature review. In several studies the main intrinsic motivation to participate in a running event is the experience, seeking enjoyment and excitement (Aicher et al., 2015; Davies et al., 2018). Other studies highlight the development of sporting activities in destinations with specific features, for example, an ecological certification (Singh et al. 2016).

Most of the respondents traveled alone to participate in the event. Only 37% of the Portuguese and 23% of the international had, at least, one companion or more. The information is of relevance because local impacts at an economic level can be extended if participants have one or more traveling companions (family or friends). On the other hand, ultimately, this would represent a higher number of people visiting and staying in the region, which could result in a less positive environmental and social impact.

The number of people traveling with the respondents will certainly influence the amount spent by each participant, particularly with travel, accommodation, and catering. According to the results, as shown in Table 5, international respondents not only spend much more with travel, but also with accommodation and catering (restaurants/meals), and shopping and leisure/cultural activities. In all the expenses, the probability value (p) is smaller than .05, so the results are significant. There are statistically significant differences in the average amount spent on different items by national and international tourists.

Table 4. Motivations to participate in the event

	Portuguese			International			Mann-Whitney test	Mann-Whitney test *U*	Z
	n	**mean**	**Standard Dev.**	**n**	**mean**	**Standard Dev.**	**(Sig.)**		
For the whole experience	114	4,51	,641	37	4,08	,862	,004	1515,000	-2,87
Familiar atmosphere of the event	115	4,32	,779	37	4,22	,886	,574	2007,500	-,56
Quality of organization	115	4,54	,717	36	4,69	,467	,437	2218,000	,77
Natural context /Landscape	115	4,89	,318	37	4,22	1,084	,000	1338,500	4,89
Cultural context	115	4,47	,692	36	3,78	1,098	,000	1273,500	-3,81
Acknowledgment of Carlos Sá Nature Events company	115	4,58	,662	36	3,86	1,099	.000	1266,000	-4,00
Gastronomy	115	4,15	,929	36	3,94	1,040	,313	1853,000	-1,00

Source: authors

Table 5. Participants' average expenditures

	Nationality		Mann-Whitney Sig.	Mann-Whitney *U*	Z
	Portuguese	International			
	Average €				
Average amount spent on travel	€27	€572	.000	3505,500	8,29
Average amount spent on accommodation	€27	€207	.000	2793,000	7,11
Average amount spent on catering	€19	€133	.000	3437,500	7,69
Average amount spent on shopping/gifts	€2	€88	.000	3163,500	8,14
Average amount spent on culture/leisure	€12	€105	.000	2749,000	7,12
Average amount spent on other things	€0	€24	.000	576,000	5,71

Source: authors

Interviews

The perception of interviewed stakeholders as described above in the methodology is that the event is, in general, very positive, with several advantages for local communities and local businesses. It attracts many participants from different countries, who travel to Peneda-Gerês with family or friends, extending their stay for longer periods. The economic impacts are therefore relevant to the local economy, both directly and indirectly, given the visibility the region gets around the world, through media coverage and participants sharing their experience through social networks. Moreover, based on their experience organizing or co-organising the event, interviewed stakeholders believe that these participants are extremely conscious and environmentally sensitive, which minimizes the environmental impacts. The community not only accepts but supports the running event.

To minimize unfavorable situations, the event jointly organized by Carlos Sá Nature Events, Adere-Peneda Gerês, and local Municipalities, in a close cooperative manner. The last two not only are relevant by facilitating logistic support, but also financial support and communication. Carlos Sá's experience and expertise are widely acknowledged in the region, a fact which facilitates to a great extent, cooperation with local businesses and organizations.

According to the results, several short, and long-term benefits for local communities are identified, both direct and indirect. As suggested in the literature (Chalip, 2004; Coleman & Ramchandani, 2010; Kennelly, 2017) runners participating at TPG are willing to pay not only the race registration fee, especially non-local ones, but also to stay longer than the event day, spending, in total, higher amounts in accommodation and catering, the most significant economic revenue for local communities. As most participants travel with family or friends and some extend their stay for up to 3 to 5 days, especially non-local ones, economic impacts are evident for local businesses, sometimes through specific arrangements with the event organizer. On the other hand, the event is also acknowledged as showcasing the region, through the participants' social networks and the extensive media coverage (at national and international level).

CONCLUSION, LIMITATIONS, IMPLICATIONS AND FURTHER RESEARCH

Sports events are increasingly important for tourism, and although they are recognized as having significant contributions to host cities, regions, and counties, impacts may not always be positive, particularly when these events take place in natural and sensitive environments, as protected areas. This study attempted to address this issue by identifying the profile and behavior of the participants in the 2019 TPG edition in Peneda-Gerês National Park (North of Portugal), and by assessing local stakeholders' perception of economic, environmental and social impacts.

The event attracted both, domestic and international participants. Although there were more domestic participants, the results of this study revealed a significant proportion of international participants, that seem to be older, highly educated and have a higher income. The whole experience, the quality of the event organization, and the natural context/landscape are the main motivations for the interviewed participants, in general. Most participants traveled alone and according to the results, international respondents spent higher amounts with travel, accommodation, and catering (restaurants/meals), as well as with shopping and leisure/cultural activities. These benefits are even of greater relevance as running activities take place during low season months.

The study found that the event is mainly perceived as positive by local stakeholders, despite some minimal concerns pointed out in terms of pollution risks and noise. Besides the immediate positive impacts, it is believed the initiative will create medium and, and even longer-term impacts, considering the participants manifested the desire to return to the region, not only to running-related activities and events but also for leisure activities, bringing their closest relatives or friends. Besides, the event will also generate a broader impact at an international level, by raising awareness about the conditions offered by the PNPG for nature and outdoor sports. In the particular case of TPNG, the Organization has adopted the motto "leave no more than footprints and take no more than photographs", which was strongly supported and widely adopted by participants.

In spite of the perceived benefits, and positive features of the region, it was widely acknowledged the need for investment in the region, creating the necessary conditions for nature tourism, qualifying the existing offer and bringing new, sustainable and responsible solutions. Although not directly perceived as resulting from running activities, in some parts of the Park, there is already some evidence of massive flows of visitors, especially during summer, a phenomenon associated with overtourism.

Therefore, to maximize the benefits of running events, and preventing from overdevelopment, the formulation of policy guidelines and management strategies are suggested, ensuring the articulation between local stakeholders and public entities. The revitalization of the rural economy depends on the regeneration of local structures and the diversification of activities as part of local entrepreneurial processes. Local populations need to embrace the concept and integrate it into their lifestyle. That is acting to condition and limit the negative effects of events, without, however, impairing their potential economic and social return.

This study has limitations and caution should be used in any generalization of the findings to other regions and events. Firstly, the reduced number of questionnaires might limit a wide-ranging perspective of participants. Secondly, data collected refers to a specific event and location and, therefore, findings may be specific to the businesses and event organizers operating in PNPG. Then, the qualitative analysis may be limited due to the small number of responses. Also, data were collected from event organizers and other relevant stakeholders in a protected area in the Northern region of Portugal. Thus, their perceptions and answers to the interview are most likely based on their experiences in this specific geographic context.

Despite its limitation, the findings of this study suggest several practical implications for the development and management of tourism-related sports events in protected areas, namely running events. In particular, the research presented here illustrates participants' profile and local stakeholders' perception of the economic, environmental and social impacts of these sports events. This study reinforces the idea that participants of this type of events seem to be higher educated and with higher income. This educational and economic profile is associated with higher levels of environmental awareness and more responsible tourists, both at the environmental and social levels when it comes to making choices for leisure activities. Thus, this study indicates that event organizers may contribute to stimulating environmental and social responsibility awareness in all participants, by adopting initiatives that underline the importance of the preservation of nature and respect for local communities. Also, the study findings allow an improved understanding of the perceived positive economic impacts, but they also emphasize existing concerns regarding tourism flows and acceptable, low-pressure level. Furthermore, and despite the overall optimistic stance of both participants, stakeholders and organizing team, the results also contribute to raising awareness of potential negative impacts, which need to be considered within effective events planning and management strategies. Thus, this study could provide an important reference to organizations, not only of the studied area but also from other regions, by considering the presented findings when promoting sport-related activities and events in the future.

Concerning a regional level, the results of this study emphasize that the Minho region, in general, and Gerês, in particular, attracts both international and international participants, which is a strong indicator that the region is perceived to have interesting characteristics for the practice of this sport. Thus, these results could motivate the organization and promotion of World Championship events. Indeed, these results can help different stakeholders in the development and promotion of other sports events in the future, and to the development of a tourist expanded offer out of this sporting event, increasing the possibility of attracting, not only athletes to the event, but also tourists to the region, enlarging their length of stay and willingness to spend money, and, therefore, contributing positively to the region.

Besides, Portugal can thus project its mountains and protected areas, starting here an assertive and permanent promotion of the various destinations similar to Peneda Gerês National Park, as competitive destinations for running, and other outdoor activities. The trail is not just another activity, it has the potential to attract a whole market that can only practice mountain sports from June to September. Tourism has grown in the wake of international trends, and the trail is yet another path that points to the future of the mountain villages and villages near abandonment and which needs to be energized. This potential is amplified as Portugal's climate in the eyes of Northern and Central Europeans resembles a long spring, and therefore, Portugal has the potential to become a winter destination for all who seek to run and discover otherwise unreached corners.

The findings of this study also present future research opportunities. Given the limited number of questionnaires and interviews applied to a specific event and geographical context, more longitudinal research is needed. Future research would allow a more comprehensive understanding of both participants' and stakeholders' perceptions and changing patterns regarding this typology of events in the protected areas. Further studies would also allow comparative analyses of participants' profile and stakeholders' perceived impacts. Also, future research could explore differences between perceptions of domestic and international and their image of the region where the event took place and the future intention to return in the future.

ACKNOWLEDGMENT

This project is carried out within the scope of the TURNOUT Project - Development of Outdoor Tourism of the Northern Portugal, POCI-01-0145-ERDF-032289 financed by the European Investment Fund Regional Development (ERDF) through the Operational Program Competitiveness and Innovation (POCI) and the Foundation for Science and Technology, I.P. (OE).

REFERENCES

Aicher, T. J., Karadakis, K., & Eddosary, M. M. (2015). Comparison of sport tourists' and locals' motivation to participate in a running event. *International Journal of Event and Festival Management*, *6*(3), 215–234. doi:10.1108/IJEFM-03-2015-0011

Chalip, L. (2004). Beyond impact: a general model for host community event leverage. In B. W. Ritchie, & D. Adair (Eds.), *Sport Tourism: Interrelationships, Impacts and Issues* (pp. 236–262). Clevedon: Channel View.

Chase, A. W., & Hobbs, N. (2010). *Ultimate Guide to Trail Running: Everything You Need To Know About* (2nd ed.). U.S.A.: Falcon Guides.

Coleman, R., & Ramchandani, G. (2010). The hidden benefits of non-elite mass participation sports events: An economic perspective. *International Journal of Sports Marketing & Sponsorship*, *12*(1), 19–31. doi:10.1108/IJSMS-12-01-2010-B004

Costa, C., & Chalip, L. (2005). Adventure sport tourism in rural revitalisation – an ethnographic evaluation. *European Sport Management Quarterly*, *5*(3), 257–279. doi:10.1080/16184740500190595

Davies, M., Hungenberg, E., & Aicher, T. (2018). The relationship between runner environmental paradigm and their motives to participate in an urban or rural marathon. *International Journal of Event and Festival Management*, *10*(1), 48–66. doi:10.1108/IJEFM-02-2018-0013

Duglio, S., & Beltramo, R. (2017). Estimating the Economic Impacts of a Small-Scale Sport Tourism Event: The Case of the Italo-Swiss Mountain Trail CollonTrek. *Sustainability*, *9*(3), 343. doi:10.3390u9030343

Gammon, S., & Robinson, T. (2003). Sport and Tourism: A Conceptual Framework. *Journal of Sport & Tourism*, *8*(1), 21–26. doi:10.1080/14775080306236

Gavrili-Alexandris, D., & Vakalakis, P. (2016). Running on Heritage, a Conceptual Discussion on the Roles of Heritage Trails and of Augmented Reality on Amateur Runner Athletes. In V. Katsoni, & A. Stratigea (Eds.), *Tourism and Culture in the Age of Innovation* (pp. 589–602). Springer Proceedings in Business and Economics. doi:10.1007/978-3-319-27528-4_40

Getz, D., & Fairley, S. (2004). Media Management At Sport Events For Destination Promotion : Case Studies And Concepts. *Event Management*, *8*(3), 127–139. doi:10.3727/1525995031436926

Getz, D., & McConnell, A. (2014). Comparing trail runners and mountain bikers: Motivation, involvement, portfolios, and event-tourist careers. *Journal of Convention & Event Tourism*, *15*(1), 69–100. do i:10.1080/15470148.2013.834807

Gibson, H. J. (1998). Sport Tourism : A Critical Analysis of Research. *Sport Management Review, 1*(1), 45–76. doi:10.1016/S1441-3523(98)70099-3

Hudson, S., & Ritchie, J. R. B. (2009). Branding a Memorable Destination Experience. The Case of "Brand Canada.". *International Journal of Tourism Research, 11*(2), 217–228. doi:10.1002/jtr.720

Instituto da Conservação da Natureza e Florestas. (2019, July 3). Retrieved from http://www2.icnf.pt/portal/icnf

International Trail Running Association. (2019, July 25). *Discovering Trail running*. Retrieved https://itra.run/page/259/Discovering_trail-running.html

Kaplanidou, K., & Vogt, C. (2010). The meaning and measurement of a sport event experience among active sport tourists. *Journal of Sport Management, 24*(5), 544–566. doi:10.1123/jsm.24.5.544

Kennelly, M. (2017). We've never measured it, but it brings in a lot of business. *International Journal of Contemporary Hospitality Management, 29*(3), 883–899. doi:10.1108/IJCHM-10-2015-0541

Koronios, K., & Kriemadis, T. (2018). An exploration of motives, constraints and future participation intention in sport and exercise events. *Sport, Business, and Management, 8*(5), 436–453. doi:10.1108/SBM-09-2017-0044

Lee, C. K., & Taylor, T. (2005). Critical reflections on the economic impact assessment of a mega-event: The case of 2002 FIFA World Cup. *Tourism Management, 26*(4), 595–603. doi:10.1016/j.tourman.2004.03.002

Marques, C., Reis, E., & Menezes, J. (2010). Profiling the segments of visitors to Portuguese protected areas. *Journal of Sustainable Tourism, 18*(8), 971–996. doi:10.1080/09669582.2010.497222

Milano, C., Novelli, M., & Cheer, J. M. (2018). Overtourism and Tourismphobia: A journey through four decades of tourism development, planning and local concerns. *Tourism Planning and Development Journal*, 1–4.

Newsome, D. (2014). Appropriate policy development and research needs in response to adventure racing in protected areas. *Biological Conservation, 171*, 259–269. doi:10.1016/j.biocon.2014.01.008

Nogueira, S., & Pinho, J. C. (2015). Stakeholder network integrated analysis: The specific case of rural tourism in the Portuguese Peneda-Gerês National Park. *International Journal of Tourism Research, 17*(4), 325–336. doi:10.1002/jtr.1989

Pan, S. Y., Gao, M., Kim, H., Shah, K. J., Pei, S. L., & Chiang, P. C. (2018). Advances and challenges in sustainable tourism toward a green economy. *The Science of the Total Environment, 635*, 452–469. doi:10.1016/j.scitotenv.2018.04.134 PMID:29677671

Perrin-Malterre, C. (2018). Tourism diversification process around trail running in the Pays of Allevard (Isère). *Journal of Sport & Tourism, 22*(1), 67–82. doi:10.1080/14775085.2018.1432410

PORDATA. (2019, July 20). Praticantes desportivos federados: total e por algumas federações desportivas. Retrieved from https://www.pordata.pt/Portugal/Praticantes+desportivos+federados+total+e+por+algumas+federações+desportivas-2226.

Preuss, H. (2005). The Economic Impact of Visitors at Major Multi-sport Events. *European Sport Management Quarterly*, *5*(3), 281–301. doi:10.1080/16184740500190710

Richards, G., & Wilson, J. (2004). The impact of cultural events on city image: Rotterdam, cultural capital of Europe 2001. *Urban Studies (Edinburgh, Scotland)*, *41*(10), 1931–1951. doi:10.1080/0042098042000256323

Richards, G., & Wilson, J. (2006). Developing creativity in tourist experiences: A solution to the serial reproduction of culture? *Tourism Management*, *27*(6), 1209–1223. doi:10.1016/j.tourman.2005.06.002

Richelieu, A. (2018). A sport-oriented place branding strategy for cities, regions and countries. *Sport, Business and Management*, *8*(4), 354–374. doi:10.1108/SBM-02-2018-0010

Ritchie, B., & Adair, D. (2009). The Growing Recognition of Sport Tourism. *Current Issues in Tourism*, *5*(1), 1–6. doi:10.1080/13683500208667903

Roche, S., Spake, D., & Joseph, M. (2013). A model of sporting event tourism as economic development. *Sport, Business, and Management*, *3*(2), 147–157. doi:10.1108/20426781311325078

Santiago, C. (2019). *Trail running: modelo e potencial territorial enquanto produto turístico* (Unpublished Master's thesis). Escola Superior de Turismo e Hotelaria do Estoril. Retrieved from https://comum.rcaap.pt/bitstream/10400.26/18234/1/2016.04.017_.pdf

Schmidt, C. E. (2013). *Reducing muscular fatigue in trail running mechanisms and strategies*, Unpublished Master's thesis). The University of Sydney & Université Nice Sophia Antipolis.

Seraphin, H., Sheeran, P., & Pilato, M. (2018). Over-tourism and the fall of Venice as a destination. *Journal of Destination Marketing and Management, 9*(September 2017), 374–376.

Singh, S., Dash, T., & Vashko, I. (2016). Tourism, ecotourism and sport tourism: The framework for certification. *Marketing Intelligence & Planning*, *34*(2), 236–255. doi:10.1108/MIP-09-2014-0180

Van Dyck, D., Cardon, G., De Bourdeaudhuij, I., De Ridder, L., & Willem, A. (2017). Who Participates in Running Events? Socio-Demographic Characteristics, Psychosocial Factors and Barriers as Correlates of Non-Participation—A Pilot Study in Belgium. *International Journal of Environmental Research and Public Health*, *14*(11), 1315. doi:10.3390/ijerph14111315 PMID:29143780

Wilson, R. (2006). The economic impact of local sport events: Significant, limited or otherwise? A case study of four swimming events. *Managing Leisure*, *11*(1), 57–70. doi:10.1080/13606710500445718

Wise, N., & Perić, M. (2017). Sports Tourism, Regeneration and Social Impacts: New Opportunities and Directions for Research, the Case of Medulin, Croatia. In N. Bellini, & C. Pasquinelli (Eds.), *Tourism in the City* (pp. 311–320). Berlin, Germany: Springer. doi:10.1007/978-3-319-26877-4_22

Chapter 19
Outdoor Solutions for the Seasonal Concentration of Tourism Demand in Northern Portugal:
An Integrated Approach Based on the Gini Index

Paula Odete Fernandes
https://orcid.org/0000-0001-8714-4901
UNIAG, Portugal & Polythecnic Institute of Bragança, Portugal

Alcina Maria Nunes
UNIAG, Portugal & Polythecnic Institute of Bragança, Portugal

Cláudia Miranda Veloso
GOVCOPP, ESTGA, University of Aveiro, Portugal

Eleonora Santos
https://orcid.org/0000-0003-4693-0804
Polythecnic Institute of Bragança, Portugal

Fernanda A. Ferreira
https://orcid.org/0000-0002-1335-7821
UNIAG, Portugal & Polythecnic Institute of Porto, Portugal

Manuel José Serra Fonseca
https://orcid.org/0000-0002-3290-8449
UNIAG, Portugal & Polytechnic Institute of Viana do Castelo, Portugal

ABSTRACT

Measures that address the seasonality, one of the identified overtourism direct causes, allow making a critical reflection on the application of control policies and monitoring regional measures crucial on a sector with such importance at the regional development analysis. The measures should stimulate or reduce tourism demand during low or peak seasons, respectively, generating a better distribution of tourism flows and eliminating potential overtourism situations. Therefore, this chapter focuses on the development of a Gini Index to analyse the distribution of the seasonality in northern Portugal and through it control current public measures in practice and suggest the implementation of different and more effective policy measures. For instance, the ones that make a clear bet on outdoor tourism activities. It will be used in the Gini Index, by the tourists' market of origin measured by the overnight stays in hotel establishments. The results showed the high seasonal concentration of tourist flows.

DOI: 10.4018/978-1-7998-2224-0.ch019

INTRODUCTION

The discussion around the concept of overtourim is not recent. Overtourism, albeit a relative new term, deals with old problems (Capocchi, Vallone, Amaduzzi, & Pierotti, 2019) and gained a renewed importance due to the increasing concern expressed by academics, policy-makers, population and social media. Such concerns include the excessive number of tourists in a place at a given moment, that may generate conflicts between residents and tourists and crowding phenomena, loss of authenticity, reduction of residents' quality of life and lower experience enjoyment by tourists (Dodds & Butler, 2019; Phi, 2019). According to Goodwin (2017) among the direct causes of overtourim is the seasonality. Indeed seasonality, related to the trend of tourism flows concentrated in a short period of the year, became one of the crucial challenge factors that the managers of the tourism sector have been facing. Its reduction and, ultimately, elimination, has not been easy to achieve by policy and decision-makers, whether at the public or private level since it presents an imbalance between demand and supply of a certain tourism destination.

Measures that address the seasonality allow making a critical reflection on the application of control policies and monitoring regional measures crucial on a sector with such importance at the regional development analysis. Such measures should stimulate tourism demand during low seasons or reduce tourism demand during peak seasons. In this regards, the current chapter focus on the development of a Gini Index to analyse the distribution of the seasonality in the North region of Portugal and through it control current public measures in practice and suggest the implementation of different and more effective policy measures like, for instance, outdoor events (which may include sports, culture and music). Such alternative tourism strategies have already been identified as strategies to manage seasonality and differentiate destination offers (Smeral, 2019). The Gini index, that is being viewed as a measure of the level of fairness of resources distribution among a group of individuals (namely to measure the level of inequality in distributions), will be applied in the present context as a measure of the dispersion of the data over the year among different regions in an innovative approach applied to the Portuguese tourism sector. With the specific application to the North region of Portugal, a region well known not only by the cultural patrimony and sea activities but also by its natural and well-preserved mountain landscapes, specific gastronomy and Winter cultural traditions, which offer multiple and different experiences all over the year, would be possible to recommend and suggest tourism activities related to outdoor tourism. Such unexplored branch of tourism might allow the Portuguese northern tourism destinations to benefit from a more uniform distribution of tourism demand over the year and optimise the existent resources causing minimal negative impacts associated with seasonal fluctuations in tourism demand and enhancing the positive impacts for the development of a region that suffers from problems as ageing, desertification, low levels of employment and regional income.

In order to reach the main objective of the study, it will be used the information of the tourism market of origin (Portugal market and the Top-5 foreign countries with the highest market share: Spain, France, United Kingdom, America and Germany), measured by the overnight stays in hotel establishments. This study allows obtaining the Gini Index for each tourism destination (8 regions), each segment share of overall tourism demand and the Gini relation of these segments with the overall tourism demand. To understand the tourism flows the analysis will be focused on the matrix of interregional tourism flows in the North region of Portugal and in the graphical techniques. The current and official data of the National Statistics of Portugal (INE) will be considered for all 8 regions that belong to the North region of Portugal, over the period 2014-2017 (the period for each could be found the most updated statistics according to the last NUT[1] revision, in Portugal).

In Portugal, namely in the North region as in many other destinations, the complex phenomenon of seasonality is a concern in the tourism sector. At the institutional level, the attempt to combat seasonality has been recognised as an important topic and it is revealed in the Portuguese National Strategic Tourism Plan as an objective to improve. At the same time, and at the regional level, the municipalities intend to implement measures directed towards encouraging efforts to mitigate seasonality as part of the growth strategy for tourism. The seasonal concentration produced by tourists visiting the North region of Portugal and its municipalities cannot be considered very high since it does not reach a high seasonality ratio, i.e. over two. This situation can be justified by the adjustments and efforts that have been made by public policymakers to reduce the seasonality problem in the tourism sector.

According to the above mentioned, it is fundamental to apply measures never used in the Portuguese economy to measure the seasonality concentration at a regional level, to present solutions to mitigate the seasonality and, by consequence, potential overtourim issues. Such measures, already presented for other international realities have proved to be efficient to monitor the movement of tourism flows all over the months of a year and regions. Indeed, with this chapter, it is believed that calculating and presenting an innovative but robust measurement of tourism seasonality is important from the perspective of developing and evaluating seasonal public policies, measures and private business strategies.

The first section of the chapter is devoted to the literature review on the concept of seasonality and the phenomena of the seasonality, in particular, in the tourism research. The section also approaches tourism development in the North region of Portugal. In the third part of the chapter, the methodology for conducting the study is presented and explained, after the goals being set and the statistical data collection and analysis are described. In the fourth part, the results obtained from the application of the method of analysis are presented and critically analysed. In conclusion, some remarks on recommendations and implications are presented.

BACKGROUND

Seasonality in Tourism Research

Touristic seasonality phenomenon denote a key topic in the fields of policymaking and practical tourism management and also in the academic literature on literature (Koenig-Lewis & Bischoff, 2005). The seasonal touristic demand variations or touristic seasonality is a phenomenon characterized by instability between supply and demand at certain times of the year. It is assumed to be the unequal distribution of tourism movements over time (concentration of tourist flows over short periods of the year) and can be expressed in terms of number of visitors, revenues, employment, occupancy rates, traffic, among other indicators. The uneven demand growth generates fluctuations and consequent imbalances for the territories and the operators (Coshall, Charlesworth & Page, 2015).

The concentration of holidays or the existence of events, weather conditions (seasons), purchasing power, social or cultural behaviours (religious or sports motivations), fashion trends and habit are factors that influence the seasonality of the touristic demand (Rosselló & Sansó, 2017). There are also issues related to the environment itself such as temperature, sunny hours, rain levels or snow (Koenig-Lewis & Bischoff, 2005).

The causes of seasonality are assorted and multifaceted, as are their effects, dealt with in some academic research about seasonality. Wall and Yan (2003) argue that the impacts are likely to have become larger with the development of mass tourism. This is due to the fact that the number of tourism-dependent enterprises has increased as well as their size and thus their ability to adapt to changing demand has been reduced (Koenig-Lewis & Bischoff, 2005). According to Wall and Yan (2003), a strict and detailed analysis of demand variations and their effects can point out profit opportunities and find out ways to reduce losses.

In these contexts, seasonality conditions the social and economic sustainability of tourist destinations, threatening their environmental and natural resources through the pressure exerted at summertime on the territory, local people, services and infrastructure (Budeanu, 2007). In economic terms, seasonality motivates low profitability of tourism facilities (occupancy rate drops), unbalanced pricing policy (high swings) and creates precarious employability (Turrión-Prats & Duro, 2018). As far as the environmental aspect is concerned, there are consequences such as the unregulated construction of infrastructures and equipment and the pollution itself. Overcrowding of spaces may also lead to an increase in crime and accidents (Cannas, 2012).

Regarding the question of tourist seasonality in Portugal, Kastenholz and Lopes de Almeida (2008) describe that, as a beach destination, its attraction is strongly associated with the climate, which contributes to the accentuation of seasonality. These flows are not only visible in sun and sea destinations such as the Algarve, but also in nature destinations such as those in the north of the country. These authors point out that Portugal needs measures to minimize this reality. In order to mitigate the seasonality are suggested strategies which are based on the design and development of new products as well as the diversification of markets and consequent actions of communication and promotion to new segments (Connell, Page & Meyer, 2015).

Being able to quantify the degree of seasonality of an observed demand pattern is extremely important in assessing its impact. However, there are few exploratory works in this area. Most of the research has been focused on longitudinal studies involving time series decomposition. In general, seasonal variables are isolated in the first step of the analysis and subsequently are compared using a variety of measures for the acuteness of the seasonal distinctions (Koenig-Lewis & Bischoff, 2005).

As an example of indicators aimed at measuring the level of seasonality, it can be highlight those based on the ratio of foreign visitor arrivals in the summer/winter months to total entries during the year; the seasonal amplitude index, which explores the relationship between summer and winter traffic variations; the tourism saturation index, based on the relationship between the number of tourists visiting a destination and its population (Cunha, 2006, 2007). Moreover, the coefficient of seasonal variation, seasonality ratio, seasonality indicator and the Gini index are the tools that allow to analyse tourism seasonality derived from seasonal factors (Koenig-Lewis, & Bischoff, 2004, 2005; Þórhallsdóttir, & Ólafsson, 2017).

Finally, effective policies for tourism in the North of Portugal may allow the generation of spillover effects which allow tourism sustainability avoiding or mitigating the negative impacts of seasonality and overtourism on the territory and its communities, maintaining the interest in both the national and international touristic flows and its long term economic profitability, social positive impact and environmental protection.

Tourism Development in North Region of Portugal

Portugal has a well-balanced tourism strategy and has several competitiveness strengths, achieving the 15th place in the Travel and Tourism Competitiveness Index Ranking 2015.

The Northern region of Portugal has 144 kilometres of Atlantic coast and is the Portuguese region with the largest border area (it borders the Spanish Autonomous Communities of Galicia and Castile-Leon). According to the *Comissão de Coordenação e Desenvolvimento Regional do Norte* (CCDRN, 2018), the North region of Portugal is one of the tourism destinations with the highest potential for internal and external growth. For several consecutive years, it has earned the praises of different foreign entities, such as the title of Best European Destination, which resulted in an increasing number of visitors. Among the various tourist attractions, four UNESCO "World Heritage Site" classifications to recover urban, historical and natural areas with strong potential for internationalisation and tourism are also part of the region's cultural assets and its potential tourism: i) the Oporto's Historic City Centre, in 1996; ii) the Prehistoric Rock Art Site of the Côa Valley; in 1998; iii) the Historic Centre of *Guimarães*, and iv) the *Alto Douro* Wine Region, in 2001, which was considered a "living evolutionary cultural landscape."

The relevance that the national government attributes to tourism and its promotion (national and international) is well expressed in the public policies that frame the current Tourism Strategy 2027.

Territory is a result of the complex interdependencies between market size, the dynamics of interactions and a political-institutional framework more or less conducive to economic and social achievements (Camagni, 2008). It is the modes of governance (shared and territorially integrated) that help territories achieve higher or lower levels of competitiveness and territorial cohesion (C&TC). Competitiveness and territorial cohesion depend on the external dynamics that condition territories, but they also depend largely on the ability to build endogenous, differentiating and community-wide solutions in the form of shared public policies and modes of governance.

Tourism is considered a sector of the economy and as any sector of the economy and its related activities has specificities that objectively condition interventions in its domain, whether public policy, enterprise policy or modes of innovation (Nordin & Hjalager, 2017). Tourism has as its main specificity to be produced and consumed in a specific territory and its added value, namely in terms of its multiplier effects - is determined, quantified and qualified territorially. Tourism can be a territorial singularity, that is, it can be a manifestation of economic ubiquity (production and consumption) in the same territory (Nunes & Sousa, 2019). The contribution of tourism to C&TC, in addition to the result of innovative and competitive tourism companies, is the result of this competitiveness integrated into an economically coherent economic framework.

METHODOLOGY

A wide range of measures has been used to describe the tourism seasonality as a temporal and spatial inequality (Cuccia, & Rizzo, 2011; Connell, Page, & Meyer, 2015; Duro, 2016; Gil-Alana, & Huijbens, 2018; Turrión-Prats, & Duro, 2018), but the definition that seems to be the most acceptable one has been provided by Butler (1994). Butler (1994) defined tourism seasonality as a temporal imbalance that can be expressed in terms of different indicators, as numbers of visitors, expenditure of visitors, traffic on high-ways and other forms of transportation, employment and admissions to touristic attractions. The basic unit used to measure seasonality is the number of visitors (Lundtorp, 2001), overnight stay data or arrivals/departure data using days, weeks or months (Cuccia, & Rizzo, 2011; Martín-Martín, Jiménez-Aguilera, & Molina-Moreno, 2014; Þórhallsdóttir, & Ólafsson, 2017).

The most used tools to analyse tourism seasonality, derived from seasonal factors, are the coefficient of seasonal variation, the seasonality ratio and the Gini index (Koenig-Lewis, & Bischoff, 2004, 2005; Þórhallsdóttir, & Ólafsson, 2017). According with these authors the:

- Coefficient of variation (*CV*) is a measure of spread and describes the amount of variability relative to the mean. It can be used instead of the standard deviation to compare the spread of data sets that have different measurement units or different means and is expressed in Eq. (1):

$$CV = \frac{s}{\bar{x}};\ \textit{where } s \textit{ is the standard deviation and } \bar{x} \textit{ is the mean value} \quad (1)$$

- Seasonality ratio (*SR*) is found by dividing the highest value of a distribution by its average value (Eq. 2). The ratio increases with the degree of seasonal variation. Its lower bound is one and its upper bound is equal to the number of periods compared - twelve when monthly data is used (Lundtorp, 2001). If the same number of visitors came every month, the ratio would be one. A seasonality ratio ^{3} 2, indicates extreme seasonality. The seasonality indicator (*SI*) is the inverse value of the seasonality ratio (Eq. 3). A seasonality indicator £ 0.5, indicates extreme seasonality.

$$SR = \frac{x_{max.}}{\bar{x}} \quad (2)$$

$$SI = \frac{\bar{x}}{x_{max.}} \Leftrightarrow SI = \frac{1}{SR} \quad (3)$$

- Gini index is the most used tool to analyse seasonality based on the well-known Lorenz curve, which is defined as twice the area between the curve and the line of perfect equality (Duro, 2016). It is derived from the Lorenz curve that displays the cumulative frequency of the ranked observations starting with the lowest number. The coefficient is varying between 0 (0%), reflecting complete equality (every month would have the same number of tourist overnight stays), and 1 (100%), indicating complete inequality (all annual tourist overnight stays would be realised in one month). This is 0£Gini index£1. The formula, adapted to the analysis of seasonal tourism concentration, can be expressed as follows (Eq. 4):

$$G = \frac{1}{2\mu}\sum_{i}\sum_{j} p_i p_j \left| y_i - y_j \right| \quad (4)$$

where p_i and p_j are the relative weights of the observations (months or seasons), y_i, is the variable for observations (in this case tourism demand), and μ is the annual mean. The values are first ordered by size - from the lowest to the highest value - and then normalised by the total number of visitors over the year; the sum of the normalised values should be one and the normalised values are then accumulated and the highest of the cumulative values should be one (Þórhallsdóttir & Ólafsson, 2017).

The Gini index is equal to the area between the Lorenz curve (area A) and the 45° line divided by the total area under the line (area B) (see Fig. 1). This means Gini index is equal to the area under the line (Area A) of perfect equality (0.5 by definition) minus the area below the Lorenz curve (Area B), divided by the area below the line of perfect equality. To find the area between the Lorenz curve and the 45° line, area B needs to be found. Mathematically, the Gini index is defined as a ratio of the areas on the Lorenz curve diagram (see Fig. 1). If the area between the line of perfect equality and the Lorenz curve is A, and the area under the Lorenz curve is B, then the Gini index is A/(A+B); as A+B=0.5, the Gini index is given by A/(0.5) (Þórhallsdóttir & Ólafsson, 2017). The higher Gini index, the greater is the inequality of distribution of tourist overnight stays, i.e., the seasonality. A Gini index ³ 0.5 shows extreme seasonality.

To understand tourism development in the North Region of Portugal time series based on two different frequencies it will be used: annual and monthly, from 2014 to 2017. As a numerical measure of tourism demand will be used the number of overnight stays in hotel establishments by the tourists' market of origin (Portugal market and the Top-5 foreign countries with the highest market share: Spain, France, United Kingdom, America and Germany). The current and official data of the National Statistics of Portugal (INE) will be considered for all 8 regions that belong to the North region of Portugal; the period for each could be found the most updated statistics according to the last NUT III revision, in Portugal. The 8 regions under analysis will be (Fig. 2): *Alto Minho* (AMin); *Área Metropolitana do Porto* (AMP); *Alto Tâmega* (ATâm); *Ave* (Ave); *Cávado* (Cáv); *Douro* (Dou); *Tâmega e Sousa* (T&S); and *Terras de Trás-os-Montes* (TTM). To understand the tourism flows the analyses will be focused on the matrix of interregional tourism flows and on the graphical technique.

Figure 1. The Lorenz curve.
Source: Adapted from Þórhallsdóttir and Ólafsson (2017, p.20).

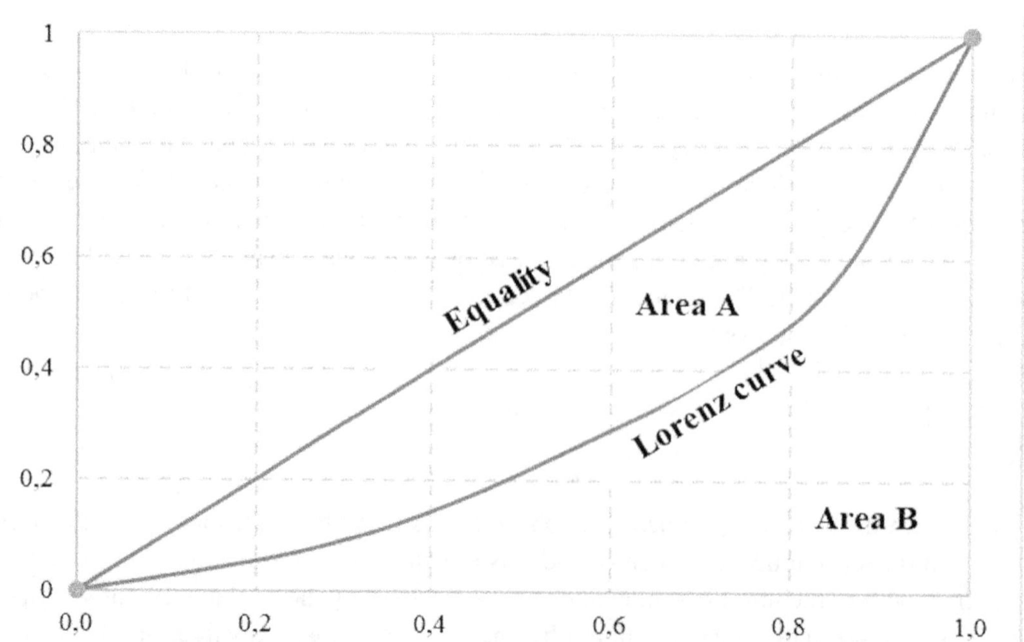

It should be noted that overnight stays data show where the visitors overnight but not which destinations they visit. The off-peak visitors have a tendency to stay in the main local destinations or cities and travel from there to other destinations as day visitors. Therefore, overnight stay data does not give accurate results about which destinations and regions they visit. In this chapter, the goal is to find a more appropriate method.

RESULTS AND ANALYSIS

In the last four years, the tourism in the North region of Portugal expanded significantly. On average, 12% per year from 2014 to 2017. During the period under analysis, there was an average annual variation of around 12% (Fig.3). Hotel establishments recorded an average of 6.5 million overnight stays, and it is obvious the region's growing trend (Fig. 3). Based on the following figure, the existing seasonality is observed. This seasonality presents high peaks in summer months and low peaks in winter months. This phenomenon and profile is typical of Mediterranean destinations. The different regions that form the north of Portugal follow exactly the same behaviour and profile presented for the region as a whole. This means that the tourist arrives at the chosen destination and present a propensity to visit other places in the region.

The internal market contributed with an average of 3.4 million overnight stays, which correspond to an average around 46% of the market share. The five main inbound markets, plus other markets, represented 51.9% of total overnight stays of non-residents and mostly presented positive results. The Top-5 inbound market has a significant expression, with the main markets in the region being Spain, America (namely Brazil) and France. The contribution of the inbound markets has been very positive, with all presenting increasing results except for the French market that decreased in 2017. All this achievements can be observed in Table 1.

Figure 2. North region of Portugal and division into NUT III.
Source: Adapted from Google.com.

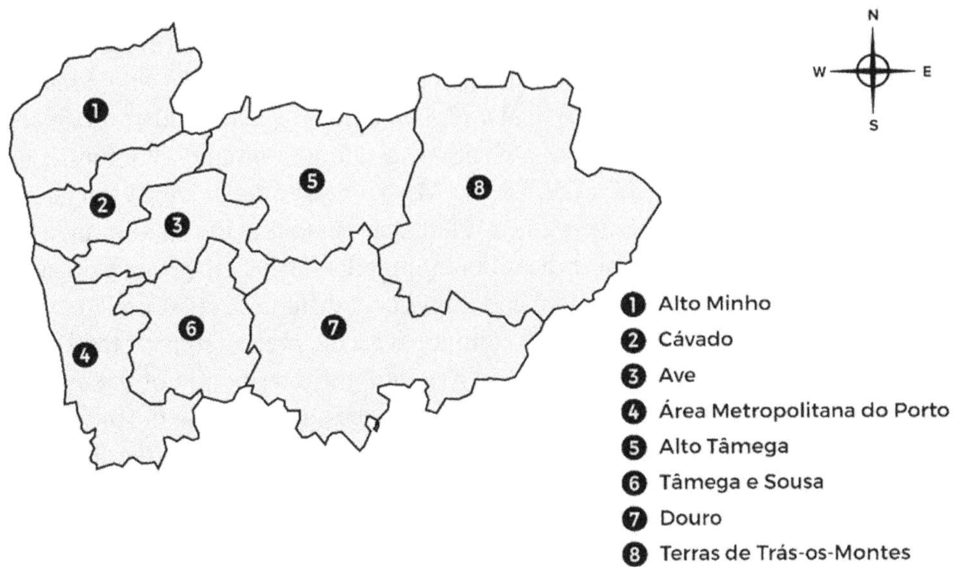

Figure 3. Monthly number of overnight stays at hotels in the North region of Portugal (2014:01-2017:12).

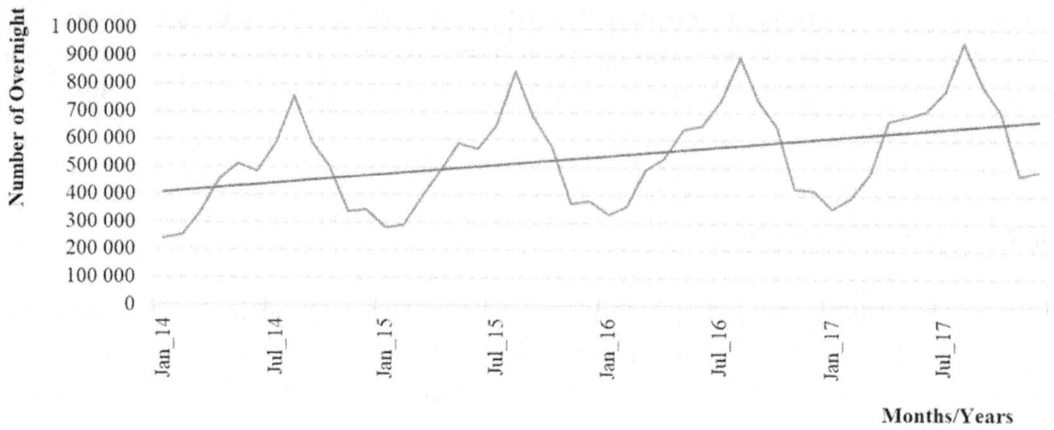

Table 1. Share of the main markets in northern Portugal.

Markets	2014	2015	2016	2017
Spain	11,6%	12,1%	12,3%	12,0%
France	8,0%	8,5%	8,8%	7,8%
United Kingdom	3,0%	3,5%	3,7%	3,7%
America	8,6%	8,0%	8,8%	10,7%
Germany	4,3%	4,5%	4,6%	4,6%
Portugal	48,1%	46,8%	44,7%	42,3%
Others	16,4%	16,6%	17,0%	18,9%
Total	**100,0%**	**100,0%**	**100,0%**	**100,0%**

Table 2 presents some indicators that allow the analysis of overnight stays in the northern region of Portugal. The calculations made were based on the monthly overnight stays for the period under analysis (2014-2017). For the northern region of Portugal, the maximum number of overnight stays was in August and the lowest in January. As already mentioned, this profile is typical of coastal countries, due to the climatic conditions that Portugal, and this particular region, enjoys. The monthly average increases over the years, which allows to state that there was a positive variation of around 12% for the 4 years under review. Looking at the values presented for CV, SR and SI, they can be considered worrying values. They are in the threshold but still below the theoretical reference. Undoubtedly, to minimize this situation, overnight stays should be decentralized over the different months. That is, to provide greater distribution throughout the year, there is a need to promote and implement different action policies.

The analysis for the North region of Portugal continue with the presentation of the Lorenz curve, for the period under analysis. The Lorenz curve (Figure 4) shows the distribution of the overnight stays for each year. The unequal distribution of overnights has yielded the curve. Due to the intense concentration of tourist arrivals in the top-performing months of the year, the gap between the line of equality and the Lorenz curve is high, pointing out inequality distribution and seasonal concentration. The shape of the Lorenz curve, within the observed period, is consistent and is pointing to the stability of seasonal

Table 2. Summary of the descriptive statistics for northern Portugal.

	2014	2015	2016	2017
Overnight stays	**5 400 591**	**6 103 053**	**6 878 367**	**7 480 137**
$Maximum_{monthly}$	754 166	845 300	897 070	952 717
$Minimum_{monthly}$	243 282	279 424	328 250	353 181
$Average_{monthly}$	450 049	508 588	573 197	623 345
$SD_{monthly}$	152 195.0	172 509.8	175 548.6	183 986.4
CV	0.338	0.339	0.306	0.295
SR	1.676	1.662	1.565	1.528
SI	0.597	0.602	0.639	0.654

Note: SD-Standard Deviation; CV- Coefficient of Variation; SR-Seasonality Ratio; SI-Seasonality Indicator.

Figure 4. Lorenz curve for the overnights for North region of Portugal (2014-2017).

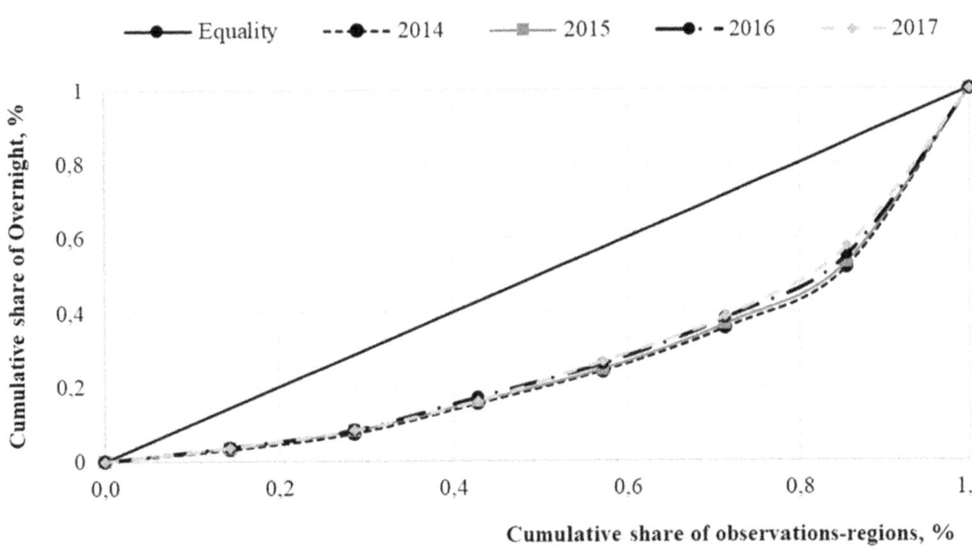

concentration of tourist activities. However, from 2014 to 2017, there has been a decrease in unevenness, that is, in concentration. The curves closest to the equality line are the one presented for 2016 and 2017.

Table 3 presents some indicators for the 8 regions that make up the northern region of Portugal. These indicators provide an overview of interregional behaviour. As for the region, as a whole, the 8 regions follow the same profile. In all regions, except the Douro region, there has been a gradual increase in the number of overnight stays in hotels. The region with the largest market share is the *Area Metropolitana do Porto*, with a very significant percentage, which averages around 76%. The region is followed by the *Cávado* and *Alto Minho* regions, with 7% and 6%, respectively. The region with a very low weight is *Terras de Trás-os-Montes* region with an average market share of 0.7%. This information shows that the regions that are most preferred due to big urban and coastal centres have relatively low market shares. Regarding the main markets, it can be said that in all regions the Top-3 are the same as those already shown for Northern Portugal, that is, in descending order, Portugal, Spain and France. The American

Table 3. Summary of the descriptive statistics for the 8 regions of the northern Portugal.

NUT III	2014	2015	2016	2017	NUT III	2014	2015	2016	2017
Alto Tâmega					*Área Metropolitana do Porto*				
Overnight stays	**42 592**	**58 104**	**66 628**	**79 489**	*Overnight stays*	**4 162 882**	**4 645 322**	**5 268 933**	**5 628 977**
Market share	0.8%	1.0%	1.0%	1.1%	*Market share*	77.1%	76.1%	76.6%	75.3%
$Maximum_{monthly}$	5 891	7 785	8 769	14 794	$Maximum_{monthly}$	580 779	644 809	686 415	726 028
$Minimum_{monthly}$	1 800	2 655	2 991	2 944	$Minimum_{monthly}$	188 154	212 089	251 873	269 059
$Average_{monthly}$	3 549	4 842	5 552	6 624	$Average_{monthly}$	346 907	387 110	439 078	469 081
$SD_{monthly}$	1 251.7	1 604.5	1 773.7	3 137.2	$SD_{monthly}$	117 108.9	131 777.9	134 105.8	146 035.3
CV	0.353	0.331	0.319	0.474	*CV*	0.338	0.340	0.305	0.311
SR	1.660	1.608	1.579	**2.233**	*SR*	1.674	1.666	1.563	1.548
SI	0.603	0.622	0.633	**0.448**	*SI*	0.597	0.600	0.640	0.646
Cávado					**Ave**				
Overnight stays	**349 399**	**406 077**	**471 775**	**552 345**	*Overnight stays*	**209 811**	**207 173**	**235 006**	**274 852**
Market share	6.5%	6.7%	6.9%	7.4%	*Market share*	3.9%	3.4%	3.4%	3.7%
$Maximum_{monthly}$	48 286	55 464	60 868	67 350	$Maximum_{monthly}$	29 595	28 801	30 749	34 198
$Minimum_{monthly}$	15 789	18 849	22 718	25 767	$Minimum_{monthly}$	9 292	9 333	11 077	12 168
$Average_{monthly}$	29 117	33 840	39 315	46 029	$Average_{monthly}$	17 484	17 264	19 584	22 904
$SD_{monthly}$	9 736.8	11 285.1	11 915.5	12 787.1	$SD_{monthly}$	6 021.9	5 954.7	6 116.9	6 597.2
CV	0.334	0.333	0.303	0.278	*CV*	0.344	0.345	0.312	0.288
SR	1.658	1.639	1.548	1.463	*SR*	1.693	1.668	1.570	1.493
SI	0.603	0.610	0.646	0.683	*SI*	0.591	0.599	0.637	0.670
Alto Minho					**Douro**				
Overnight stays	**306 971**	**376 212**	**389 410**	**478 689**	*Overnight stays*	**190 980**	**225 356**	**258 214**	**255 000**
Market share	5.7%	6.2%	5.7%	6.4%	*Market share*	3.5%	3.7%	3.8%	3.4%
$Maximum_{monthly}$	42 079	50 765	49 391	58 062	$Maximum_{monthly}$	27 795	32 002	35 421	33 025
$Minimum_{monthly}$	14 157	17 951	19 260	23 315	$Minimum_{monthly}$	8 177	10 148	11 514	10 950
$Average_{monthly}$	25 581	31 351	32 451	39 891	$Average_{monthly}$	15 915	18 780	21 518	21 250
$SD_{monthly}$	8 324.4	10 074.6	9 357.9	10 628.7	$SD_{monthly}$	5 693.7	6 536.0	7 079.8	6 317.8
CV	0.325	0.321	0.288	0.266	*CV*	0.358	0.348	0.329	0.297
SR	1.645	1.619	1.522	1.456	*SR*	1.746	1.704	1.646	1.554
SI	0.608	0.618	0.657	0.687	*SI*	0.573	0.587	0.607	0.643

continues on following page

Table 3. Continued

NUT III	2014	2015	2016	2017		NUT III	2014	2015	2016	2017
Alto Tâmega						*Área Metropolitana do Porto*				
Terras de Trás-os-Montes						**Tâmega e Sousa**				
Overnight stays	**37 072**	**52 877**	**45 188**	**54 143**		*Overnight stays*	**100 884**	**131 932**	**146 647**	**159 625**
Market share	0.7%	0.9%	0.7%	0.7%		*Market share*	1.9%	2.2%	2.1%	2.1%
$Maximum_{monthly}$	4 993	7 192	5 746	13 172		$Maximum_{monthly}$	14 748	18 482	19 711	19 661
$Minimum_{monthly}$	1 683	2 408	2 090	2 018		$Minimum_{monthly}$	4 230	5 991	6 727	6 960
$Average_{monthly}$	3 089	4 406	3 766	4 512		$Average_{monthly}$	8 407	10 994	12 221	13 302
$SD_{monthly}$	1 033.2	1 494.2	1 178.2	2 931.7		$SD_{monthly}$	3 074.4	3 836.9	3 952.3	4 028.4
CV	0.334	0.339	0.313	0.650		*CV*	0.366	0.349	0.323	0.303
SR	1.616	1.632	1.526	**2.919**		*SR*	1.754	1.681	1.613	1.478
SI	0.619	0.613	0.655	**0.343**		*SI*	0.570	0.595	0.620	0.677

market, namely Brazil, has been growing in all regions in a sharp way. In addition, seasonality is manifested by high peaks in the summer months, namely August, and low values in the winter months, such as December, January and February, for all regions and every year.

Continuing to observe the values presented in Table 3, and comparing the month with the highest tourist overnight stays, August, and the lowest tourist overnights, January, enormous uneven proportions are identified. This evidence shows the levels of concentration and seasonality that continue to exist. According to the calculated seasonality ratio and seasonality indicator each region is experiencing seasonality problems, almost on the threshold of extreme seasonality, as it can be verified by the values of the seasonality indicators. In 2017, *Alto Tâmega* e *Terras de Trás-os-Montes* regions already presented perturbing values, since the value presented for the seasonality ratio is higher than 2 and for the seasonality indicator is smaller than 0.5. Both values indicate extreme seasonality.

The last applied measurement method is the Gini index. As presented in Table 4, Gini index, in the period under analysis, is very high, express the seasonal concentration of tourist flows and correspondingly unequal distribution. With the exception of *Área Metropolitana do Porto*, where there is a more adequate and balanced distribution over the months, all other regions of northern Portugal have high values (greater than 0.5 indicating extreme seasonality). Regional tourism destinations that need more attention are *Terras de Trás-os-Montes* ($GI_{Average}$=0.74) and *Alto Tâmega* ($GI_{Average}$=0.72). However, from 2016 to 2017, the indicator declined in almost all regions. This satisfactory situation may be the consequence of the policies implemented in the different municipalities where it is intend to implement a more equal distribution regarding the northern region of Portugal.

CONCLUSION

With tourism gaining global importance and presenting increasingly greater volatility, there is the need to study its evolution over time. One of the main problems intrinsically linked to this economic activity is the seasonality, which can have economic, social and environmental repercussions and is considered

Table 4. Gini Index Evolution, 2014-2017.

Regions (NUT III)	Years				Evolution of Gini Index (2014-2017) 2014 2015 2016 2017	Average	SD	CV
	2014	2015	2016	2017				
North	0.465	0.452	0.432	0.425		0.444	0.018	0.042
AMin	0.578	0.571	0.584	0.567		0.575	0.008	0.013
AMP	0.400	0.378	0.356	0.350		0.371	0.023	0.062
Cáv	0.579	0.577	0.563	0.548		0.567	0.014	0.025
Ave	0.556	0.563	0.538	0.563		0.555	0.012	0.021
ATâm	0.739	0.731	0.709	0.704		0.721	0.017	0.023
T&S	0.624	0.606	0.600	0.591		0.605	0.014	0.023
Dou	0.599	0.586	0.560	0.553		0.575	0.022	0.038
TTM	0.753	0.719	0.739	0.736		0.737	0.014	0.019

Note: *Alto Minho* (AMin); *Área Metropolitana do Porto* (AMP); *Alto Tâmega* (ATâm); *Ave* (Ave); *Cávado* (Cáv); *Douro* (Dou); *Tâmega e Sousa* (T&S); and *Terras de Trás-os-Montes* (TTM).

one of the main direct causes of overtourim. Seasonality is one of the main phenomena affecting tourism. It depends on the characteristics of both tourism demand and tourism destinations in terms of location and services supplied. Analysing seasonal patterns helps to understand and predict their impacts on tourism production and consumption (Turrión-Prats, & Duro, 2018), facilitating the formulation and implementation of strategic and destination development policies.

The main purpose of this chapter was to describe a method to find and analyse tourism distribution in time and space, in the North region of Portugal, to identify and understand existing seasonality. For that, the current and official data of the National Statistics of Portugal (INE) was considered for all 8 regions that belong to the North region of Portugal, over the period 2014-2017. Four indicators were calculated, seasonality ratio, seasonality indicator, coefficient of variation seasonal and the Gini index complemented by the Lorenz curve. For the North region of Portugal, it was important to understand how seasonality varies in the 8 different regions that compose it.

The results were consistent, allowing to conclude about the high concentration. The region with the largest market share is the *Area Metropolitana do Porto*, with a very significant percentage (76%), followed by the *Cávado* and *Alto Minho* regions, with 7% and 6%, respectively. *Terras de Trás-os-Montes* is the region with the lower average market share (just 0.7%). This results shows the regions that are ultraperipheric regarding big urban and coastal centres have relatively low market shares. Regarding the main markets, it can be said that in all regions the Top-3 markets are Portugal, Spain and France.

The Gini Index applied in this chapter allowed to calculate the level of distribution of several tourism indicators among North Portuguese regions to understand that the phenomena of seasonality - that here it is believed has a strong tie with the overtourism - is indeed a reality and which regions are being suffering from it over the period in analysis. The knowledge and understanding of the reality and its (no)evolution over time allows to identify and recommend the most adequate public (both national and regional) policies to combat the problems caused by seasonality and consequently overtourism and the ones that are the most appropriate to generate sustainability on the tourism sector in all the regions that compose the North of Portugal. If the public policies are fundamental to address both issues mentioned above it is also important that the private agents and the communities, in general, understand the problem and how to overcome it in order it could be possible the tourism sector generates employment and income for all the regions in an equality way. The seaside and the proximity to the most important means of transportation – as the airport, the cruise port or the second most import railway station in the country – influences the existence of overtourism in some regions of the North of Portugal, namely the ones nearest the city of Oporto. However, all the northern region presents a multiplicity of tourism supply that should be more well-known, explored and publicized.

The diversification of tourism supply, namely with a focus on the outdoor tourism for which the less developed North regions are well prepared with the beautiful and almost unexplored landscapes, the cultural and religious heritage and the tasty gastronomy, could be a suitable and sustainable solution for a better distribution of tourism indicators over the entire northern region. The offer of an outdoor tourism plan according with the most suitable rules and supplied adequately by public and private agents is a solution that needs extensively knowledge, proper analysis and public investment on more depressed and less touristic demanded regions. More measures of spatial planning, based on well-defined policies and indicators, like the one here proposed by the application of the Gini Index, will allow a suitable management of all forms of tourism supply and consequently may allow the generation of spillover effects like the tourism sustainability - by avoiding or mitigating the negative impacts of seasonality and overtourism on the territory and populations - the raise of economic productivity, the enhancement of positive social impacts (namely the maintenance of populations in less populated regions) and the environmental protection.

In considering the persistent nature of seasonality in the tourism market, its measurement and analysis are still challenging issues in nowadays tourism research. It is believed that calculating and presenting a robust innovative measurement of tourism seasonality is important to develop/evaluate seasonal public policies and private business alternative strategies. Analysing the distribution of visitors at destinations is valuable for a deeper understanding of seasonality in tourism since gives planners and managers of tourism destinations data that is appreciated when planning services and infrastructures.

ACKNOWLEDGMENT

This research is a part of a project title "TURNOUT: Desenvolvimento do Turismo Outdoor da Região Norte de Portugal", with the reference POCI-01-0145-FEDER-032289; funded by the European Regional Development Fund (FEDER) (through the Operational Programme 'Innovation and competitiveness') and by the Portuguese Foundation for the Development of Science and Technology (FCT), of the Ministry of Science, Technology and Higher Education. This work is, also, funded by National Funds through the Foundation for Science and Technology under the project UID/GES/04752/2019.

REFERENCES

Þórhallsdóttir, G., & Ólafsson, R. (2017). A method to analyse seasonality in the distribution of tourists in Iceland. *Journal of Outdoor Recreation and Tourism, 19*, 17–24. doi:10.1016/j.jort.2017.05.001

Budeanu, A. (2007). Sustainable tourist behaviour–a discussion of opportunities for change. *International Journal of Consumer Studies, 31*(5), 499–508. doi:10.1111/j.1470-6431.2007.00606.x

Butler, R. W. (1994). Seasonality in Tourism: Issues and implications. In A. Seaton (Ed.), Tourism: The state of the art (pp. 332-339). Chichester: Wiley.

Camagni, R. (2008). Regional Competitiveness: Towards a Concept of territorial capital. In R. Capello, R. Camagni, B. Chizzolini, & U. Fratesi (eds.), Modelling Regional Scenarios for the Enlarged Europe: European Competitiveness and Global Strategies, 33-48, Berlin, Germany: Springer Verlag. doi:10.1007/978-3-540-74737-6_3

Cannas, R. (2012). An overview of tourism seasonality: Key concepts and policies. *Almatourism-Journal of Tourism, Culture, and Territorial Development, 3*(5), 40–58.

Capocchi, A., Vallone, C., Amaduzzi, A., & Pierotti, M. (2019). Is 'overtourism'a new issue in tourism development or just a new term for an already known phenomenon? *Current Issues in Tourism, 1-5*. doi:10.1080/13683500.2019.1638353

CCDRN. (2018). Comissão de Coordenação e Desenvolvimento Regional do Norte. Retrieved from http://www.ccdrn.pt/sites/default/files/ficheiros_ccdrn/af_ccdrn_brochura_regiao_ing.pdf

Connell, J., Page, S. J., & Meyer, D. (2015). Visitor attractions and events: Responding to seasonality. *Tourism Management, 46*, 283–298. doi:10.1016/j.tourman.2014.06.013

Coshall, J., Charlesworth, R., & Page, S. J. (2015). Seasonality of overseas tourism demand in Scotland: A regional analysis. *Regional Studies, 49*(10), 1603–1620. doi:10.1080/00343404.2013.859666

Cuccia, T., & Rizzo, I. (2011). Tourism seasonality in cultural destinations: Empirical evidence from Sicily. *Tourism Management, 32*(3), 589–595. doi:10.1016/j.tourman.2010.05.008

Cunha, L. (2006). *Economia e política do turismo*. Lisboa, Portugal: Editorial Verbo.

Cunha, L. (2007). *Introdução ao turismo. 3ª edição*. Lisboa, Portugal: Editorial Verbo.

Dodds, R., & Butler, R. (2019). Introduction. In R. Dodds, & R. Butler (Eds.), *Overtourism: Issues, Realities, and Solutions* (Vol. 1). Walter de Gruyter GmbH & Co KG. doi:10.1515/9783110607369-001

Duro, J. A. (2016). Seasonality of hotel demand in the main Spanish provinces: Measurements and decomposition exercises. *Tourism Management, 52*, 52–63. doi:10.1016/j.tourman.2015.06.013

Gil-Alana, L. A., & Huijbens, E. H. (2018). Tourism in Iceland: Persistence and seasonality. *Annals of Tourism Research, 68*, 20–29. doi:10.1016/j.annals.2017.11.002

Goodwin, H. (2017). *The challenge of overtourism*. Responsible Tourism Partnership.

Kastenholz, E., & Lopes de Almeida, A. (2008). Seasonality in rural tourism–the case of North Portugal. *Tourism Review*, *63*(2), 5–15. doi:10.1108/16605370810883905

Koenig-Lewis, N., & Bischoff, E. E. (2004). Analyzing seasonality in Welsh room occupancy data. *Annals of Tourism Research*, *31*(2), 374–392. doi:10.1016/j.annals.2003.12.006

Koenig-Lewis, N., & Bischoff, E. E. (2005). Seasonality research: The state of the art. *International Journal of Tourism Research*, *7*(4/5), 201–219. doi:10.1002/jtr.531

Lundtorp, S. (2001). Measuring tourism seasonality. In T. Baum, & S. Lundtorp (Eds.), *Seasonality in tourism*. Oxford, UK: Pergamon. doi:10.1016/B978-0-08-043674-6.50006-4

Martín-Martín, J. M., Jiménez-Aguilera, J. D. D., & Molina-Moreno, V. (2014). Impacts of seasonality on environmental sustainability in the tourism sector based on destination type: An application to Spain's Andalusia region. *Tourism Economics*, *20*(1), 123–142. doi:10.5367/te.2013.0256

Nordin, S., & Hjalager, A. M. (2017). Doing, Using, Interacting – Towards a New Understanding of Tourism Innovation Processes. In A. Kiralova (Ed.), *Driving tourism through creative destinations and activities*. IGI Global. doi:10.4018/978-1-5225-2016-0.ch008

Nunes, S., & Sousa, V. (2019). Perfect Resources, Tourism and Territorial Singularities: Contributions to the Development of a Scientific Tourism Line in Golegã. *Revista Portuguesa de Estudos Regionais*, *50*, 27–47.

Phi, G. T. (2019). Framing overtourism: A critical news media analysis. *Current Issues in Tourism*, *1-5*. doi:10.1080/13683500.2019.1618249

Rosselló, J., & Sansó, A. (2017). Yearly, monthly and weekly seasonality of tourism demand: A decomposition analysis. *Tourism Management*, *60*, 379–389. doi:10.1016/j.tourman.2016.12.019

Smeral, E. (2019). Overcrowding of tourism destinations: Some suggestions for a solution. In H. Pechlaner, E. Innerhofer, & G. Erschbamer (Eds.), *Overtourism: Tourism Management and Solutions* (pp. 163–173). Routledge. doi:10.4324/9780429197987-12

Turrión-Prats, J., & Duro, J. A. (2018). Tourist seasonality and the role of markets. *Journal of Destination Marketing & Management*, *8*, 23–31. doi:10.1016/j.jdmm.2016.11.004

Wall, G., & Yan, M. (2003). Disaggregating visitor flows - the example of China. *Tourism Analysis*, *7*(3/4), 191–205.

ENDNOTE

1 Nomenclature of Territorial Units for Statistics (NUT).

Compilation of References

(2019). LET [Laboratórios Estratégicos para o Turismo] (2016). *Estratégia Turismo 2027* – Portugal (Portugal–Laboratório Estratégico Lisboa. Lestegás, I., Seixas, J. & González, R. (2019). *Commodifying Lisbon: A Study on the Spatial Concentration of Short-Term Rentals. Soc. Sci., 8*, 33. Retrieved from https://www.researchgate.net/publication/330640248_Commodifying_Lisbon_A_Study_on_the_Spatial_Concentration_of_Short-Term_Rentals

A21mx. (2017, November 30). *Airbnb grows 165% in Mexico and goes for national tourists*. A21mx online newspaper. Retrieved from a21.com.mx/innovación/2017/11/30/airbnb-crece-165-en-mexico-y-va-por-turistas-nacionales

Aalbers, M. (2016). *The Financialization of Housing: a Political Economy Approach*. London, UK: Routledge. doi:10.4324/9781315668666

Aalbers, M. (Ed.). (2012). *Subprime Cities: The Political Economy of Mortgage Markets*. Oxford, UK: Wiley Blackwell. doi:10.1002/9781444347456

Aall, C., & Koens, K. (2019). The Discourse on Sustainable Urban Tourism: The need for Discussing More Than Overtourism. *Sustainability, 11*(15), 4228. doi:10.3390u11154228Andereck, K. L., & Becker, R. H. (1993). Perceptions of carry-over crowding in recreation environments. *Leisure Sciences, 15*(1), 25–35. doi:10.1080/01490409309513184

Abril-Sellarés, M., Azpelicueta, M. D. C., & Sánchez-Fernández, M. D. (2015). Turismo sostenible: Lugareños frente a turistas. El caso de la ciudad de Barcelona. *HOLOS, 3*, 331–337. doi:10.15628/holos.2015.2809

Advisors, H. (2018, February 7). *Operatsionnye pokazateliotelej Sankt Peterburga za 2017 god [Performance indicators of St. Petersburg hotels for the year of 2017]*. Retrieved from https://hoteladvisors.ru/blog-2/Operatsionnye-pokazateli-oteley-Sankt-Peterburga-za-2017-god/

Aena. (2018). *Aeropuerto de Palma de Mallorca*. Retrieved from http://www.aena.es/es/aeropuerto-palma-mallorca/presentacion.html

Agostinelli, A. (2018). Culture & Tourism in Rome: What is the Future of the Eternal City? A New Storytelling to Save the Old History [PDF document]. Retrieved from https://www.researchgate.net/publication/323855568_Culture_Tourism_in_Rome_What_is_the_Future_of_the_Eternal_City_A_New_Storytelling_to_Save_the_Old_History

Aguiló, E., & Sastre, A. (1984). La medición de la estacionalidad del turismo: El caso de Baleares. *Estudios Turísticos, 81*, 79–88.

Agyeiwaah, E., McKercher, B., & Suntikul, W. (2017). Identifying core indicators of sustainable tourism: A path forward? *Tourism Management Perspectives, 24*, 26–33. doi:10.1016/j.tmp.2017.07.005

AHP. (2016). *O Alojamento Local em Portugal: qual o fenómeno?* Lisboa, Portugal: Nova School of Business and Economics e Faculdade de Direito da Universidade Nova de Lisboa.

Aicher, T. J., Karadakis, K., & Eddosary, M. M. (2015). Comparison of sport tourists' and locals' motivation to participate in a running event. *International Journal of Event and Festival Management*, *6*(3), 215–234. doi:10.1108/IJEFM-03-2015-0011

Airbnb.com. (2018). *Earn money as a host on Airbnb*. Colorado. Retrieved from https://www.airbnb.mx/host/homes?from_nav=1. https://web.archive.org/web/20120729194252/http://www.airbnb.es/info/ten_million

Airdna.co. (2018a). *Airbnb data & insights to succeed in the sharing economy*. Retrieved from https://airdna.co

Airdna.co. (2018b). *Vacation rental data Puerto Vallarta*. Retrieved from https://www.airdna.co/vacation-rental-data/app/mx/jalisco/puerto-vallarta/overview

Airdna.co. (2018c). *Vacation rental data Puerto Vallarta*. Retrieved from https://www.airdna.co/vacation-rental-data/app/mx/jalisco/puerto-vallarta/overview

Airdna.co. (2018d). *Vacation rental data Puerto Vallarta*. Retrieved from https://www.airdna.co/vacation-rental-data/app/mx/jalisco/puerto-vallarta/overview

Ajuntament de València. (2010). *Revisión simplificada del plan general de Valencia. Catálogo de bienes y espacios protegidos*. Valencia: Ajuntament de València.

Aledo, A. M. (2007). La insostenibilidad del turismo residencial. *Antropología y turismo: claves culturales y disciplinares*, 185-208.

Aledo, A. M. (2016). Turismo residencial y vulnerabilidad en el interior del Levante español. (Ed.) Turismo residencial y gentrificación rural. (pp. 37-60) Alicante. Colección PASOS 16 & Foro de Turismo Responsable.

ALEP [Associação de Alojamento Local em Portugal]. (2019). *Alojamento local em Portugal*. Retrieved from http:www.alep.pt

Ali, R. (2016). *Exploring the Coming Perils of Over tourism*. Skift. Retrieved July, 7, 2019 from https://skift.com/2016/08/23/exploring-the-coming-perils-of-overtourism/

All-Russian Center for Researching General Public's Opinion. (2019, March 13). *Hostely v zhilyhdomah: zapretit' iliostavit' [Hostels in residential buildings: ban them or keep them]*. Retrieved from https://wciom.ru/index.php?id=236&uid=9592

Alonso, M., & Miranda, A. (2017). *The collaborative company. The new economic revolution*. Spain: Pearson.

Alquiler Seguro. (2018). *II Estudio comparativo sobre alquiler turístico vs. residencial*. Madrid, Spain.

Altuzarra, A., & Esteban, M. (2011). Land prices and housing prices: The case of Spain. *Journal of Housing and the Built Environment*, *26*(4), 397–409. doi:10.100710901-011-9235-8

Álvarez-Sousa, A. (2018). The Problems of Tourist Sustainability in Cultural Cities: Socio-Political Perceptions and Interests Management. *Sustainability*, *10*(2), 503. doi:10.3390u10020503

Alves, R., Pereira, M., & Rafeiro, M. (2015). O Memorando de Entendimento e as reformas no mercado de arrendamento e na reabilitação urbana. In *M. Lurdes Rodrigues, & P. Adão e Silva (Orgs.), Governar com a Troika: Políticas Públicas em Tempo de Austeridade* (pp. 237–259). Coimbra: Almedina.

Alves, S. (2017). Requaliðcação e reabilitação do centro histórico do Porto. *Scripta Nova. Revista Electrónica de Geografía y Ciencias Sociales*, *21*(557), 1–32.

Alvesson, M., & Sandberg, J. (2013). *Constructing Research Questions: Doing Interesting Research*. London, UK: Sage. doi:10.4135/9781446270035

Andereck, K. L., & Nyaupane, G. P. (2011). Exploring the nature of tourism and quality of life perceptions among residents. *Journal of Travel Research*, *50*(3), 248–260. doi:10.1177/0047287510362918

Andereck, K. L., Valentine, K. M., Knopf, R. C., & Vogt, C. A. (2005). Residents' perceptions of community tourism impacts. *Annals of Tourism Research*, *32*(4), 1056–1076. doi:10.1016/j.annals.2005.03.001

Andereck, K. L., & Vogt, C. A. (2000). The relationship between residents' attitudes toward tourism and tourism development options. *Journal of Travel Research*, *39*(1), 27–36. doi:10.1177/004728750003900104

Andereck, K., & Vogt, C. (2000). The Relationship between Residents' Attitudes toward Tourism and Tourism Development Options. *Journal of Travel Research*, *39*(1), 27–36.

Andriotis, K. (2005). Community groups' perceptions and preferences to tourism development. Evidence from Crete. *Journal of Hospitality & Tourism Research (Washington, D.C.)*, *29*(1), 67–90. doi:10.1177/1096348004268196

Anthena Insight. (2016). *Airbnb Market Report – Lisbon August 2016*. Retrieved from http://insideairbnb.com/lisbon/?neighbourhood=&filterEntireHomes=false&filterHighlyAvailable=false&filterRecentReviews=false&filterMultiListings=false#

Arbulu, I., Lozano J. & Rey-Maquieira J. (2016). The challenges of municipal solid waste management systems provided by public-private partnerships in mature tourist destinations: The case of Mallorca. *Waste Management*, (51), 252-258.

Ardura Urquiaga, A. (2017). La incidencia de la proliferación del alojamiento turístico en viviendas en el centro de Madrid. In *Vivir las ciudades en Europa*. Madrid, Spain: II Encuentro Europeo de Asociaciones de Vecinos.

Ardura Urquiaga, A., & Sorando Ortin, D. (2016). *First we take Manhattan. La destrucción creativa de las ciudades*. Madrid, Spain: Editorial Catarata.

Arenas, S. (2018, July 31). *The residents of Marina Vallarta are worried about building with excess height*. Observatoriobahía.com.mx. Retrieved from http://observatoriobahia.mx/preocupa-a-residentes-marina-vallarta-edificacion-altura-excedente/

Arias Sans, A., & Dominguez, Q. (2016). Unravelling Airbnb: Urban perspectives from Barcelona. *Reinventing the local in tourism: Producing, consuming, and negotiating place*. 73, 209-228.

Arias-Sans, A. (2018). Turismo y Gentrificación. Apuntes desde Barcelona. *Papers. Regió Metropolitana de Barcelona. Territori, estratègies, planejament*, 60, 130–139.

Armstrong, E. K., & Kern, C. L. (2011). Demarketing manages visitor demand in the Blue Mountains National Park. *Journal of Ecotourism*, *10*(1), 21–37. doi:10.1080/14724040903427393

Arrazola, M. (2015). Long-run Supply and Demand Elasticities. *Journal of European Real Estate Research,* 37, 371-404.

Arteaga, J., & Hernández, A. (2017). *Citizen Engagement versus Tourism-phobia: The Role of Citizens in a Country's Brand.* Retrieved from Developing Ideas by Llorente & Cuenca. website: http://www.desarrollando-ideas.com/wp-content/uploads/sites/5/2017/09/170905_DI_Informe_Marca_Pa%C3%ADs_ENG.pdf

Ashworth, G. J., & Tunbridge, J. E. (2000). *The tourist-historic city. Retrospect and the prospect of managing the heritage city*. Oxford, UK: Pergamon.

Ashworth, G., & Page, S. (2011). Urban tourism research: Recent progress and current paradoxes. *Tourism Management*, *32*(1), 1–15. doi:10.1016/j.tourman.2010.02.002

Ashworth, G., & Tunbridge, J. (1990). *The tourist-historic city*. London, UK: Belhaven.

Asia, G. G. R. (2019). *Macau's assembly to debate city tourist capacity*, available at https://www.ggrasia.com/macaus-assembly-to-debate-city-tourist-capacity/ (Accessed on May 30, 2019).

ASOCIACIÓN EMPRESARIAL HOTELERA DE MADRID. (2018). *Análisis estratégico de los riesgos y escenarios de futuro del sector turístico y de hospedaje de la Comunidad de Madrid. Propuestas de actuación 2018.* Available at https://aehm.es/wp-content/uploads/2018/04/Propuestas_AEHM_2018vf.pdf

Association of Collective Investment Institutions and Pension Funds (Inverco). (2018). *Statistics,* Retrieved from http://www.inverco.es/38/0/43

Auken, M. (2009). *Informe sobre el impacto de la urbanización extensiva en España en los derechos individuales de los ciudadanos europeos, el medio ambiente y la aplicación del Derecho comunitario.* Parlamento Europeo, Comisión de Peticiones, 20. Estrasburgo, Francia. Retrieved from http://www.europarl.europa.eu/sides/getDoc.do?pubRef=-//EP//TEXT+REPORT+A6-2009-0082+0+DOC+XML+V0//ES

Austrian Hotelier Association & Roland Berger. (2018). *Protecting your city from overtourism. European city tourism study 2018.* Munich, Germany: Roland Berger GmbH.

Bærenholdt, J. O., Haldrup, M., Larsen, J., & Urry, J. (2004). *Performing tourist places*. London, UK: Ashgate.

Ballester, F. (28 de 08 de 2014). *El turismo urbano de la ciudad de Valencia se enfrenta a un problema de masificación y poco gasto.* Retrieved April 10, 2019 from Valencia plaza: http://epoca1.valenciaplaza.com/ver/137689/el-turismo-urbano-de-la-ciudad-de-valencia-se-enfrenta-a-un-problema-de-masificacion-y-poco-gasto.html

Bank of Spain. (2018). *Financial balance of non-bank financial institutions*. Madrid, Spain.

Bank, U. B. S. (2019). *UBS Global Real Estate Bubble Index 2019.* Retrieved from https://www.ubs.com/global/en/ubs-news/r-news-display-ndp/en-20190930-grebi-global.html

Barata Salgueiro, T. (2017). Alojamentos Turísticos em Lisboa. *Scripta Nova. Revista Electrónica de Geografía y Ciencias Sociales*, *21*(578), 1–43.

Barata Salguiero, T., Mendes, L., & Guimarães, P. (2017). Tourism and Urban changes: Lessons from Lisbon. In *M. Gravari-Barbas, & S. Guinand, Tourism and Gentrification in contemporary monopolises* (pp. 255–275). London, UK: Routledge. doi:10.4324/9781315629759-11

Bardin, L. (2015). *Análise de Conteúdo. Editora: Edições 70.* Coimbra: Extra Coleção.

Barrera, A. P., & Cesario, M. (2017). Factors influencing the choice of the Algarve region as a tourist destination: Does season matter? *International Journal of Tourism Research*, 1–10. doi:10.1002/jtr.2207

Barreto, M. (2013). *Manual de iniciação ao estudo do turismo* (13th ed.). Campinas, SP: Editora Papirus.

Bauman, Z. (1998). *Work, Consumerism, and the New Poor*. Philadelphia: Open University Press.

Bec, A., McLennan, C. L., & Moyle, B. D. (2016). Community resilience to long-term tourism decline and rejuvenation: A literature review and conceptual model. *Current Issues in Tourism*, *19*(5), 431–457. doi:10.1080/13683500.2015.1083538

Becken, S., & Simmons, D. G. (2019). Stakeholders management: different interest and different action. In R. Dodds, & R. W. Butler (Eds.), *Overtourism. Issues, realities, and solutions* (pp. 234–249). Berlin, Germany: De Gruyter; doi:10.1515/9783110607369-016

Belhassen, Y., Uriely, N., & Assor, O. (2014). The Touristification of a Conflict Zone: The Case of Bil'in. *Annals of Tourism Research*, *49*, 174–189. doi:10.1016/j.annals.2014.09.007

Belisle, F., & Hoy, D. (1980). The Perceived Impact of Tourism by Residents. *Annals of Tourism Research*, *7*(1), 83–101. doi:10.1016/S0160-7383(80)80008-9

Beni, M. C. (2019). *Análise estrutural do turismo* (14th ed., p. 558). São Paulo, Brazil: Editora Senac.

Benner, M. (2019). *From overtourism to sustainability: A research agenda for qualitative tourism development in the Adriatic*. [PDF document]. Retrieved from https://mpra.ub.uni-muenchen.de/92213/

Berger, P., & Luckmann, T. (1966). *The Social Construction of Reality. A Treatise in the Sociology of Knowledge*. New York: Anchor Books.

Berges, A., & Ontiveros, E. (2007). La nueva Ley de Suelo desde la perspectiva económica. *Ciudad y Territorios. Estudios Territoriales*, *152-153*, 259–275.

Bernardo, E. (2013). *Uma introdução ao turismo: conceitos, classificações e tipologias*. CIES E-Working Paper no. 164, Instituto Universitário de Lisboa, Lisboa.

Berrueta, J. (13 de 08 de 2017). *Venecia se ahoga de turistas*. Retrieved May 27, 2019, from ctxt revista contexto: https://ctxt.es/es/20170809/Politica/14394/venecia-turismo-gentrificacion-ctxt-alquiler.htm

Biagi, B., Brandamo, M., & Lambiri, D. (2015). Does Tourism Affect House Prices? Evidence from Italy. *Growth and Change*, *46*(3), 501–528. doi:10.1111/grow.12094

Bianchini, F., & Parkinson, M. (1993). *Cultural Policy and Urban Regeneration: The West European Experience*. Manchester, UK: Manchester University Press.

BJH Advisors L.L.C. (2016). *Short Changing New York City The impact of Airbnb on New York City's housing market. Nueva York: Housing Conservation Coordinators Inc*. MFY Legal Services Inc.

Blanco Izquierdo, R., & Arroyo Romero-Salazar, L. (2018). Geografía humana del Casco Histórico de Toledo. Estilos de vida y su transformación. In J. Manzano (Ed.), *Casco Histórico de Toledo. Construir la convivencia* (pp. 7–9). Toledo: Toletum Revolutum.

Blanco-Romero, A., Blázquez-Salom, M., & Cànoves, G. (2018). Barcelona, Housing Rent Bubble in a Tourist City. Social Responses and Local Policies. *Sustainability*, *10*(6), 2043. doi:10.3390u10062043

Blau, P. M. (1964). *Exchange and Power in Social Life*. Hoboken, NJ: J. Wiley.

Bocadosdemadriz, E. (01 de 09 de 2018). *La "turistificación" de los barrios de Madrid: Llegaron las chanclas*. Retrieved May 27, 2019 from Huffpost: https://www.huffingtonpost.es/evabocadosdemadriz/la-turistificacion-de-los-barrios-de-madrid-llegaron-las-chanclas_a_23513335/

Boissevaint, J. (Ed.). (1996). *Coping with tourists: European reactions to mass tourism* (Vol. 1). Providence: Berghahn Books.

Boivin, M., & Tanguay, G. A. (2019). Analysis of the determinants of urban tourism attractiveness: The case of Québec City and Bordeaux. *Journal of Destination Marketing & Management*, *11*, 67–79. doi:10.1016/j.jdmm.2018.11.002

Boluk, K., Cavaliere, C. T., & Higgins-Desbiolles, F. (2017). Critical thinking to realize sustainability in tourism systems: Reflecting on the 2030 sustainable development goals. *Journal of Sustainable Tourism*, *25*(9), 1201–1204. doi:10.1080/09669582.2017.1333263

Boluk, K., Cavaliere, C. T., & Higgins-Desbiolles, F. (2019). A critical framework for interrogating the United Nations Sustainable Development Goals 2030 Agenda in tourism. *Journal of Sustainable Tourism*, *27*(7), 847–864. doi:10.1080/09669582.2019.1619748

Boniface, P. (2013). *Managing quality cultural tourism*. London, UK: Routledge. doi:10.4324/9780203424230

Borg, J. V. D., Costa, P., & Gotti, G. (1996). Tourism in European heritage cities. *Annals of Tourism Research*, *23*(2), 302–321.

Bossel, H. (1999). Indicators for sustainable development: theory, method applications. A report to the Balaton group. Winnipeg, Canada: International Institute for Sustainable Development (IISD).

Botsman, R., & Rogers, R. (2010). *What´s mine is yours. The rise of collaborative consumption*. USA. Harper Collins e-books.

Boukas, N., & Ziakas, V. (2016). Tourism policy and residents' well-being in Cyprus: Opportunities and challenges for developing an inside-out destination management approach. *Journal of Destination Marketing & Management*, *5*(1), 44–54. doi:10.1016/j.jdmm.2015.12.004

Boulding, J. (2018). Overtourism a sign of the times, *Canadian Travel Press*, 1. Retrieved from: https://www.travelpress.com/subscription/IDEA/100017_010118_ctp/F912647B177BF06BB74179E561EA228E/Canadian%20Travel%20Press%20-%20January%204,%202018.pdf

Boyer, M. (2003). *História do turismo de massa*. Bauru, SP: EDUSC.

Boyer, R. (2011). Is a finance-led growth regime a viable alternative to Fordism? A preliminary analysis. *Economy and Society*, *29*(1), 111–145. doi:10.1080/030851400360587

Brakus, J. J., Schmitt, B., & Zarantonello, L. (2009, May). Brand Experience: What Is It? How Is It Measured? Does It Affect Loyalty? *Journal of Marketing*, *73*(3), 52–68. doi:10.1509/jmkg.73.3.052

Brandajs, F., & Russo, A. P. (2019). Whose is that square? Cruise tourists' mobilities and negotiation for public space in Barcelona. *Applied Mobilities*, 1–25. doi:10.1080/23800127.2019.1576257

Brandão, C. DO N.; Barbieri, J C., & Reyes Junior, E. (2015). Análise da sustentabilidade do turismo: um estudo em comunidades indígenas no Estado de Roraima, Brasil. *Revista Brasileira de Pesquisa em Turismo. São Paulo*, *9*(2), 500-518.

Bresson, G., & Lagossah, K. (2011, February). Crowding-out effects of cruise tourism on stay-over tourism in the Caribbean: Non-parametric panel data evidence. *Tourism Economics*, *17*(1), 127–158. doi:10.5367/te.2011.0028

Briggs, D. (2013). *Deviance and Risk on Holiday: An Ethnography of British Tourists in Ibiza*. London, UK: Palgrave MacMillan. doi:10.1057/9781137022400

Brito, M. (2007). Las ciudades históricas como destinos patrimoniales: Potencialidades y requisitos. Fundación Cátedra Iberoamericana (Ed.) Palma de Mallorca.

Brossat, I. (2018). *La ville ubérisée*. France: Éditions La Ville Brule.

Brossat, I. (2019). *Airbnb. La ciudad Uberizada*. Madrid, Spain: Editorial Katakrak.

Brougham, J. E., & Butler, R. W. (1981). A Segmentation Analysis of Resident Attitudes to Social Impact of Tourism. *Annals of Tourism Research*, *8*(4), 569–590. doi:10.1016/0160-7383(81)90042-6

Brunt, P., & Courtney, P. (1999). Host Perceptions of Sociocultural Impacts. *Annals of Tourism Research*, *26*(3), 493–515. doi:10.1016/S0160-7383(99)00003-1

Buckley, R. (2012). Sustainable Tourism: Research & Reality. *Annals of Tourism Research*, *39*(2), 528–546. doi:10.1016/j.annals.2012.02.003

Budeanu, A. (2007). Sustainable tourist behaviour–a discussion of opportunities for change. *International Journal of Consumer Studies, 31*(5), 499–508. doi:10.1111/j.1470-6431.2007.00606.x

Bulencea, P., & Egger, R. (2015). *Gamification in tourism: Designing memorable experiences.* Norderstedt: Bod. Books on demand.

Burgen, S. (2018). 'Tourists go home, refugees welcome': Why Barcelona chose migrants over visitors. Retrieved 16 May, 2019, from https://www.theguardian.com/cities/2018/jun/25/tourists-go-home-refugees-welcome-why-barcelona-chose-migrants-over-visitors

Burgen, S. (2018, June 25). *Tourists go home, refugees welcome': why Barcelona chose migrants over visitors.* Retrieved from The Guardian: https://www.theguardian.com/cities/2018/jun/25/tourists-go-home-refugees-welcome-why-barcelona-chose-migrants-over-visitors

Burkart, A., & Medlik, S. (1981). *Tourism, Past Present and Future.* Oxford, UK: Heinemann.

Burns, P. M., & Novelli, M. (Eds.). (2008). *Tourism and Mobilities. Local-Global Connections.* Wallingford: CABI. doi:10.1079/9781845934040.0000

Burns, P., & Holden, A. (1995). *Tourism: a New Perspective.* Harlow: Prentice Hall.

Butler, R. W. (1994). Seasonality in Tourism: Issues and implications. In A. Seaton (Ed.), Tourism: The state of the art (pp. 332-339). Chichester: Wiley.

Butler, R. (1980). The concept of a tourist area cycle of evolution: Implications for management of resources. *The Canadian Geographer. Geographe Canadien, 24*(1), 5–12. doi:10.1111/j.1541-0064.1980.tb00970.x

Butler, R. (1995). Introduction. In R. Butler, & D. Pearce (Eds.), *Change in Tourism: People, Places, and Processes* (pp. 1–11). New York: Routledge.

Butler, R. (2018). Challenges and opportunities. *Worldwide Hospitality and Tourism Themes, 10*(6), 635–641. PMID:30591573

Butler, R. (2018). Sustainable Tourism in Sensitive Environments: A Wolf in Sheep's Clothing? *Sustainability, 10*(6), 1789. doi:10.3390u10061789

Cabrerizo, C. (2016). *La ciudad negocio: turismo y movilización social en pugna.* Madrid, Spain: Cisma.

Calgaro, E., Lloyd, K., & Dominey-Howes, D. (2014). From vulnerability to transformation: A framework for assessing the vulnerability and resilience of tourism destinations. *Journal of Sustainable Tourism, 22*(3), 341–360. doi:10.1080/09669582.2013.826229

Calleja, D. (2017, September 14). *What are unicorn companies?* Entrepreneur Magazine. Retrieved from https://www.entrepreneur.com/article/268414

Camagni, R. (2008). Regional Competitiveness: Towards a Concept of territorial capital. In R. Capello, R. Camagni, B. Chizzolini, & U. Fratesi (eds.), Modelling Regional Scenarios for the Enlarged Europe: European Competitiveness and Global Strategies, 33-48, Berlin, Germany: Springer Verlag. doi:10.1007/978-3-540-74737-6_3

Campos Echevarría, J. L. (2008). *La burbuja inmobiliaria española.* Madrid, Spain: Marcial Pons.

Cañada, E. (2018). Too Precarious to be Inclusive? Hotel Maid Employment in Spain. *Tourism Geographies, 20*(4), 653–674. doi:10.1080/14616688.2018.1437765

Canalis, X. (16 de 05 de 2017). *Turistificación: la palabra de moda para demonizar al turismo.* Retrieved May 1, 2019 from Hosteltour: https://www.hosteltur.com/122062_turistificacion-palabra-moda-demonizar-al-turismo.html

Canarias7. (2019, Feb. 2). *Guanarteme, contra la desregulación del alquiler vacacional.* Retrieved from https://www.canarias7.es/siete-islas/gran-canaria/las-palmas-de-gran-canaria/guanarteme-contra-la-desregulacion-del-alquiler-vacacional-GY6504932

Cañigueral, A. (2014). *Live better with less. Discover the advantages of the new collaborative economy.* Barcelona, Spain: Editorial Connect.

Cannas, R. (2012). An overview of tourism seasonality: Key concepts and policies. *Almatourism-Journal of Tourism, Culture, and Territorial Development, 3*(5), 40–58.

Capocchi, A., Vallone, C., Amaduzzi, A., & Pierotti, M. (2019). Is 'overtourism'a new issue in tourism development or just a new term for an already known phenomenon? *Current Issues in Tourism, 1-5.* doi:10.1080/13683500.2019.1638353

Capocchi, A., Vallone, C., Pierotti, M., Amaduzzi, A., & Capocchi. (2019). Overtourism: A literature review to assess implications and future perspectives. *Sustainability, 11*(12), 3303. doi:10.3390u11123303

Capo, J., Font, A., & Nadal, J. (2007). Dutch disease in tourism economies: Evidence from the Balearics and the Canary Islands. *Journal of Sustainable Tourism, 15*(6), 615–627. doi:10.2167/jost698.0

Cardoso, C., & Silva, M. (2018). Residents' perceptions and attitudes towards future tourism development A challenge for tourism planners. *Worldwide Hospitality and Tourism Themes, 10*(6), 688–697.

Casanovas, O. (2017). Turismo y convivência ciudadana. In L. *Coma, & J. Santacana, (Coords.), Ciudad educadora y turismo responsable* (pp. 135–145). Ediciones Trea.

Castells, M. (1986). *La ciudad y las masas: sociologia de los movimientos urbanos.* Madrid, Espana: Alianza Editorial.

Cavus, S., & Tanrisevdi, A. (2003). Residents' Attitudes toward Tourism Development: A Case Study in Kusadasi, Turkey. *Tourism Analysis, 7*(3), 259–269. doi:10.3727/108354203108750102

CCDRN. (2018). Comissão de Coordenação e Desenvolvimento Regional do Norte. Retrieved from http://www.ccdrn.pt/sites/default/files/ficheiros_ccdrn/af_ccdrn_brochura_regiao_ing.pdf

CEC (Commission of the European Communities). (2002). *Recommendation of the European Parliament and Council of 30 May 2002 concerning the implementation of integrated coastal zone management in Europe 2002/413/EC.*

Céron, J. P., & Dubois, G. (2003). Tourism and sustainable development indicators: The gap between theoretical demands and practical achievements. *Current Issues in Tourism, 6*(1), 54–75. doi:10.1080/13683500308667944

Chalip, L. (2004). Beyond impact: a general model for host community event leverage. In B. W. Ritchie, & D. Adair (Eds.), *Sport Tourism: Interrelationships, Impacts and Issues* (pp. 236–262). Clevedon: Channel View.

Chase, A. W., & Hobbs, N. (2010). *Ultimate Guide to Trail Running: Everything You Need To Know About* (2nd ed.). U.S.A.: Falcon Guides.

Chaudhry, P., Cesareo, L., & Pastore, A. (2019). Resolving the jeopardies of consumer demand: Revisiting demarketing concepts. *Business Horizons, 65*(5), 663–677. doi:10.1016/j.bushor.2019.05.002

Cheer, J. M., & Lew, A. A. (2018). Understanding tourism resilience: Adapting to social, political, and economic change. In J. M. Cheer, & A. A. Lew (Eds.), Routledge advances in tourism: Vol. 42. Tourism, resilience, and sustainability: Adapting to social, political and economic change (pp. 3–17). London, UK: Routledge.

Cheer, J. M., Milano, C., & Novelli, M. (2018, Jul 18). Overtourism: a growing global problem. *The Conversation.*

Cheer, J. M., & Lew, A. A. (Eds.). (2018). *Tourism, Resilience and Sustainability: Adapting to Social, Political and Economic Change*. London, UK: Routledge.

Cheer, J. M., Milano, C., & Novelli, M. (2019). Tourism and community resilience in the Anthropocene: Accentuating temporal overtourism. *Journal of Sustainable Tourism*, *27*(4), 554–572. doi:10.1080/09669582.2019.1578363

Chen, Ch.-F., & Chen, P.-Ch. (2011). Resident attitudes toward heritage tourism development. *Tourism Geographies*, *12*(4), 525–545. doi:10.1080/14616688.2010.516398

Cheng, C., Li, X. R., Petrick, J. F., & O'Leary, J. T. (2011). An examination of tourism journal development. *Tourism Management*, *32*(1), 53–61. doi:10.1016/j.tourman.2009.11.004

Cheung, K. S., & Li, L. H. (2019). Understanding visitor–resident relations in overtourism: Developing resilience for sustainable tourism. *Journal of Sustainable Tourism*, *27*(8), 1197–1216. doi:10.1080/09669582.2019.1606815

Chiou, C. (2008). The effect of concept mapping on students' learning achievements and interests. *Innovations in Education and Teaching International*, *45*(4), 375–387. doi:10.1080/14703290802377240

Choi, H. C., & Sirakaya, E. (2006). Sustainability indicators for managing community tourism. *Tourism Management*, *27*(6), 1274–1289. doi:10.1016/j.tourman.2005.05.018

Chossudovsky, M. (2003). *The Globalization of Poverty and The New World Order*. Ontario, Canada: Global Outlook.

Christophers, B. (2011). Revisiting the Urbanization of Capital. *Annals of the Association of American Geographers*, *101*(6), 1347–1364. doi:10.1080/00045608.2011.583569

Churchill, G. A. Jr. (1979). A paradigm for developing better measures of marketing constructs. *JMR, Journal of Marketing Research*, *16*(1), 64–73. doi:10.1177/002224377901600110

Ciedes Foundation. (2017). *Málaga in figures*. Retrieved from http://ciedes.es/gestion-y-seguimiento/malaga-en-cifras.html

CIES. (1974). Introducción a un estudio socioeconómico del turismo en la provincia de Las Palmas. Las Palmas de Gran Canaria: CIES (Centro de Investigación Económico y Social).

CML [Câmara Municipal de Lisboa]. (2019). *Regulamento Municipal do Alojamento Local*. Retrieved from http://www.cm-lisboa.pt/viver/urbanismo/alojamento-local

CML. (2009). *Estratégias para a Cultura em Lisboa, CML – Pelouro da Cultura – Site da CML: Estudos de Caracterização*. Retrieved from http://www.cm-lisboa.pt/fileadmin/VIVER/Urbanismo/urbanismo/Documentos/estrategias_para_a_cultura.pdf

CML. (2011). *Estratégia de Reabilitação Urbana de Lisboa – 2011/2024*. Site da CML. Retrieved from http://www.cm-lisboa.pt/fileadmin/VIVER/Urbanismo/urbanismo/Reabilitacao/estrat.pdf

CML. (2013). *Programa de Governo da Cidade de Lisboa 2013/2017*. Retrieved from http://issuu.com/camara_municipal_lisboa/docs/programa_gov_cidade_de_lisboa_2013

CML. (2016). *Historic Lisbon, Global City and Pombalina Lisbon – Heritage of Humanity Proposal*. Retrieved from http://www.cm-lisboa.pt/viver/urbanismo/planeamento-urbano/planodiretor-municipal/pdm-em-vigor

CML. (2018). *Visão Estratégica para o Ordenamento Urbanístico do Turismo em Lisboa – 2018* [Strategic Vision for the Urban Planning of Tourism in Lisbon - 2018 CML - City Planning]. Retrieved from http://www.cm-lisboa.pt/fileadmin/VIVER/Urbanismo/urbanismo/alojamentolocal/4_Vis%C3%A3o_estrat%C3%A9gica_Zonas_Tur%C3%ADsticas_Homog%C3%A9neas_Cidade_de_Lisboa_2018-10-15.pdf

Cochrane, A. (2007). *Understanding Urban Policy*. Oxford, UK: Blackwell.

Cocola Gant, A. (2016). Holiday Rentals: The New Gentrification Battlefront. *Sociological Research Online*, *21*(3), 10. doi:10.5153ro.4071

Cocola-Gant, A. (2015). Tourism and commercial gentrification. *RC21 International Conference on "The Ideal City: between myth and reality. Representations, policies, contradictions, and challenges for tomorrow's urban life"*. Urbino (Italy), August 27-29, 2015.

Cocola-Gant, A. (2018). Tourism gentrification. In L. Lees, & M. Phillips (Eds.), *Handbook of Gentrification Studies* (pp. 281–293). Cheltenham, UK: Edward Elgar Publishing. doi:10.4337/9781785361746.00028

Coffey, H. (2017). *How Magaluf is shedding its party-town image and appealing to families.* Retrieved from: https://www.independent.co.uk/travel/news-and-advice/magaluf-holidays-clubbing-drinking-family-appeal-upmarket-image-change-mallorca-melia-hotels-calvia-a8080371.html

Cohen, E., & Cohen, S. A. (2015). A mobilities approach to tourism from emerging world regions. *Current Issues in Tourism*, *18*(1), 11–43. doi:10.1080/13683500.2014.898617

Cohen, E., & Cohen, S. A. (2015). Beyond Eurocentrism in tourism: A paradigm shift to mobilities. *Tourism Recreation Research*, *40*(2), 157–168. doi:10.1080/02508281.2015.1039331

Cohen, S. A., Duncan, T., & Thulemark, M. (2015). Lifestyle mobilities: The crossroads of travel, leisure and migration. *Mobilities*, *10*(1), 155–172. doi:10.1080/17450101.2013.826481

Colaninno, N., & Roca Cladera, J. N. (2012). Un desarrollo metropolitano policéntrico: dispersión y compactación en los procesos de crecimiento urbano, a escala territorial. In F. Gaja-Díaz (Ed.), Deconstruction. La desconfiguración del litoral mediterráneo español. Colección Manual de referencia (pp. 14-75). Valencia, Spain: Editorial UPV.

Coldwell, D. (2017). *First Venice and Barcelona: now anti-tourism marches spread across the Europe.* Retrieved from: https://www.theguardian.com/travel/2017/aug/10/anti-tourism-marches-spread-across-europe-venice-barcelona

Coleman, R., & Ramchandani, G. (2010). The hidden benefits of non-elite mass participation sports events: An economic perspective. *International Journal of Sports Marketing & Sponsorship*, *12*(1), 19–31. doi:10.1108/IJSMS-12-01-2010-B004

Cole, S. (2012). Synergy and congestion in the tourist destination life cycle. *Tourism Management*, *33*(5), 1128–1140. doi:10.1016/j.tourman.2011.11.018

Collins Online Dictionary. (2019). Overtourism. Retrieved May 16, 2019, from https://www.collinsdictionary.com/submission/19794/Overtourism

Colomb, C., & Novy, J. (2016). *Protest and Resistance in the Tourist City*. London, UK: Routledge. doi:10.4324/9781315719306

Colomb, C., & Novy, J. (Eds.). (2017). *Protest and resistance in the tourist city*. New York, NY: Routledge.

Comisión Nacional del Mercado de Valores (CNMV). "Series Estadísticas de las Instituciones de Inversión Colectivas. 2016. www.cnmv.es

Confidencial Imobiliário. (2016). Portuguese Housing Market Survey. Monthly Report, October. Lisboa.

Conley, Ch. (2014, August 29). *Airbnb is an old idea with a new tech twist*. Hotel News Now. Retrieved from http://www.hotelnewsnow.com/articles/23687/Airbnb-is-an-old-idea-with-a-new-tech-twist

Connell, J., Page, S. J., & Meyer, D. (2015). Visitor attractions and events: Responding to seasonality. *Tourism Management*, *46*, 283–298. doi:10.1016/j.tourman.2014.06.013

Cooper, C., Fletcher, J., Fyall, A., Gilbert, D., & Wanltill, S. (2008). *Turismo: princípios e práticas* (3rd ed.). Bookman, São Paulo: Editora.

Corneo, G., & Jeanne, O. (1999). Segmented communication and fashionable behaviour. *Journal of Economic Behavior & Organization*, *39*(4), 371–385. doi:10.1016/S0167-2681(99)00046-3

Coshall, J., Charlesworth, R., & Page, S. J. (2015). Seasonality of overseas tourism demand in Scotland: A regional analysis. *Regional Studies*, *49*(10), 1603–1620. doi:10.1080/00343404.2013.859666

Costa, C., & Chalip, L. (2005). Adventure sport tourism in rural revitalisation – an ethnographic evaluation. *European Sport Management Quarterly*, *5*(3), 257–279. doi:10.1080/16184740500190595

Costa, J., Montenegro, M., & Gomes, J. (2018). What challenges and opportunities will lead to success? *Worldwide Hospitality and Tourism Themes*, *10*(6), 631–634. doi:10.1108/WHATT-08-2018-0053

Crespi Vallbona, M., & Domínguez Pérez, M. (2016). Los mercados de abastos y las ciudades turísticas. *Pasos. Revista de Turismo y Patrimonio Cultural*, 401-416.

Crommelin, L., Troy, L., Martin, C., & Parkinson, S. (2018). *Technological Disruption in Private Housing Markets: The Case of Airbnb. AHURI Final Report 305*. Melbourne, Australia: Australian Housing and Urban Research Institute Limited.

Cuccia, T., & Rizzo, I. (2011). Tourism seasonality in cultural destinations: Empirical evidence from Sicily. *Tourism Management*, *32*(3), 589–595. doi:10.1016/j.tourman.2010.05.008

Cullwick, D. (1975). Positioning Demarketing Strategy: Marketers must integrate the changes in the business environment into effective new product, pricing, distribution, and promotion strategies. *Journal of Marketing*, *39*(2), 51–57.

Cummings, J. (2015). Confronting Favela Chic: The Gentrification of Informal Settlements in Rio de Janeiro, Brazil. In L. Lees, H. B. Shin, & E. Lopez-Morales (Eds.), *Global Gentrifications: Uneven Development and Displacement* (pp. 81–99). Bristol: Policy Press. doi:10.2307/j.ctt1t894bt.11

Cummins, R. A. (1996). Assessing quality of life. *Quality of life for people with disabilities: model, research and practice*, 116–150.

Cunha, L. (2006). *Economia e política do turismo*. Lisboa, Portugal: Editorial Verbo.

Cunha, L. (2007). *Introdução ao turismo. 3ª edição*. Lisboa, Portugal: Editorial Verbo.

Cunha, L. (2013). *Economia e política do Turismo*. Lisboa, Portugal: Lidel Edições Técnica.

Cunha, L., & Abrantes, A. (2013). *Introdução ao Turismo, 5ª Edição*. Lidel.

Curry, B., & Mounitnho, L. (1992). Diseño de los aspectos medioambientales del turismo. *Papers de Turisme*, *10*, 7–22.

Davies, M., Hungenberg, E., & Aicher, T. (2018). The relationship between runner environmental paradigm and their motives to participate in an urban or rural marathon. *International Journal of Event and Festival Management*, *10*(1), 48–66. doi:10.1108/IJEFM-02-2018-0013

De Frantz, M. (2018). Tourism marketing and urban politics: Cultural planning in a European capital. *Tourism Geographies*, *20*(3), 481–503. doi:10.1080/14616688.2017.1354392

De Juan-Vigaray, M. D., Sarabia-Sánchez, F. J., & Garau-Vadell, J. B. (2013). The acculturation of international residential tourists and their shopping behaviours. *Tourism Management, 36*, 115–118. doi:10.1016/j.tourman.2012.11.017

De la Calle Vaquero, M., García Hernández, M., & Yubero Bernabé, C. (2017). Las dinámicas del sector del alojamiento en la ciudad de Donostia-San Sebastián: implicaciones turísticas y urbanas. In *Actas XXV Congreso de la AGE. 50 años de congresos de Geografía. Naturaleza, territorio y ciudad en un mundo global* (pp. 1553-1562). Madrid, Spain: Asociación de Geógrafos Españoles.

De la Casa Martínez, C. (2000). El turismo cultural: una alternativa al desarrollo económico. En F. d. León, Turismo cultural: el patrimonio histórico como fuente de riqueza (pp. 97-107). Valladolid: Fundación del patrimonio histórico de Castilla y León.

Deery, M., Jago, L., & Fredline, L. (2012). Rethinking social impacts of tourism research: A new research agenda. *Tourism Management, 33*(1), 64–73. doi:10.1016/j.tourman.2011.01.026

Del Romero, L. (2018). Tourisitification, sharing economies and the new geography of urban conflicts. *Urban Sci, 2*(4), 104. doi:10.3390/urbansci2040104

Delgadillo, V. (2015). Patrimonio urbano, turismo y gentrificación. In V. Delgadillo, I. Díaz, & L. Salinas (Orgs.), Perspectivas del Estudio de la Gentrificacíon en México y América Latina (pp. 113-132). Coyoacán: Instituto de Geografía, UNAM.

Delgado, M. (s.f.). La cultura de las calles. El espacio público como patrimonio sociocultural. En T. Soldino Nicolás, M. Delgado, & A. Montiel Márquez (Eds.), Patrimoni i societat (pp. 61-77). Valencia, Spain: Universitat de València.

DeVerteuil, G. (2011). Evidence of Gentrification-induced Displacement among Social Services in London and Los Angeles. *Urban Studies (Edinburgh, Scotland), 48*(8), 1563–1580. doi:10.1177/0042098010379277

Deyà Tortella, B., & Tirado, D. (2011). Hotel water consumption at a seasonal mass tourist destination. The case of the island of Mallorca. *Journal of Environmental Management, 92*(10), 2568–2579. doi:10.1016/j.jenvman.2011.05.024 PMID:21696882

Di Grino, B. N. (1986). Community surveys: Purchasing confidence and accuracy. *Journal of Park and Recreation Administration, 4*(1), 61–69.

Díaz Hernández, R. P. (2017). La vivienda secundaria y vacacional en la diversificación de la oferta alojativa de Canarias. *Revista Atlántida: Revista Canaria de Ciencias Sociales, 8*, 243–264.

Díaz-Armas, R. J., & Gutiérrez-Taño, D. (2017). Turismofobia. In E. A. González, & F. C. García (Eds.), *La actividad turística española en 2016* (pp. 407–410). Madrid, Spain: Editorial Síntesis.

Dichter, A., & Manzo, G. (Orgs.) (2017). *Coping with Success. Managing Overcrowding in Tourism Destinations.* London, UK: McKinsey & Company and World Travel & Tourism Council. [Online], [Accessed July 5, 2018], Available from https://www.wttc.org/-/media/files/reports/policy-research/coping-with-success---managing-overcrowding-in-tourism-destinations-2017.pdf

DICJ (Direcção de Inspecção e Coordenação de Jogos). (2017). *Gaming Statistics*, available at http://www.dicj.gov.mo/web/en/information/DadosEstat/2017/content.html#n1 (Accessed on March 15, 2018).

Dicken, P. (1992). *Global shift: The internationalization of economic activity*. London, UK: Chapman Publishing.

Diedrich, A., & García-Buades, E. (2009). Local Perceptions of Tourism as Indicators of Destination Decline. *Tourism Management, 30*(4), 512–521. doi:10.1016/j.tourman.2008.10.009

Dines, N. (2016). An irreconcilable first-place: The precarious life of tourism and heritage in a southern European historic-centre. *International Journal of Heritage Studies*, *24*(2), 142–153. doi:10.1080/13527258.2016.1195428

Dioko, L. (2017). The problem of rapid tourism growth – an overview of the strategic question. *Worldwide Hospitality and Tourism Themes*, *9*(3), 252–259. doi:10.1108/WHATT-02-2017-0005

Directorate-General of Tributes. (2014). *Recaudación y Estadísticas del Sistema Tributario Español 2006-2016*. Madrid, Spain: Ministry of Finance and Public Administration.

Ditton, R. B., Fedler, A. J., & Graefe, A. R. (1983). Factors contributing to perceptions in recreational crowding. *Leisure Sciences*, *5*(4), 273–288. doi:10.1080/01490408309513009

Dixón. E. (2019, April 5). *Family finds hidden camera streaming live from their Airbnb in Ireland.* CNN News in Spanish. Retrieved from https://cnnespanol.cnn.com/2019/04/05/famiia-encuentra-camara-oculta-transmitiendo-en-vivo-desde-en-su-airbnb-en-irlanda/

Dodds, R., & Butler, R. (2019). *Overtourism: Issues, realities and solutions. The Gruyter Studies in Tourism.* Walter de Gruyter GmbH Editors, Berlin/Boston. Retrieved from https://books.google.pt/books?id=qoKVDwAAQBAJ&pg=PT11&hl=pt-PT&source=gbs_toc_r&cad=3#v=onepage&q&f=false

Dodds, R., & Butler, R. W. (2019a). Conclusion. In R. Dodds, & R. W. Butler (Eds.), *Overtourism. Issues, realities, and solutions* (pp. 262–276). Berlin, Germany: De Gruyter. doi:10.1515/9783110607369-018

Dodds, R., & Butler, R. W. (2019b). Introduction. In R. Dodds, & R. W. Butler (Eds.), *Overtourism. Issues, realities, and solutions* (pp. 1–5). Berlin, Germany: De Gruyter. doi:10.1515/9783110607369-001

Dolnicar, S., Yanamandram, V., & Cliff, K. (2012). The contribution of vacations to quality of life. *Annals of Tourism Research*, *39*(1), 59–83. doi:10.1016/j.annals.2011.04.015

Domblas, N. (2018). *Las carreteras de Baleares soportan casi tres millones de desplazamientos diarios.* Retrieved from: https://ultimahora.es/noticias/local/2018/04/18/994177/carreteras-baleares-soportan-casi-tres-millones-desplazamientos-diarios.html

Domínguez-Mujica, J., González-Pérez, J. M., & Parreño-Castellano, J. M. (2008). Transformaciones recientes en barrios turísticos maduros. Los casos de Palma de Mallorca y Las Palmas de Gran Canaria, España. *Scripta Nova. XII, 270 (93).*

Domínguez-Mujica, J., González-Pérez, J. M., & Parreño-Castellano, J. M. (2011). Tourism and human mobility in Spanish Archipelagos. *Annals of Tourism Research*, *38*(2), 586–606. doi:10.1016/j.annals.2010.11.016

Doxey, G. (1975). A causation theory of visitor-resident irritants, methodology and research inferences. The Impact of Tourism. In *Sixth Annual Conference Proceeding of the Travel Research Association*, (pp. 195-198). San Diego, CA.

Doxey, G. (1975). A causation theory of visitor-resident irritants: Methodology and research inferences. In *Proceedings from Travel and Tourism Research Association's (TTRA) 6th Annual Conference,* 195-198. San Diego, CA: TTRA.

Doxey, G. (1975). A Causation Theory of Visitor–Resident Irritants: Methodology and Research Inferences. In *Proceedings of the Impact of Tourism Sixth Annual Conference* (pp.195-198). San Diego, CA: The Travel Research Association.

Doxey, G. V. (1975). A causation theory of visitor-resident irritants: Methodology and research inferences. In *Proceedings of Sixth annual conference of the Travel Research Association*, 195-198. San Diego, CA: Travel Research Association Conference.

Doxey, G. V. (1975, September). A causation theory of visitor-resident irritants: Methodology and research inferences. The Travel Research Association. In *Sixth Annual Conference Proceedings, 'The Impact of Tourism', San Diego, CA* (pp. 8-11).

Dredge, D. (2017). "Overtourism" Old wine in new bottles? Retrieved from https://www.linkedin.com/pulse/overtourism-old-wine-new-bottles-dianne-dredge

Dredge, D., Gyimothy, S., & Birkbak, A. (2016). The impact of regulatory approaches targeting collaborative economy in the tourism accommodation sector: Barcelona, Berlin, Amsterdam, and Paris. Copenhagen: Paper 9 prepared for the European Commission DG GROWTH. Aalborg University.

Dredge, D., Gyimóthy, S., Birkbak, A., Elgaard Jensen, T., & Madsen, A. (2016). *The impact of regulatory approaches targeting collaborative economy in the tourism accommodation sector: Barcelona, Berlin, Amsterdam and Paris.* Copenhagen, Denmark: European Commission DG GROWTH, Aalborg University.

DSEC. (Direcção dos Serviços de Estatística e Censos). (2016). *Intercensos 2016 resultados globais*, available at http://www.dsec.gov.mo/getAttachment/bfa0112a-eaf3-49a9-9168-b5add 46e9d65/C_ICEN_PUB_2016_Y.aspx (Accessed on May 8, 2018) (In Portuguese and Chinese).

DSEC. (Direcção dos Serviços de Estatística e Censos). (2017). *Environmental statistics 2016*, available at http://www.dsec.gov.mo/getAttachment/5399842e-51c4-462c-81eb-942cf94778 6f/E_AMB_PUB_2016_Y.aspx?disposition=attachment (Accessed on March 15, 2018).

DSEC. (Direcção dos Serviços de Estatística e Censos). (2018a). *Tourism Statistics 2018*, available at https://www.dsec.gov.mo/Statistic.aspx?NodeGuid=7b23463a-d253-4750-bd12-958030df5ccb#Pba46f0c9-8b5b-46b6-b346-8cd1ded3c1f5 (Accessed on May 6, 2019).

DSEC. (Direcção dos Serviços de Estatística e Censos). (2018b). *Demographic statistic (4th Quarter/2017)*, available at http://www.dsec.gov.mo/getAttachment/9ea0efa4-6dfc-4e55-b82 9-743acd429c17/E_DEM_FR_2017_Q4.aspx?disposition=attachment (Accessed on March 15, 2018).

DSEC. (Direcção dos Serviços de Estatística e Censos). (2018c). *Employment survey* (January - March/2018), available at http://www.dsec.gov.mo/getAttachment/3579e9e4-5a5c-4717-b fbc-cccc8b9e1365/C_IE_FR_2018_M01.aspx (Accessed on May 8, 2018).

DSEC. (Direcção dos Serviços de Estatística e Censos). (2019). *Visitor arrivals*, available at http://www.dsec.gov.mo/TimeSeriesDatabase.aspx (Accessed on May 15, 2019).

Du Cros, H. (2012). *Tourism and intangible cultural heritage.* Madrid, Spain: UNWTO.

Duarte, A. D., & Nyanjom, J. (2016). Tourism, quality of life, and residents of a rural town. *Tourism Analysis*, *21*(6), 617–629. doi:10.3727/108354216X14713487283165

Duglio, S., & Beltramo, R. (2017). Estimating the Economic Impacts of a Small-Scale Sport Tourism Event: The Case of the Italo-Swiss Mountain Trail CollonTrek. *Sustainability*, *9*(3), 343. doi:10.3390u9030343

Dumbrovská, V. (2017). Urban Tourism Development in Prague: From Tourist Mecca to Tourist Ghetto. *Tourism in the City: Towards an integrative agenda on urban tourism*, 275-283.

Duro, J. A. (2016). Seasonality of hotel demand in the main Spanish provinces: Measurements and decomposition exercises. *Tourism Management*, *52*, 52–63. doi:10.1016/j.tourman.2015.06.013

EBD [European Best Destinations]. (2019). *European Best Destinations Awards.* Retrieved from https://www.europeanbestdestinations.com/destinations/search/search-results/#e=external&q=lisbon%20awards

EC. (2019). *Reflection paper:* Towards a Sustainable Europe by 2030, Highlights its determination to achieve the 2030 SDGs (Agenda 2030), COM(2019)22.

ECM-European Cities Marketing. (2018). *Managing Tourism Growth In Europe The ECM Toolbox*, 30.

Editors, C. N. T. (2018, October 30). *Parts of the world we love are straining under too much travel. Some aren't handling it well. But among others, there are signs of hope.* Retrieved from Condé Nast Traveler: https://www.cntraveler.com/story/loving-the-world-to-death-the-good-bad-and-ugly-of-overtourism

Edufinet Financial Education (Ed.). (2018). Guía Financiera. Málaga: Aranzadi S.A. European Commission (2019) Communication from the Commission to the European Parliament, the European Council, the Council, the European Central Bank, the European Economic and Social Committee, the Committee of the Regions and the European Investment Bank. European Semester: country-specific recommendations. Brussels.

EGEAC. (2018). *Relatório e Contas – 2018*. Retrieved from http://www.egeac.pt

Egresi, I. (2018). "Tourists Go Home!" – Tourism Overcrowding And "Tourismophobia" In European Cities (Can Tourists and Residents Still Co-Habitate In The City?). En TAW2018 International Scientific Conference (p. 23).

Eldiario.es. (2018). *¿Cuántas viviendas de tu barrio están en Airbnb? Descúbrelo en este mapa, manzana a manzana*? Retrieved from https://www.eldiario.es/economia/concentracion-Airbnb-manzana-viviendas-vacacional_0_809119651.html

Emerson, R. M. (1976). Social Exchange theory. *Annual Review of Sociology*, *2*(2), 335–362. doi:10.1146/annurev.so.02.080176.002003

Erschbamer, G., Innerhofer, E., & Pechlaner, H. (2018). *Overtourism: How Much Tourism is Too Much*. Bolzano, Italy: Eurac Research.

Escudero Gómez, L. A. (2018). Realities and problems of a major cultural tourist destination in Spain, Toledo. *Pasos. Revista de Turismo y Patrimonio Cultural*, *16*(3), 617–636. doi:10.25145/j.pasos.2018.16.045

España, G. (2018). *A toda costa*. Retrieved from https://es.greenpeace.org/es/sala-de-prensa/informes/a-toda-costa/

Espiner, S., Orchiston, C., & Higham, J. (2017). Resilience and sustainability: A complementary relationship? Towards a practical conceptual model for the sustainability–resilience nexus in tourism. *Journal of Sustainable Tourism*, *25*(10), 1385–1400. doi:10.1080/09669582.2017.1281929

Esping-Andersen, G. (1990). *The Three Worlds of Welfare Capitalism*. Cambridge, UK: Polity Press.

Etxezarreta Etxarri, A., Hoekstra, J., Dol, K., & Cano Fuentes, G. (2012). De la burbuja inmobiliaria a las ejecuciones hipotecarias. *Ciudad y Territorio*, *174*, 597–613.

Euromonitor International. (2018). *Top 100 City Destinations 2018*. Retrieved from https://go.euromonitor.com/white-paper-travel-2018-100-cities#download-link

Euromonitor International. (2018). Top 100 City Destinations 2018. Retrieved from https://go.euromonitor.com/white-paper-travel-2018-100-cities.html?utm_source=blog&utm_medium=blog&utm_campaign=CT_WP_18_12_04_100%20Cities&utm_content=organic

Europe Press. (2016). *Palma es la segunda Ciudad española con más atascos de tráfico*. Retrieved from: http://www.elmundo.es/baleares/2016/03/22/56f17536e2704e430a8b45b3.html

European Cities Marketing. (2018). *Managing tourism growth in Europe: the ECM toolbox*. Dijon, France.

EUROPEAN COMMISSION. (2013). *The European tourism indicator system. Toolkit for sustainable destinations*. Luxembourg: European Union.

EUROPEAN COMMISSION. (2016). *The European Tourism Indicator System. ETIS toolkit for sustainable destination management*. Luxembourg: Publications Office of the European Union; doi:10.2873/983087

European Commission. EURES. (2018). *Labour market information – Illes Baleares*. Retrieved from: https://ec.europa.eu/eures/printLMIText.jsp?lmiLang=en&regionId=ES5&catId=457

European Parliament TRAN Committee. (2018). *Overtourism: Impact and Possible Policy Responses*. Brussels, Belgium: European Parliament.

European Parliament's Committee on Transport and Tourism (TRAN Committee). (2018). Overtourism: impact and possible policy responses. Brussels, Belgium. European Parliament, Policy Department for Structural and Cohesion Policies.

Eusébio, C., Vieira, A. L., & Lima, S. (2018). Place attachment, host–tourist interactions, and residents' attitudes towards tourism development: The case of Boa Vista Island in Cape Verde. *Journal of Sustainable Tourism*, *26*(6), 890–909. doi:10.1080/09669582.2018.1425695

Exceltur. (2015). *Alojamiento turístico en viviendas de alquiler: Impactos y retos asociados*. Madrid, Spain. Exceltur.

Fairbnb.coop. (2019). Community Powered Tourism. Retrieved October 5, 2019, from https://fairbnb.coop/

Farias, S. A. (2010). Internal Marketing (IM): A literature review and research propositions for service excellence. *Brazilian Business Review*, *7*(2), 99–115. doi:10.15728/bbr.2010.7.2.6

Faulkner, B., & Tideswell, C. (1997). A Framework for Monitoring Community Impacts of Tourism. *Journal of Sustainable Tourism*, *5*(1), 3–28. doi:10.1080/09669589708667273

Feliziani, V., & Miarelli, M. (2012). How many visitors should there be in the city? The case of Rome. *Review of European Studies*, *4*(2), 179. doi:10.5539/res.v4n2p179

Femenia-Serra, F., Neuhofer, B., & Ivars-Baidal, J. (2019). Towards a conceptualisation of smart tourists and their role within the smart destination scenario. *Service Industries Journal*, *39*(2), 109–133. doi:10.1080/02642069.2018.1508458

Fernandes, J. R. (2011). Area-Based Initiatives and Urban Dynamics. The Case of the Porto City Centre. *Urban Research & Practice*, *4*(3), 285–307. doi:10.1080/17535069.2011.616747

Fessler Vaz, L., & Berenstein Jacques, P. (2006). Contemporary urban spectacularisation. In J. Monclús, & M. Guàrdia (Eds.), Culture, Urbanism, and Planning. Routledge. AZW ebook.

Fields, D., & Uffer, S. (2014). The financialisation of rental housing: A comparative analysis of New York City and Berlin. *Urban Studies (Edinburgh, Scotland)*, *53*(7), 1486–1502. doi:10.1177/0042098014543704

Fink, A. (2005). *Conducting Research Literature Reviews: From the Internet to Paper* (2nd ed.). Thousand Oaks, CA: Sage.

Fistola, R., & La Rocca, R. A. (2017). Driving Functions for Urban Sustainability: The Double-Edged Nature of Urban Tourism. *International Journal of Sustainable Development and Planning*, *12*(3), 425–434. doi:10.2495/SDP-V12-N3-425-434

Font, A. (2005). Problemas urbanos y paradigmas disciplinares en los territorios de la urbanística actual. In A. Font, M. Corominas, & J. Sabaté (Eds.), *Los territorios del urbanista, 10 años del Master en Proyectación Urbanística*. Barcelona, Spain: Ediciones UPC.

Forrest, R., Koh, S. Y., & Wissink, B. (2017). Hyper-divided cities and the 'immoral' super-rich: Five parting questions. In *Cities and the Super-Rich* (pp. 273–287). New York: Palgrave Macmillan. doi:10.1057/978-1-137-54834-4_14

Forster, J. (1964). The Sociological Consequences of Tourism. *International Journal of Comparative Sociology*, *5*(2), 217–227. doi:10.1177/002071526400500208

Fotiadis, A. K. (2016). Modifying and applying time and cost blocks: The case of E-Da theme park, Kaohsiung, Taiwan. *Tourism Management*, *54*, 34–42. doi:10.1016/j.tourman.2015.10.013

Fotocasa. (2019). *Real estate index*, retrieved from https://www.fotocasa.es/indice/#/filter/eyJ0cmFuc2FjdGlvbiI6ImJ1eSJ9

Freeman, R. E. (1984). *Strategic Management: A Stakeholder Approach*. Boston, MA: Pitman.

Freytag, T., & Bauder, M. (2018). Bottom-up touristification and urban transformations in Paris. *Tourism Geographies*, *20*(3), 443–460. doi:10.1080/14616688.2018.1454504

Fuller, H., & Michel, B. (2014). Stop being a tourist! New dynamics of urban tourism in Berlin-Kreuzberg. *International Journal of Urban and Regional Research*, *38*(4), 1304–1318. doi:10.1111/1468-2427.12124

Gaja i Díaz, F. (2008). El "tsunami urbanizador" en el litoral mediterráneo. El ciclo de hiperproducción inmobiliaria 1996-2006. *Scripta Nova, XII, 270 (66).*

Gaja-Díaz, F. (Ed.). (2012). *Deconstruction. La desconfiguración del litoral mediterráneo español. Colección Manual de referencia*. Valencia, Spain: Editorial UPV.

Galvín, A. (2018). *Baleares marca el camino del adiós a las cápsulas de café, las pajitas y los bastoncillos de usar y tirar.* Retrieved from: https://www.eldiario.es/sociedad/Baleares-capsulas-bastoncillos-mecheros-reciclables_0_730027687.html

Gammon, S., & Robinson, T. (2003). Sport and Tourism: A Conceptual Framework. *Journal of Sport & Tourism*, *8*(1), 21–26. doi:10.1080/14775080306236

Gao, W., Zhang, Q., Lu, Z., Wu, D., & Du, X. (2018). „Modelling and application of fuzzy adaptive minimum spanning tree in tourism agglomeration area division. *Knowledge-Based Systems*, *143*, 317–326. doi:10.1016/j.knosys.2017.06.007

Garau-Vadell, J. B., Gutierrez-Taño, D., & Diaz-Armas, R. (2018). Economic crisis and residents' perception of the impacts of tourism in mass tourism destinations. *Journal of Destination Marketing & Management*, *7*, 68–75. doi:10.1016/j.jdmm.2016.08.008

García Hernández, M. (2003). *Turismo y conjuntos monumentales. Capacidad de acogida turística y gestión de flujos de visitantes*. Valencia, Spain: Tirant lo Blanch.

García Montalvo, J. (2000). El precio del suelo. La polémica interminable. *Instituto Valenciano de Investigaciones Económicas y Universidad de Valencia*, 1-47.

García Peña, M. (2017). *El impacto del turismo sobre la disponibilidad y precio de la vivienda*. Málaga: Fundación CIEDES.

García, J. G. (2019, May 21). *"Dinkis", the sector that spends more than a family when it comes to Puerto Vallarta.* A. Z News. Retrieved from https://aznoticias.mx/index.php/puerto-vallarta-movil/32365_di...-que-gasta-mas-que-una-familia-cuando-viene-a-puerto-vallarta

García, F. A., Vázquez, A. B., & Macía, R. C. (2015). Resident's attitudes towards the impacts of tourism. *Tourism Management Perspectives*, *13*, 33–40. doi:10.1016/j.tmp.2014.11.002

García-Hernández, M., de la Calle-Vaquero, M., & Yubero, C. (2017). Cultural heritage and urban tourism: Historic city centres under pressure. *Sustainability*, *9*(8), 1346. doi:10.3390u9081346

Garcia-López, M., Jofre-Monseny, A. J., Martínez Mazza, R., & Segú, M. (2019). Do Short-Term Rental Platforms Affect Housing Markets? Evidence From Airbnb in Barcelona. IEB Working Paper, 2019/05.

Garda, J. C. (2018, September 22). *Demonstration and blockades in Marina Vallarta*. Vallarta Daily Review. Retrieved from http://vallartaopina.net/2018/09/22/local/ciudad/manifestacion-y-bloqueo-en-marina-vallarta/

Garijo, M. (in press 2017). Estas son las empresas que explotan el negocio del alquiler turístico al calor de Airbnb. *Eldiario.es*

Garousi, V. (2015). A bibliometric analysis of the Turkish software engineering research community. *Scientometrics*, *105*(1), 23–49.

Garrigós Simón, F. J., Narangajavana, Y., & Marqués, D. P. (2004). Carrying capacity in the tourism industry: A case study of Hengistbury Head. *Tourism Management*, *25*(2), 275–283. doi:10.1016/S0261-5177(03)00089-X

Gavrili-Alexandris, D., & Vakalakis, P. (2016). Running on Heritage, a Conceptual Discussion on the Roles of Heritage Trails and of Augmented Reality on Amateur Runner Athletes. In V. Katsoni, & A. Stratigea (Eds.), *Tourism and Culture in the Age of Innovation* (pp. 589–602). Springer Proceedings in Business and Economics. doi:10.1007/978-3-319-27528-4_40

Gerritsma, R., & Vork, J. (2017). Amsterdam residents and their attitude towards tourists and tourism. *Coactivity: Philosophy. Communication*, *25*(1), 85–98. doi:10.3846/cpc.2017.274

Getz, D., & Fairley, S. (2004). Media Management At Sport Events For Destination Promotion : Case Studies And Concepts. *Event Management*, *8*(3), 127–139. doi:10.3727/1525995031436926

Getz, D., & McConnell, A. (2014). Comparing trail runners and mountain bikers: Motivation, involvement, portfolios, and event-tourist careers. *Journal of Convention & Event Tourism*, *15*(1), 69–100. doi:10.1080/15470148.2013.834807

Gibson, H. J. (1998). Sport Tourism : A Critical Analysis of Research. *Sport Management Review*, *1*(1), 45–76. doi:10.1016/S1441-3523(98)70099-3

Giddens, A. (1993). *Consecuencias de la modernidad*. Madrid, Spain: Alianza.

Gil, J., & Sequera, J. (2018). Expansión de la ciudad turística y nuevas resistencias. El caso de Airbnb en Madrid. *EMPIRIA. Revista de Metodología de Ciencias Sociales*, 41, 15-32.

Gil-Alana, L. A., & Huijbens, E. H. (2018). Tourism in Iceland: Persistence and seasonality. *Annals of Tourism Research*, *68*, 20–29. doi:10.1016/j.annals.2017.11.002

Gil, J., & Sequera, J. (2018). The expansion of the tourist city and new resistances. The case of Airbnb in Madrid. *Revista de Metodología de Ciencias Sociales*, *41*, 15–32.

Glaesser, D., Kester, J., Paulose, H., & Abbas, A. (2017, April). Global Travel Patterns: An overview. *Journal of Travel Medicine*, *24*(4), 1–5. doi:10.1093/jtm/tax007 PMID:28637267

Glaser, E., Gyourko, J., & Saiz, A. (2008). Housing supply and housing bubbles. *Journal of Urban Economics*, *64*(2), 198–217. doi:10.1016/j.jue.2008.07.007

Glasson, J., Godfrey, K., Goodey, B., & Absalom, H. (1995). *Carrying Capacity and Management Responses in Europe's Historic Towns and Cities*. Aldershot: Avebury.

Glass, R. (1964). *London: aspects of Change*. London, UK: Centre for Urban Studies & MacGibbon and Kee.

GOB-Mallorca. (2018). *Grupo Ornitológico Balear* (GOB-Mallorca). Retrieved June 11, 2019, from https://www.gobmallorca.com/

Goeldner, C. R., & Ritchie, J. R. B. (2011). *Tourism: Principles, Practices, Philosophies* (12th ed.). New York: John Wiley.

Gogonea, R. M., Baltalunga, A. A., Nedelcu, A., & Dumitrescu, D. (2017). Tourism Pressure at the Regional Level in the Context of Sustainable Development in Romania. *Sustainability*, *9*(5), 698.

Goldstein, M. (2017). *Mallorca joins destinations saying #Touristgohome.* Retrieved from: https://www.forbes.com/sites/michaelgoldstein/2017/09/28/mallorca-joins-destinations-saying-touristgohome/#614be1106f6e

González Pérez, J. M. (2019). The dispute over tourist cities. Tourism gentrification in the historic centre of Palma (Majorca, Spain). *Tourism Geographies*, 1–21. doi:10.1080/14616688.2019.1586986

Goodwin, H. (2017). *The challenge of overtourism.* Responsible Tourism Partnership Working Paper 4. October 2017 [Online], [Accessed 5 July 2018], Available at http://haroldgoodwin.info/pubs/RTP'WP4Overtourism01'2017.pdf

Goodwin, H. (2017). *The challenge of overtourism. Responsible Tourism Partnership* Working Paper 4. October.

Goodwin, H. (2017). The challenge of overtourism. *Responsible Tourism Partnership Working Paper 4*. Retrieved from https://haroldgoodwin.info/pubs/RTP'WP4Overtourism01'2017.pdf.

Goodwin, H. (2017, October). The challenge of overtourism. *Responsible Tourism Partnership*. Retrieved from https://haroldgoodwin.info/pubs/RTP'WP4Overtourism01'2017.pdf

Goodwin, H. (2017, October). *The challenge of overtourism.* Retrieved from https://haroldgoodwin.info/pubs/RTP'WP4Overtourism01'2017.pdf

Goodwin, H. (2018, October). *Managing Tourism in Barcelona.* Retrieved from https://haroldgoodwin.info/RTPWP/012ManagingTourisminBarcelona.pdf

Goodwin, H. (2016). *Responsible tourism: Using tourism for sustainable development.* London, UK: Goodfellow Publishers. doi:10.23912/978-1-910158-84-5-3101

Goodwin, H. (2017). *The challenge of overtourism.* Responsible Tourism Partnership.

Goodwin, H. (2019). Barcelona – crowding out the locals: a model for tourism management? In R. Dodds, & R. W. Butler (Eds.), *Overtourism. Issues, realities, and solutions* (pp. 125–138). Berlin, Germany: De Gruyter. doi:10.1515/9783110607369-009

Gotham, K. (2005). Tourism Gentrification: The Case of New Orleans' Vieux Carre (French Quarter). *Urban Studies (Edinburgh, Scotland)*, *42*(7), 1099–1121. doi:10.1080/00420980500120881

Gotham, K. F. (2009). Creating liquidity out of spatial fixity: The secondary circuit of capital and the subprime mortgage crisis. *International Journal of Urban and Regional Research*, *33*(2), 355–371. doi:10.1111/j.1468-2427.2009.00874.x

Gravari-Barbas, M., & Guinand, S. (2017b). Introduction: Addressing tourism-gentrification processes in contemporary metropolises. In M. Gravari-Barbas, & S. Guinand (Eds.), *Tourism and gentrification in contemporary metropolises: International perspectives* (pp. 1–22). New York, NY: Routledge. doi:10.4324/9781315629759-1

Gravari-Barbas, M., & Jacquot, S. (2016). No conñict? Discourses and management of tourism-related tensions in Paris. In C. Colomb, & J. Novy (Eds.), *Protest and Resistance in the Tourist City* (pp. 45–65). London, UK: Routledge.

Gravary-Barbas, M., & Guinand, S. (2017). *Tourism and Gentrification in Contemporary Metropolises: International Perspectives*. London, UK: Routledge. doi:10.4324/9781315629759

Gregson, J. (2017, March 1). *The richest countries in the world*, available at https://www.gfmag.com/global-data/economic-data/richest-countries-in-the-world?page=9 (Accessed on June 1, 2018).

Groundwater, B. (2019). Ten destinations to avoid in 2019. Retrieved May 16, 2019, from https://www.stuff.co.nz/travel/themes/109938533/ten-destinations-to-avoid-in-2019

Gunn, C. A., & Var, T. (2002). *Tourism planning: Basics, concepts, cases*. Psychology Press.

Gurran, N. (2017). When Tourists Move In: How Should Urban Planners Respond to Airbnb? Journal of the American Planning Association, 80-92.

Gürsoy, I. (2019). Beauty and the Beast: A Fairy Tale of Tourismphobia. *Tourism Planning & Development*, *16*(4), 434–451. doi:10.1080/21568316.2019.1582086

Gutiérrez, J., García-Palomares, J. C., Romanillos, G., & Salas-Olmedo, M. H. (2017). The eruption of Airbnb in tourist cities: Comparing spatial patterns of hotels and peer-to-peer accommodation in Barcelona. *Tourism Management*, *62*, 278–291. doi:10.1016/j.tourman.2017.05.003

Gyimóthy, S. (2017). Business Models of the Collaborative Economy. In D. Dredge, & S. Gyimóthy (Eds.), *Collaborative Economy and Tourism. Tourism on the Verge*. Cham, Switzerland: Springer. doi:10.1007/978-3-319-51799-5_3

Habermas, J. (1973). *Legitimationsprobleme im Spätkapitalismus*. Frankfurt, Germany: Suhrkamp.

Hair, J. F., Black, W. C., Babin, B. J., & Anderson, R. E. (2018). *Multivariate data analysis* (8th ed.). London, UK: Cengage Learning EMEA.

Hall, C. M. (2015). On the mobility of tourism mobilities. *Current Issues in Tourism*, *18*(1), 7–10. doi:10.1080/13683500.2014.971719

Hall, P. (1970). Horizon of hotels. *New Society*, *15*(389), 389–445.

Hamnett, C. (1994). Socio-economic change in London: Professionalisation not polarization. *Built Environment*, *31*(3), 192–203.

Hanafiah, M.H.; Azman, I.; Jamaluddin, M.R., & Aminuddin, N. (2016). Responsible Tourism Practices and Quality of Life: Perspective of Langkawi Island communities. *Procedia - Social 128 and Behavioral Sciences*, 222, 406–413.

Hannam, K., Butler, G., & Paris, C. M. (2014). Developments and key issues in tourism mobilities. *Annals of Tourism Research*, *44*, 171–185. doi:10.1016/j.annals.2013.09.010

Harrill, R. (2004). Residents' Attitudes toward Tourism Development: A Literature Review with Implications for Tourism Planning. *Journal of Planning Literature*, *18*(3), 251–266. doi:10.1177/0885412203260306

Harrill, R., & Potts, T. (2003). Tourism Planning in Historic Districts: Attitudes toward Tourism Development in Charleston. *Journal of the American Planning Association*, *69*(3), 233–244. doi:10.1080/01944360308978017

Hartman, S. (2018). Resilient tourism destinations? Governance implications of bringing theories of resilience and adaptive capacity to tourism practice. In Innerhofer, Fontanari, & Pechlaner (Eds.), Destination Resilience – Challenges and Opportunities for Destination Management and Governance. 66-75. Abingdon, UK: Routledge doi:10.4324/9780203701904-5

Harvey, D. (1989). From Managerialism to Entrepreneurialism: The Transformation in Urban Governance in Late Capitalism. *Geografiska Annaler. Series B, Human Geography*, *71*(1), 3–17. doi:10.1080/04353684.1989.11879583

Harvey, D. (2003). *The 'New' Imperialism: Accumulation by dispossession*. Oxford, UK: Oxford University Press.

Harvey, D. (2005). *A Brief History of Neoliberalism*. New York: Oxford University Press.

Henriques, C. (2003). *Turismo, Cidade e Cultura – Planeamento e Gestão Sustentável*. Edições Silabo.

Hernández Cordero, A. (2014). Gentrificación comercial y mercados públicos: El mercado de Santa Caterina, Barcelona. *Working paper series. CONTESTED_CITIES*, 2341-2755.

Hernàndez i Martí, G.-M. (2005). La difusión del patrimonio cultural y el turismo. En G.-M. Hernàndez i Martí, B. Santamarina Campos, A. Moncusí Rodrigo, & M. Albert Rodrigo (Eds.), La memoria construida. Patrimonio cultural y modernidad (pp. 159-192). Valencia, Spain: Titant lo Blanch.

Hernàndez i Martí, G.-M. (2005). La globalización y el patrimonio cultural. En G.-M. Hernàndez i Martí, B. Santamarina Campos, A. Moncusí Ferré, & M. Albert Rodrigo (Eds.), La memoria construida. Patrimonio cultural y modernidad (pp. 123-158). Valencia, Spain: Tirant lo Blanch.

Hernàndez i Martí, G.-M., Santamarina Campos, B., Moncusí Ferré, & Albert Rodrigo M. (Eds.), *La memoria construida. Patrimonio cultural y modernidad.* Valencia, Spain: Tirant lo blanch.

Hernandez-Pezzi, C. (2018). *Turismo: ¿truco o trato?: políticas urbanas para el turismo de masas*. Madrid, Spain: Los Libros de la Catarata.

Hiernaux, D., & González, C. (2014). Turismo y gentrificación: Pistas teóricas sobre una articulación. *Revista de Geografía Norte Grande*, *58*(58), 55–70. doi:10.4067/S0718-34022014000200004

Higgins-Desbiolles, F. (2017). 'Sustainable Tourism' is not working – here's how we can change that. Retrieved May 16, 2019, from https://theconversation.com/sustainable-tourism-is-not-working-heres-how-we-can-change-that-76018

Higgins-Desbiolles, F. (2018a). Why Australia might be at risk of 'overtourism'. Retrieved May 16, 2019, from https://theconversation.com/why-australia-might-be-at-risk-of-overtourism-99213

Higgins-Desbiolles, F. (2019). Rethinking tourism so the locals actually benefit from hosting visitors. Retrived October 5, 2019, from https://theconversation.com/rethinking-tourism-so-the-locals-actually-benefit-from-hosting-visitors-116066

Higgins-Desbiolles, F. (2018b). Sustainable tourism: Sustaining tourism or something more? *Tourism Management Perspectives*, *25*, 157–160. doi:10.1016/j.tmp.2017.11.017

Higgins-Desbiolles, F., Carnicelli, S., Krolikowski, C., Wijesinghe, G., & Boluk, K. (2019). Degrowing tourism: Rethinking tourism. *Journal of Sustainable Tourism*, *27*(12), 1926–1944. doi:10.1080/09669582.2019.1601732

Hodkinson, S. (2013). The new urban enclosures. *City*, *16*(5), 500–518. doi:10.1080/13604813.2012.709403

Hoffman, L., Fainstein, S., & Judd, D. (2003). *Cities and Visitors: Regulating People, Markets and City Space*. Oxford, UK: Blackwell. doi:10.1002/9780470773673

Homeaway. (2018). *V Barómetro del alquiler Vacacional en España*. Salamanca. Universidad de Salamanca.

Horm, K., & Merante, M. (2017). Is home sharing driving up rents? Evidence from Airbnb in Boston. *Journal of Housing Economics*, *38*, 14–24. doi:10.1016/j.jhe.2017.08.002

Hosteltur. (2018). *España alcanzó los 82 millones de turistas internacionales en 2017 | Economía.* Retrieved June 11, 2019, from Hosteltur: Toda la información de turismo website: https://www.hosteltur.com/126008_espana-alcanzo-82-millones-turistas-internacionales-2017.html

Hudson, S., & Ritchie, J. R. B. (2009). Branding a Memorable Destination Experience. The Case of "Brand Canada.". *International Journal of Tourism Research*, *11*(2), 217–228. doi:10.1002/jtr.720

Huete, R., & Mantecón, A. (2018). El auge de la turismofobia ¿hipótesis de investigación o ruido ideológico? *Pasos – Revista de Turismo y Patrimonio Cultural, 16*, 9-19.

Huete, R., & Mantecón, A. (2018). El auge de la turismofobia ¿hipótesis de investigación o ruido ideológico? *Pasos. Revista de Turismo y Patrimonio Cultural, 16*(1), 9–19. doi:10.25145/j.pasos.2018.16.001

Hughes, N. (2018). 'Tourists go home': Anti-tourism industry protest in Barcelona. *Social Movement Studies, 17*(4), 471–477. doi:10.1080/14742837.2018.1468244

Huízar, M. A., Baños, J. A., & López, J. L. (2018). The impact of LGBT tourism in Puerto Vallarta. In S. M. Arnaiz, & A. B. Gómez (Eds.), *Tourism against the objectives of Sustainable Development*. Mexico: University of Guadalajara.

Idoate, E. Z. (2008). Políticas de vivienda en el Estado español. In Seminario de Economía Crítica Taifa, (pp. 66-80).

IFT (Institude for Tourism Studies). (2015). *Findings of the annual study of Macao's Tourism Carrying Capacity, 2013 and 2014,* available at https://www.ift.edu.mo/Membership/ Download/TCC]_[2013and2014/2013-2014_Press_Release.pdf (Accessed on June 1, 2019).

Ignarra, L. R. (2000). *Fundamentos do Turismo*. São Paulo: Pioneira.

IMPACTUR [Indicators for the Monitoring and Forecast of Tourist Activity]. (2019). *Informação estatística turística de Lisboa*. Retrieved from http://www.ciitt.ualg.pt/impactur/dt_anuais.asp?nut=17

INE [Instituto Nacional de Estatística]. (2018). Estatísticas de Portugal. Retrieved from https://www.ine.pt/xportal/xmain?xpgid=ine_main&xpid=INE&xlang=pt

INE. (2018a). *Resultados Preliminares sobre Actividade Turística 2017*. Lisboa, Portugal: Instituto Nacional de Estatística.

INE. (2018b). *Estatísticas de Preços da Habitação Local*. Lisboa, Portugal: Instituto Nacional de Estatística.

Infobae.com. (2018, October 11). *The house of horrors: they rented on Airbnb and have not recovered from the nightmare*. Online publication. Retrieved from https://www.infobae.com/america/eeuu/2018/10/11/la-casa-de-los-horrores-rentaron-en-airbnb-y-no-se-han-recuperado-de-la-pesadilla/

Insideairbnb (2019, 2018). *Inside Airbnb. Adding Data to the Debate*. Retrieved from http://insideairbnb.com/lisbon/

InsideAirbnb. (2019). *Sobre Airbnb en Málaga*. Retrieved from http://insideairbnb.com/malaga/

Inskeep, E. (1991). *Tourism Planning: An Integrated and Sustainable Development Approach*. New York: Van Nostrand Reinhold.

Instituto da Conservação da Natureza e Florestas. (2019, July 3). Retrieved from http://www2.icnf.pt/portal/icnf

Instituto Nacional de Estadística (INE). (2018). PIB per capita - Illes Balers, 2000-2017. Retrieved from: http://www.ine.es/prensa/cre_2017_1.pdf, http://www.ine.es/dyngs/INEbase/es/operacion.htm?c=Estadistica_C&cid=1254736167628&menu=resultados&idp=1254735576581# Interreg

International Conference on Responsible Tourism in Destinations. (2002). *The Cape Town Declaration, Cape Town*. Retrieved from http://responsibletourismpartnership.org/cape-town-declaration-on-responsible-tourism/

International Financial Analysts. AFI data laboratory. (in press 2016). ¿Cuánto bajaron los precios de la vivienda en tu provincia con la crisis? *Diario El País*.

International Trail Running Association. (2019, July 25). *Discovering Trail running*. Retrieved https://itra.run/page/259/Discovering_trail-running.html

International, I. P. K. (2018). *ITB World Travel Trends Report*. Berlin, Germany: Messe Berlin GmbH.

Inurrieta Berruete, A. (2007). *Mercado de vivienda en alquiler en España: más vivienda social y más mercado profesional*. Madrid, Spain: Fundación Alternativas.

Ivars-Baidal, J., Femenia-Serra, F., Giner Sánchez, D., & Celdrán-Bernabeu, M. (2018a). *Las OGDs ante el ecosistema turístico inteligente: oportunidades y riesgos*. Paper presented at the 2nd UNWTO World Conference on Smart Destinations. Oviedo (Spain), June 25-27, 2018.

Ivars-Baidal, J., Navarro, E., Celdrán, M., Perles, J., & Pérez-Medina, M. (2018b). El enfoque integral de los destinos turísticos inteligentes (DTI) en áreas litorales: alcance, progresos y limitaciones. In *Actas IV Congreso Ciudades Inteligentes* (pp. 471-476) Madrid, Spain: Tecma.

Ixová, M., & Sládek, J. (2017). Touristification and awakening civil society in post-socialist Prague. In C. Colomb & J. Novy (Eds.), *Protest and resistance in the tourist city* (pp. 73–89). London, UK: Routledge.

Jackson, M. S., & Inbakaran, R. J. (2006). Evaluating Residents' Attitudes and Intentions to Act toward Tourism Development in Regional Victoria, Australia. *International Journal of Tourism Research*, *8*(5), 355–366. doi:10.1002/jtr.584

Jacobsen, J. (2000). *Tourism in Peripheral Areas: Case Studies*. Frances Brown, Derek D. Hall, Derek R. Hall.

Jacobsen, J. K. S., Iversen, N. M., & Hem, L. E. (2019). Hotspot crowding and over-tourism: Antecedents of destination attractiveness. *Annals of Tourism Research*, *76*, 53–66. doi:10.1016/j.annals.2019.02.011

Jamal, T., & Camargo, B. (2017). Tourism governance and policy: Whither justice? *Tourism Management Perspectives*. doi:10.1016/j.tmp.2017.11.009

Jansen-Verbeke, M., & Lievois, E. (1999). Analysing heritage resources for urban tourism in European cities. In D. G. Pearce, & R. W. Butler (Eds.), *Contemporary Issues in Tourism Development* (pp. 81–107). New York: Routledge. doi:10.4324/9780203380307_chapter_6

Jennings, G., & Nickerson, N. (Eds.). (2006). *Quality tourism experiences*. Burlington, MA: Elsevier Butterworth-Heinemann.

JN [Jornal de Negócios]. (2018). *As freguesias de Lisboa mais pressionadas pelo alojamento local*. Retrieved from https://www.jornaldenegocios.pt/economia/rendas/detalhe/mapa-quais-sao-as-freguesias-de-lisboa-mais-pressionadas-pelo-alojamento-local. https://www.publiturishotelaria.pt/2018/12/20/alojamento-local-teve-um-impacto-economico-de-412-milhoes-de-euros-em-2018/

Jogaratnam, G., Mccleary, K., Mena, M., & Yoo, J. J. (2005). An Analysis of Hospitality and Tourism Research: Institutional Contributions. *Journal of Hospitality & Tourism Research (Washington, D.C.)*, *29*(3), 356–371. doi:10.1177/1096348005276929

Joppe, M. (2018). Tourism policy and governance: Quo vadis? *Tourism Management Perspectives*, *25*, 201–204. doi:10.1016/j.tmp.2017.11.011

Joppe, M. (2019). The roles of policy, planning and governance in preventing and managing overtourism. In R. Dodds, & R. Butler (Eds.), *Overtourism: Issues, Realities, and Solutions* (pp. 250–261). Berlin, Germany: De Gruyter. doi:10.1515/9783110607369-017

Jordan, E. J., Spencer, D. M., & Prayag, G. (2019). Tourism impacts, emotions and stress. *Annals of Tourism Research*, *75*, 213–226. doi:10.1016/j.annals.2019.01.011

Jordan, E. J., Vogt, C. A., & DeShon, R. P. (2015). A stress and coping framework for understanding resident responses to tourism development. *Tourism Management*, *48*, 500–512. doi:10.1016/j.tourman.2015.01.002

JRC. (2018). *The Cultural and Creative Cities Monitor*. JRC/European Commission.

Jurowski, C., & Gursoy, D. (2004). Distance Effects on Residents' Attitudes toward Tourism. *Annals of Tourism Research*, *31*(2), 296–312. doi:10.1016/j.annals.2003.12.005

Kádár, B. (2014). Measuring tourist activities in cities using geotagged photography. *Tourism Geographies*, *16*(1), 88–104. doi:10.1080/14616688.2013.868029

Kádár, B., & Gede, M. (2013). Where do tourists go? Visualizing and analyzing the spatial distribution of geotagged photography. *Cartographica*, *48*(2), 78–88. doi:10.3138/carto.48.2.1839

Kandampully, J., Zhang, T. C., & Jaakkola, E. (2018). Customer experience management in hospitality: A literature synthesis, new understanding and research agenda. *International Journal of Contemporary Hospitality Management*, *30*(1), 21–56. doi:10.1108/IJCHM-10-2015-0549

Kaplanidou, K., & Vogt, C. (2010). The meaning and measurement of a sport event experience among active sport tourists. *Journal of Sport Management*, *24*(5), 544–566. doi:10.1123/jsm.24.5.544

Kaspar, C. (1973). *"Tourism Ecology – New Dimension of Tourism Science". Festschrift zur Vollendung des 65. Lebensjahres von P. Bernecker. Beitraege zur Fremdenverkehrsforschung* (pp. 139–143). Wien: Ender W.

Kastenholz, E., & Lopes de Almeida, A. (2008). Seasonality in rural tourism–the case of North Portugal. *Tourism Review*, *63*(2), 5–15. doi:10.1108/16605370810883905

Katarzyna, J., Six, S., & Vanneste, D. (2017). Building tourism-resilient communities by incorporating residents' perceptions? A photo-elicitation study of tourism development in Bruges. *Journal of Tourism Future*, *3*(2), 127–143. doi:10.1108/JTF-04-2017-0011

Kato, K. (2018). Debating Sustainability in Tourism Development: Resilience, Traditional Knowledge and Community: A Post-disaster Perspective. *Tourism Planning & Development*, *15*(1), 55–67. doi:10.1080/21568316.2017.1312508

Kennelly, M. (2017). We've never measured it, but it brings in a lot of business. *International Journal of Contemporary Hospitality Management*, *29*(3), 883–899. doi:10.1108/IJCHM-10-2015-0541

Kent State University. (2018). *SPSS tutorials: independent samples t test*, available at https://libguides.library.kent.edu/SPSS/IndependentTTest (Accessed on Jun 2, 2018).

Kharas, H. (2017, February). *The Unprecedented Expansion of the Global Middle Class: An Update.* Global Economy and Development. Washington, DC: The Brookings Institution. Retrieved from https://think-asia.org/handle/11540/7251

Kim, K., Uysal, M., & Sirgy, M. J. (2013). How does tourism in a community impact the quality of life of community residents? *Tourism Management*, *36*, 527–540. doi:10.1016/j.tourman.2012.09.005

King, B., Pizam, A., & Milman, A. (1993). Social impacts of tourism: Host perceptions. *Annals of Tourism Research*, *20*(4), 650–665. doi:10.1016/0160-7383(93)90089-L

King, W., & He, J. (2005). Understanding the Role and Methods of Meta-Analysis in IS Research. *Communications of the Association for Information Systems*, *16*, 665–686. doi:10.17705/1CAIS.01632

Klein, R. (2009). Keeping the Cruise Tourism Responsible: The Challenge for Ports to Maintain High Self Esteem. Available at www.cruisejunkie.com/icrt2.pdf

Koenig-Lewis, N., & Bischoff, E. E. (2004). Analyzing seasonality in Welsh room occupancy data. *Annals of Tourism Research, 31*(2), 374–392. doi:10.1016/j.annals.2003.12.006

Koenig-Lewis, N., & Bischoff, E. E. (2005). Seasonality research: The state of the art. *International Journal of Tourism Research, 7*(4/5), 201–219. doi:10.1002/jtr.531

Koens, K., Postma, A., & Papp, B. (2018). Is Overtourism Overused? Understanding the Impact of Tourism in a City Context, *Sustainability*, 10(12), 4384.

Koens, K., Postma, A., & Papp, B. (2018). Is overtourism overused? Understanding the impact of tourism in a city context. *Sustainability, 10*(12), 4384. doi:10.3390u10124384

Kommersant. (2018, June 25). *Alye parusa 2018 posetili 1,2 mln chelovek [Scarlet Sails 2018 were visited by 1,2 million of people].* Retrieved from https://www.kommersant.ru

Koronios, K., & Kriemadis, T. (2018). An exploration of motives, constraints and future participation intention in sport and exercise events. *Sport, Business, and Management, 8*(5), 436–453. doi:10.1108/SBM-09-2017-0044

Körössy, N. (2008). Do "Turismo predatório" ao "Turismo sustentável": Uma revisão sobre a origem e a consolidação do discurso da sustentabilidade na atividade turística. *Caderno Virtual de Turismo, 8*(2), 56–68.

Kotler, P., & Levy, S. (1971, November-December). Demarketing- Yes, Demarketing. *Harvard Business Review*, pp. 74-80.

Kotler, P. (2000). *Marketing Management: The Millennium Edition.* Upper Saddle River, NJ: Prentice-Hall.

Kotler, P., Armstrong, G., Saunders, J., & Wong, V. (2001). *Principles of Marketing*. Upper Saddle River, NJ: Prentice Hall.

Krähmer, K. (2017). *Gentrification without gentry? Tourism and Real Estate Investment in Lisbon.* Master's thesis. Torino: Politecnico di Torino, Corso di laurea magistrale in Pianificazione Territoriale, Urbanistica e Paesaggistico-Ambientale.

Krippendorf, J. (1999). *The Holliday Makers Understanding the Impact of Leisure and Travel.* London, UK: Routledge.

Kumar, G., & Srivastav, A. (2019). Sustainable Tourism Marketing in Andaman and Nicobar Islands. *International Journal of Scientific Research and Review, 7*(3), 663–671.

Kuščer, K., & Mihalič, T. (2019). Residents' Attitudes towards Overtourism from the Perspective of Tourism Impacts and Cooperation – The Case of Ljubljana. *Sustainability, 11*(6), 1823. doi:10.3390u11061823

Laborda, J. (in press 2018). Ajuste de cuentas de las élites: pasado, presente y futuro. *Vozpopuli.com*

Laerd Statistics. (2018a). *Independent t-test using SPSS Statistics*, available at https:// statistics.laerd.com/spss-tutorials/independent-t-test-using-spss-statistics.php (Accessed on June 2, 2018).

Laerd Statistics. (2018b). *Kruskal-Wallis H test using SPSS Statistics*, available at https://statistics.laerd.com/spss-tutorials/kruskal-wallis-h-test-using-spss-statistics.php (Accessed on Jun 2, 2018).

Lai, I. K. W., & Hitchcock, M. (2017). Local reactions to mass tourism and community tourism development in Macau. *Journal of Sustainable Tourism, 25*(4), 451–470. doi:10.1080/09669582.2016.1221413

Lalicic, L., & Onder, I. (2018). Residents' Involvement in Urban Tourism Planning: Opportunities from a Smart City Perspective. *Sustainability, 10*(6). doi:10.3390u10061852

Lamsfus, C., Martín, D., Alzua-Sorzabal, A., & Torres-Manzanera, E. (2015). Smart Tourism Destinations: An Extended Conception of Smart Cities Focusing on Human Mobility. In I. Tussyadiah, & A. Inversini (Eds.), *Information and Communication Technologies in Tourism. Proceedings of the International Conference in Lugano, Switzerland*, February 3-6, 2015 (pp. 363-375). Springer International Publishing. 10.1007/978-3-319-14343-9_27

Lankford, S. V., & Howard, D. R. (1994). Developing a tourism impact attitude scale. *Annals of Tourism Research*, *21*(1), 121–139. doi:10.1016/0160-7383(94)90008-6

Larsen, S., Wolff, K., Marnburg, E., & Øgaard, T. (2013). Belly full, purse closed: Cruise line passengers' expenditures. *Tourism Management Perspectives*, *6*, 142–148. doi:10.1016/j.tmp.2013.02.002

LaSalle, J. L. (2017). Portugal Real State, Market 360º. Lisboa, Portugal.

LaSalle, J. L. (2015). *Apartamentos Turísticos em Lisboa*. Lisboa, Portugal: Um Mercado em Crescimento.

Law (Lei) n.º 62/2018- 22 august (changes the regime of Alojamento Local units exploration authorization).

Law, Ch. (2002). *Urban Tourism. The Visitor Economy and the Growth of Large Cities*. London, UK: Mansell.

Lee, D. (2016). How Airbnb short-term rentals exacerbate Los Angeles's affordable housing crisis: Analysis and policy recommendations. *Harv. L. & Pol'y Rev.*, 10, 229. Retrieved from https://harvardlpr.com/wp-content/uploads/sites/20/2016/02/10.1_10_Lee.pdf

Lee, C. K., & Taylor, T. (2005). Critical reflections on the economic impact assessment of a mega-event: The case of 2002 FIFA World Cup. *Tourism Management*, *26*(4), 595–603. doi:10.1016/j.tourman.2004.03.002

Lee, D. (2016). How Airbnb short-term rentals exacerbate Los Angeles's affordable housing crisis: Analysis and policy recommendations. *Harvard Law & Policy Review*, *10*, 229–253.

Lees, L., Shin, H. B., & López-Morales, E. (2016). *Planetary gentrification*. Cambridge, MA: John Wiley & Sons.

Leiper, N. (1990). Partial industrialization of tourism systems. *Annals of Tourism Research*, *17*, 600–605.

Leite, N. K. (2008). Turismo e territorio: um estudo sobre a turistificacao de Portimao (Algarve/Portugal) a partir da geografia do turismo. Master's thesis. Universidade de Lisboa.

Lerario, A., & Di Turi, S. (2018). Sustainable Urban Tourism: Reflections on the Need for Building-Related Indicators. *Sustainability*, *10*(6), 1981. doi:10.3390u10061981

Lestegás, I. (2019). Lisbon after the crisis: From credit-fuelled suburbanization to tourist-driven gentrification. *International Journal of Urban and Regional Research*, *43*(4), 705–723. doi:10.1111/1468-2427.12826

Lestegás, I., & Lois González, R. C. (2018). The global rent gap of Lisbon's historic centre. *International Journal of Sustainable Development and Planning*, *13*(4), 683–694. doi:10.2495/SDP-V13-N4-683-694

Leung, Y. F., Spenceley, A., Hvenegaard, G., & Buckley, R. (Eds.). (2018). *Tourism and Visitor Management in Protected Areas: Guidelines for Sustainability. Best Practice Protected Area Guidelines Series No. 27*. Gland, Switzerland: IUCN. doi:10.2305/IUCN.CH.2018.PAG.27.en

Lew, A. A., & Cheer, J. M. (Eds.). (2018). *Tourism Resilience and Adaptation to Environmental Change*. London, UK: Routledge.

Lew, A. A., & Cheer, J. M. (Eds.). (2018). *Tourism resilience and adaptation to environmental change: Definitions and frameworks*. London, UK: Routledge.

Ley, D. (1986). Alternative Explanations for Inner-City Gentrification: A Canadian Assessment. *Annals of the Association of American Geographers*, *74*(4), 521–535. doi:10.1111/j.1467-8306.1986.tb00134.x

Liang, Z., & Hui, T. (2016). Residents' quality of life and attitudes toward tourism development in China. *Tourism Management*, *57*, 56–67. doi:10.1016/j.tourman.2016.05.001

Lingard, L. (2018). Mapping the gap. *Perspectives on Medical Education*, *7*(1), 47–49. doi:10.100740037-017-0401-x PMID:29260402

Lipietz, A. (1996). Warp, woof, and regulation: A tool for social science. In G. Benko, & U. Strohmayer (Eds.), *Space and Social Theory* (pp. 250–283). Cambridge, MA: Blackwell.

Lipovčan, L. K., Brajša-Žganec, N., & Poljanec-Borić, S. (2014). What is Good for Tourists Should Be Good for Residents Too: The Relationship Between the Quality of the Touristic Offer and Subjective Well-Being of Residents. *Tourism Analysis*, *19*(6), 719–730. doi:10.3727/108354214X14146846679448

Litchfield, R. C. (2008). Brainstorming Reconsidered: A goal-based view. *Academy of Management Review*, *33*(3), 649–668. doi:10.5465/amr.2008.32465708

Little, V., Lee, C., & Nair, S. (2019). Macro-demarketing: The Key to unlocking unsustainable production and consumption systems? *Journal of Macromarketing*, *39*(2), 166–187. doi:10.1177/0276146718823885

Liu, C., Zhang, R., Wang, M., & Xu, J. (2018). Measurement and Prediction of Regional Tourism Sustainability: An Analysis of the Yangtze River Economic Zone, China. *Sustainability*, *10*(5), 1321.

Liu, J. C., Sheldon, P. J., & Var, T. (1987). Resident perception of the environmental impacts of tourism. *Annals of Tourism Research*, *14*(1), 17–37. doi:10.1016/0160-7383(87)90045-4

Liu, J., Liu, N., Zhang, Y., Qu, Z., & Yu, J. (2019). Evaluation of the non-use value of beach tourism resources: A case study of Qingdao coastal scenic area, China. *Ocean and Coastal Management*, *168*, 63–71. doi:10.1016/j.ocecoaman.2018.10.030

Liu, J., & Var, T. (1986). Residents Attitudes toward Tourism Impacts in Hawaii. *Annals of Tourism Research*, *13*(2), 193–214. doi:10.1016/0160-7383(86)90037-X

Li, X., & Wan, Y. K. P. (2017). Residents' support for festivals: Integration of emotional solidarity. *Journal of Sustainable Tourism*, *25*(4), 517–535. doi:10.1080/09669582.2016.1224889

Lohman, M., & Beer, H. (2013). Fundamentals of tourism: What makes a person a potential tourist and a region a potential tourism destination? Poznan University of Economic Review, 13(4), 83–96.

López Andion, M. (1998). Modelos econométricos de la vivienda. *Econometrics Working Paper Series Economic Development de la Universidad de Santiago de Compostela*, 59, 1-46.

López Palomeque, F., Torres-Delgado, A., Font Urgell, X., & Serrano Miracle, D. (2018). Gestión sostenible de destinos turísticos: La implementación de un sistema de indicadores de turismo en los destinos de la provincia de Barcelona. *Boletín de la Asociación de Geógrafos Españoles*, *77*(77), 428–461. doi:10.21138/bage.2547

Los turistas solo entrarán al Mercado Central con guías oficiales de la Comunitat Valenciana. (02 de 05 de 2018). Retrieved February 2, 2019 from Valencia plaza: https://valenciaplaza.com/los-turistas-solo-entraran-al-mercado-central-con-guias-oficiales-de-la-comunitat-valenciana

Loudis, J. (2018). *'Overtourism' Is Driving Europeans Crazy.* Retrieved from https://www.thenation.com/article/overtourism-is-driving-europeans-crazy/

Lu, J., & Nepal, S. K. (2009). Sustainable tourism research: An analysis of papers. *Journal of Sustainable Tourism*, *17*(1), 5–16. doi:10.1080/09669580802582480

Lundtorp, S. (2001). Measuring tourism seasonality. In T. Baum, & S. Lundtorp (Eds.), *Seasonality in tourism.* Oxford, UK: Pergamon. doi:10.1016/B978-0-08-043674-6.50006-4

Luque-Gil, A. M., Gómez-Moreno, M. L., & Peláez-Fernández, M. A. (2018). Starting to enjoy nature in Mediterranean mountains: Crowding perception and satisfaction. *Tourism Management Perspectives*, *25*, 93–103. doi:10.1016/j.tmp.2017.11.006

Luthe, T., & Wyss, R. (2014). Assessing and planning resilience in tourism. *Tourism Management*, *44*, 161–163. doi:10.1016/j.tourman.2014.03.011

Macau Daily Times. (2018, February 26). *More tourists, less satisfaction in 2017*, available at https://macaudailytimes.com.mo/tourists-less-satisfaction-2017.html (Accessed on May 17, 2018).

Macau Daily Times. (2019, February 1). *Lawmaker calls for action to avoid 'over-Tourism'*, available at https://macaudailytimes.com.mo/lawmaker-calls-for-action-to-avoid-over-tourism.html (Accessed on May 5, 2019).

Macau News. (2016, May 11). *Macau records the highest population density per sq km*, available at https://macaunews.mo/macau-records-highest-population-density-per-sq-km/ (Accessed on May 17, 2018).

Macleod, D. V. L. (2005). *Alternative Tourism: A comparative analysis of meaning and impact*. Oxford, UK: Butterworth-Heinemann.

Maitland, R. (2010). Everyday life as a creative experience in cities. *Journal of Culture Tourism and Hospitality Research*, 176-185.

Maitland, R., & Newman, P. (2008). Visitor-host relationships: conviviality between visitors and host communities. In B. Hayllar, T. Griffin, & D. Edwards (Eds.), *City Spaces–Tourist Places: Urban Tourism Precincts* (pp. 223–242). New York: Elsevier. doi:10.1016/B978-0-7506-8195-7.00011-1

Malighetti, P. P., Paleari, S., & Redondi, R. (2009). Pricing strategies of low-cost airlines: The Ryanair case study. *Journal of Air Transport Management*, *15*(4), 195–203. doi:10.1016/j.jairtraman.2008.09.017

Mannheim, K. (1929). *Ideologie und Utopie*. Bonn, Germany: Cohen.

Mansfeld, Y. (1992). Group-differentiated Perceptions of Social Impacts Related to Tourism Development. *The Professional Geographer*, *44*(4), 377–392. doi:10.1111/j.0033-0124.1992.00377.x

Mansilla, J. (2019). Gentrificación, turistificación y clases socials en las ciudades del Mediterráneo. *Alba Sud*, *9*, 62–65.

Mantecón, A. (2010). Tourist Modernisation and Social Legitimation in Spain. *International Journal of Tourism Research*, *12*(5), 617–626.

Mantecón, A. (2016). La opinión pública sobre el turismo. Crítica y propuesta para una nueva orientación. *Estudios y Perspectivas en Turismo*, *25*(2), 110–123.

Mantecón, A., & Huete, R. (2011). Sociological Insights on Residential Tourism: Host Society Attitudes in a Mature Destination. *European Journal of Tourism Research*, *4*(2), 109–122.

Maoz, D. (2006). The mutual gaze. *Annals of Tourism Research*, *33*(1), 221–239. doi:10.1016/j.annals.2005.10.010

Marcuse, P. (1985). Gentrification, abandonment, and displacement: Connections, causes, and policy response in New York City. *Journal of Urban and Contemporary Law*, *28*, 195–240.

Marques, C., Reis, E., & Menezes, J. (2010). Profiling the segments of visitors to Portuguese protected areas. *Journal of Sustainable Tourism*, *18*(8), 971–996. doi:10.1080/09669582.2010.497222

Marrero Rodríguez, J. R. (2006). "El discurso de rechazo al turismo en Canarias: Una aproximación cualitativa". *PASOS Revista de Turismo y Patrimonio Cultural, 4*(3), 327-341. Instituto de Investigación Social y Turismo, 2006.

Marrero, J. R., & Huete, R. (2013). La opinión pública sobre el empleo turístico en la Comunidad Valenciana. *Cuadernos de Turismo*, *32*, 189–206.

Marsiglio, S. (2018). On the implications of tourism specialization and structural change in tourism destinations. *Tourism Economics*, *24*(8), 945–962. doi:10.1177/1354816618784788

Martín Martín, J., Guaita Martínez, J., & Salinas Fernández, J. (2018). An analysis of the factors behind the citizen's attitude of rejection towards tourism in a context of overtourism and economic dependence on this activity. *Sustainability*, *10*(8), 2851. doi:10.3390u10082851

Martín Plaza, A. (10 de 08 de 2017). RTVE. La turistificación, una "burbuja" que amenaza no solo a los vecinos sino al propio sector. Barcelona, Spain.

Martinez Pagés, J., & Maza, L. (2003). Análisis del precio de la vivienda en España. *Documento de Trabajo. Servicios de Estudios del Banco de España*, *307*, 1–46.

Martin, J. M. M., Guaita, J. M. M., & Salinas, J. A. F. (2018). An Analysis of the Factors behind the Citizen's Attitude of Rejection towards Tourism in a Context of Overtourism and Economic Dependence on This Activity. *Sustainability*, *10*(8), 2851.

Martín-Martín, J. M., Jiménez-Aguilera, J. D. D., & Molina-Moreno, V. (2014). Impacts of seasonality on environmental sustainability in the tourism sector based on destination type: An application to Spain's Andalusia region. *Tourism Economics*, *20*(1), 123–142. doi:10.5367/te.2013.0256

Mastercard. (2018). *The 'Global Destination Cities Index – 2018'* (Mastercard's 2018 Global Destination Cities Index). Retrieved from https://newsroom.mastercard.com/press-releases/big-cities-big-business-bangkok-london-and-paris-lead-the-way-in-mastercards-2018-global-destination-cities-index/

Mastercard. (2019). *The Global Destination Cities Index 2019*. Retrieved from https://newsroom.mastercard.com/wp-content/uploads/2019/09/GDCI-Global-Report-FINAL-1.pdf

Mathew, P. V., & Sreejesh, S. (2017). Impact of responsible tourism on destination sustainability and quality of life of the community in tourism destinations. *Journal of Hospitality and Tourism Management*, *31*, 83–89.

Mathieson, A., & Wall, G. (1982). Tourism, economic, physical and social impacts. *Journal of Travel.*

Mathieson, A., & Wall, G. (1982). *Tourism economic, physical and social impacts*. Harlow: Longman.

Mathieson, A., & Wall, G. (1982). *Tourism, economic, physical, and social impacts*. Harlow: Pearson Education.

Mathieson, A., & Wall, G. (1982). *Tourism: Economic, Physical and Social Impacts*. London, UK: Longman.

Mavrič, M., & Urry, J. (2009). Tourism studies and the new mobilities paradigm. In T. Jamal, & M. Robinson (Eds.), *The Sage Handbook of Tourist Studies* (pp. 645–657). Los Angeles, CA: Sage. doi:10.4135/9780857021076.n36

Maxim, C. (2017). Challenges Faced by World Tourism Cities–London's Perspective. *Current Issues in Tourism*, 1006–1024.

McCool, S. F., & Lime, D. W. (2001). Tourism carrying capacity: Tempting fantasy or useful reality? *Journal of Sustainable Tourism*, *9*(5), 372–388. doi:10.1080/09669580108667409

McIntyre, G. (1993). Sustainable tourism development: guide for local planners. Madrid, Spain: World Tourism Organization (UNWTO).

McKercher, B. (1993). Some fundamental truths about tourism: Understanding tourism's social and environmental impacts. *Journal of Sustainable Tourism*, *1*(1), 6–16. doi:10.1080/09669589309450697

McKinsey & Co., & WTTC. (2017, December). *Coping with success: Managing Overcrowding in tourism destinations.* London, UK: McKinsey & Company, & World Travel & Tourism Council. Retrieved from World's Travel and Tourism Council website.

McKinsey & Company and World Travel & Tourism Council. (2017). *Coping with success Managing Overcrowding in Tourism Destinations.* McKinsey & Company and World Travel & Tourism Council. Retrieved from https://www.wttc.org/-/media/files/reports/policy-research/coping-with-success---managing-overcrowding-in-tourism-destinations-2017.pdf

McKinsey & Company, & World Travel & Tourism Council. (2017). *Coping with success. Managing overcrowding in tourism destinations.* Retrieved from https://www.mckinsey.com/industries/travel-transport-and-logistics/our-insights/coping-with-success-managing-overcrowding-in-tourism-destinations

Mediterranean BlueIslands. (2018). *Seasonal concentration and characterization of marine litter in selected beaches of 8 Mediterranean islands.* Retrieved from: https://blueislands.interreg-med.eu/fileadmin/user_upload/Sites/Sustainable_Tourism/Projects/BLUEISLANDS/Library/Deliverables/WP3/D3.4.4_Seasonal_concentration_and_characterization_of_ML.pdf

Medway, D., Warnaby, G., & Dharni, S. (2010). Demarketing places: Rationales and strategies. *Journal of Marketing Management, 27*(1-2), 124–142. doi:10.1080/02672571003719096

Mellinas, J. P. (2017). La relativa importancia de la velocidad en las conexiones WI-FI. In J. F. Vera-Rebollo, J. A. Ivars-Baidal, & M. A. Celdrán Bernabeu (Eds.), Actas del Seminario Internacional Destinos Turísticos Inteligentes: nuevos horizontes en la investigación y gestión del turismo. (pp. 38-51) Sant Vicent del Raspeig: Publicacions de la Universitat d'Alacant. doi:10.14198/Destinos-Turisticos-Inteligentes.2017.02

Mendes, L. (2008). *A Nobilitação Urbana no Bairro Alto: Análise de um Processo de Recomposição Sócio-espacial.* Master's thesis. University of Lisbon, Portugal.

Mendes, L. (2016a). Tourism gentrification: touristification as Lisbon´s new urban frontier of gentrification. Master Class *Tourism Gentrification and City Making*, Stadslab and Academia Cidadã, Lisbon, April 16.

Mendes, L. (2018). Tourism gentrification in Lisbon: The panacea of touristification as a scenario of a post-capitalist crisis. In I. David (Eds.), Crisis, Austerity, and Transformation: How Disciplinary Neoliberalism is Changing Portugal (pp. 25-48). London, UK: Lexington.

Mendes, L. (2006). A nobilitação urbana no Bairro Alto: Análise de um processo de recomposição socio-espacial. *Finisterra, 41*(81), 57–82.

Mendes, L. (2013). Public policies on urban rehabilitation and their effects on gentrification in Lisbon. *AGIR. Revista Interdisciplinar de Ciências Sociais e Humanas, 1*(5), 200–218.

Mendes, L. (2014). Gentrificação e políticas de reabilitação urbana em Portugal: Uma análise crítica à luz da tese rent gap de Neil Smith. *Cadernos Metrópole, 16*(32), 487–511. doi:10.1590/2236-9996.2014-3209

Mendes, L. (2016b). What can be done to resist or mitigate tourism gentrification in Lisbon? Some Policy Findings & Recommendations. In M. Glaudemans, & I. Marko (Eds.), *City Making & Tourism Gentrification* (pp. 34–41). Tilburg: Stadslab.

Mendes, L. (2017). Gentrificação turística em Lisboa: Neoliberalismo, financeirização e urbanismo austeritário em tempos de pós-crise capitalista 2008-2009. *Cadernos Metrópole, 19*(39), 479–512. doi:10.1590/2236-9996.2017-3906

Merton, R. K. (1938). Social Structure and Anomie. *American Sociological Review, 3*(5), 672–682. doi:10.2307/2084686

Mexican Institute of Competitiveness (IMCO). (2017). *Airbnb in Mexico City: Extends the economic benefits of tourism and reduces the wage gap. Case study for the capital of the country*. Mexico: IMCO.

Mihalic, T. (2016). "Sustainable-responsible tourism discourse–Towards 'responsustable' tourism". *Journal of Cleaner Production*, 111(B), 461–470.

Mihalic, T. (2016). Sustainable-responsible tourism discourse – Towards 'responsustable' tourism. *Journal of Cleaner Production*, *111*, 461–470. doi:10.1016/j.jclepro.2014.12.062

Miklós-Thal, J., & Zhang, J. (2011, June 29). Strategic Demarketing. *MIT Sloan School Working Paper 4814-10*, 1-52.

Milano, C. (2017). *Overtourism and Tourismphobia: Global trends and local context*. [PDF document]. Retrieved from https://www.academia.edu/35911428/Overtourism_and_Tourismphobia_Global_trends_and_local_contexts?auto=download

Milano, C. (2018). Overtourism, Social Unrest and Tourismphobia. A controversial debate, PASOS-REVISTA DE TURISMO Y PATRIMONIO CULTURAL, 16(3), 551-564.

Milano, C., & Mansilla, J. A. (Eds.). (2018). Ciudad de vacaciones. Conflictos urbanos en espacios turísticos. Barcelona, Spain: Pol·len Edicions.

Milano, C., Novelli, M., & Cheer, J. M. (2018). Overtourism and Tourismphobia: A journey through four decades of tourism development, planning and local concerns. *Tourism Planning and Development Journal*, 1–4.

Milano, C. (2017). *Overtourism y Turismofobia. Tendencias globales y contextos locales*. Barcelona, Spain: Ostelea School of Tourism & Hospitality.

Milano, C. (2018). Overtourism, Social Unrest, and Tourismphobia. A controversial debate. *Pasos: Revista de Turismo y Patrimonio Cultural*, *16*(3), 551–564. doi:10.25145/j.pasos.2018.16.041

Milano, C. (2018b). *Tendencias en viajes: "live like a local" y saturación turística*. Barcelona, Spain: Ostelea School of Tourism & Hospitality.

Milano, C. (2019). *Ciudades y turismo urbano: ¿una relación indestructible?* Barcelona, Spain: Ostelea School of Tourism & Hospitality.

Milano, C., Cheer, J., & Novelli, M. (Eds.). (2019). *Overtourism: Excesses, Discontents and Measures in Travel & Tourism*. Wallingford: CABI. doi:10.1079/9781786399823.0000

Milano, C., Novelli, M., & Cheer, J. (2019). Overtourism and Tourismphobia: A Journey through Four Decades of Tourism Development, Planning and Local Concerns. *Tourism Planning & Development*, *16*(4), 353–357.

Milano, C., Novelli, M., & Cheer, J. M. (2018). *Overtourism and Tourismphobia: A journey through four decades of tourism development, planning and local concerns*. Tourism Planning and Development Journal; doi:10.1080/21568316.2019.1599604

Milano, C., Novelli, M., & Cheer, J. M. (2019). Overtourism and degrowth: A social movements perspective. *Journal of Sustainable Tourism*, *27*(12), 1857–1875. doi:10.1080/09669582.2019.1650054

Milano, C., Novelli, M., & Cheer, J. M. (2019). Overtourism and Tourismphobia: A Journey Through Four Decades of Tourism Development, Planning and Local Concerns. *Tourism Planning & Development*, *16*(4), 353–357.

Miller, G. (2001). The development of indicators for sustainable tourism: Results of a Delphi survey of tourism researchers. *Tourism Management*, *22*(4), 351–362. doi:10.1016/S0261-5177(00)00067-4

Milman, A., & Asli, D. A. (2018). Exploring the experiential and sociodemographic drivers of satisfaction and loyalty in the theme park context. *Journal of Destination Marketing & Management*, *8*, 385–395. doi:10.1016/j.jdmm.2017.06.005

Mincotur. (2018). *Previsiones Turísticas 2018 (p. 2)*. Retrieved from Mincotur website: https://www.tourspain.es/en-us/Conozcanos/NotasPrensaTemp/2018%2007%2005%20Espa%C3%B1a%20recibir%C3%A1%20este%20verano%20 30,4%20millones%20de%20turistas%20internacionales.pdf

Minetad - Ministerio de energía, turismo y agenda digital. (2018.). *España logra un récord de llegadas en 2017 con 82 millones de turistas internacionales*. Retrieved from http://www.minetad.gob.es/es-es/gabineteprensa/notasprensa/2017/documents/180110%20np%20balance%20turismo%202017.pdf

Ministry of development. (2019). *Protected housing and rehabilitation*. Retrieved from https://apps.fomento.gob.es/BoletinOnline2/?nivel=2&orden=31000000

Miralles, J. L. (2014). *El darrer cicle immobiliari al País Valencià o El progrés de la misèria*. Valencia, Spain: Fundació Nexe.

Misachi, J. (2019). *What Is Overtourism? Overtourism occurs when a destination receives too many visitors*. Retrieved from https://www.worldatlas.com/articles/what-is-overtourism.html

Moher, D., Liberati, A., Tetzlaff, J., & Altman, D. G.The PRISMA Group. (2009). Preferred Reporting Items for Systematic Reviews and Meta-Analyses: The PRISMA Statement. *PLoS Medicine*, *6*(7). doi:10.1371/journal.pmed.1000097

Mohn, T. (2015). Co-Working on Vacation: A Desk in Paradise. The New York Times (January 19). Retrieved from https://www.nytimes.com/2015/01/20/business/co-working-on-vacation-a-desk-in-paradise.html?action=click&contentCollection=Business%20Day&pgtype=imageslideshow&module=RelatedArticleList%C2%AEion=CaptionArea&version=SlideCard-4

Montanari, A., & Staniscia, B. (2010). Rome: A difficult path between tourist pressure and sustainable development. *Rivista di Scienze del Turismo*, *1*, 301–316.

Montezuma, J., & McGarrigle, J. (2018). What motivates international homebuyers? Investor to lifestyle 'migrants' in a tourist city. *Tourism Geographies*. doi:10.1080/14616688.2018.1470196

Mora, M. (2017). *8 of the 24 constructions of the Romantic Zone were authorized in this administration*. Independent Vallarta, Online Journal. Recovered from https://vallartaindependiente.com/2017/11/20/8-las-24-construc...s-la-zona-romantica-fueron-autorizadas-en-esta-administracion/

Moreno-Izquierdo, L., Ramón-Rodríguez, A., & Such-Devesa, M. J. (2018). The challenge of long-term tourism competitiveness in the age of innovation: Spain as a case study. *Investigaciones Regionales – Journal of Regional Research*, 42, 13-34.

Moreno-Izquierdo, L., Ramon-Rodriguez, A., & Jesus, M. S. D. (2018). The challenge of long-term tourism competitiveness in the age of innovation: Spain as a case study, Investigaciones Regionales-. *Journal of Regional Research*, *42*, 13–34.

Morteza, Z., Reza, F., Seddiq, M., Sharareh, P., & Jamal, G. (2016). Selection of the optimal tourism site using the ANP and fuzzy TOPSIS in the framework of Integrated Coastal Zone Management: A case of Qeshm Island. *Ocean and Coastal Management*, *130*, 179–187. doi:10.1016/j.ocecoaman.2016.06.012

Muler Gonzalez, V., Coromina, L., & Gali, N. (2018). Overtourism: Residents' perceptions of tourism impact as an indicator of resident social carrying capacity – Case study of a Spanish heritage town. *Tourism Review*, *73*(3), 277–296. doi:10.1108/TR-08-2017-0138

Muler, V. G., Coromina, L., & Gali, N. U. (2018). Overtourism: Residents' perceptions of tourism impact as an indicator of resident social carrying capacity - case study of a Spanish heritage town. *Tourism Review*, *73*(3), 277–296.

Mullins, P. (1991). Tourism Urbanization. *International Journal of Urban and Regional Research*, *15*(3), 326–342. doi:10.1111/j.1468-2427.1991.tb00642.x

Munguía, C. (1997). *Historic panorama of Puerto Vallarta and Banderas Bay*. Mexico: University of Guadalajara.

Municipal Málaga Housing Institute. (2019). *Promotions*. Retrieved from http://smv.malaga.eu/es/proyectos/proyectos/#!tab2

Munro, J., Allison, H., & Moore, S. (2010). Using resilience concepts to investigate the impacts of protected area tourism on communities. *Annals of Tourism Research*, *37*(2), 499–519. doi:10.1016/j.annals.2009.11.001

Munsters, W. (2005). Cultural Tourism in Belgium. In G. Richards (Ed.), *Cultural Tourism in Europe* (pp. 80–92). Association for Tourism and Leisure Education; Retrieved from http://www.tram-research.com/cultural_tourism_in_europe.PDF

Namberger, P., Jackisch, S., Schmude, J., & Karl, M. (2019). Overcrowding, Overtourism and Local Level Disturbance: How Much Can Munich Handle? *Tourism Planning & Development*, *16*(4), 452–472. doi:10.1080/21568316.2019.1595706

Naredo Pérez, J. M. (2013). Patrimonio familiar y crisis inmobiliaria. La apuesta por el alquiler. *Fundación de Estudios Financieros*, 169-186.

Naredo Pérez, J. M. (2019). Diagnóstico del panorama inmobiliario actual. ¿Rebrota una burbuja inmobiliaria comparable a la anterior? *Papeles de relaciones ecosociales y cambio global*, 145, 129-140.

National Competition Commission. (2013). *Problemas de competencia en el mercado del suelo en España*. Madrid, Spain.

National Statistical Institute. (2011) Census of Population and Housing (2011). Retrieved from https://www.ine.es/censos2011_datos/cen11_datos_inicio.htm

National Statistical Institute. (2019) Statistics of tourist movements at the border. Frontur, Arrival of international tourists according to access road. Retrieved from https://www.ine.es/dyngs/INEbase/es/operacion.htm?c=Estadistica_C&cid=1254736176996&menu=ultiDatos&idp=1254735576863

National Statistical Institute. (2019). *Rent per household by autonomous communities*. Retrieved from http://www.ine.es/jaxiT3/Tabla.htm?t=9949

National Statistics Institute. (2019). *Hotel Occupancy Survey*. Retrieved from http://www.ine.es/dynt3/inebase/es/index.htm?padre=238&dh=1

National Stock Market Commission (CNMV). (2019). *Statistical Series of Collective Investment Institutions*. Retrieved from https://www.cnmv.es/portal/Publicaciones/Estadisticas-IIC.aspx

Navarro Castelló, C. (20 de 09 de 2018). *Vecinos del Carme exigen más medidas contra la sobreocupación de las zonas peatonales y la turistificación*. Retrieved May 5, 2019 from El Diario.es: https://www.eldiario.es/cv/Vecinos-Carme-turistificacion-reduccion-terrazas_0_816469343.html

Navarro Jurado, E., Tejada Tejada, M., Almeida García, F., Cabello González, J., Cortés Macías, R., Delgado Peña, J., ... Solís Becerra, F. (2012). Carrying capacity assessment for tourist destinations. Method for the creation of synthetic indicators applied in a coastal area. *Tourism Management*, *33*(6), 1337–1346. doi:10.1016/j.tourman.2011.12.017

Nawijn, J., & Mitas, O. (2012). Resident attitudes to tourism and their effect on subjective well-being: The case of Palma de Mallorca. *Journal of Travel Research*, *51*(5), 531–541. doi:10.1177/0047287511426482

Netto, L. G. P. (2017). *O que é turismo*. São Paulo, Brazil: Editora Brasileiense.

Newsome, D. (2014). Appropriate policy development and research needs in response to adventure racing in protected areas. *Biological Conservation*, *171*, 259–269. doi:10.1016/j.biocon.2014.01.008

Nichols, M., Stitt, B. G., & Giacopassi, D. (2002). Community assessment of the effects of casinos on quality of life. *Social Indicators Research*, *57*(3), 229–262. doi:10.1023/A:1014777323534

Nofre, J., Martins, J. C., Vaz, D., Fina, R., Sequera, J., & Vale, P. (2018). *The 'Pink Street' in Cais do Sodré: urban change and liminal governance in a nightlife district of Lisbon*. Urban Research & Practice; doi:10.1080/17535069.2018.1449010

Nogueira, S., & Pinho, J. C. (2015). Stakeholder network integrated analysis: The specific case of rural tourism in the Portuguese Peneda-Gerês National Park. *International Journal of Tourism Research*, *17*(4), 325–336. doi:10.1002/jtr.1989

Nogués, A. M. (2007). Lugares y territorios: la segregación social y política en contextos turísticos. (Ed) Antropología y Turismo (pp. 165-184). México DF: Plaza y Valdés.

Nolan, E., & Séraphin, H. (2019). Venice: capacity and tourism. In R. Dodds, & R. W. Butler (Eds.), *Overtourism. Issues, realities, and solutions* (pp. 139–151). Berlin, Germany: De Gruyter. doi:10.1515/9783110607369-010

Nordin, S., & Hjalager, A. M. (2017). Doing, Using, Interacting – Towards a New Understanding of Tourism Innovation Processes. In A. Kiralova (Ed.), *Driving tourism through creative destinations and activities*. IGI Global. doi:10.4018/978-1-5225-2016-0.ch008

Novy, J. (2011). Berlin does not love you. *The Berlin Reader*, 223-233.

Novy, J. (2018). Urban tourism as a bone of contention: four explanatory hypotheses and a caveat. *International Journal of Tourism Cities*.

Novy, J. (2019). Urban tourism as a bone of contention: Four explanatory hypotheses and a caveat. *International Journal of Tourism Cities*, *5*(1), 63–74. doi:10.1108/IJTC-01-2018-0011

Nozari, E. (2017). *Mallorca locals put pressure on the government to clamp down on tourism.* Retrieved from: https://www.cntraveler.com/story/mallorca-locals-put-pressure-on-government-to-clamp-down-on-tourism?mbid=nl_092817_Intel&CNDID=45233894&spMailingID=12026653&spUserID=MTU4NzkwNjAyOTY0S0&spJobID=1242432308&spReportId=MTI0MjQzMjMwOAS2

Numbeo. (2018) *Cost of living in Spain.* Retrieved from: https://www.numbeo.com/cost-of-living/country_result.jsp?country=Spain

Nunes, S., & Sousa, V. (2019). Perfect Resources, Tourism and Territorial Singularities: Contributions to the Development of a Scientific Tourism Line in Golegã. *Revista Portuguesa de Estudos Regionais*, *50*, 27–47.

Nunkoo, R. (2016). Toward a more comprehensive use of social exchange theory to study residents' attitudes to tourism. *Procedia Economics and Finance*, *39*, 588–596. doi:10.1016/S2212-5671(16)30303-3

Nunkoo, R., & Ramkissoon, H. (2011). Residents' satisfaction with community attributes and support for tourism. *Journal of Hospitality & Tourism Research (Washington, D.C.)*, *35*(2), 171–190. doi:10.1177/1096348010384600

Nunkoo, R., & Ramkissoon, H. (2012). Power, trust, social exchange and community support. *Annals of Tourism Research*, *39*(2), 997–1023. doi:10.1016/j.annals.2011.11.017

Nunnally, J. C. (1978). *Psychometric theory*. New Jersey: McGraw-Hill Book Company.

O'Brien, R. G. (1979). A general ANOVA method for robust tests of additive models for variances. *Journal of the American Statistical Association*, *74*(368), 877–880. doi:10.1080/01621459.1979.10481047

O'Regan, D. M., & Choe, J. (2017). Airbnb: Turning the Collaborative Economy into a Collaborative Society. In D. Dredge, & S. Gyimóthy (Eds.), Collaborative Economy and Tourism. Tourism on the Verge. Cham, Switzerland: Springer.

O'Reilly, K. (2007). Emerging tourism futures: residential tourism and its implications. In: C. Geoffroy, & R. Sibley (Eds.), Going Abroad: Travel, Tourism, and Migration. Cross-Cultural Perspectives on Mobility. Newcastle upon Tyne: Cambridge Scholars Publishing, pp. 144-157.

O'Reilly, A. M. (1986). Tourism carrying capacity: Concept and issues. *Tourism Management*, *7*(4), 254–258. doi:10.1016/0261-5177(86)90035-X

Observatory of the Urban Environment of Malaga. (2016). *Annual rent per household in Malaga, 2015*. Retrieved from http://www.omau-malaga.com/2/com1_md3_cd-1585/renta-anual-hogar-malaga

Ocaña, M. A. (April 10, 2017). *The Romantic Zone loses its identity*. Vallarta Daily Review. Retrieved from http://vallartaopina.net/2017/04/10/s-gente-pv/la-zona-romantica-pierde-identidad/

OECD-ORGANISATION FOR ECONOMIC CO-OPERATION AND DEVELOMPENT. (1993). OECD core set of indicators for environmental performance reviews. *Environment Monographs*, 38. Paris, France: Organisation for Economic Co-operation and Develompent.

Official College of Architects of Malaga. (2019). *Data concerning the files of new work of housing approvals in the COA Málaga*. Retrieved from http://coamalaga.es/estadisticas/

Oklevik, O., Gössling, S., Hall, C. M., Steen Jacobsen, J. K., Grøtte, I. P., & McCabe, S. (2019). Overtourism, optimisation, and destination performance indicators: A case study of activities in Fjord Norway. *Journal of Sustainable Tourism*, 1–21. doi:10.1080/09669582.2018.1533020

Olcina Cantos, J., & Vera-Rebollo, J. F. (2016). Adaptación del sector turístico al cambio climático en España. La importancia de las acciones a escala local y en empresas turísticas. *Anales de Geografía de la Universidad Complutense*, *36*(2), 321–352.

Oliveira, E. S. (2008). Impactos socioambientais e econômicos do turismo e suas repercussões no desenvolvimento local: o caso de Itacare – Bahia / Elton Silva Oliveira. – Ilheus, BA: UESC, 153 fls.

OMT – Organização Mundial de Turismo. (1999). *Conta Satélite do Turismo*. Madrid, Spain: Quadro Conceptual.

OMT - Organização Mundial de Turismo. (2003). *Guia de desenvolvimento do turismo sustentável*. Porto Alegre, Brazil: Bookman.

Ordiz, E. (2017). "Turismofobia en España: Dónde se producen las protestas y cuál es su objetivo". *20 Minutos*. Available at https://www.20minutos.es/noticia/3111276/0/turismofobia-en-espana-claves/#xtor=%20AD-15andxts=467263

Oskam, J., & Boswijk, A. (2016). Airbnb: The future of networked hospitality businesses. *Journal of Tourism Futures*, *2*(1), 22–42. doi:10.1108/JTF-11-2015-0048

Oxford Dictionary. (2019). *Word of the Year 2018: shortlist,* available at https://en.oxforddictionaries.com/word-of-the-year/shortlist-2018 (Accessed on June 10, 2019).

Ozomek, A. (2018). Cronomapa del Expolio en el BIC Centro Histórico de Málaga 1957-2016. Edifeicios: Paisaje Urbano Patrimonio BIC Centro Histórico de Málaga.

Page, S. (2003). *Tourism Management: Managing for Change*. Oxford, UK: Elsevier.

Palacios García, A. & Vinuesa Angulo, J. (2010). Un análisis cualitativo sobre la política de vivienda en España. La opinión de los expertos. *Anales de Geografía de la Universidad Complutense,* 101-118.

Palos-Sanchez, P. R., & Correia, M. B. (2018). The Collaborative Economy Based Analysis of Demand: Study of Airbnb Case in Spain and Portugal. *Journal of Theoretical and Applied Electronic Commerce Research, 13*(3), 85–98. doi:10.4067/S0718-18762018000300105

Panayiotopoulos, A., & Pisano, C. (2019). Overtourism dystopias and socialist utopias: Towards an urban armature for Dubrovnik. *Tourism Planning & Development, 16*(4), 393–410. doi:10.1080/21568316.2019.1569123

Pan, S. Y., Gao, M., Kim, H., Shah, K. J., Pei, S. L., & Chiang, P. C. (2018). Advances and challenges in sustainable tourism toward a green economy. *The Science of the Total Environment, 635*, 452–469. doi:10.1016/j.scitotenv.2018.04.134 PMID:29677671

Papathanassis, A. (2016). Curing the 'Beach Disease': Corruption and the Potential of Tourism-led Transformation for Developing Countries and Transitional Economies. *Ovidius University Annals Economic Sciences Series, 16*(1), 75–80.

Papathanassis, A. (2016b). Combating tourism-related corruption: Effective countermeasures derived from analysing tourists' perceptions and experiences. *Ovidius University Annals Economic Sciences Series, 16*(2), 248–255.

Papathanassis, A. (2017). Cruise tourism management: State of the art. *Tourism Review, 72*(1), 104–119. doi:10.1108/TR-01-2017-0003

Papathanassis, A. (2017). Over-tourism and anti-tourist sentiment: An exploratory analysis and discussion. *Ovidius University Annals. Economic Sciences Series, 17*(2), 288–293.

Papathanassis, A. (2017b). Over-Tourism and Anti-Tourist Sentiment: An Exploratory Analysis and Discussion. "Ovidius" University Annals. *Economic Sciences Series., XVII*(2), 288–293.

Papathanassis, A. (2019). The growth and development of the cruise sector: A perspective article. *Tourism Review*, TR-02-2019-0037. doi:10.1108/TR-02-2019-0037

Pappalepore, I., Maitland, R., & Smith, A. (2014). Presuming creative urban areas. Evidence from East London. *Annals of Tourism Research, 44*, 227–240.

Pappalepore, I., Maitland, R., & Smith, A. (2014). Prosuming creative urban areas. Evidence from East London. *Annals of Tourism Research, 44*, 227–240. doi:10.1016/j.annals.2013.11.001

Paré, G., Trudel, M.-C., Jaana, M., & Kitsiou, S. (2015). Synthesizing information systems knowledge: A typology of literature reviews. *Information & Management, 52*(2), 183–199. doi:10.1016/j.im.2014.08.008

Paris, C. (2013). The homes of the super-rich: Multiple residences, hyper-mobility and decoupling of prime residential housing in global cities. In I. Hay, & M. Flinders (Eds.), *Geographies of the super-rich* (pp. 94–109). Cheltenham, UK: Edward Elgar. doi:10.4337/9780857935694.00013

Parreño-Castellano, J. M., Domínguez-Mujica, J., Armengol-Martín, M. T., Pérez-García, T., & Boldú-Hernández, J. (2018). Foreclosures and Evictions in Las Palmas de Gran Canaria during the Economic Crisis and Post-Crisis Period in Spain. *Urban Science, 2*(4), 109, 1-15.

Parreño-Castellano, J. M., Domínguez-Mujica, J., Armengol-Martín, M. T., Boldú-Hernández, J., & Pérez-García, T. (2019). Real estate dispossession and evictions in Spain: A theoretical geographical approach. *Boletín de la Asociación de Geógrafos Españoles, 80*(2602), 1–25.

Pavel, F. (2017). Turismo e mercantilização da cidade. *Revista espaços vividos e espaços construídos: estudos sobre a cidade, 5*(1), 21-28.

Pearce, P. L. (2018). Limiting overtourism; the desirable new behaviours of the smart tourist. In *Proceedings of the Tourism Intelligence Forum (t-Forum) Global Exchange Conference 2018,* Palma de Mallorca, Spain, March 11-13.

Peattie, K., Peattie, S., & Newcombe, R. (2016). Unintended consequences in demarketing antisocial behaviour: Project Bernie. *Journal of Marketing Management, 32*(17-18), 1588–1618. doi:10.1080/0267257X.2016.1244556

Peeters, P., Gössling, S., Klijs, J., Milano, C., Novelli, M., Dijkmans, C., ... Postma, A. (2018). *Research for TRAN Committee - Overtourism: impact and possible policy responses.* Brussels, Belgium: European Parliament, Policy Department for Structural and Cohesion Policies.

Peeters, P., Gössling, S., Klijs, J., Milano, C., Novelli, M., Dijkmans, C., & Postma, A. (2018). *Research for TRAN Committee - Overtourism: Impact and Possible Policy Responses.* Brussels, Belgium: European Parliament, Policy Department for Structural and Cohesion Policies.

Peetrs, P., Gossling, S., Klijs, J., Milano, C., Novelli, M., Dijkmans, C., & Postma, A. (2018). *Research for TRAN Committee - Overtourism: impact and possible policy responses.* Brussels, Belgium: European Parliament, Policy Department for Structural and Cohesion Policies.

Perdue, R. R., Long, P. T., & Allen, L. (1990). Resident support for tourism development. *Annals of Tourism Research, 17*(4), 586–599. doi:10.1016/0160-7383(90)90029-Q

Pereira, H., & Teixeira, N. (2017). *(coord.). O impacto económico do Alojamento Local na Área Metropolitana de Lisboa 2016-2020.* Lisboa, Portugal: AHRESP.

Pérez, C. (2013). Patrimonialización, turistificación y autenticidad en exaltación de la Cruz, Argentina. *Estudios y Perspectivas en Turismo, 22,* 785–804.

Perkumienė, D., & Pranskūnienė, R. (2019). Overtourism: Between the Right to Travel and Residents' Rights. *Sustainability, 11*(7), 2138. doi:10.3390u11072138

Perrin-Malterre, C. (2018). Tourism diversification process around trail running in the Pays of Allevard (Isère). *Journal of Sport & Tourism, 22*(1), 67–82. doi:10.1080/14775085.2018.1432410

Pfeffer, J. (2007). A Modest Proposal: How We Might Change the Process and Product of Managerial Research. *Academy of Management Journal, 50*(6), 1334–1345. doi:10.5465/amj.2007.28166117

Phi, G. T. (2019). Framing overtourism: A critical news media analysis. *Current Issues in Tourism,* 1–5. doi:10.1080/13683500.2019.1618249

Picornell, M. (2014). Insular identity and urban contexts: Representations of the local in the construction of an image of Palma (Mallorca, Balearic Islands). *Island Studies Journal, 9*(2), 223–238.

Pigram, J., & Wahab, S. (1997). *Tourism, Development, and Growth.* Abingdon, UK: Routledge.

Pile, T. (2017). Who are the world's worst tourists? Six nations that stand out - you may be surprised. [Online] Available at http://www.sohu.com/a/157741590_644346, Accessed date: October 30, 2019.

Pinke-Sziva, I., Smith, M., Olt, G., & Berezvai, Z. (2019). Overtourism and the night-time economy: A case study of Budapest. *International Journal of Tourism Cities, 5*(1), 1–16.

Pinke-Sziva, I., Smith, M., Olt, G., & Berezvai, Z. (2019). Overtourism and the night-time economy: A case study of Budapest. *International Journal of Tourism Cities, 5*(1), 1–16. doi:10.1108/IJTC-04-2018-0028

Pinkster, F., & Boterman, W. (2017). When the spell is broken: Gentrification, urban tourism and privileged discontent in the Amsterdam canal district. *Cultural Geographies*, *24*(3), 457–472. doi:10.1177/1474474017706176 PMID:29278248

Pizam, A. (1978). Tourism's Impacts: The Social Costs to the Destination Community as Perceived by Its Residents. *Journal of Travel Research*, *16*(4), 8–12. doi:10.1177/004728757801600402

Pizam, A., & Milman, A. (1986). The social impacts of tourism. *Tourism Recreation Research*, *11*(1), 29–33. doi:10.1080/02508281.1986.11014414

Platges de Balears. (2018). *Caló des Màrmols*. Retrieved from: http://www.platgesdebalears.com/caplaibFront/playa_basico.ct.jsp?cPla=65017&cMun=1&cIsl=MA

Plichta, G. (2018). The Problem of Short-term Rental as a Manifestation of the Overtourism Effect on the Housing Market in the Historical City of Krakow, *Świat Nieruchomości, 4*(106) 57-64.

Podilchak, W. (1991). Distinctions of fun, enjoyment and leisure. *Leisure Studies*, *10*(2), 133–148. doi:10.1080/02614369100390131

Poll, B. (2019, July 25). *What we want to do to help combat overtourism*. Retrieved from Spotted by Locals: www.spottedbylocals.com/blog/overtourism

Popova, T. (2016, December 10). *Knight Frank: Rynok gostinichnoj nedvizhimosti. Sankt-Peterburg - 2016 [Knight Frank: The hotel real estate marketplace. St. Petersburg - 2016]*. Retrieved from http://www.hotelscatalog.ru/novosti-gostinits-sankt-peterburga/knight-frank-rynok-gostinichnoj-nedvizhimosti-sankt-peterburg-2016.html

Popp, M. (2012). Positive and negative urban tourist crowding: Florence, Italy. *Tourism Geographies*, *14*(1), 50–72. doi:10.1080/14616688.2011.597421

Popp, M. (2009). Der touristische Blick im St€adtetourismus der Postmoderne. Das Beispiel der italienischen Stadt Florenz [The tourist gaze in postmodern urban tourism. The example of the Italian city of Florence]. *Geographische Rundschau*, *61*(2), 42–48.

Population and Housing Census of The National Statistical Institute. (2011). Retrieved from https://www.ine.es/censos2011_datos/cen11_datos_inicio.htm

PORDATA. (2019, July 20). Praticantes desportivos federados: total e por algumas federações desportivas. Retrieved from https://www.pordata.pt/Portugal/Praticantes+desportivos+federados+total+e+por+algumas+federações+desportivas-2226.

Postma, A. (2013). When the tourists flew in: Critical encounters in the development of tourism. The Netherlands: Groningen.

Postma, A., & Schmuecker, D. (2017). Understanding and overcoming negative impacts of tourism in city destinations: Conceptual model and strategic framework. *Journal of Tourism Futures*, *3*(2), 144–156. doi:10.1108/JTF-04-2017-0022

Prakash, S. L., Perera, P., Newsome, D., Kusuminda, T., & Walker, O. (2019). Reasons for visitor dissatisfaction with wildlife tourism experiences at highly visited national parks in Sri Lanka. *Journal of Outdoor Recreation and Tourism*, *25*, 102–112. doi:10.1016/j.jort.2018.07.004

Prats, Ll. (1997). *Antropología y patrimonio*. Barcelona, Spain: Ariel.

Prayag, G. (2017). Symbiotic relationship or not? Understanding resilience and crisis management in tourism. *Tourism Management Perspectives*. doi:10.1016/j.tmp.2017.11.012

Preuss, H. (2005). The Economic Impact of Visitors at Major Multi-sport Events. *European Sport Management Quarterly*, *5*(3), 281–301. doi:10.1080/16184740500190710

Project DataHippo. (2019). Retrieved from https://datahippo.org/es/

Publituris Hotelaria. (2018). Alojamento Local teve um impacto económico de 412 milhões de euros em 2018. *Publituris Hotelaria*, 20-12-2018

Puczkó, L., & Smith, M. (2011). Tourism-specific quality-of-life index: the Budapest model. In M. Budruk, & R. Phillips (Eds.), *Quality-of-life community indicators for parks, recreation and tourism management* (Vol. 43, pp. 163–183). The Netherlands: Springer. doi:10.1007/978-90-481-9861-0_9

Puertos. (2017). *Spanish ports exceeded 9.2 million cruise goers in 2017* (p. 2). Retrieved from Spanish Ministery of Public Works and Transport website: http://www.puertos.es/Documents/Notas%20de%20Prensa/NPSeatradeGlobal_02032018[ENG].pdf

Quadros, M. (2017). *Percepções dos residentes sobre os impactos do turismo na comunidade local. 77f. Dissertação (Mestrado em Gestão do turismo internacional). Universidade dos Açores faculdade de economia.* Ponta Delgada, Portugal.

R Development Core Team. (2008). R: A language and environment for statistical computing. R Foundation for Statistical Computing, Vienna, Austria. Retrieved from http://www.R-project.org

Rabadan Formies, M. (2009). La inversión colectiva: concepto y evolución. Sus funciones. Situación actual y perspectivas. *Fundación de Estudios Financieros*, 27-42.

Ramón, A., & Abellán, M. J. (1995). Estacionalidad de la demanda turística en España. *Papers de Turisme*, *17*, 45–73.

Ram, Y., Hall, C., & Michael, W. (2018). Score and tourist accommodation. *International Journal of Tourism Cities*, *4*(3), 369–375.

Rasoolimanesh, S. M., Jaafar, M., Kock, N., & Ramayah, T. (2015). A Revised Framework of Social Exchange Theory to Investigate the Factors Influencing Residents' Perceptions. *Tourism Management Perspectives*, *16*, 335–345. doi:10.1016/j.tmp.2015.10.001

Rees, W. E. (1997). Is sustainable city an oxymoron? *Local Environment*, *2*(3), 303–310. doi:10.1080/13549839708725535

Regoeczi, W. C. (2008). Crowding in context: An examination of the differential responses of men and women to high-density living environments. *Journal of Health and Social Behavior*, *49*(3), 254–268. doi:10.1177/002214650804900302 PMID:18771062

Regueira, M. (2019). *1 in 10 Airbnb users have found hidden cameras in their accommodations.* The computer group.com. Retrieved from https://www.elgrupoinformatico.com/cada-usuarios-airbnb-encontrado-camaras-ocultas-sus-alojamientos-t73574.html

Renda, A. I., Mendes, J. C., & Valle, P. O. (2011). A structural model approach of residents' perception of tourism impacts in their own quality of life: the municipality of Loule, Algarve. In *Proceedings International Conference on Tourism & Management Studies - Algarve, Vol. II.* October 26-29, 2011, Faro, Algarve, Portugal.

Resort Company International [RCI]. (2017). Results 2016. EU. Retrieved from https://www.rci.com

Responsible Tourism. (2019). *Overtourism,* available at http://responsibletourismpartnership .org/ overtourism/ (Accessed on June 10, 2019).

Rezende, C. F., & Rezende, D. D. (2005). *Impactos do turismo: uma análise sob a ótica da população receptora.* Anais do Encontro Nacional de Pós-Graduação e Pesquisa em Administração-ENANPAD.

RIA novosti. (2019, June 8). *PMEF v etom godu posetili bolee 19 tysjach uchastnikov iz 125 stran mira [PMEF was visited this year by more than 19 participants from 125 countries of the world].* Retrieved from https://ria.ru

Richards, G., & Marques, L. (2018). *Creating synergies between cultural policy and tourism for permanent and temporary citizens*. Retrieved from http://www.agenda21culture.net/our-cities/barcelona

Richards, G. (2007). *Cultural Tourism: Global and local perspectives*. Binghamton: Haworth Hospitality Press.

Richards, G. (2011). *Cultural tourism: global and local perspectives*. Abingdon, UK: Routledge.

Richards, G. (2017). The new geographies of tourism: Space, place and locality. In D. Dredge, & S. Gyimóthy (Eds.), *Tourism and Collaborative Consumption: Perspectives, Politics, Policies and Prospects* (pp. 169–184). Springer. doi:10.1007/978-3-319-51799-5_10

Richards, G. (2018). Cultural tourism: A review of recent research and trends. *Journal of Hospitality and Tourism Management, 36*, 12–21. doi:10.1016/j.jhtm.2018.03.005

Richards, G., & Marques, L. (2018). *Creating synergies between cultural policy and tourism for permanent and temporary citizens*. Barcelona, Spain: UCLG Committee.

Richards, G., & Wilson, J. (2004). The impact of cultural events on city image: Rotterdam, cultural capital of Europe 2001. *Urban Studies (Edinburgh, Scotland), 41*(10), 1931–1951. doi:10.1080/0042098042000256323

Richards, G., & Wilson, J. (2006). Developing creativity in tourist experiences: A solution to the serial reproduction of culture? *Tourism Management, 27*(6), 1209–1223. doi:10.1016/j.tourman.2005.06.002

Richardson, D. (2017). Suffering the strain of tourism. Retrieved from: TTG@wtm

Richelieu, A. (2018). A sport-oriented place branding strategy for cities, regions and countries. *Sport, Business and Management, 8*(4), 354–374. doi:10.1108/SBM-02-2018-0010

Rifkin, J. (2000). *The era of access. The revolution of the new economy*. Spain: Paidós.

Ritchie, B., & Adair, D. (2009). The Growing Recognition of Sport Tourism. *Current Issues in Tourism, 5*(1), 1–6. doi:10.1080/13683500208667903

Rivano, H., Augé-Blum, I., Bechkit, W., Boussetta, K., Fiore, M., Stanica, R., & Valois, F. (2018). Wireless Access Networks for Smart Cities. In I. Management Association (Ed.), Smart Technologies: Breakthroughs in Research and Practice (pp. 476-507). Hershey, PA: IGI Global. doi:10.4018/978-1-5225-2589-9.ch022

RNAL [Registo Nacional de Turismo]. (2019, 2018). *Alojamento Local - Registo Nacional de Turismo*. Retrieved from https://tdpindicadores.maps.arcgis.com/apps/MapSeries/index.html?appid=d2e682edee1248a79d15b7b43a68ca2d

Roberts, D. (2017, December 1). *Wealthier people produce more carbon pollution — even the "green" ones*. (Vox, Producer, & Vox Media, Inc.) Retrieved from Vox: https://www.vox.com/energy-and-environment/2017/12/1/16718844/green-consumers-climate-change

Robertson, R. (1995). Glocalization: Time - Space and Homogenity - Heterogeneity. En M. Featherstone, S. Lash, & R. Robertson (Eds.), Global Modernities (pp. 25-44). London, UK: Sage.

Roche, S., Spake, D., & Joseph, M. (2013). A model of sporting event tourism as economic development. *Sport, Business, and Management, 3*(2), 147–157. doi:10.1108/20426781311325078

Rodríguez López, J. (2017). Las viviendas que pudieron hundir la economía española. La caída del mercado de vivienda y sus consecuencias. *Cuadernos de Relaciones Laborales*, 71-99.

Rodríguez López, J. (2007). Urbanismo, vivienda y economía en España. *Arquitectura. Ciudad y Entorno*, *3*, 80–85.

Rodríguez-Salmones, N. A. (2003). El alojamiento turístico privado en el contexto del sistema de estadísticas del turismo. *Estudios Turísticos*, n.° 155-156, 7-32.

Roland Berger/ERTRL/TL. (2014). *The Strategic Plan for Tourism in the Lisbon Region 2015–2019* [Plano Estratégico para o Turismo na Região de Lisboa 2015-2019 - Posicionar a Região de Lisboa num novo patamar de excelência turística]. Retrieved from http://ertrl/TL%20

Romera, J., Brualia, A., & García, C. S. (2017, May 29). *Airbnb faces a wave of fraud due to false floors.* Online newspaper El Economisa. Retrieved from https://www.eleconomista.es/vivienda/noticias/8390344/05/17/airbnb-afronta-una-oleada-de-fraudes-con-pisos-falsos.html

Romero, J., Jiménez, F., & Villoria, M. (2012). (Un) sustainable territories: Causes of the speculative bubble in Spain (1996–2010) and its territorial, environmental, and socio-political consequences. *Environment and Planning. C, Government & Policy*, *30*(3), 467–486. doi:10.1068/c11193r

Rončák, M. (2019). Prague and the impact of low-cost airlines. In R. Dodds, & R. W. Butler (Eds.), *Overtourism. Issues, realities, and solutions* (pp. 152–1638). Berlin, Germany: De Gruyter. doi:10.1515/9783110607369-011

Ros, M. (11 de 05 de 2012). *El Mercado Central estudia cómo cobrar a los turistas por emplear los aseos.* Retrieved May 05, 2019 from Levante-EMV: https://www.levante-emv.com/valencia/2012/05/11/mercado-central-estudia-cobrar-turistas-aseos/904208.html

Rosen, E. (2018, September 6). *New Rankings of The World's Fastest-Growing Tourism Destinations.* (Forbes Media LLC.) Retrieved from Forbes: https://www.forbes.com/sites/ericrosen/2018/09/06/new-rankings-of-the-worlds-fastest-growing-tourism-destinations/#725d17f457ea

Rosselló, J., & Saenz-de-Miera, O. (2011). Road accidents and tourism: The case of the Balearics Islands. *Accident; Analysis, and Prevention*, *43*(3), 675–683. doi:10.1016/j.aap.2010.10.011 PMID:21376854

Rosselló, J., & Sansó, A. (2017). Yearly, monthly and weekly seasonality of tourism demand: A decomposition analysis. *Tourism Management*, *60*, 379–389. doi:10.1016/j.tourman.2016.12.019

Routledge, P. (2001). 'Selling the rain' resisting the sale: Resistant identities and the conflict over tourism in Goa. *Social & Cultural Geography*, *2*(2), 221–240. doi:10.1080/14649360120047823

Royle, S. A. (2009). Tourism changes on a Mediterranean island: Experience from Mallorca. *Island Studies Journal*, *4*(2), 225–240.

Royo, M., & Ruiz, M. E. (2009). Actitud del residente hacia el turismo y el visitante: Factores determinantes en el turismo y excursionismo rural-cultural. *Cuadernos de Turismo*, *23*, 217–236.

Royo, M., & Ruíz, M. E. (2009). Actitud del residente hacia el turismo y el visitante: Factores determinantes en el turismo y excursionismo rural-cultural. *Cuadernos de Turismo*, *23*, 217–236.

Ruiz Baudrihaye, J.-A. (1997). El turismo cultural: Luces y sombras. *Estudios Turísticos*, 43–54.

Russo, A. P. (2002). The "vicious circle" of tourism development in heritage cities. *Annals of Tourism Research*, *29*(1), 165–182. doi:10.1016/S0160-7383(01)00029-9

Ryan, C., & Montgomery, D. (1994). The Attitude of Bakewell Residents to Tourism and Issues in Community Response Tourism. *Tourism Management*, *15*(5), 358–369. doi:10.1016/0261-5177(94)90090-6

Saenz-de-Miera, O., & Rosselló, J. (2014). Modeling tourism impacts on air pollution: The case study of PM10 in Mallorca. *Tourism Management*, *40*(C), 273–281. doi:10.1016/j.tourman.2013.06.012

Sakshi, P. (2017). Do social media effect tourists' behaviour? *International Journal of Advance Research and Innovative Ideas in Education, 3*(4), 38-45.

Saladié, O. A.-J. (2014). La influencia de las rutas de vuelos de bajo coste en la elección del destino turístico. *Cuadernos de Turismo*, *34*, 287–312.

Salas-Olmedo, H., Moya-Gómez, B., García-Palomares, J. C., & Gutiérrez, J. (2018). Tourists' digital footprint in cities: Comparing big data sources. *Tourism Management*, *66*, 13–25. doi:10.1016/j.tourman.2017.11.001

Sánchez Rivero, M., & Pulido Fernández, I. (2008). *Medida de la sostenibilidad turística. Propuesta de un índice sintético*. Madrid, Spain: Editorial Universitaria Ramón Areces.

Sánchez, R., & Ordaz, A. (in press 2018). ¿Cuántas viviendas de tu barrio están en Airbnb? *Eldiario.es*

Sanchez, P. M., & Adams, K. M. (2008). The Janus-faced character of tourism in Cuba. *Annals of Tourism Research*, *35*(1), 27–46. doi:10.1016/j.annals.2007.06.004

Sánchez-Quiles, D., & Tovar-Sánchez, A. (2015, October). Are sunscreens a new environmental risk associated to coastal tourism? *Environment International*, *83*, 158–170. doi:10.1016/j.envint.2015.06.007 PMID:26142925

Sanfelici, D., & Halbert, L. (2018). Financial market actors as urban policy-makers: The case of real estate investment trusts in Brazil. *Urban Geography*, *40*(1), 83–103. doi:10.1080/02723638.2018.1500246

Sans, A. A., & Quaglieri, A. (2016). Unravelling Airbnb: Urban Perspectives from Barcelona. In *G. R. A. P. Russo (Ed.), Reinventing the local in tourism: Producing, consuming and negotiating place* (pp. 209–228). New York: Channel View Publications.

Santamarina Campos, B. (2005). Una aproximación al patrimonio cultural. En G.-M. Hernànez i Martí, B. Santamarina Campos, A. Moncusí Ferré, & M. Albert Rodrigo, La memoria construida (pp. 21-51). Valencia: Tirant lo Blanch.

Santana Talavera, A. (2003). Turismo cultural, culturas turísticas. *Horizontes Antropológicos*, *20*(9), 31–57. doi:10.1590/S0104-71832003000200003

Santiago, C. (2019). *Trail running: modelo e potencial territorial enquanto produto turístico* (Unpublished Master's thesis). Escola Superior de Turismo e Hotelaria do Estoril. Retrieved from https://comum.rcaap.pt/bitstream/10400.26/18234/1/2016.04.017_.pdf

Sarantakou, E., & Terkenli, T. S. (2019). Non-Institutionalized Forms of Tourism Accommodation and Overtourism Impacts on the Landscape: The Case of Santorini, Greece. *Tourism Planning & Development*, *16*(4), 411–433. doi:10.1080/21568316.2019.1569122

Sastre, B. (2012). Linguistic integration of the resident German population in Mallorca. *Treballs de Sociolinguística Catalana*, *22*, 249–269.

Saura, J. R., Palos-Sánchez, P. R., & Reyes-Menéndez, A. (2017). Marketing a través de aplicaciones móviles de turismo (m-tourism): Un estudio exploratorio. *International Journal of World of Tourism*, *4*(8), 45–56. doi:10.12795/IJWT.2017.i08.04

Saveriades, S. (2000). Establishing the social tourism carrying capacity for the tourist resorts of the east coast of the Republic of Cyprus. *Tourism Management*, *21*(2), 147–156. doi:10.1016/S0261-5177(99)00044-8

Scalabrini, E., Remoaldo, P., & Lourenço, J. (2014). Percepções de residentes a respeito dos impactos da atividade turística: Uma análise das publicações brasileiras sobre o tema. *Tourism and Hospitality International Journal, 2*(2), 12–31.

Schäfer, P., & Hirsch, J. (2017). Do urban tourism hotspots affect Berlin housing rents? *International Journal of Housing Markets and Analysis, 10*(2), 231–255. doi:10.1108/IJHMA-05-2016-0031

Schianetz, K., & Kavanagh, L. (2008). Sustainability indicators for tourism destinations: A complex adaptive systems approach using systemic indicator systems. *Journal of Sustainable Tourism, 16*(6), 601–628. doi:10.1080/09669580802159651

Schmidt, C. E. (2013). *Reducing muscular fatigue in trail running mechanisms and strategies*, Unpublished Master's thesis). The University of Sydney & Université Nice Sophia Antipolis.

Schoenberger, E. (1988). From Fordism to Flexible Accumulation: Technology, Competitive Strategies and International Location. *Environment and Planning. D, Society, & Space, 6*(3), 245–262. doi:10.1068/d060245

Secretaría de Turismo del estado de Jalisco [SECTURJAL]. (1994). *Statistical Yearbook 1993.* Mexico. Gobierno de Jalisco.

Secretaría de Turismo del estado de Jalisco [SECTURJAL]. (2017). *Statistical Yearbook 2016.* Mexico. Gobierno de Jalisco.

Segittur. (2015). *Informe destinos turísticos inteligentes: construyendo el futuro.* Retrieved from: https://www.segittur.es

Sennett, R. (2018). *Building and Dwelling: Ethics for the City.* New York: Farrar, Straus, and Giroux.

SEPE. (2018). Informe del Mercado de Trabajo Estatal. Datos 2017. *Catálogo general de publicaciones de la Administración General del Estado* http://publicacionesoficiales.boe.es, 148. Retrieved from http://www.sepe.es/contenidos/observatorio/mercado_trabajo/3017-1.pdf

Sequera, J., & Nofre, J. (2018). *Shaken, not stirred.* City; doi:10.1080/13604813.2018.1548819

Seraphin, H., Sheeran, P., & Pilato, M. (2018). Over-tourism and the fall of Venice as a destination. *Journal of Destination Marketing and Management, 9*(September 2017), 374–376.

Séraphin, H., Gowreesunkar, V., Zaman, M., & Bourliataux-Lajoin, S. (2019). Community based festivals as a tool to tackle tourismphobia and antitourism movements. *Journal of Hospitality and Tourism Management, 39*, 219–223. doi:10.1016/j.jhtm.2018.12.001

Seraphin, H., Sheeran, P., & Pilato, M. (2018). Over-tourism and the fall of Venice as a destination. *Journal of Destination Marketing & Management, 9*, 374–376. doi:10.1016/j.jdmm.2018.01.011

Séraphin, H., Yallop, A., Capatina, A., & Gowreesunkar, V. (2018). Heritage in tourism organisations' branding strategy: The case of a post-colonial, post-confiict and post disaster destination. *International Journal of Culture, Tourism and Hospitality Research, 12*(1), 89–105. doi:10.1108/IJCTHR-05-2017-0057

Séraphin, H., Zaman, M., Olver, S., Bourliataux-Lajoinied, S., & Dosquete, F. (2019). Destination branding and overtourism. *Journal of Hospitality and Tourism Management, 38*, 1–4. doi:10.1016/j.jhtm.2018.11.003

Servihabitat. (2018). *Mercado residencial en España. Segundo semestre 2017.* Barcelona, Spain: Caixabank.

Shafiee, G., & Hasanzadeh, J. (2019). Developing a model for sustainable smart tourism destinations: A systematic review. *Tourism Management Perspectives, 31*, 287–300. doi:10.1016/j.tmp.2019.06.002

Sharpley, R. (2014). Host Perceptions of Tourism: A Review of the Research. *Tourism Management, 42*, 37–49. doi:10.1016/j.tourman.2013.10.007

Shea, B. J., Grimshaw, J. M., Wells, G. A., Boers, M., Andersson, N., Hamel, C., ... Bouter, L. M. (2007). Development of AMSTAR: A measurement tool to assess the methodological quality of systematic reviews. *BMC Medical Research Methodology*, *7*(1), 10. doi:10.1186/1471-2288-7-10 PMID:17302989

Shelby, B., Heberlein, T. H., Vaske, J. J., & Alfano, G. (1983). Expectations, preferences, and feeling crowded in recreation activities. *Leisure Sciences*, *11*, 269–291. doi:10.1080/01490408909512227

Sheng, L., Li, T., & Wang, J. (2017). Tourism and externalities in an urban context: Theoretical model and empirical evidence. *Cities (London, UK)*, *70*, 40–45. doi:10.1016/j.cities.2017.06.012

Sheppard, V., & Williams, P. (2016). Factors that strengthen tourism resort resilience. *Journal of Hospitality and Tourism Management*, *28*, 20–30. doi:10.1016/j.jhtm.2016.04.006

Shin, H. J., Kim, H. N., & Son, J. Y. (2017). Measuring the economic impact of rural tourism membership on local economy: A Korean case study. *Sustainability*, *9*(4), 639. doi:10.3390u9040639

Shoval, N. (2018). Urban planning and tourism in European cities. *Tourism Geographies*, *20*(3), 371–376. doi:10.1080/14616688.2018.1457078

Simancas Cruz, M., & Zaragoza, M. P. P. (2019). Analysis of the Accommodation Density in Coastal Tourism Areas of Insular Destinations from the Perspective of Overtourism. *Sustainability*, *3031*(11), 3031. doi:10.3390u11113031

Simancas, M., & Peñarrubia, M. P. (2019). Analysis of the Accommodation Density in Coastal Tourism Areas of Insular Destinations from the Perspective of Overtourism. *Sustainability, MDPI. Open Access Journal*, *11*(11), 1–19.

Simpson, F. (1999). Tourism impact in the historic center of Prague: Resident and visitor perceptions of the historic built environment. *The Geographical Journal*, *165*(2), 173–183. doi:10.2307/3060415

Singh, S., Dash, T., & Vashko, I. (2016). Tourism, ecotourism and sport tourism: The framework for certification. *Marketing Intelligence & Planning*, *34*(2), 236–255. doi:10.1108/MIP-09-2014-0180

Sirgy, M. J., & Lee, D. J. (2006). Macro measures of consumer well-being (CWB): A critical analysis and a research agenda. *Journal of Macromarketing*, *26*(1), 27–44. doi:10.1177/0276146705285669

Skeli, S., & Schmid, M. (2019). Mitigating overtourism with the help of smart technology solutions–a situation analysis of European city destinations. In *ISCONTOUR 2019 Tourism Research Perspectives: Proceedings of the International Student Conference in Tourism Research*, 13. Norderstedt, Germany: BoD–Books on Demand.

Slater, T. (2006). The eviction of critical perspectives from gentrification research. *International Journal of Urban and Regional Research*, *32*(1), 192–194.

Slee, T. (2017). *Yours is mine. Against the collaborative economy*. Mexico: Taurus. doi:10.2307/j.ctv62hf03

Smeral, E. (2019). Overcrowding of tourism destinations: Some suggestions for a solution. In H. Pechlaner, E. Innerhofer, & G. Erschbamer (Eds.), *Overtourism: Tourism Management and Solutions* (pp. 163–173). Routledge. doi:10.4324/9780429197987-12

Smith, M., Sziva, I., & Olt, G. (2019). Overtourism and Resident Resistance in Budapest. *Tourism Planning & Development*, *16*(4), 376–392. doi:10.1080/21568316.2019.1595705

Smith, N. (1979). Toward a theory of gentrification. A back to the city movement by capital, not people. *Journal of the American Planning Association*, *45*(4), 538–548. doi:10.1080/01944367908977002

Smith, N. (2002). New globalism, new urbanism: Gentrification as global urban strategy. *Antipode*, *34*(3), 427–450. doi:10.1111/1467-8330.00249

Snaith, T., & Haley, A. (1999). Residents' opinions of tourism development in the historic city of York, England. *Tourism Management, 20*(5), 595–603. doi:10.1016/S0261-5177(99)00030-8

Soares, N., Vairiinhos, M., Beça, P., & Ribeiro, T. F. (2019). Real-Time Notifications: Marketing Applied to the Internet of Things. In D. Simões, B. Barbosa, & S. Filipe (Eds.), *Smart Marketing with the Internet of Things* (pp. 114–122). Hershey, PA: IGI Global; doi:10.4018/978-1-5225-5763-0.ch006

Social Barometer. (2019). *Net disposable household income per capita.* Retrieved from https://barometrosocial.es/renta-y-patrimonio

Sociedad de Tasación. (2018). *El perfil del comprador de vivienda 2018*, Informe planner exhibitions - st sociedad de tasación Octubre 2018.

Soja, E. W. (1989). *Postmodern geographies: The reassertion of space in critical social theory.* London, UK: Verso.

Soldino Nicolás, T. (2004). Patrimonio y Sociedad. Acción ciudadana y compromiso por el patrimonio. En T. Soldino Nicolás, M. Delgado, & A. Montiel Márquez, Patrimoni i Societat (pp. 35-59). Valencia: Universitat de València.

Soliman, D. (2010). Managing Visitors Via Demarketing in The Egyptian World Heritage Site: Giza Pyramids. *Journal of Association of Arab Universities for Tourism and Hospitality*, *7*(1), 15–20.

Sommer, C., & Helbrecht, I. (2017). Seeing like a tourist city: how administrative constructions of conflictive urban tourism shape its future. *Journal of Tourism Futures, 3*(2), 157-170.

Sommer, C., & Helbrecht, I. (2017). Seeing like a tourist city: How administrative constructions of conflictive urban tourism shape its future. *Journal of Tourism Futures*, *3*(2), 157–170. doi:10.1108/JTF-07-2017-0037

Sorando, D., & Ardura, A. (2018). Procesos y dinámicas de gentrificación en las ciudades españolas. *Papers. Regió Metropolitana de Barcelona. Territori, estratègies, planejament*, 60, 35–47. Retrieved from https://www.raco.cat/index.php/PapersIERMB/article/view/339239/430199

Sotiriadou, P., Le, A., & Brouwers, J. (2014). Choosing a qualitative data analysis tool: A comparison of NVivo and Leximancer. *Annals of Leisure Research*, *17*(2), 218–234. doi:10.1080/11745398.2014.902292

Spanish Tax Agency. (2019). *Statistics of IRPF declarants by municipalities,* retrieved from https://www.agenciatributaria.es/AEAT.internet/datosabiertos/catalogo/hacienda/Estadistica_de_los_declarantes_del_IRPF_por_municipios.shtml

Stefanidaki, E., & Lekakou, M. (2014). Cruise carrying capacity: A conceptual approach. *Research in Transportation Business & Management*, *13*, 43–52. doi:10.1016/j.rtbm.2014.11.005

Steven, R., Pickering, C., & Castley, J. G. (2011). A review of the impacts of nature-based recreation on birds. *Journal of Environmental Management*, *92*(10), 2287–2294. doi:10.1016/j.jenvman.2011.05.005 PMID:21640470

STHDA. (2018). *Kruskal-Wallis test in R*, available at http://www.sthda.com/english/ wiki/kruskal-wallis-test-in-r (Accessed on Jun 2, 2018).

Stiglitz, J. (2010). *Caída libre*. Madrid, Spain: Taurus.

Straaten, J. V. (2000). Can sustainable tourism positively influence rural regions? In G. Richards, & D. Hall (Eds.), *Tourism and Sustainable Community Development* (pp. 221–232). London, UK: Routledge.

Street, F. (2018, October 3). *Can the world be saved from overtourism?* Available at https://edition.cnn.com/travel/article/overtourism-solutions/index.html (Accessed on June 10, 2019).

Su, M. M., & Wall, G. (2013). Community involvement at Great Wall World Heritage sites. Beijing, China. *Current Issues in Tourism, 18*(2), 137–157. doi:10.1080/13683500.2013.823917

Swarbooke, J. (1999). *Sustainable Tourism Management*. London, UK: CABI. doi:10.1079/9780851993140.0000

Tague-Sutcliffe, J. (1992). An introduction to infometrics. *Information Processing & Management, Oxford, 28*(1), 1–3.

Tao, K., Li, X., Zhou, Q., Moher, D., Ling, C., & Yu, W. (2011). From QUOROM to PRISMA: A Survey of High-Impact Medical Journals' Instructions to Authors and a Review of Systematic Reviews in Anesthesia Literature. *PLoS One, 6*(11), 1–5. doi:10.1371/journal.pone.0027611 PMID:22110690

Tasaciones InmobiliariasS. A. (2019). Retrieved from .https://www.tinsa.es/precio-vivienda

Taylor, I. (2018). *Overtourism 'highest risk' warns Macao chief*, available at http://www.travelweekly.co.uk/articles/304025/over-tourism-highest-risk-warns-macao-chief (Accessed on June 1, 2019).

TCV. (2019). *Afrontando los retos del overtourism a través de open/big data. Proyecto Territorial Integrado "HERIT-DATA"*. Retrieved from http://www.turisme.gva.es/turisme/es/files/pdf/HD_Informe_ejecutivo_BENCHMARKING.pdf

Templier, M., & Pare, G. (2015). A Framework for Guiding and Evaluating Literature Reviews. *Communications of the Association for Information Systems, 37*(6), 112–137.

Teye, V., Sönmez, S. F., & Sirakaya, E. (2002). Residents' attitudes toward tourism development. *Annals of Tourism Research, 29*(3), 668–688. doi:10.1016/S0160-7383(01)00074-3

The official site of St. Petersburg Authorities. (2018, October 25). *V Peterburge podveli itogi navigatsii 2018 goda [Marine navigation in St. Petersburg in 2018 was assessed]*. Retrieved from https://www.gov.spb.ru/gov/otrasl/c_tourism/news/149312/

The State Duma of the Russian Federation. (2019, April 4). *Zakon o hostelah prinjat v okonchatel'noj redaktsii [The law on hostels has been ratified in its most recent iteration]*. Retrieved from http://duma.gov.ru/news/30244/

Þórhallsdóttir, G., & Ólafsson, R. (2017). A method to analyse seasonality in the distribution of tourists in Iceland. *Journal of Outdoor Recreation and Tourism, 19*, 17–24. doi:10.1016/j.jort.2017.05.001

Timothy, D. (2007). Empowerment and stakeholder participation in tourism destination communities. In A. Church, & T. Coles (Eds.), Tourism, Power, and Space, 199-216. London: Routledge.

Timothy, D. J. (2011). *Cultural heritage and tourism: an introduction*. Bristol: Channel View Publications.

Tinsa. (2019). *Interactive Imie, The price of housing in Spain*. Retrieved from https://www.tinsa.es/precio-vivienda/

Titman, S. (1985). Urban land prices under uncertainty. *The American Economic Review, 75*, 505–514.

TL [Turismo de Lisboa]. (2016). *Site do Turismo de Lisboa. Lisbon Tourism Survey*. Retrieved from http://www.visitlisboa.com/see-and-do/heritage-and-culture

TL OTL [Turismo de Lisboa /Observatório de Turismo de Lisboa]. (2018). *Inquérito Motivacional 2018 – Cidade de Lisboa*.

TL OTL [Turismo de Lisboa /Observatório de Turismo de Lisboa]. (2019). *Inquérito Motivacional 2018 – Cidade de Lisboa*. Retrieved from https://www.visitlisboa.com

TL OTL. (2018). *Inquérito Motivacional – Cidade de Lisboa*. Retrieved from https://www.visitlisboa.com

TL OTL. (2018). *The Satisfaction and Image Survey - 2018*. Retrieved from https://www.visitlisboa.com

TL. (2015 a). *Visitor Activities and Information Survey – Lisbon Region - 2015*. Observatório de Turismo de Lisboa. Retrieved from http://www.visitlisboa.com/sites/default/files/2016-10/VISITOR%20ACTIVITIES%20AND%20INFORMATION%20SURVEY%202015.pdf

TL. (2015). *Satisfaction and Image Survey - 2015 – Lisbon Region*. Observatório de Turismo de Lisboa. Retrieved from http://www.visitlisboa.com/sites/default/files/2016-10/SATISFACTION%20AND%20IMAGE%20SURVEY%202015.pdf

TomTom. (2019). *Traffic congestion ranking | TomTom Traffic Index [Index]*. Retrieved from https://tomtom.com/en_gb/traffic-index/ranking/

Torre, O. (1992). *El Turismo, Fenómeno Social* (1st ed.). Mexico: Fondo de Cultura Económica.

Torres-Delgado, A., & López Palomeque, F. (2014). Measuring sustainable tourism at the municipal level. *Annals of Tourism Research*, *49*, 122–137. doi:10.1016/j.annals.2014.09.003

Torres-Delgado, A., & Saarinen, J. (2013). Using indicators to assess sustainable tourism development: A review. *Tourism Geographies*, *16*(1), 31–47. doi:10.1080/14616688.2013.867530

Tourism, R. (n.d.). *OverTourism*. Retrieved from http://responsibletourismpartnership.org/over tourism.

TourMIS. (n.d.). Retrieved from https://www.tourmis.info/index_e.html

TP/Governo de Portugal. (2015). *Turismo 2020: Plano de Ação para o Desenvolvimento do Turismo em Portugal*. Retrieved from http://www.turismodeportugal.pt/Portugu%C3%AAs/turismodeportugal/apresentacoes/Documents/Turismo-2020-Sintra.pdf

Trancoso, A. V. G. (2018). The problem of overtourism and the impact of cruises, Investigaciones Regionales. *Journal of Regional Research*, *42*, 35–51.

Treserras, J. J. (2004). La tematización cultural de las ciudades como estrategia de desarrollo a través del turismo. *Portal Iberoamericano de Gestión Cultural*, 1-19.

TripAdvisor. (2018). *Mallorca Cathedral*. Retrieved from: https://www.tripadvisor.com/Attraction_Review-g187463-d244037-Reviews-or10-Catedral_de_Mallorca-Palma_de_Mallorca_Majorca_Balearic_Islands.html

Troitiño Vinuesa, M. (2005). Retos y oportunidades en la gestión del turismo cultural y de ciudad. En M. Gómez Borja, J. Mondéjar Jiménez, & C. Sevilla Sevilla (Eds.), Gestión del turismo cultural y de ciudad (pp. 15-27). Cuenca: Ediciones de la Universidad de Castilla-La Mancha.

Tscheu, F., & Buhalis, D. (2016). Augmented Reality at Cultural Heritage sites. In A. Inversini, & R. Schegg (Eds.), *Information and Communication Technologies in Tourism 2016*. Cham, Switzerland: Springer. doi:10.1007/978-3-319-28231-2_44

Tulumello, S. (2015). Reconsidering neoliberal urban planning in times of crisis: Urban regeneration policy in a "dense" space in Lisbon. *Urban Geography*. doi:10.1080/02723638.2015.1056605

Turrión-Prats, J., & Duro, J. A. (2018). Tourist seasonality and the role of markets. *Journal of Destination Marketing & Management*, *8*, 23–31. doi:10.1016/j.jdmm.2016.11.004

UNEP & UNWTO. (2005). *Making Tourism More Sustainable - A Guide for Policy Makers*, p. 11-12. Retrieved from: http://www.unep.fr/shared/publications/pdf/DTIx0592xPA-TourismPolicyEN.pdf

UNESCO. (2005). *Mostar, Macao and Biblical vestiges in Israel are among the 17 cultural sites inscribed on UNESCO's World Heritage List*, available at http://whc.unesco.org/en/news/135/ (Accessed on March 15, 2018).

United Nations, World Tourism Organization, (UNWTO). (2017). *Yearbook of Tourism Statistics. Madrid, UNWTO* (p. 15). Madrid, Spain: Calle Poeta Joan Maragall.

United Nations. World Tourism Organization. (1994). Recommendations on Tourisme Statistics, Serie M, nº 83, United Nations, New York.

UNWTO - World Tourism Organization. (2001). *Global Code of Ethics for Tourism*. Madrid, Spain: UNWTO.

UNWTO - World Tourism Organization. (2019). Overtourism? – Understanding and Managing Urban Tourism Growth beyond Perceptions: Vol. 2. *Case Studies*. Madrid, Spain: UNWTO.

UNWTO Committee on Tourism Statistics and TSA. (2018). *Statistical Framework for Measuring Sustainable Tourism: Consultation draft*. UNWTO.

UNWTO Communications Department. (2019, January 21). *International Tourist Arrivals Reach 1.4 billion Two Years Ahead of Forecasts.* Retrieved from UNWTO: https://www2.unwto.org/press-release

UNWTO. (1981). *Saturation of Tourist Destinations*. Madrid, Spain: General Secretary, World Tourism Organisation.

UNWTO. (1981). *Saturation of Tourist Destinations. Report of the Secretary General*. Madrid, Spain: World Tourism Organisation.

UNWTO. (2017). UNWTO & WTM Minister' Summit Overtourism: "growth is not the enemy, it is how we manage it". Retrieved October 5, 2019, from https://www2.unwto.org/unwto-wtm-ministers-summit-2017

UNWTO. (2018). '*Overtourism'? Understanding and Managing Urban Tourism Growth beyond Perceptions,* available at https://www.e-unwto.org/doi/pdf/10.18111/9789284420070 (Accessed on May 5, 2019).

UNWTO. (2019). *International Tourism Highlights 2019*. Edition UNWTO (2019) International Tourism Highlights 2019 Edition.

Urban, T. U. R. (2016). *Monitor de competitividad turística de los destinos urbanos españoles*. Retrieved from https://www.exceltur.org/urbantur/#

Urrestarazu-Capellán. (2019). *Los factores determinantes de la oferta y precio de la vivienda en alquiler: una visión desde el turismo* (Unpublished doctoral dissertation). University of Málaga.

Urry, J. (1990). *The Tourist Gaze. Leisure and Travel in Contemporary Societies*. London, UK: Sage.

Urtasun, A., & Gutierrez, I. (2006). Tourism agglomeration and its impact on social welfare: An empirical approach to the Spanish case. *Tourism Management*, *27*(5), 901–912. doi:10.1016/j.tourman.2005.05.004

Uysal, M., Sirgy, M. J., Woo, E., & Kim, H. (2016). Quality of life (QoL) and well-being research in tourism. *Tourism Management*, *53*, 244–261. doi:10.1016/j.tourman.2015.07.013

Valencia, A. V. (Dirección). (2005). *Mercado Central. Mil años de historia* [Película].

Valencia, A. G. (2017). *El eje económico de la sostenibilidad turística*. México.

Valenciana, G. T. (2019). Afrontando los retos del overtourism en destinos turísticos patrimoniales y naturales a través de open/big data. Valencia: Generalitat Valenciana.

Vallés, M. (2017). No es turismofobia, es turismofilia, *Diario de Mallorca*, 2 de agosto del 2017.

van der Borg, J. (1998). Tourism management in Venice, or how to deal with success. In D. Tyler, Y. Guerrier, & M. Robertson (Eds.), *Managing tourism in cities. Policy, Process, and Practice*. West Sussex: John Wiley & Sons.

van Doren, C. S., Koh, Y. K., & McCahill, A. (1994). Tourism research: a state-of-the art citation analysis (1971 -1990). In A. V. Seaton (Ed.), Tourism: The state of the art (308-315). Chichester, UK: John Wiley &Sons.

Van Dyck, D., Cardon, G., De Bourdeaudhuij, I., De Ridder, L., & Willem, A. (2017). Who Participates in Running Events? Socio-Demographic Characteristics, Psychosocial Factors and Barriers as Correlates of Non-Participation—A Pilot Study in Belgium. *International Journal of Environmental Research and Public Health*, *14*(11), 1315. doi:10.3390/ijerph14111315 PMID:29143780

Van Niekerk, M. (2016). Community perceptions on the impacts of art festivals and its impact on overall quality of life: A case study of the innibos National Art festivals, South Africa. In C. Newbold, & J. Jordan (Eds.). Focus of world festivals: Contemporary case studies and perspectives, 152–163. Oxford: Goodfellow Publishers.

Vargas-Sánchez, A., Oom-do-Valle, P., da-Costa-Mendes, J., & Silva, J. A. (2015). Residents' Attitude and Level of Destination Development: An International Comparison. *Tourism Management*, *48*, 199–210. doi:10.1016/j.tourman.2014.11.005

Varma, A., Jukic, N., Pestek, A., Shultz, C., & Nestorov, S. (2016). Airbnb: Exciting innovation or passing fad? *Tourism Management Perspectives*, *20*, 228–237. doi:10.1016/j.tmp.2016.09.002

Veiga, C., Santos, M. C., Águas, P., & Santos, J. A. C. (2018). Sustainability as a key driver to address challenges. *Worldwide Hospitality and Tourism Themes*, *10*(6), 662–673. doi:10.1108/WHATT-08-2018-0054

Velasco, M. (2016). Dynamics of Spanish Tourism Policy: The Political System as a Driver and Policy Instruments as Indicators of Change (1952–2015). *International Journal of Tourism Policy*, *6*(3/4), 256–272. doi:10.1504/IJTP.2016.081526

Vera, J. F. (2001). Planificación y gestión del desarrollo turístico sostenible: Propuestas para la creación de un sistema de indicadores. Alicante: Instituto Universitario de Alicante.

Vidal-Folch, I. (2018). *La «fobia» al turismo en Barcelona* | Opinion Home | EL MUNDO. Elmundo. Recuperado de https://www.elmundo.es/opinion/2017/08/08/5988933a268e3e15708b45be.html

Vives-Miró, S., & Rullan, O. (2017). ¿Desposesión de vivienda por turistización? Revalorización y desplazamientos en el Centro Histórico de Palma (Mallorca). *Revista de Geografía Norte Grande*, *67*(67), 53–71. doi:10.4067/S0718-34022017000200004

Wachsmuth, D., & Weisler, A. (2018). Airbnb and the rent gap: Gentrification through the sharing economy. *Environment and Planning A. Economy and Space*, *50*(6), 1147–1170.

Wagar, J. A. (1964). The carrying capacity of wild lands for recreation. *Forest Science*, *10*, a0001–a0024.

Wagar, J. A. (1964). The Carrying Capacity of wild lands for recreation. *Forest Science*, 1–24.

Waligo, V. M., Clarke, J., & Hawkins, R. (2013). Implementing sustainable tourism: A multi-stakeholder involvement management framework. *Tourism Management*, *36*, 342–353. doi:10.1016/j.tourman.2012.10.008

Wall, G. (1997). Is ecotourism sustainable? *Environmental Management*, *21*(4), 483–491. doi:10.1007002679900044 PMID:9175538

Wall, G., & Yan, M. (2003). Disaggregating visitor flows - the example of China. *Tourism Analysis*, *7*(3/4), 191–205.

Walmsley, A. (2017). *Overtourism and underemployment: a modern labour market dilemma*. Paper presented at Responsible Tourism in Destinations 13 – Tackling Overtourism – Local Responses, September 29-30, 2017 – Hannesarholt, Reykjavik, Iceland.

Wang, D., Li, X., & Li, Y. (2013). China's "smart tourism destination" initiative: A taste of the service-dominant logic. *Journal of Destination Marketing & Management*, *2*(2), 59–61. doi:10.1016/j.jdmm.2013.05.004

Wan, Y. K. P., & Li, X. (2013). Sustainability of tourism development in Macao, China. *International Journal of Tourism Research*, *15*(1), 52–65. doi:10.1002/jtr.873

Watson, G. L., & Kopachevsky, J. P. (1996). Tourist carrying capacity: A critical look at the discursive dimension. *Progress in Tourism and Hospitality Research*, *2*(2), 169–179. doi:10.1002/(SICI)1099-1603(199606)2:2<169::AID-PTH35>3.0.CO;2-R

Wearing, S., Schweinsberg, S., & Tower, J. (2016). *Marketing national parks for sustainable tourism*. Channel View Publications. doi:10.21832/9781845415594

Weaver, D. B., & Lawton, L. J. (2007). Twenty Years On: The State of Contemporary Ecotourism Research. *Tourism Management*, *28*(5), 1168–1179.

Weber, F., Stettler, J., Priskin, J., Rosenberg-Taufer, B., Ponnapureddy, S., Fux, S., . . . Barth, M. (2017). Tourism destinations under pressure. Challenges and innovative solutions. Lucerne, Switzerland.

Weber, J. (2014). Augmented Reality Gaming: A new Paradigm for Tourist Experience. In *Information and communication technologies in tourism, eProceedings of the ENTER 2014 PhD Workshop* (pp. 57-67), Dublin, Ireland: Springer.

Weber, F., Stettler, J., Priskin, J., Rosenberg-Taufer, B., Ponnapureddy, S., Camp, M.-A., & Bath, M. (2017). *Tourism destinations under pressure: Challenges and Innovative solutions*. Institute of Tourism ITW, Lucerne University of Applied Sciences and Arts, Roesslimatte.

Weber, F., Stettler, J., Priskin, J., Rosenberg-Taufer, B., Ponnapureddy, S., Fux, S., & Barth, M. (2017). *Tourism destinations under pressure*. Luzern, Switzerland: Lucerne University of Applied Sciences and Arts, Institute of Tourism ITW.

Wegmann, J., & Jiao, J. (2017). Taming Airbnb: Toward guiding principles for local regulation of urban vacation rentals based on empirical results from five US cities. *Land Use Policy*, *69*, 494–501. doi:10.1016/j.landusepol.2017.09.025

Weidenfeld, A. (2018). Tourism Diversification and Its Implications for Smart Specialisation. *Sustainability*, *10*(2), 319.

Wen, J., Meng, F., Ying, T., Qi, H., & Lockyer, T. (2018). Drug tourism motivation of Chinese outbound tourists: Scale development and validation. *Tourism Management*, *64*, 233–244. doi:10.1016/j.tourman.2017.08.001

Which city's populations grow the most due to tourism? (2019). [Graph illustration the relationship between total native population and annual visitor numbers in cities]. *Data Access from World Economic Forum and Mastercard Global Destination Index*. Retrieved from https://www.onthegotours.com/city-populations-that-grow-most-from-tourism/

Whitehad, C., & Williams, P. (2011). Causes and consequences? Exploring the shape and direction of the housing system in the UK post the financial crisis. *Housing Studies*, *26*(7-8), 1157–1169. doi:10.1080/02673037.2011.618974

Williams, J., & Lawson, R. (2001). Community Issues and Resident Opinions of Tourism. *Annals of Tourism Research*, *28*(2), 269–290. doi:10.1016/S0160-7383(00)00030-X

Williams, P. W., Penrose, R. W., & Hawkes, S. (1998). Shared decision-making in tourism land use planning. *Annals of Tourism Research*, *25*(4), 860–889. doi:10.1016/S0160-7383(98)00037-1

Williams, S. (1998). *Tourism Geography*. London, UK: Routledge.

Williams, S. W. (2009). *Tourism Geography: a new synthesis*. New York: Routledge.

Wilson, J., & Tallon, A. (2012). Geographies of gentrification and tourism. In J. Wilson (Ed.), *The Routledge Handbook of Tourism Geographies* (pp. 103–112). London, UK: Routledge. doi:10.4324/9780203859742

Wilson, R. (2006). The economic impact of local sport events: Significant, limited or otherwise? A case study of four swimming events. *Managing Leisure, 11*(1), 57–70. doi:10.1080/13606710500445718

Wise, N., & Perić, M. (2017). Sports Tourism, Regeneration and Social Impacts: New Opportunities and Directions for Research, the Case of Medulin, Croatia. In N. Bellini, & C. Pasquinelli (Eds.), *Tourism in the City* (pp. 311–320). Berlin, Germany: Springer. doi:10.1007/978-3-319-26877-4_22

Wöhler, K. (2011). *Touristifizierung von Räumen. Kulturwissenschaftliche und soziologische Studien zur Konstruktion von Räumen* [Touristification of spaces. Cultural and social studies research on the construction of spaces]. Wiesbaden: VS.

Wohlmuther, C., & Wintersteiner, W. (2013). Peace Sensitive Tourism: How Tourism Can Contribute to Peace. In C. Wohlmuther, & W. Wintersteiner (Eds.), International Handbook on Tourism and Peace. Klagenfurt: Drava, 1-379.

Woo, E., Kim, H., & Uysal, M. (2015). Life satisfaction and support for tourism development. *Annals of Tourism Research, 50*, 84–97. doi:10.1016/j.annals.2014.11.001

Woo, E., Uysal, M., & Sirgy, M. J. (2018). Tourism impact and stakeholders' quality of life. *Journal of Hospitality & Tourism Research (Washington, D.C.), 42*(2), 260–286. doi:10.1177/1096348016654971

Woosnam, K. M. (2012). Using emotional solidarity to explain residents' attitudes about tourism and tourism development. *Journal of Travel Research, 51*(3), 315–327. doi:10.1177/0047287511410351

World Economic Forum. (2019). *Making Affordable Housing a Reality in Cities.* Cities, Urban Development & Urban Services Platform In Collaboration with PwC. Cologny/Geneva.

World Tourism Organization (UNWTO). (2018). *'Overtourism'? Understanding and managing urban tourism growth beyond perceptions*. Madrid, Spain: UNWTO.

World Tourism Organization (UNWTO). Centre of Expertise Leisure, Tourism, & Hospitality, NHTV Breda University of Applied Sciences & NHL Stenden University of Applied Sciences. (2018). 'Overtourism'? – Understanding and Managing Urban Tourism Growth beyond Perceptions, Executive Summary, Madrid, Spain: UNWTO.

World Tourism Organization (UNWTO). *Tourism Highlights*, 2016 Edition. Downloaded on 6/05/2019 from https://www.e-unwto.org/doi/pdf/10.18111/9789284418145

World Tourism Organization (UNWTO). *Tourism Towards 2030 / Global Overview* - Advance edition presented at UNWTO 19th General Assembly - October 10, 2011.

World Tourism Organization (UNWTO); Centre of Expertise Leisure, Tourism & Hospitality; NHTV Breda University of Applied Sciences; and NHL Stenden University of Applied Sciences. (2018). 'Overtourism'? – Understanding and Managing Urban Tourism Growth beyond Perceptions, Executive Summary, UNWTO, Madrid, Doi:10.18111/9789284420070

World Tourism Organization. (1983). *Risks of saturation of tourist carrying capacity overload in holiday destinations. Report of the secretary-general on the execution of the general programme of work for the period 1982-1983.* New Delhi, India: UNWTO.

World Tourism Organization. (2019). *UNWTO Tourism Definitions*. Madrid, Spain: UNWTO; doi:10.18111/9789284420858

World Tourism Organization. Centre of Expertise Leisure, Tourism & Hospitality, NHTV Breda University of Applied Sciences & NHL Stenden University of Applied Sciences (Eds.). (2019). 'Overtourism'? – Understanding and Managing Urban Tourism Growth beyond Perceptions, Volume 2: Case Studies. Madrid, Spain: UNWTO. doi:10.18111/9789284420629

World Travel & Tourism Council (WTTC) and McKinsey & Company. (2017). Coping with Success: Managing Overcrowding in Tourism Destinations. Retrieved August 14, 2019, from https://www.wttc.org/priorities/sustainable-growth/destination-stewardship/

WTO-ORGANIZACIÓN MUNDIAL DEL TURISMO. (1995). *Lo que todo gestor turístico debe saber. Guía práctica para el desarrollo y uso de indicadores de turismo sostenible*. Madrid, Spain: UNWTO.

WTO-WORLD TOURISM ORGANIZATION. (2004). *Indicators of Sustainable Development for Tourism Destinations. A Guidebook*. Madrid, Spain: UNWTO.

WTO-WORLD TOURISM ORGANIZATION. (2005). *Tourism Congestion Management at Natural and Cultural Sites*. Madrid, Spain: UNWTO.

WTO-WORLD TOURISM ORGANIZATION. (2018). *'Overtourism'? – Understanding and Managing Urban Tourism Growth beyond Perceptions*. Madrid, Spain: UNWTO; doi:10.18111/9789284419999

WTTC & McKinsey & Company. (2017). *Coping with Success: Managing Overcrowding in Tourism Destinations*. London, UK: World Travel & Tourism Council.

WTTC [World Travel & Tourism Council]. (2019). *Travel & Tourism Economic Impact - World*. Retrieved from https://www.wttc.org/economic-impact/country-analysis/country-data/

WTTC. (2018). *City Travel & Tourism Impact 2018*. Retrieved from http:www.wttc.org.com

WTTC. (2018). *City Travel & Tourism Impact 2018*. Retrieved from https://wttc.org/-/media/files/reports/economic-impact-research/cities-2018/city-travel--tourism-impact-2018final.pdf

WTTC. (2018). *Travel & tourism economic impact 2018 - Macau*, available at https://www.wttc.org/-/media/files/reports/economic-impact-research/countries-2018/macau2018.pdf (Accessed on May 29, 2018).

WTTC. (2018). *Travel and Tourism Economic Impact 2018 Portugal Report*. London, UK: World Travel & Tourism Council.

WTTC. (2019, March 20). *One in five Thai baht spent is in Travel & Tourism, according to new WTTC research*. Retrieved from World Travel & Tourism Council: https://www.wttc.org/about/media-centre/press-releases/press-releases/2019/one-in-five-thai-baht-spent-is-in-travel-and-tourism-according-to-new-wttc-research/

Xian Liang, Z., & Gang Bao, J. (2015). Tourism gentrification in Shenzhen, China: Causes and socio-spatial consequences. *Tourism Geographies*, *17*(3), 461–481. doi:10.1080/14616688.2014.1000954

Xiao, H., & Smith, S. (2006). The maturation of tourism research: Evidence from a content analysis. *Tourism Analysis*, *10*(4), 335–348. doi:10.3727/108354206776162796

Xu, F., Buhalis, D., & Weber, J. (2017). Serious games and the gamification of tourism. *Tourism Management*, *60*, 244–256. doi:10.1016/j.tourman.2016.11.020

Yamada, N., Heo, J., King, C., & Fu, Y. (2009). Life satisfaction of urban residents: do health perception, wealth, safety, community pride and, and cultural tourism matter? In International CHRIE Conference-Refereed Track. Paper 24.

Yetiskul, E., & Demirel, S. (2018). Assembling gentrification in Istanbul: The Cihangir neighbourhood of Beyoğlu. *Urban Studies (Edinburgh, Scotland)*, *55*(15), 3336–3352. doi:10.1177/0042098017746623

Yetter, G., & Capaccioli, K. (2010). Differences in responses to web and paper surveys among school professionals. *Behavior Research Methods*, *42*(1), 266–272. doi:10.3758/BRM.42.1.266 PMID:20160305

Yrigoy, I. (2018). Rent gap reloaded: Airbnb and the shift from residential to tourist rental housing in the Palma Old Quarter in Mallorca, Spain. *Urban Studies (Edinburgh, Scotland)*. doi:10.1177/0042098018803261

Zanirato, S. H., & Tomazzoni, E. L. (2014). A sustentabilidade do turismo em Fernando de Noronha (PE-Brasil). *Turydes. Malaga*, *7*(17), 1–24.

Zerva, K., Palou, S., Blasco, D., & Donaire, J. A. (2019). Tourism-philia versus Tourism-phobia: Residents and Destination Management Organization's Publicly Expressed Tourism Perceptions in Barcelona. *Tourism Geographies*, *21*(2), 306–329. doi:10.1080/14616688.2018.1522510

Zhang, C., Pearce, P., & Chen, G. (2019). Not losing our collective face: Social identity and Chinese tourists' reflections on uncivilised behaviour. *Tourism Management*, *73*, 71–82. doi:10.1016/j.tourman.2019.01.020

Zhang, Y., Xiang, L., & Qin, S. (2017b). Does spatial layout matter to theme park tourism carrying- capacity? *Tourism Management*, *61*, 82–95. doi:10.1016/j.tourman.2017.01.020

Zhang, Y., Xiang, L., Qin, S., & Hu, X. (2017). Exploring a theme park's tourism carrying capacity: A demand-side analysis. *Tourism Management*, *59*, 564–578. doi:10.1016/j.tourman.2016.08.019

Zukin, S. (1989). *Loft Living. Culture and Capital in Urban Change*. New Brunswick: Rutgers University Press.

Zwiersa, M., Gideon, B., Van Ham, M., & Van Kempen, R. (2016). The global financial crisis and neighborhood decline. *Urban Geography*, *37*(5), 664–684. Doi:10.1080/02723638.2015.1101251.

About the Contributors

Cláudia Ribeiro de Almeida is Professor of Tourism at School of Management, Hospitality and Tourism of the University of Algarve. She is the Tourism Course Director since 2017. She holds a PhD and a Post Doctoral Degree in Tourism (University of Aveiro), a MA in Total Quality Management (Universidade Politécnica da Cataluña) and a BSc in Marketing (University of Algarve). Cláudia Ribeiro de Almeida's current research interests include Air transport & Tourism dynamics; Second home tourism and Decision making process. She is a member of the Research Centre for Tourism, Sustainability and Well-Being (CinTurs) of University of Algarve. She is a member of several research projets. Professor Cláudia is reviewer of several national and international journals.

Alfred Quintano is a Malta based senior lecturer and management consultant specialising in the Tourism, Aviation and Hospitality sectors. He holds a PhD from the University of Portsmouth, an MBA from Maastricht School of Management and a BA (Honours) degree in Economics from the University of Malta. He has published academic and trade papers and a book on decision choice analysis in the Hospitality industry.

Moisés Simancas is a Senior Lecturer of Department of Geography at La Laguna University. PhD in Geography. Master's Degree in Environmental Impact Assessment. Coordinator of the research group of the ULL "ReinvenTUR: renovation and tourist innovation". Deputy Director of the Tourism Chair CajaCanarias-Ashotel-University of La Laguna. The current lines of research are analyzes of public policies for renovation of mature tourist areas and tourism territorial planning.

Raquel Huete is an Associate Professor in Tourism Studies at the University of Alicante (Spain). She holds a PhD in Sociology from the UA and a BA in Tourism Management from the Official Tourism School in Madrid. She is a researcher at the University Institute for Tourism Research at UA. She has taught at the UA delivering classes on sociology of tourism, social research and local development. In addition to giving worldwide lectures on second home tourism and urban sustainability, she has guest lectured at events and conferences worldwide and contributes to the tourism community through a number of boards and memberships. Dr Huete has been a researcher in about 50 R & D projects and research contracts. She has been a principal researcher in several projects. This work has been reflected in more than 60 articles published in scientific journals and chapters in collective books. Prior to teaching at the university she worked as a travel agent and tourist guide and also she held several positions related to Higher Education Management. She held a position as Director-General for Tourism at the Region of Valencia Government from 2016 to 2018. Her expertise areas are: public opinion in tourism, relationship between tourism and migrations, sustainability and smart tourism destinations.

Zélia Breda holds a Ph.D. in Tourism, a MA in Chinese Studies (Business and International Relations) and a BSc in Tourism Management and Planning from the University of Aveiro, where she is Assistant Professor, at the Department of Economics, Management, Industrial Engineering, and Tourism (DEGEIT), holding the position of Director of the Master's in Tourism Management and Planning. She is a senior member of the Research Unit 'Governance, Competitiveness and Public Policies' (GOVCOPP) of the University of Aveiro, a founding member and vice-president of the Observatory of China, and a senior consultant of Idtour, a tourism R&D spin-off company. She is also a member of the editorial, scientific and review boards of a few academic international peer-reviewed journals, as well as a member of the organizing and scientific committees of international tourism conferences. She has authored and co-authored several national and international papers and communications on tourism development, networks, tourism in China and Goa (India), gender and tourism, and internationalization of the tourism economy. She has also been taking part in several research projects in the tourism field, both as a member of the team and as a consultant.

* * *

Ruben Arnandis-i-Agramunt, Tourismologist, specialist in Tourism Public Management. Bachelor in Tourism, postgraduate in Tourism Management and Planning, in Environmental and Territorial Management Techniques and PdD in Tourism. Expertise, since 2001, in tourism planning and management at the local and regional levels as a technical professional and consultant. Teaching and scientific research fields are also linked to tourism planning and management (strategic planning, stakeholders' participation, public policies evaluation), especially, in assessing and applying adaptation processes to tourism use in cultural and natural heritage.

José Baños, Architecture degree, Master in Sustainable development and tourism, PhD. in City, territory and sustainability. Professor in Universidad de Guadalajara.

Luciana Barbieri da Rosa, Bachelor's degree in Business Administration from the Faculty of Education of São Luis (Bachelor) and a full degree from the Special Teacher Education Graduation Program for Professional Education from the Federal University of Santa Maria. Specialist in Environmental Education, and in Public Management from the Federal University of Santa Maria. Master in Business Administration from the Postgraduate Program in Business Administration of the Federal University of Santa Maria. She served as a substitute professor at the Department of Administrative Sciences of the Federal University of Santa Maria (2013-2015) and Tutor of the Specialization in Public Management (2017-2018), and professor at Faculdade Palotina - FAPAS (2018). She is currently a Ph.D. student in Business Administration from the Postgraduate Program in Business Administration. Operates in the areas of Innovation, Sustainability, Strategy, Marketing, and Production. She is part of the Editorial Advisor of Management Review from the Federal University of Santa Maria.

Alexandra Bol-Esteve, Tourismologist, specialist in local sustainable development and heritage for human development. Bachelor in Tourism.

Irina Borovskaia received her PhD in Economics from the St. Petersburg State University of Economics (Russia) in 2007. She is currently working for her Alma mater as Associate Professor in International Business department. Besides, she is a visited Associate Professor and researcher in National Research University Higher School of Economics, Russia. Having diverse experiences across several different industries, she writes among others on issues of cultural tourism, creativity in hospitality industry, creative gastronomy, event management and volunteering.

Kadir Çakar, PhD, is currently working as an Assistant Professor at the Faculty of Tourism at Mardin Artuklu University. His PhD examined the motivations and experiences of travellers visiting Gallipoli Peninsula within the context of dark tourism. His main research areas include Dark Tourism, Tourist Destination Governance, Tourism Education, Information Communication Technologies, Destination Marketing and Management, Crisis Management and Sustainable Tourism.

Fatima Carvalho, Political scientist (Phd Government, Essex, 2000) and senior researcher working in project 'Tourist experiences' impacts on the destination image: searching for new opportunities to the Algarve'. Track record of interdisciplinary studies in tourism studies, discourse analysis, policy-making, and regulation. Her most recent research interest are – Tourism & customer experience and models of governance to the flourishing of sustainable tourism:

Nimit Chowdhary is a professor of tourism. He heads the Department of Tourism and Hospitality Management at Jamia Millia Islamia, a top-ranked university in India. He has more than 25 years of teaching, training, consulting and research experience. He has authored 9 books and publications around 115 papers and chapters. He has supervised 15 P.hDs. His research interests are in tourism, marketing, destination management, rural tourism, and tour guiding and leadership. He has significant experience in teaching in different regions of India and around the world. He ha successfully administered academic programs and institutions.

Rafael Temes-Cordovez, Architect by Polytechic University of Valencia (UPV) (1994-99). Master in Urban projects by Politecnical University of Cataluña (UPC) (1999-2001). PhD by UPV, 2007. Assistant Director of Urban Design and Regional Plannning Department UPV. Director of the Postgraduate Studies "Geographic Information System for the regional planning, urban design and landscape" (UPV). Dr. Temes works on a variety of GIS archives related with cities, tourism, landscape and regional planning. He runs a research team (www.vlcurbanbigdata.com) focused on different topics based on GIS: city and health; urban vulnerability; urban renewal in residential and tourist environments; Big Data and apps development.

Alexandra Correia, Associate Professor of Tourism Studies at School of Technology and Management (ESTG), of Polytechnic Institute of Viana do Castelo (IPVC), and integrated member Centre for Tourism Research, Development, and Innovation (CiTUR). She also collaborates as a researcher on Applied Management Research Unit (UNIAG), and on Applied Research Centre for Digital Transformation (ARC4DigIT). She is the Head of the Master in Innovative Tourism Development and her main interests include wine tourism, inter-business cooperation, tourism experiences, soft skills in tourism, outdoor tourism, digital marketing, with a special emphasis on experiential marketing, and on social media marketing in Tourism. She is currently involved in several research projects, both at national and international level, namely Project INCOME: Innovative Cooperation Business – HEI learning model for Tourism, co-funded by Erasmus + (KA2); and Project TURNOUT aiming at understanding the development of Outdoor Tourism in the North region (Portugal), co-financed by FEDER.

Luana Damke, Master in Business Administration PPGA UFSM and PhD student in Administration PPGA UFSM

Juan José Tuset Davó is an architect by Polytechnic University of Valencia (UPV) (1994-2001). Master in Preservation of Architectural Heritage (UPV) (2011). Ph.D. by UPV (2008). Lecturer at Architectural Design Department UPV since 2009. Dr. Tuset works on a variety of themes: housing design, garden design and GIS related to tourism, landscape, and regional planning. He is a member of the Contemporary Art and Architecture research team (UPV) and collaborates with vlcurbanbigdata focused on different topics based on GIS.

Manuel De la Calle-Vaquero, PhD in Geography, degree in Geography and degree in Sociology (1996). Associate Professor of Human Geography (Department of Geography). Member of the Research Group "Tourism, Heritage and Development", with special dedication to topics such as cultural tourism, urban tourism, tourism planning, etc.

Ruth De León-Rodríguez, Technical Architect (1996) by University La Laguna, Civil Engineering (2002) and Architecture (2008) by Polytechnic University of Valencia (UPV). PhD Student in Architecture from 2018. She has worked in the field of Civil Engineer, in the department of Urban Development and Urbanization Projects as well as in bidding processes for works and projects. Since he joined the European University of Valencia (UEV) four years ago, he combines his external professional work with teaching. His area of study is Urban Planning, specialising in Geographic Information Systems applied to Territorial Planning, Urban Planning and Landscape.

Josefina Domínguez-Mujica is Professor of Human Geography at the University of Las Palmas de Gran Canaria (Spain). Her career as a researcher has long been linked to Population Geography and, more specifically, to International Migration processes as well as Urban and Tourism Geography. She leads the Atlantic Societies and Spaces Research Group at the University of Las Palmas de Gran Canaria (Spain). She currently chairs the Commission 16.17 Global Change and Human Mobility (Globility) of the International Geographical Union. Josefina has participated in numerous research projects and published articles in national and international prestigious reviews.

Ekaterina Fedorova in 2019 received master degree in the Higher School of Economics Saint Petersburg at the academic program 'Cultural and Event Tourism Management'. Currently working for Roscongress Foundation, non-commercial institution of social development, the largest organizer of governmental and international events in Russia.

Paula Fernandes, PhD in Economics and Management, Professor of Management at the Polytechnic Institute of Bragança (IPB) - Portugal. She is a researcher and Scientific Coordinator of UNIAG (Applied Management Research Unit, since 2013) and was a researcher at the Research Unit in Business Sciences (NECE-UBI, 2010-2017). She participated in 15 international and national projects I&D (6 as

Responsible Researcher), supervised one PhD thesis and several master students (over 95 master theses). She published more than 200 publications in peer-reviewed journals, book chapters and proceedings, and has several communications at international scientific conferences and won 8 scientific awards. She serves as a member of Program Board and Organizing Committees for some Scientific Committees of International Conferences and has collaborated as a reviewer with several journals. Her current research interest is in Tourism Research, Importance-Performance Analysis, Management, Artificial Neural Network, Entrepreneurship, Econometric Modelling, Marketing and Strategic Management, Market Research, Corporate Social Responsibility and Sustainable, Higher Education Quality and Applied Research Methods.

Fernanda A. Ferreira is a full-time associate professor in the Department of Mathematics at the School of Hospitality and Tourism of Polytechnic Institute of Porto, Portugal. She holds a BS in Mathematics and a PhD in Applied Mathematics from the University of Porto. She obtained also a Diploma of Advanced Studies in Statistics and Operations Research from Vigo University. A member and coordinator of the Applied Management Research Unit (UNIAG), her publications, more than 100, in peer-reviewed journals, book chapters and proceedings, cover the research interest areas of industrial organization, game theory and tourism (ORCID ID: orcid.org/0000-0002-1335-7821). Co-author of two books published in the Mathematics area. Supervised master thesis in the areas of Hospitality, Game Theory and Management. She also organizes Symposium on "Operational Research and Applications" in several International Conferences and has collaborated as reviewer with several journals. She has several communications at international scientific conferences and she serves as a member of Program Board and Organizing Committees for some Scientific Committees of International Conferences. Her current research interest is in Tourism Research, Hospitality Management, Industrial Organization and Operational Research.

Elena Ferreiro-Calzada, Degree in Geography. Master on Tourism Management Destinations. Training researcher on Tourism Geography. PhD student at the Complutense University of Madrid

Manuel José Fonseca, PhD in Information Sciences, Manuel José Serra da Fonseca was born in Lisbon in 1971. He played professional activities such as marketing communications technician and professor of higher education. He is currently Adjunct Professor at the Polytechnic Institute of Viana do Castelo (Portugal) and Coordinator of Marketing Master's degree at its Business School (ESCE-IPVC). He is a researcher at UNIAG (Applied Management Research Unit, since 2014). He participated in 3 national projects I&D and supervised about 40 undergraduate and masterworks in the areas of Communication Sciences and Business Sciences. He published several publications in peer-reviewed journals, book chapters and proceedings and has several communications at national and international scientific conferences. He collaborates as a reviewer with several journals. He acts in the Social Sciences field with an emphasis in Communication and Marketing Sciences. In his academic curriculum, the most frequent terms in the context of scientific production are: attitude towards advertising, attitude towards communication skills, road safety communication campaigns, consumer behavior, integrated marketing communications, social marketing, public service advertising, viral marketing and organizational communication.

María García-Hernández, PhD in Geography. Associate Professor of Human Geography at the Complutense University of Madrid. Degree in Geography and degree in Documentation Co-director of the research group "Tourism, Heritage and Development" (www.ucm.es/geoturis)

María Antonia García-Sastre, An associate professor of marketing at the University of the Balearic Islands, Spain. Her research includes tourism, sports tourism, and management.

Tais Pentiado Godoy, Atualmente, Doutoranda em Administração pela Universidade Federal de Santa Maria (2016). Graduada em Administração na UFSM (2017). Mestrado em Engenharia de Produção pela Universidade Federal de Santa Maria (2013). Possui Técnico em Administração; Eixo Tecnológico: Gestão e Negócios pela UFSM (2013). Bacharel em Direito pela Faculdade Metodista de Santa Maria (2010). Tem experiência na área de Direito, com ênfase em Direito Previdenciário.

Clandia Maffini Gomes, Associated Professor at Santa Maria Federal University Editor in Chief of Brazilian Journal of Management.

Luis Alfonso Escudero Gómez, born in Aarberg (Switzerland) in 1971, has a PhD in Geography from the University of Santiago de Compostela since 2000. He is a lecturer in the Department of Geography and Spatial Planning of the University of Castilla-La Mancha since 2000, Senior Lecturer of Human Geography since 2007. He has carried out research on cultural tourism and the urban image. He has published several articles in journals such as Estudios Geográficos,Boletín de la Asociación de Geógrafos Españoles, Pasos, Ciudad y Territorio, Anales de Geografía, Urban Science or Sustainability.

Valery Gordin, Head of Laboratory of Economics of Culture, professor of Department of Management in National Research University Higher School of Economics. Head of researches in the field of culture and tourism, carried out for the Nordic Council of Ministers, the Ministry of Education and Science of the Russian Federation, the Committee for Culture of St. Petersburg. Under his leadership researches for leading Russian museums have been done: The State Hermitage Museum, the State Russian Museum, The Tsarskoe Selo State Museum and Heritage Site; researchers of the project "Museum Night" are being conducted for several years. Author of articles on cultural and creative tourism.

Maria Manuela Guerreiro is Assistant Professor at the Faculty of Economics, University of Algarve (Portugal). She is the co-coordinator of the Research Centre for Tourism, Sustainability and Well-Being (CinTurs) and Director of the Master of Marketing Management. She has a PhD in Economic and Management Sciences, and a Msc in Cultural Management. Her current research interests include consumer behaviour, marketing management and design intersections, brand management, e-marketing and social media marketing, consumer experiences, arts and events marketing, especially in tourism. She publishes academic papers, books, and book chapters.

Cláudia Henriques has a PhD in Economics (University of Algarve, 2002), thesis under the subject of "Management and Planning of Sustainable Tourism in cities and historic city quarters. The case of Lisbon". She is lecturer of Tourism Economics, Cultural and Creative Sector and Tourism Planning and

Managing at University of Algarve (ESGHT, UALG) from 1993 until today. Her research interests are in Cultural Tourism, Tourism Planning and Management; Cultural and Tourism Policies and Cultural heritage (Historic Centres). She is member of CIEO (Centre for Spatial and Organizational Dynamics).

María de los Angeles Huízar, Tourism Degree, Master in Sustainable development and tourism, PhD in City, Territory and Sustainability. Research professor in Universidad de Guadalajara.

Snigdha Kainthola is a doctoral researcher in the Department of Tourism and Hospitality Management, Jamia Millia Islamia, a top-ranked university in India. She is a graduate in history and has completed Masters in tourism. She has worked in the tourism sector for 2 years as a tour leader. She is an avid traveler. She decided to pursue academics and make a career in the tourism education sector. Her research inclination is in tourism marketing, destination marketing and management, and spiritual tourism.

Hio Kuan Lai is a Master's degree holder in Tourism Economics & Regional Development in the Faculty of Economics at University of Algarve, Portugal. Hio Kuan's research interests concern the residents' attitude in tourism destinations and sustainable tourism development. For over 5 years, Hio Kuan has been professionally active in hospitality industry in Macau.

Jorge López, Law degree, Master in Sustainable development and tourism, PhD. in Sustainable development. Bussiness Consultant.

Alejandro Mantecón holds a Ph.D. in Sociology from University of Alicante (UA). He is Senior Lecturer in the Department of Sociology I at the UA (Spain). His research interests are: Tourism-migration-urbanization nexus, leisure-oriented migrations and, tourism and social change in Spain. He has investigated and attended at courses and seminars at several universities in Spain, Portugal, Switzerland, United Kingdom, United States, Brazil and Ecuador. He has participated in 24 Research Projects and Groups funded by different institutions.

Nataša Marković, MA Tourism Destination Management, is a promoter of cultural heritage and sustainable tourism in the Adriatic-Ionian Macro-Region. As a researcher and consultant, her work is focused on a holistic approach to innovations in tourist destinations.

Maria Carolina Martins-Rodrigues, International PhD in Management, University of Extremadura, Spain Post-graduate in Knowledge management, University of Belgrano, Buenos Aires, Argentina, and a research fellow of CIEO (Algarve University). Her research interests currently are Sustainability, innovation, Knowledge Management, Entrepreneurship, Social Responsibility and Business Management. Has co-authored over 20 articles and book chapters and published in several scientific journals. She has also organized Iberian Conference on Entrepreneurship and international Conferences of Sustainability and Innovation. Founder member and vice-president of the direction of Empreend - Portuguese Association for Entrepreneurship. Was Professor in University Aberta (UAb) and University Moderna. In TAX Administration was Tax Administration Technician Advisor acting as coordinator in the International Relations Services Directorate (DSRI). and was Head of the Tax Service.

Nelson Matos, Invited Adjunct Professor (University of Algarve). PhD in Tourism (2013), degree in Tourism (2002). Especialises in Consumer Behaviour, Marketing Experience, Destination Image, Tourism Experience, Destination Marketing, and Digital Marketing.

Luís Mendes, Geographer. Invited Professor at the School of Architecture of the Technical University of Lisbon (2005/2006), at the Lisbon School of Education (2010 / ...) and the Institute of Geography and Spatial Planning of the University of Lisbon (IGOT-UL) (2012 / ...). Member of the Portuguese Association of Geographers, the Geography Teachers Association and the Lisbon Tenants Association, integrating since 2012 and 2018, respectively, the board of the latter two. Since 2003, carries Permanent Researcher functions in the Centre for Geographical Studies of the University of Lisbon (CEG/UL) of IGOT-UL, in which has developed research in the fields of Urban Studies (notably gentrification and urban regeneration), Geographical Education and Teaching of Geography. He is author of more than one hundred and fifty titles: numerous papers, presentations, book chapters, books, technical opinions and reports, in addition to other publications in the area of research topics mentioned above. It was also awarded with the Amilcar Patrício Award 2005 by the Portuguese Association of Geographers, and more recently received the Diploma of Merit, recognition award for activities with relevance to the community, for the three-year period 2016/2017/2018 awarded by the Polytechnic Institute of Lisbon. He has also worked in the last three years as an activist in the social movement Morar em Lisboa.

Sofía Mendoza de Miguel, Degree in Tourism. Master on Tourism Management Destinations. Training researcher on Tourism Geography. PhD student at the Complutense University of Madrid.

Claudio Moreno-Medina is PhD in Geography and Lecturer of Regional Geographical Analysis at the University of Las Palmas de Gran Canaria. He is member of the research team "Atlantic Societies and Spaces" and of the Spanish Association of Geography as well as of the Educational Innovative Group EDUGEDESIS. His research has been conducted in the fields of spatial analysis and he has published papers in impact journals and books.

Alcina Nunes is a Professor at the Polytechnic Institute of Bragança, Portugal, where she teaches several courses in the area of Economics and Management, is the responsible for the bachelor degree of International Business Management (European Degree) and a research member of UNIAG, a research unit funded by FCT. With a PhD in Mathematical Economics and Econometric Models her actual research concerns the econometric evaluation of public policies, namely labour market, health, gender and development policies, entrepreneurship and business demography. She also collaborated with the Group of Economic Studies of the Portuguese Economics Minister, the Group of Economics and Monetary Studies of the University of Coimbra and with several universities as an invited lectured at the same time she published several scientific articles and book chapters.

Juan Manuel Parreño-Castellano is Senior Lecturer of Human Geography at the University of Las Palmas de Gran Canaria and the Director of the Department of Geography. He is member of the research team "Atlantic Societies and Spaces" and of the group "Urban Geography" in the Association of Spanish Geographers. His research career has been focused on Urban Studies, Migrations and Tourism. He has participated in some research projects on housing models, residential space, production systems of the urban-tourist space, mechanisms of social participation and migrations. He is the author of 30 papers published in journals refereed in SSCI and SCOPUS and of about 15 research books and chapters in prestigious publishers.

Pedro Pintassilgo holds a PhD in Economics from the New University of Lisbon. He is Assistant Professor at the Faculty of Economics, University of Algarve, Portugal, where he got the Aggregation title. He is also Director of the Master in Tourism Economics and Regional Development. He has published in influential journals in the fields of Environmental Economics and Tourism.

Patrícia Pinto holds a PhD in quantitative methods applied to economics and management. She is associate professor at the Faculty of Economics, University of Algarve, and an active member of board of the PhD program in Tourism in the same institution. Patrícia is the coordinator of the Research Centre for Tourism, Sustainability and Well-Being (CinTurs). She teaches data analysis (statistics and econometrics) and research methods and her current research interests include applications of statistical methods, especially Structural Equation Modeling (SEM), in Tourism, Marketing and Psychology. She is the editor in chief of Journal of Spatial and Organizational Dynamics and leading or participant researcher in funded research projects.

Alejandro Garcia Pozo, PhD in Economic and Business Sciences and Extraordinary Doctorate Award.by the University of Malaga, he is Professor of the University of Malaga at Applied Economics (Economic Structure) Department. He has published a large number of scientific papers in journals of impact in the economic scientific field and has been part of the scientific committee of many international congresses, on issues related to tourism and the environment. Head of Department of Applied Economics (Economic Structure) at the University of Malaga from 10/04/2013, and Lab Head of SEJ139 research group at Andalusian research system.

Francisco Rejón-Guardia is an associate professor in the Department of Business and Economics at the University of The Balearic Islands. His research interest includes consumer behavior, internet advertising, tourism, and social media. He published on these topics in journals such as Internet Research, Psychology & Marketing, Industrial Management & Data Systems, Journal of Business Research, Tourism Economics, Journal of International Consumer Marketing, Journal of Technology and Science Education as well as books and book chapters on international publishers (Springer, IGIglobal, Taylor & Francis Group) His papers have appeared in conferences such as Academy Marketing Science (AMS), European Marketing Academy (EMAC), the Spanish Academic and Professional Marketing Association (AEMARK). He is chief executive of the TURNETMK research group.

José Luis Sanchez-Ollero, Doctor in Economic and Business Sciences from the University of Malaga, he is Professor of the University of Malaga in the Department of Applied Economics (Economic Structure). He has published a large number of scientific papers in journals of great impact in the international scientific field and has been part of the scientific committee of many international congresses; having assumed university management responsibilities as Vice-Chancellor of the Technological Campus of the International University of Andalusia (UNIA) from 15/07/2015 to 30/09/2017, or General Director Of Sports UMA (2004-2008) and Deputy Director of the Campus in Malaga of the UNIA (2014-2015).

Eleonora Santos is a researcher at the School of Technology and Management (ESTiG), at the Polytechnic Institute of Bragança (IPB). She has been working at the IPB since February 2019 in the project TURNOUT - Development of Outdoor Tourism in the Northern Region of Portugal". She holds a PhD in Economics, from the Faculty of Economics, University of Coimbra (2018); a MSc in Economics (2002) and a degree in Economics (1994) from the Faculty of Economics of the University of Coimbra. She is a collaborating researcher at the Center for Business and Economics Research (CeBER) at the Faculty of Economics, University of Coimbra; and a consultant at the Interdisciplinary Center for Marine and Environmental Research (CIIMAR) of the University of Porto. She was Adjunct Professor at the D. Dinis Higher Education Institute (ISDOM) in Marinha Grande from 2005 to 2014, where she coordinated and taught 14 Courses in Accounting and Business Administration, Solicitor, Human Resources Management and Management and Informatics. She was also the Director of the Free Course of Preparation for the Chartered Accountants Examination and a representative of the Adjunct Professors in the Scientific Council of ISDOM, as well as a member of the Pedagogical Council and the General Council. She has participated in 5 national and international R&D projects (1 still in progress) at the universities of Coimbra, Aveiro and Porto and at the Polytechnic of Bragança. She was awarded with 4 scholarships and 2 scientific merit awards from the University of Coimbra, Calouste Gulbenkian Foundation (Portugal), Oxford University (UK) / General Electric Healthcare Budapest (Hungary) and Tokyo Foundation (Japan). Her scientific areas of interest and research are: Tourism; Marine Economics, Foreign Direct Investment, Economic Growth, Productivity, Industrial Organization and Health Economics. In addition, she has published several articles in the field of Medical Physics; has been nominated as a participant in the Fourth Technical Meeting on Modelling and Data for Radiological Impact Assessments (MODARIA) at the Headquarters of the International Atomic Energy Agency (IAEA) in Vienna; and has participated in several national and worldwide congresses.

Begoña Serrano-Lanzarote, Architect and PhD by Polytechnic University of Valencia (UPV). Professor at the School of Architecture of Valencia (UPV). Director of the Valencian Institute of Building (IVE) that works on 36 research European Projects (INTERREG Programmes, Intelligent Energy, Horizon 2020 and Climate KIC) focused on inclusive and sustainable buildings and cities, urban regeneration, nearly zero energy building, energy retrofit, housing buildings characterization, changing user behaviour to conserve resources, community based research and environmental assessment. Coordinator of the Postgraduate Studies "Buildings Renovation and Urban Regeneration". She has directed 6 PhD theses.

Goretti Silva is Associate Professor in Tourism Studies at the Polytechnic Institute of Viana do Castelo, Portugal. She obtained her PhD degree at Bournemouth University (UK), about the role of relationships, conceptualized as social capital and social support, in the creation and management of Tourism related small family businesses in rural areas. Her research interests include tourism as a strategy for rural development, particularly in peripheral regions, SME's in Tourism, tourism destinations' sustainable development, and tourism planning at regional, local and project level. Prior to being a member of academia she has collaborated with several organisations as tourism consultant, namely the Peneda Gerês National Park in the North of Portugal. She is currently involved in several research projects, both at national and international level, namely Project INCOME: Innovative Cooperation Business – HEI learning model for Tourism, co-funded by Erasmus + (KA2); and Project TURNOUT aiming at understanding the development of Outdoor Tourism in the North region (Portugal), co-financed by FEDER.

André Soares is a former researcher in the Project TURNOUT that aims at understanding the development of Outdoor Tourism in the North region (Portugal), and is co-financed by FEDER. He is graduated in Tourism by the Polytechnic Institute fo Viana do Castelo.

Pinaz Tiwari is a research scholar in the Department of Tourism and Hospitality in Jamia Millia Islamia, India. She has worked in the tourism sector for two years. She has expertise in customer management and has been involved in travel agents' capacity development projects in her organisation. She has completed her Masters in Tourism with specialization in International Tourism Business and has been awarded as the topper of the batch. She looks forward to making a career in teaching and creating a difference in the tourism industry with her research works. Her interest areas are overtourism, destination marketing, management, stakeholders' capacity development, and responsible tourism.

Ricardo Urrestarazu, Doctor from the University of Malaga in the interuniversity tourism program. Degree in business sciences from the University of Malaga, graduated in political science from UNED. and Master in MBA from ESESA. Employee of Unicaja Bank since 1997, where I have held branch address positions for 10 years, investment analyst and commercial.

María Velasco holds a Bachelor of Law and Political Science and Administration, from the Complutense University and a Doctorate of Political Science. She is a lecturer in the Department of Political Science and Administration of the Faculty of Political Science (Universidad Complutense de Madrid). Her research interests are public policies, as theoretical field, and tourism, urban and cultural policy as the object of specific analysis.

Cláudia Veloso, European PhD in Social Sciences (Economics and Management) and she is graduates in Management from the ISEG University of Lisboa. She is currently a professor at ESTGA – School of Technology and Management, University of Aveiro, where she lectures several courses in the areas of management, both at master and graduate levels. She is a researcher at Research Unit on Governance, Competitiveness and Public Policies - GOVCOPP. Her current research interest is in Management, Tourism Research, Entrepreneurship, Marketing and Strategic Management, Market Research, Corporate Social Responsibility and Sustainable and Higher Education Quality. She has published more than a dozen scientific articles in international journals, 8 of which are indexed in the WoS and SCOPUS databases, 4 book chapters, and about five dozen articles published in proceeding of international conferences, one of which are indexed in the WoS and 1 book. She participated in 3 scientific research projects. She has collaborated as a reviewer of articles, presented at journals and international conferences, indexed at WoS and moderating international congresses. Since 2017, he has been a member of the Scientific Committee of Economic and Social Developments (EDS) Conferences. In addition, she has worked in the business context, with more than 10 years of experience in public and business management, standing out as Administrator of the Northeast Hospital Center, EPE.

Elvira Vieira has a diverse experience in economics and management, firstly in a business environment and after the finishing of her Ph.D. in Applied Economy at the Santiago de Compostela University (USC), back in 2007, as a researcher and teacher, collaborating in institutions like the Portuguese Institute of Marketing Management (IPAM), Minho University (UM), Vigo University (UV), Polytechnic Institute of Viana do Castelo (IPVC) and Institute of Administration and Management (ISAG). In 2010 she was invited to take the position of Director of the European Grouping of Territorial Cooperation Galicia – North of Portugal, for a period of 2 years. In 2012 she returns to ISAG where she is now full time Professor, CEO and member of the Scientific and Technical Council, along with the position of part time Professor at IPVC. She also develops research projects in various disciplines (Human Capital, Innovation, Tourism, Regional Development, among others), collaborating with UNIAG.

Index

L

M

N

O

P

Q

R

S

T

U

V

W

www.ingramcontent.com/pod-product-compliance
Lightning Source LLC
LaVergne TN
LVHW060355300926
849162LV00006B/1272